131/132
Advances in Polymer Science

A. Abe · H.-J. Cantow · P. Corradini · K. Dušek · S. Edwards
H. Fujita · G. Glöckner · H. Höcker · H.-H. Hörhold
H.-H. Kausch · J. P. Kennedy · J. L. Koenig · A. Ledwith
J. E. McGrath · L. Monnerie · S. Okamura · C. G. Overberger
H. Ringsdorf · T. Saegusa · J. C. Salamone · J. L. Schrag · G. Wegner

Springer
Berlin
Heidelberg
New York
Barcelona
Budapest
Hong Kong
London
Milan
Paris
Santa Clara
Singapore
Tokyo

Rotational Isomeric State Models in Macromolecular Systems

By M. Rehahn, W. L. Mattice, U. W. Suter

Springer

Authors

Dr. M. Rehahn
University of Karlsruhe
Institute of Polymers
D-76128 Karlsruhe, FRG
*E-mail: rehahn@polyibm2.chemie.
uni-karlsruhe.de*

Prof. U. W. Suter
Dept. of Materials
Institute of Polymers
ETH, CNB E92
CH-8092 Zürich, Switzerland
E-mail: suter@ifp.mat.ethz.ch

Prof. W. L. Mattice
University of Akron
Dept. of Polymer Science
Akron OH 44325-3909, USA
E-mail: wlm@frank.polymer.uakron.edu

This series presents critical reviews of the present and future trends in polymer and biopolymer science including chemistry, physical chemistry, physics and materials science. It is addressed to all scientists at universities and in industry who wish to help abreast of advances in the topics covered.

As a rule, contributions are specially commissioned. The editors and publishers will, however, always be pleased to receive suggestions and supplementary information. Papers are accepted for „Advances in Polmyer Science" in English. In references Advances in Polymer Science is abbreviated Adv. Polym. Sci. and is cited as a journal.

Springer WWW home page: http://www.springer.de

ISSN 0065-3195
ISBN 3-540-62487-2
Springer-Verlag Berlin Heidelberg NewYork

Library of Congress Catalog Card Number 61642

This work is subject to copyright. All rights are reserved, whether the whole or part of the material is concerned, specifically the rights of translation, reprinting, re-use of illustrations, recitation, broadcasting, reproduction on microfilms or in other ways, and storage in data banks. Duplication of this publication or parts thereof is only permitted under the provisions of the German Copyright Law of September 9, 1965, in its current version, and permission for use must always be obtained from Springer-Verlag. Violations are liable for prosecution under the German Copyright Law.

© Springer-Verlag Berlin Heidelberg 1997
Printed in Germany

The use of registered names, trademarks, etc. in this publication does not imply, even in the absence of a specific statement, that such names are exempt from the relevant protective laws and regulations and therefore free for general use.

Typesetting: Macmillan India Ltd., Bangalore-25
SPIN: 10548351   02/3020 - 5 4 3 2 1 0 - Printed on acid-free paper

Editors

Prof. Akihiro Abe
Department of Industrial Chemistry
Tokyo Institute of Polytechnics
1583 Iiyama, Atsugi 243-02, Japan

Prof. Hans-Joachim Cantow
Freiburger Materialforschungszentrum
Stefan Meier-Str. 31a
D-79104 Freiburg i. Br., FRG

Prof. Paolo Corradini
Università di Napoli
Dipartimento di Chimica
Via Mezzocannone 4
80134 Napoli, Italy

Prof. Karel Dušek
Institute of Macromolecular Chemistry, Czech
Academy of Sciences
16206 Prague 616, Czech Republic

Prof. Sam Edwards
University of Cambridge
Department of Physics
Cavendish Laboratory
Madingley Road
Cambridge CB3 OHE, UK

Prof. Hiroshi Fujita
35 Shimotakedono-cho
Shichiku, Kita-ku
Kyoto 603, Japan

Prof. Gottfried Glöckner
Technische Universität Dresden
Sektion Chemie
Mommsenstr. 13
D-01069 Dresden, FRG

Prof. Dr. Hartwig Höcker
Lehrstuhl für Textilchemie
und Makromolekulare Chemie
RWTH Aachen
Veltmanplatz 8
D-52062 Aachen, FRG

Prof. Hans-Heinrich Hörhold
Friedrich-Schiller-Universität Jena
Institut für Organische
und Makromolekulare Chemie
Lehrstuhl Organische Polymerchemie
Humboldtstr. 10
D-07743 Jena, FRG

Prof. Hans-Henning Kausch
Laboratoire de Polymères
Ecole Polytechnique Fédérale
de Lausanne, MX-D
CH-1015 Lausanne, Switzerland

Prof. Joseph P. Kennedy
Institute of Polymer Science
The University of Akron
Akron, Ohio 44 325, USA

Prof. Jack L. Koenig
Department of Macromolecular Science
Case Western Reserve University
School of Engineering
Cleveland, OH 44106, USA

Prof. Anthony Ledwith
Pilkington Brothers plc. R & D
Laboratories, Lathom Ormskirk
Lancashire L40 SUF, UK

Prof. J. E. McGrath

Polymer Materials and Interfaces Lab.
Virginia Polytechnic and State University
Blacksburg
Virginia 24061, USA

Prof. Lucien Monnerie

Ecole Superieure de Physique et de Chimie
Industrielles
Laboratoire de Physico-Chimie
Structurale et Macromoléculaire
10, rue Vauquelin
75231 Paris Cedex 05, France

Prof. Seizo Okamura

No. 24, Minamigoshi-Machi Okazaki
Sakyo-Ku, Kyoto 606, Japan

Prof. Charles G. Overberger

Department of Chemistry
The University of Michigan
Ann Arbor, Michigan 48109, USA

Prof. Helmut Ringsdorf

Institut für Organische Chemie
Johannes-Gutenberg-Universität
J.-J.-Becher Weg 18-20
D-55128 Mainz, FRG

Prof. Takeo Saegusa

KRI International
Inc. Kyoto Research Park 17
Chudoji Minamima-chi
Shimogyo-ku Kyoto 600, Japan

Prof. J. C. Salamone

University of Lowell
Department of Chemistry
College of Pure and Applied Science
One University Avenue
Lowell, MA 01854, USA

Prof. John L. Schrag

University of Wisconsin
Department of Chemistry
1101 University Avenue
Madison, Wisconsin 53706, USA

Prof. G. Wegner

Max-Planck-Institut für Polymerforschung
Ackermannweg 10
Postfach 3148
D-55128 Mainz, FRG

Foreword

The Rotational Isomeric State (RIS) method has become a popular tool for the analysis of the conformational behavior of macromolecules when proper attention to the details of the chemical structure - constitution and configuration - of the unperturbed chain is required. It has been implemented in several commercial software packages and is simple enough that many researchers have coded their own programs to carry out the necessary computations. Several review articles and two books have been dedicated to this topic and it seemed as if there was no need for a further publication. The practitioner of the art of RIS calculations will quickly find out, however, that while it is one thing to be able to carry out the calculations and thereby obtain a wealth of information on the conformational behavior of a macromolecule, it is quite another matter to devine the parameters necessary for the calculations. And when he resorts to the literature, he finds it difficult to find the necessary publications since little standardization exists in the field and, once the appropriate sources are located, he will often be confused when more than one source is found that provides different, seemingly contradictory values and models. Even a researcher intimately familiar with the method and its foundation, is often faced with the task of finding RIS parameters from the literature for unfamiliar constitutions and configurations, a task that can prove formidable.

We set out, in consequence, to ease the task of finding and comparing RIS models and parameters in the literature. A review seemed to be the most appropriate method, but in the process of its compilation it became rapidly clear that a complete representation of the published work since Volkenstein´s first article on the subject in 1951 was not possible. Nevertheless we are confident that we were able to provide a sufficiently complete listing of the most relevant publications on fundamental aspects of RIS theory as well as a comprehensive overview of the published models available to the end of 1994. More than 680 models for structures of symmetric and asymmetric synthetic polymers, polysaccharides, polypeptides, polynucleotides, rigid chains, branched, comb, and star molecules, and others are compiled here.

We would like to thank many of the authors of the reviewed publications for patiently checking the accuracy and completeness of their contributions and hope for the understanding of those researchers whose work we have inadvertently overlooked.

December 1996

Matthias Rehahn
Wayne L. Mattice
Ulrich W. Suter

Table of Contents

1 Motivation 1
2 The Rotational Isomeric State Method 2
3 Publications on Fundamental Aspects of RIS Theory 4
4 Notation Used and Structure of the Collection of RIS Models ... 5
5 Index of Compounds 7
6 Collection of RIS Models 19

    Simple Symmetric Chains .... S 001 - S 188
    (Simple) Asymmetric Chains .... A 001 - A 111
    Chains with Rigid Moieties and Virtual Bonds .... V 001 - V 147
    Copolymers .... C 001 - C 023
    Branched Polymer Molecules, Combs, Stars .... X 001 - X 057
    Naturally Occuring Polymers .... N 001 - N 157

7 Author Index Volumes 101 - 131/132 .... AI 1

Rotational Isomeric State Models in Macromolecular Systems

M. Rehahn[1] · W. L. Mattice[2] · U. W. Suter[3]

[1] University of Karlsruhe, Institute of Polymers, D-76128 Karlsruhe, FRG
 E-mail: rehahn@polyibm2.chemie.uni-karlsruhe.de
[2] University of Akron, Dept. of Polymer Science, Akron OH 44325-3909, USA
 E-mail: wlm@frank.polymer.uakron.edu
[3] Dept. of Materials, Institute of Polymers, ETH, CNB E92, CH-8092 Zurich, Switzerland
 E-mail: suter@ifp.mat.ethz.ch

1
Motivation

The Rotational Isomeric State method was created to make possible the description of the conformational behavior of macromolecules with proper attention to the details of the chemical structure of the chain – constitution and configuration – albeit neglecting the effects of excluded volume, i.e. for unperturbed chains. It is based on the generator matrix techniques pioneered by Kramers and Wannier[1] and was introduced to macromolecules by Volkenstein.[2] At first, the approach was of considerable complexity but a number of significant developments[3-7] have yielded a method that allowed ready application even to complex chemical structures. In 1969, Flory published his classic book[8] on RIS theory, making the method available to a larger audience. He further improved it by recasting the formalism into a particularly compact and simple form which he published in 1974.[9] Many incremental improvements followed these major steps. The method today presents an attractive and computationally extremely efficient way of assessing the conformational properties of most macromolecular systems that assume a significant number of different conformations, including branched, star, and cyclic chains, and macromolecules with almost any composition and structure. Today the approach has been implemented in several commercial software packages and is readily available. It was reviewed in a recent book[10] to which we refer the reader for details.

---

[1] Kramers HA, Wannier GH, (1941) Phys Rev 60:252
[2] Volkenstein MV (1951) Dokl Akad Nauk 78:879
[3] Gotlib, Yu Ya (1959) Zh Tekhn 29:523
[4] Birshtein TM, Ptitsyn OB (1959) Zh Tekhn 29:1048
[5] Lifson SJ (1959) J Chem Phys 30:964
[6] Nagai KJ (1959) J Chem Phys 31:1169
[7] Hoeve CAJ (1960) J Chem Phys 32:888
[8] Flory PJ (1969) Statistical mechanics of chain molecules. Wiley-Interscience, New York (reprinted with the same title by Hanser, Munich)
[9] Flory PJ (1974) Macromolecules 7:381
[10] Mattice WL, Suter UW (1994) Conformational theory of large molecules; the rotational isomeric state Model in macromolecular systems. Wiley, New York

Over the past four decades, RIS models for hundreds of polymer structures, and sometimes many different ones for one and the same macromolecule, have been developed and published. The casual user of RIS techniques, even one intimately familiar with the method and its foundation, is often faced with the task of finding RIS parameters from the literature for unfamiliar constitutions and configurations. It is this task that the present review is aimed at easing by providing a comprehensive overview of the published models available to date. While completeness cannot be claimed, care has been taken to find as many models as possible and to report them in as brief a space as practical.

2
The Rotational Isomeric State Method

This RIS method can be stated in concise form as a recipe:

1. Identify all skeletal bonds around which rotation is possible under the chosen conditions (a variation in bond angles can also be taken as the basis for an RIS scheme).
2. For each of these bonds, analyze the interactions of short range that determine its conformational behavior. Usually one finds that the torsion angles of the bonds in question and their adjoining bond angles can assume only relatively narrow domains and one identifies these domains with Rotational Isomeric States. The location of these states determines the geometry of those "rotational isomers" selected to represent the totality of conformations available to the macromolecule.
3. Compute the energy or free energy of the selected rotational isomers (there are v_i states for bond i) and attribute these interactions to molecular causes. Usually it is sufficient to partition the interactions into contributions of "first order," i.e. those that depend on one torsion angle only, and those of "second order," i.e. interactions that depend on two neighboring torsion angles (there are alternatives to this approach in which the conformer populations are directly estimated and an assignment to first and second order interactions can be circumvented,[10] albeit with other disadvantages). Determine the appropriate "statistical weights" that describe the propensity for population of a particular local conformation compared to its competitors, i.e. the other conformations of the same bond or bonds. Typically, these statistical weights take the form of Boltzmann factors, $\xi_j = \xi_{0,j} \exp(-E_j/kT)$, where for the statistical weight ξ_j, the prefactor $\xi_{0,j}$ comprises "entropic" contributions while E_j denotes a mean energy associated with the particular weight.
4. Collect the statistical weights into "statistical weight matrices" (of dimension $v_{i-1} \times v_i$, where v_i is the number of rotational isomeric states of bond i), one per skeletal bond subject to conformational change, indexing rows and columns of the matrices with the RIS of the bond. The generic statistical weight matrix for bond i is termed $\mathbf{U}_i$.

Once the statistical weight matrices and the local geometry of the individual conformers is determined, the RIS approach allows for the extremely efficient estimation of global characteristics. The essential simplifying assumption of the

RIS approach consists of a factorization of the conformational partition function into local contributions. The conformational partition function of a simple linear chain consisting of skeletal bonds $1 \leq i \leq n$ is given in most cases by

$$Z = \mathbf{U}_1 \mathbf{U}_2 \mathbf{U}_3 \ldots \mathbf{U}_n = \prod_{i=1}^{n} \mathbf{U}_i \qquad (1)$$

with $\mathbf{U}_1 = [1\ 0 \ldots 0]$ and $\mathbf{U}_n = \begin{bmatrix} 1 \\ 1 \\ \vdots \\ 1 \end{bmatrix}$ of appropriate dimensions.

In addition, factorization of the partition function provides a vehicle for the efficient estimation of many other properties; all that is required is that the property for a single chain conformation be expressible in factorial form similar to that for the partition function. If a conformation-dependent property f can be computed for a given (fixed) chain conformation by a product of "generator matrices" $\mathbf{F}_i$, one for each bond,

$$f = \mathbf{F}_1 \mathbf{F}_2 \mathbf{F}_3 \ldots \mathbf{F}_n = \prod_{i=1}^{n} \mathbf{F}_i \qquad (2)$$

then the exact average of f over all conformations can be obtained by a related product of matrices.

$$\langle f \rangle = \frac{1}{Z} F_1 F_2 F_3 \ldots F_n = \prod_{i=1}^{n} F_i \qquad (3)$$

where

$$F_i = (\mathbf{U}_i \otimes \mathbf{I}_f) \begin{bmatrix} \mathbf{F}_{\text{state 1}} & 0 & \cdots & 0 \\ 0 & \mathbf{F}_{\text{state 2}} & \cdots & 0 \\ \cdots & \cdots & \cdots & \cdots \\ 0 & 0 & \cdots & \mathbf{F}_{\text{state } v} \end{bmatrix}_i$$

and $\mathbf{I}_f$ is the identity matrix of the same order as the generator matrix $\mathbf{F}$, $\otimes$ denotes the direct (Kronecker) product, and the different generator matrices in the block diagonal matrix are those for the different rotational isomers for bond i, in the same sequence as the indexing of the columns in $\mathbf{U}_i$.

The virtue of the RIS approach lies in the form of Eq. (3): A global exact average (subject, of course to the simplification of factorizability as set forth in Eq. (1), Eq. (2), and Eq. (3)) over a very large number of chain conformations can be obtained with a simple matrix product, calculable with a trivial computational effort that is provided by any modern Personal Computer. And since each $\mathbf{U}_i$ and each $\mathbf{F}_i$ can be defined separately, practically any chemical structure will yield to an RIS treatment.

Today generator matrices $\mathbf{F}$ are known for many properties,[10] among them the population of different conformers, the relative stability of macromolecular diastereoisomers, the mean-square end-to-end distance, the radius of gyration, the molecular dipole moment, the molecular optical anisotropy (and, with it, the stress-optical coefficient, the Kerr effect, depolarized light scattering, and the

Table 1. Corresponding concepts between the continuum and RIS representation[a]

	continuum[b]	RIS
conformational partition function	$z = \int_0^{2\pi} \exp(-\widehat{U}/kT)\, d\varphi$	$z = \sum u_i$ where $u_i = \exp(-G_i/RT)$ $= \exp(S_i/R)\exp(-E_i/RT)$
population	$p_i = z^{-1} \int_{\text{domain}} \exp(-\widehat{U}/kT)\, d\varphi$	$p_i = z^{-1} u_i$
representative energy	$\langle \widehat{U} \rangle_i = z^{-1} \int_{\text{domain}} \widehat{U} \exp(-\widehat{U}/kt)\, d\varphi$	$E_i \approx H_i$
representative entropy	$\langle S \rangle_i = z^{-1} \int_{\text{domain}} \dfrac{\widehat{U}}{T} \exp(-\widehat{U}/kt)\, d\varphi$	S_i
representative angle	$\langle \varphi \rangle_i = z^{-1} \int_{\text{domain}} \varphi \exp(-\widehat{U}/kT)\, d\varphi$	φ_i

[a] Adapted from ref. 10.
[b] $\widehat{U}$ stands for an appropriate potential of mean force.

Cotton-Mouton effect), NMR chemical shift and coupling constants, the optical rotation of polarized light and correlation coefficients between different properties. Extensions to incorporate long-range interactions have also been elaborated[11] and it has even been possible to adapt RIS theory for the description of the dynamics of transitions between rotational isomeric states.[12,13]

It is prudent to be aware of the implications of the replacement of continuous geometrical degrees of freedom by discrete states.[10] The (hypothetical) rotational isomers are though of as distinct chemical species that have associated thermodynamic state functions just like any compound. Nevertheless, there is a one-to-one correspondence between concepts in the continuous representation of conformation and the RIS model (see Table 1). The partition functions are of substantially different character, however, one being a definite integral over a continuous domain and the other a sum of arbitrarily many terms, and care must be taken in manipulating the RIS partition function: While the usual derivatives of ln Z are relatively harmless (e.g. $\partial \ln Z/\partial T$), the magnitude of ln Z itself is strongly dependent on arbitrary technical features and should, in general, not be employed for the estimation of thermodynamic state functions such as entropy.

---

[11] Mattice WL (1981) Macromolecules 14:1491
[12] Jernigan RL, Karasz FE (ed) (1972) In: Dielectric properties of polymers. Plenum, New York, p 99
[13] Bahar I, Erman B, Monnerie L (1994) Adv Polym Sci. 116:145

3
Publications on Fundamental Aspects of RIS Theory

Abe A (1970) J Am Chem Soc 92:1136
Abe A (1970) Polymer J 1:232
Allegra G (1968) Makromol Chem 117:12
Allegra G (1968) Makromol Chem 117:24
Allegra G, Immirzi A (1969) Makromol Chem 124:70
Bahar I, Erman B, Monnerie L (1994) Adv Polym Sci 116:145
Birshtein TM, Ptitsyn OB (1959) Zh Tekhn Fiz 29:1048
Flory PJ, Mark JE, Abe A (1966) Am Chem Soc 88:639
Flory PJ (1969) Statistical mechanics of chain molecules. Wiley-Interscience New York; reprinted with the same title by Hanser, München 1989
Flory PJ (1974) Macromolecules 7:381
Flory PJ, Fujiwara Y (1969) Macromolecules 2:315
Flory PJ, Sundararajan PR, DeBold LC (1974) J Am Chem Soc 96:5015
Fujiwara Y, Flory PJ (1970) Macromolecules 3:280
Fujiwara Y, Flory PJ (1970) Macromolecules 3:288
Gotlib YuYa (1959) Zh Tekhn Fiz 29:523
Hartmann M (1989) J Macromol Sci-Phys, B 28:389
Hoeve CAJ (1960) J Chem Phys 32:888
Jernigan RL (1972) In: Karasz FE (ed) Dielectric Properties of Polymers. Plenum, New York, p 99
Kramers HA, Wannier GH (1941) Phys Rev 60:252
Lifson S (1958) J Chem Phys 29:80
Lifson S (1958) J Chem Phys 29:89
Lifson SJ (1959) J Chem Phys 30:964
Mansfield ML (1983) Macromolecules 16:1863
Mark JE (1974) Acc Chem Res 7:218
Mark JE (1976) J Polym Sci; Symp 54:91
Mark JE (1979) Acc Chem Res 12:49
Mattice WL (1981) Macromolecules 14:1491
Mattice WL, Suter UW (1994) Conformational theory of large molecules; The rotational isomeric state model in macromolecular systems, Wiley, New York
Moritani T, Fujiwara Y (1973) J Chem Phys 59:1175
Nagai KJ (1959) J Chem Phys 31:1169
Nagai KJ (1962) J Chem Phys 37:490
Natta G, Corradini P, Ganis P (1962) J Polym Sci 58:1191
Suter UW (1981) Macromolecules 14:523
Viswanadhan VN, Mattice WL (1987) Macromolecules 20:685
Volkenstein M (1958) J Polym Sci 29:441
Volkenstein MV (1951) Dokl Akad Nauk SSSR 78:879

4
Notation Used and Structure of the Collection of RIS Models

The most pertinent features of several hundred RIS models are collected on the following pages. For all properties amenable to RIS treatment, the information contained in the statistical weight matrices ist mandatory. For most properties, geometrical information, such as bond lengths, bond angles, and torsion angles are also necessary. The following collection consequently focuses on those parameters. If further details are required, the reader should consult the original literature.

Table 2. Labeling scheme for the RIS models

class label	class comprises	remarks
S	simple symmetric chains	maximum 1 relevant torsion angle in side chains
A	(simple) asymmetric chains	including most vinyl homopolymers maximum 1 relevant torsion angle in side chains
V	chains with rigid moieties and virtual bonds	e.g. polyesters, polyamides, and polybutadienes
C	copolymers	
X	branched polymer molecules, combs, stars	including chains with articulated side chains (more than 1 relevant torsion angle in side chains)
N	naturally occurring polymers	e.g. polysaccharides, polypeptides, polynucleotides

The models presented in this review are grouped into classes according to common traits of the macromolecules described by the models. Every model is identified by a unique label consisting of the "main label", according to Table 2, and a serial number.

Within one class of chains (**S**, **A**, **V**, **C**, **X**, or **N**), the models are grouped according to increasing complexity of their constitution: First, those that can be described with one constitutional repeat unit are listed [poly(A)], then those with two constitutional repeat units [poly(AB), then poly(AAB), then poly(AAAB), etc.], those with three constitutional repeat units [e.g., poly(ABC)], and so on. Wherever practical, the Organic Chemistry priorities of substituents was employed. To a certain extent, this structure is arbitrary, of course.

In general, the values for all pertinent parameters given in the original literature are reported; such numbers are printed in roman (upright) type. In order to facilitate comparison between models, however, some values have been *deduced* by the present authors from the original literature, and such numbers are typed in *italic* print.

Most RIS models follow a simple numbering scheme for the skeletal bonds of the main chain, as was implied above. There are classes of macromolecules, however, where an alternative numbering scheme may be beneficial; foremost among these are the vinyl polymers where it has become customary to group the two consecutive skeletal bonds between substituent-carrying carbon atoms into *diads* of *meso* or a *racemo* diastereoisomeric character and to use for the definition of torsion angles *right-* as well as *left-handed* coordinate systems in order to simplify the description of the conformational behavior of these chains. Consequently, the statistical weight matrices are alternately one for the first bond in a diad ($\mathbf{U}_p$) and one for the second in a diad ($\mathbf{U}_m$ or $\mathbf{U}_r$, depending on the diastereoisomeric character of the diad). Depending on the original reference, other schemes are also employed for such asymmetric chains where it seemed appropriate, e.g. subscripts *i* for *iso* and *s* for *syndio* (sometimes *h* for *hetero*) or *ll* and *dl* and *ld* and *dd*. For details one should refer to a specialized text.[8,10]

5
Index of Compounds

This index contains references to those RIS models that were explicitly constructed for specific compounds. It is neither comprehensive nor complete, but is given here as an aid for the location of models for particular compounds.

Substance	Model numbers
agar	N 137
alginic acid	N 143, N 144, N 150
n-alkanes	S 001 – S 076
amylose	N 131, N 132, N 134, N 136, N 138, N 139, N 142, N 146, N 152
amylose tricarbanilate	N 153
1,4-BBM	S 162
bisphenol A polycarbonate	V 109 – V 115
bisphenol A polythiocarbonate	V 116
1,4-butanediol dibenzoate	V 064
carrageenan	N 137
5CB	V 089, V 092
8CB	V 090, V 092
cellobiose	N 133
cellulose	N 131, N 132, N 133, N 140, N 141, N 146, N 148, N 153
cellulose tricarbanilate	N 153
chitin	N 141
chondroitin	N 137
chondroitin sulfate	N 137
cyclo-(glycyl glycyl glycyl prolyl prolyl)	N 030
cyclohexaamylose	N 139, N 142
DCP	A 002
DDP	S 180
dermatan sulfate	N 137
α,ω-dibromo-n-alkanes	S 002, S 011, S 013, S 021
α,ω-dibromoparafines	S 002, S 011, S 013, S 021
2,4-dichloro-n-pentane	A 002
diethylene glycol dibenzoate	V 044
α,ω-dihalo-n-alkanes	S 082
α,ω-dihydroperfluoro-n-alkanes	S 082
α,ω-diiodo polyoxyethylene	S 145
1,4-dimethoxybutane	S 162
1,3-dimethoxy-2,2-dimethylpropane	S 180
1,2-dimethoxyethane	S 146
1,2-dimethoxy-2-methylpropane	S 178
1,3-dimethoxypropane	S 158

Substance	Model numbers
dimethyl adipate	V 034, V 037
2,4-dimethylglutaric esters	A 078, A 080, A 082, A 084, A 086, A 088 – A 090, V 034, V 036
dimethyl malonate	V 034
dimethyl sebacate	V 034
dimethyl succinate	V 034, V 035, X 021
di(2-naphtyl)adipate	V 072
di(2-naphtyl)glutarate	V 071
di(2-naphtyl)pimelate	V 073
di(2-naphtyl)suberate	V 074
di(2-naphtyl)succinate	V 070
DMMP	S 178
1,3-DMP	S 158
elastin	N 101
β-endorphin	N 080
1,2-ethanediol dibenzoate	V 061
4-ethoxy-4'-biphenylyl cyanide	V 088
gellan	N 156
glucan	N 155
glucomannan	N 150
n-hexane	S 017
1,6-hexanediol dibenzoate	V 068
hyaluronic acid	N 137, N 148
keratan sulfate	N 137
laminaran	N 148
lecithin	N 118
β-lipotropin	N 080
maltose	N 138
mannan	N 147
Myelin Basic Protein	N 080
2OCB	V 088
4-n-octyl-4'-cyanobiphenyls	V 090, V 092
ODB	V 044
PACA	V 027, V 028
P1B	X 019, X 038
1,4-PBD	V 001, V 003, V 005, V 007, V 009 – V 011, V 014, V 016 – V 018, V 021 – V 024
PBO	V 130, V 138
PBT	V 130, V 138
PC	V 109 – V 115
PCA	A 079
PCCHA	X 020
PCCS	V 084

Index of Compounds

Substance	Model numbers
PCDO	V 085
PDA	V 076
PDAS	X 027 – X 030, X 053
PDBzI	X 021
PDCMI	X 023
PDCP	S 134 – S 136
PDEI	V 053
PDEP	V 054
PDES	X 028
PDET	V 043 – V 046
PDG	A 080
PDMS	S 116 – S 130, X 027, X 031
PDMSM	S 113 – S 115
PDO	S 181, S 182
PDPT	V 055
PDS	S 184, V 077
PDTC	S 175
PDTT	V 049
PE	S 001 – S 076, X 031
pectic acid	N 131, N 150
PEMA	X 022
1,5-pentanediol dibenzoate	V 066
4-n-pentyl-4'-cyanobiphenyls	V 089, V 092
PEO	S 137 – S 156, S 171
PEP	A 099
PES	S 169 – S 171
PET	V 040 – V 042, V 060, V 126
PEVB	C 003
PFM	A 001
PHB	V 118, V 125, V 126, V 129
PHBA	V 118, V 125, V 126, V 129
PHNA	V 121
PIB	S 088 – S 097
1,4-PIP	V 002, V 004, V 006, V 008, V 012, V 013, V 015 – V 017
PiPMA	X 022
PMA	A 074 – A 078
PMBzI	X 021
PMCPA	A 085
PMMA	A 092 – A 098, X 022
PM$_2$O	S 137 – S 156
P4MP	X 019
PMPS	A 111
PMS	A 091

Substance	Model numbers
PM$_2$S	S 169 – S 171
PM$_3$S	S 172, S 173
PM$_5$S	S 174
PMTC	V 116
PMTHF	A 102
PNA	V 039
PNS	V 038
POCPA	A 083
PODM$_3$	S 183
PODME	S 179
PODTD	S 188
POE	S 137 – S 156, S 171
POLA	V 059
poly(L-alanine)	N 006, N 010, N 012, N 014 – N 017, N 020, N 027, N 028, N 033, N 037, N 040 – N 042, N 051, N 058, N 059, N 064, N 065, N 077, N 079, N 080, N 084, N 085, N 89
poly(L-alanine-co-D-alanine)	N 041
poly(alkyl vinyl ether)s	X 011, X 015
polyamide 6	V 027, V 028
polyamide 66	V 029
poly(ε-aminocaproamide)	V 027, V 028
poly(6-aminocaproamide)	V 027, V 028
poly(arabinopyranose)	N 145
poly(L-arginine)	N 064, N 065, N 077, N 079, N 080, N 084, N 089
poly(L-asparagine)	N 064, N 065, N 077, N 079 – N 081
poly(L-aspartic acid)	N 022, N 064, N 065, N 077, N 079, N 080, N 084, N 089
poly(p-benzamide)	V 123, V 124
polybenzobisoxazole	V 130, V 138
polybenzobisthiazole	V 130, V 138
poly(β-benzyl-L-aspartate)	N 027, N 028
poly(γ-benzyl-L-glutamate)	N 002, N 025, N 104
poly{2,2-bis[4-(2-hydroxyethoxy)phenyl]-propane adipate}	V 076
poly{2,2-bis[4-(2-hydroxyethoxy)phenyl]-propane sebacate}	V 077
poly(p-bromoostyrene)	A 061
1,4-polybutadiene	V 001, V 003, V 005, V 007, V 009 – V 011, V 014, V 016 – V 018, V 021 – V 024
poly(n-butene-1)	A 042, X 019

Substance	Model numbers
poly(butylene terephthalate)	V 065
poly(*tert*-butyl vinyl ketone)	A 068
polycarbonate from 2,2-bis (4-hydroxyphenyl)propane	V 109 – V 115
polycarbonates	V 115
poly(2-chlorocyclohexyl acrylate)	X 020
poly(*o*-chlorophenyl acrylate)	A 083
poly(*m*-chlorophenyl acrylate)	A 085
poly(*p*-chlorophenyl acrylate)	A 087
poly(*p*-chlorostyrene)	A 058 – A 061
poly(*p*-chlorostyrene-*co-p*-methylstyrene)	C 010, C 011
polycyclobutene	V 001, V 003, V 005, V 007, V 009 – V 011, V 014, V 016 – V 018, V 021 – V 024
polycyclododence	V 020
poly(*cis/trans*-1,4-cyclohexane dimethanol-*alt*-formaldehyde)	V 085
poly(cyclohexane sulphone)	V 140
poly(cyclohexyl acrylate)	A 079
poly(*trans*-1,4-cyclohexylene-dimethylene-oxymethylene oxide)	V 141
polycyclooctene	V 019
poly(L-cysteine)	N 064, N 065, N 077, N 079, N 080, N 084, N 089
poly(decamethylene oxide)	S 165
poly(dialkyl siloxane)s	X 027 – X 030, X 053
poly(dibenzyl itaconate)	X 021
poly(1,1-dibromoethylene)	S 101
poly(di-*tert*-butyl siloxane)	X 056
poly(1,1-dichloroethylene)	S 098, S 099
polydichlorophosphazane	S 134 – S 136
catena-poly[(dichlorophosphorous)-μ-nitrido]phosphate	S 134 – S 136
poly(dicyclohexylmethylene itaconate)	X 023
poly(diethylene glycol isophthalate)	V 053
poly(diethylene glycol phthalate)	V 054
poly(diethylene glycol terephthalate)	V 043 – V 046, V 048
poly(diethylsiloxane)	X 028
poly(1,1-difluoroethylene)	S 100, S 102
poly(di-*n*-hexylsilane)	X 049
poly(di-*n*-hexylsilyene)	X 046
polydihydrogenesiloxane	S 131
poly(di-isopropyl siloxane)	X 030
poly(1,1-dimethylethylene)	S 088 – S 097

Substance	Model numbers
poly[(S)-3,7-dimethyl-1-octene]	X 037
poly(3,3-dimethyl oxetane)	S 181, S 182
poly(2,6-dimethyl-1,4-phenylene oxide)	V 112, V 122
polydimethylsilaethylene	S 113 – S 115
polydimethylsilmethylene	S 113 – S 115
polydimethylsiloxane	S 116 – S 130, X 027, X 031
polydimethylsilylene	S 087
poly(3,3-dimethylthiethane)	S 184
poly[(2,2-dimethyl)trimethylene oxide]	S 183
poly(1,3-dioxa-6-thiocane)	S 186
poly(1,3-dioxocane)	S 168
poly(1,3-dioxolane)	S 166 – S 167
poly(1,3-dioxonane)	S 168
catena-poly[(dioxophosphorous)-µ-oxo]phosphate	S 132, S 133
poly(2,6-diphenyl-1,4-phenylene oxide)	V 122
poly(dipropylene glycol terephthalate)	V 055
poly(di-*n*-propylsiloxane)	X 029
poly(1,3-dithiocane)	S 175
poly(ditrimethylene glycol terephthalate)	V 049
polyesters from 2,6-naphtalene dicarboxylic acid	V 075
polyethylene	S 001 – S 076, X 031
poly(ethylene-*co*-1-butene)	X 001, X 047
poly(ethylene-*co*-carbon monoxide)	C 001
polyethyleneglycol dimethyl ether	S 139, S 140, S 142, S 149
poly(ethylene oxide)	S 137 – S 156, S 171
poly(ethylene-*alt*-propylene)	X 051
poly(ethylene-*co*-propylene)	C 012, C 017, C 019, C 022
poly(ethylene sulfide)	S 169 – S 171
poly(ethylene terephthalate)	V 040 – V 042, V 060, V 126
poly(ethylene-*co*-vinyl acetate)	C 008
poly(ethylene-*co*-vinyl bromide)	C 003
poly(ethylene-*co*-vinyl chloride)	C 002, C 004, C 018
polyehtylethylene	A 042
poly(ethyl methacrylate)	X 022
polyfluoromethylene	A 001
poly(galactopyranose)	N 145
poly(glucopyranose)	N 145
poly(L-glutamic acid)	N 022, N 046, N 064, N 065, N 077, N 079, N 080, N 084, N 089
poly(L-glutamic acid-*co*-L-alanine)	N 046
poly(L-glutamine)	N 064, N 065, N 077, N 079, N 080, N 084, N 089

Substance	Model numbers
polyglycine	N 012, N 014 – N 016, N 020, N 033, N 040 – N 042, N 051, N 059, N 064, N 065, N 077, N 079, N 080, N 084, N 089
poly(guluronic acid)	N 143
poly(hexamethylene adipamide)	V 029
poly(hexamethylene oxide)	S 164
poly(hexamethylene terephthalate)	V 069
poly(L-histidine)	N 064, N 065, N 077, N 079, N 080, N 084, N 089
poly(4-hydroxybicyclo[2.2.2]octane-1-carboxylate)	V 126
poly[N^5-(4-hydroxybutyl)-L-glutamine]	N 035, N 075, N 077
poly(hydroxybutylglutamine-*ran*-L-arginine	N 071
poly(hydroxybutylglutamine-*ran*-L-asparagine)	N 073
poly(hydroxybutylglutamine-*ran*-L-aspartic acid)	N 072
poly(hydroxybutylglutamine-*ran*-L-glutamic acid)	N 055
poly(hydroxybutylglutamine-*ran*-glycine)	N 036
poly(hydroxybutylglutamine-*ran*-L-histidine)	N 092
poly(hydroxybutylglutamine-*ran*-L-leucine)	N 045
poly(hydroxybutylglutamine-*ran*-L-lysine)	N 066
poly(hydroxybutylglutamine-*ran*-L-methionine)	N 070
poly(hydroxybutylglutamine-*ran*-L-serine)	N 047
poly(hydroxybutylglutamine-*ran*-L-threonine)	N 076
poly(hydroxybutylglutamine-*ran*-L-tyrosine)	N 061
poly(hydroxybutylglutamine-*ran*-L-valine)	N 052
poly(hydroxyethyl-L-glutamine)	N 086
poly[N^5-ω-hydroxyethyl-L-glutamine]	N 043
poly(γ-hydroxy-L-proline)	N 033, N 053
poly[N^5-(3-hydroxypropyl)-L-glutamine]	N 035
poly(hydroxypropylglutamine-*ran*-L-alanine)	N 044
poly(hydroxypropylglutamine-*ran*-hydroxybutylglutamine)	N 035
poly(hydroxypropylglutamine-*ran*-L-isoleucine)	N 081
poly(hydroxypropylglutamine-*ran*-L-leucine)	N 045
poly(hydroxypropylglutamine-*ran*-L-phenylalanine)	N 050
poly(hydroxypropylglutamine-*ran*-L-valine)	N 052, N 082
polyisobutylene	S 088 – S 087
poly(L-isoleucine)	N 064, N 065, N 077, N 079, N 080, N 084, N 089
1,4-polyisoprene	V 002, V 004, V 006, V 008, V 012, V 013, V 015 – V 017
poly(L-lactic acid)	V 032, V 033

Substance	Model numbers
poly(isopropyl methacrylate)	X 022
poly(L-leucine)	N 064, N 065, N 079, N 080, N 084, N 089
poly(L-lysine)	N 064, N 065, N 077, N 079, N 080, N 084, N 089
poly(mannopyranose)	N 145
poly(mannuronic acid)	N 143
polymers from (+)-catechin and (−)-epicatechin	V 137
poly(L-methionine)	N 027, N 064, N 065, N 079, N 080, N 084, N 089
poly(methyl acrylate)	A 074 – A 078
poly(methyl acrylate-co-vinyl chloride)	C 014
poly(methyl acrylate-co-vinylidene chloride)	C 014
poly(N-methyl-L-alanine)	N 040, N 041, N 046
poly(N-methyl-L-alanine-co-N-methyl-D-alanine)	N 040, N 041
poly(β-methyl-L-aspartate)	N 017
polymethylene	S 001 – S 076
poly(γ-methyl-L-glutamate)	N 017
poly(N-methylglycine)	N 040, N 041
poly[(S)-5-methyl-1-heptene]	X 037
poly[(S)-4-methyl-1-hexene]	X 019, X 037
poly(methyl methacrylate)	A 092 – A 098, X 022
poly(methyl methacrylate-co-styrene)	C 015
poly[(S)-6-methyl-1-octene]	X 037
polymethyloxirane	A 101
poly[(S)-3-methyl-1-pentene]	X 037
poly(4-methyl-1-pentene)	X 019
poly(2-methyl-6-phenyl-1,4-phenylene oxide)	V 122
polymethylphenylsiloxane	A 111
polymethylphenylsilylene	A 108, A 109
poly(α-methyl styrene)	A 091
poly(3-methyl tetrahydrofurane)	A 102
poly(methyl vinyl ketone)	A 067
poly(monobenzyl itaconate)	X 021
poly(neopentyl glycol succinate)	V 038
poly(trans-7-oxabicyclo[4.3.0]nonane)	X 052
poly(3-oxa-1,5-dithiadecamethylene)	S 188
poly(1-oxa-3-thiacyclopentane)	S 187
poly(3-oxy-1-benzoate)	V 119
poly(4-oxy-1-benzoate)	V 118, V 125, V 126, V 129
poly(oxy[1-tert-butyl ethylene)]	X 018, X 026
poly(oxy-1,1-dimethylethylene)	S 179
polxxyethylene	S 137 – S 156, S 171
polyoxythylene glycol dimethyl ether	S 139, S 140, S 142, S 149

Substance	Model numbers
poly(oxy[1-ethyl ethylene])	X 016, X 024
poly(oxy[1-isopropyl ethylene])	X 017, X 025
polyoxymethylene	S 103–S 112
poly(oxymethylene-1,4-*trans*-cyclohexylenemethyleneoxysebacoyl)	V 083
poly(oxymethylene-1,4-*cis*-cyclohexylenemethyleneoxysebacoyl)	V 084
poly(6-oxy-2-naphthoate)	V 121
polyoxyneopentyleneoxyadipoyl	V 039
poly(4'-oxyphenyl-carboxylate)	V 120
polyoxypropylene	A 101
poly(pentamethylene sulfide)	S 174
poly(pentamethylene terephthalate)	V 067
poly(*n*-pentene-1)	A 043, X 019
poly(4-phenoxyphenyl acrylate)	A 090
poly(phenyl acrylate)	A 081
poly(L-phenylalanine)	N 064, N 065, N 079
poly(*p*-phenylene isophthalate)	V 127
poly(1,4-phenylene oxide)	V 122
poly(*p*-phenylene terephthalamide)	V 123, V 124
poly(*p*-phenylene terephthalate)	V 127
polyphosphate	S 132, S 133
poly(L-proline)	N 018, N 020, N 023, N 024, N 026, N 032, N 033, N 046, N 049, N 054, N 059, N 064, N 065, N 077, N 079, N 083
polypropylene	A 020–A 041
poly(propylene-*co*-1-butene)	X 047
poly(propylene glycol terephthalate)	V 056
poly(propylene oxide)	A 101
poly(propylene-*co*-1-pentene)	C 005, C 009
poly(propylene-*co*-styrene)	C 006
poly(propylene sulfide)	A 103–A 105
poly(propylene terephthalate)	V 063
poly(propylene-*co*-vinyl chloride)	C 007, C 021
polypropylethylene	A 043
poly(2,6-pyridinediyl sulfide)	V 143
polypyrrole	V 144
poly(2,3-quinoxaline)	V 147
poly(rU)	N 122
polysarcosine	N 049
catena-poly(selenium)	S 084, S 085
poly(L-serine)	N 064, N 065, N 077, N 079, N 080, N 089

Substance	Model numbers
polysilane	S 086
polysilapropylene	A 110
polysilastyrerne	A 106, A 107
polystyrene	A 044–057
poly(styrene-*alt*-acrylonitrile)	C 013
poly(styrene-*co*-*p*-bromostyrene)	C 020
poly(styrene-*co*-*p*-chlorostyrene)	C 010, C 011
catena-poly(sulfur)	S 084, S 085
polytetrafluoroethylene	S 077–S 081, S 083
poly(tetramethylene oxide)	S 160–S 163
poly(tetramethylene terephthalate)	V 065
poly(tetramethyl-*p*-silphenylene-siloxane)	V 145
polythiocarbonate from 2,2-bis (4-hydroxyphenyl)propane	V 116
polythiocarbonates	V 117
poly(thiodiethylene glycol)	S 185
poly(thiodiethylene glycol terephthalate)	V 050
polythioethylene	S 169–S 171
poly(thiomethylene-1,4-*trans*-cyclohexylenemethylenethiomethylene)	V 142
polythiopropylene	A 103–A 105
poly(L-threonine)	N 064, N 065, N 077, N 079, N 080, N 084, N 089
poly(triethylene glycol terephthalate)	V 047, V 052
polytrifluoroethylene	A 011
poly(trimethylene oxide)	S 157–S 159
poly(trimethylene sulfide)	S 172, S 173
poly(trimethylene terephthalate)	V 063
poly(L-tryptophane)	N 064, N 065, N 079, N 080, N 084, N 089
poly(L-tyrosine)	N 017, N 064, N 065, N 077, N 079, N 080, N 084, N 089
poly(L-valine)	N 017, N 064, N 065, N 077, N 079, N 080, N 084, N 089
poly(vinyl acetate)	A 071–A 073
poly(vinyl alcohol)	A 069, A 070
poly(vinyl bromide)	A 012, A 013, A 015, A 016, A 018, A 019
poly(*N*-vinylcarbazole)	A 065, A 066
poly(vinyl chloride)	A 002–A 008, A 014, A 016–A 019
poly(vinyl fluoride)	A 009, A 010
poly(vinylidene bromide)	S 101
poly(vinylidene chloride)	S 098, S 099

Index of Compounds

Substance	Model numbers
poly(vinylidene fluoride)	S 100, S 102
poly(2-vinylpyridine)	A 062
poly(n-vinylpyrrolidone)	A 063, A 064
poly(xylopyranose)	N 145
POM	S 103 – S 112
POM_3	S 157 – S 159
POM_4	S 160 – S 163
POM_6	S 164
POM_{10}	S 165
POP	A 101
POTC	S 187
PP	A 020 – A 041
P1P	X 019, X 038
PPA	A 081
PPCPA	A 087
PPCS	A 058 – A 061
PPIA	V 127
PPO	A 101, V 112, V 122
PPOA	A 090
PPT	V 056
PPTA	V 123, V 124, V 127
PPyS	V 143
1,3-propanediol dibenzoate	V 062
PS	A 044 – 057
PSDET	V 050
PS4MH	X 019, X 037
PSP	A 110
PTCDM	V 141
PTCMT	V 142
PTCS	V 083
PTE	S 169 – S 171
PTET	V 047, V 052
PTF_3	A 011
PTFE	S 077 – S 081, S 083
PTMPS	V 145
PTP	A 103 – A 105
pullulan	N 157
PVA	A 069, A 070
PVAc	A 071 – A 073
PVB	A 012, A 013, A 015, A 016, A 018, A 019
PVC	A 002 – A 008, A 014, A 016 – A 019
PVDB	S 101
PVDC	S 098, S 099

Substance	Model numbers
PVDF	S 100, S 102
PVF	A 009, A 010
PVK	A 065, A 066
PVP	A 063, A 064
P2VP	A 062
PXS	S 176
rhamsan	N 156
SDB	V 051
TCH	A 002
thiodiethylene glycol dibenzoate	V 051
triacetin	X 009
2,4,6-trichloro-n-heptane	A 002
tryglycerides	X 010
tRNA^{Phe*}	N 130
tropomyosin	N 096, N 125, N 129
welan	N 156
xylan	N 133, N 135, N 141, N 147

6
Collection of RIS Models

S 001

polyethylene, polymethylene, PE

Hoeve, C. A. J. *J. Chem. Phys.* **1961**, *35*, 1266.

Bond Length [pm]	Valence Angles [°]	Torsion Angles [°]	ξ (for 433 K)	ξ_0	E_ξ [kJ·mol^{-1}]		
C-C : 153	C-C-C : 112	t : 180	σ = 0.56	1	2.1		
		g$^+$: 60	ω [a] = 0.092 or 0.125	1	8.6 or 7.5	$U = \begin{bmatrix} 1 & \sigma & \sigma \\ 1 & \sigma & \sigma\omega \\ 1 & \sigma\omega & \sigma \end{bmatrix}$	rows and columns: t, g$^+$, g$^-$
		g$^-$: −60					

[a] In the original paper, combined values of matrix elements $u_{2,3} = u_{3,2}$ (= $\sigma\omega$) are given.

Calcd. quantities: $<r^2>_0 / nl^2$ = 6.75 or 6.55   (*Exptl.*: 6.55)

(433 K) $d (\ln <r^2>_0) / d (\ln T)$ = −0.44 or −0.48   (*Exptl.*: −0.44)

S 002

α,ω-dibromo-n-alkanes

Leonard, Jr., W. J.; Jernigan, R. L.; Flory, P. J. *J. Chem. Phys.* **1965**, *43*, 2256.

Bond Length [pm]	Valence Angles [°]	Torsion Angles [°]	ξ (for 298 K)	ξ_0	E_ξ [kJ·mol^{-1}]			
C-C : 153	C-C-C : 112	t : 180	σ = 0.43	1	2.1			
C-Br : 194		g$^+$: 60	σ' = 1	1	0 [a]	$U = \begin{bmatrix} 1 & \sigma & \sigma \\ 1 & \sigma & 0 \\ 1 & 0 & \sigma \end{bmatrix}$	$U_2 = U_{n-1} = \begin{bmatrix} 1 & \sigma' & \sigma' \\ 1 & \sigma' & 0 \\ 1 & 0 & \sigma' \end{bmatrix}$	rows and columns: t, g$^+$, g$^-$
		g$^-$: -60						

[a] Changes of ± 0.4 kJ·mol^{-1} in E(σ') alter μ by less than 0.5 %.

Calcd. quantities: $\mu \equiv \langle \mu^2 \rangle^{1/2}$: μ varries from 2.2 D to 2.6 D for x = 0 to 7. (*Exptl.*: 1.95 D for 1,3-dibromopropane, 2.5 D for 1,10-dibromodecane)

$|d(\ln\mu)/dT|$: < 5 × 10^{-5} K^{-1} (*Exptl.*: 0 (to 5) × 10^{-4} K^{-1})

S 003

polyethylene, polymethylene, PE

Flory, P. J.; Jernigan, R. L. *J. Chem. Phys.* **1965**, *42*, 3509.

Bond Length [pm]	Valence Angles [°]	Torsion Angles [°]	ξ (for 433 K)	ξ_0	E_ξ [kJ · mol^{-1}]			
C-C : 153	C-C-C : 112	t : 180	σ = 0.56	1	2.1			
		g$^+$: 60				$U = \begin{bmatrix} 1 & \sigma & \sigma \\ 1 & \sigma & 0 \\ 1 & 0 & \sigma \end{bmatrix}$		rows and columns: t, g$^+$, g$^-$
		g$^-$: −60						

Comments:
Exact methods are developed for calculating second and fourth moments of chain molecules. The RIS approximation is adopted to represent the effects of hinderance potentials affecting bond rotations. No other approximations are invoked in deriving the statistical-mechanical averages.

Calcd. quantities: $<r^2>_0 / nl^2$ = 7.30 (for n = 1 031) $<r^4> / <r^2>^2$ = 1.65 (for n = 1027)

                                    = 7.34 (for n = 16 391)

S 004   (see also S 005 and S 006)

polyethylene, polymethylene, PE

Abe, A.; Jernigan, R. L.; Flory, P. J. *J. Am. Chem. Soc.* **1966**, *88*, 631.

Model A:

Bond Length [pm]	Valence Angles [°]	Torsion Angles [°]	ξ (for 413 K)	ξ_0	E_ξ [kJ·mol^{-1}]
C-C : 153	C-C-C : 112	t : 180	σ = 0.73 - 0.58	1	1.1 - 1.9
C-H : 110	H-C-H : 109	g$^+$: 67.5 (±3)	ω = 0.21 - 0.14	1	5.4 - 6.7
	H-C-C : 109	g$^-$: −67.5 (±3)			

$$U = \begin{bmatrix} 1 & \sigma & \sigma \\ 1 & \sigma & \sigma\omega \\ 1 & \sigma\omega & \sigma \end{bmatrix}$$

rows and columns: t, g$^+$, g$^-$

S 005   (see also S 004 and S 006)

Model B:

Bond Length [pm]	Valence Angles [°]	Torsion Angles [°]	ξ (for 413 K)	ξ_0	E_ξ [kJ·mol^{-1}]
C-C : 153	C-C-C : 112	t : 180	σ = 0.59 - 0.49	1	1.8 - 2.5
C-H : 110	H-C-H : 109	g$^+$: 60 (±3)	ω = 0.13 - 0.10	1	7.1 - 8.0
	H-C-C : 109	g$^-$: −60 (±3)			

$$U = \begin{bmatrix} 1 & \sigma & \sigma \\ 1 & \sigma & \sigma\omega \\ 1 & \sigma\omega & \sigma \end{bmatrix}$$

rows and columns: t, g$^+$, g$^-$

S 006 (see also S 004 and S 005)

Conformations: t:1, g*⁻:σ*, g⁻:σ, g⁺:σ, g*⁺:σ*

Model C:

Bond Length [pm]	Valence Angles [°]	Torsion Angles [°]	ξ	ξ₀ (for 413 K)	E_ξ [kJ·mol⁻¹]
C-C : 153	C-C-C : 112	t : 180	σ = 0.54	1	2.1
C-H : 110	H-C-H : 109	g*⁺ : 100	σ* = 0.087 [a]	1	8.4
	H-C-C : 109	g⁺ : 65	ω* = 0.54	1	2.1
		g⁻ : −65			
		g*⁻ : −100			

$$U = \begin{bmatrix} 1 & \sigma^* & \sigma & \sigma & \sigma^* \\ 1 & \sigma^* & \sigma & \sigma\omega^* & 0 \\ 1 & \sigma^* & \sigma & 0 & \sigma^*\omega^* \\ 1 & \sigma^*\omega^* & 0 & \sigma & \sigma^* \\ 1 & 0 & \sigma\omega^* & \sigma & \sigma^* \end{bmatrix}$$

rows and columns: t, g*⁺, g⁺, g⁻, g*⁻

[a] There is a misprint in the original paper (σ* = 0.048). Also, the elements 1,5 and 2,5 are exchanged in the original publication.

Comments:
Conformation energies of *n*-butane and *n*-pentane are calculated as functions of their C—C bond rotation angles using semiempirical expressions for the repulsive and attractive energies between nonbonded atom pairs and an intrinsic threefold torsion potential. The mean dimensions of PE chains are interpreted in light of the features of the conformation energy of first neighbor bonds in lower *n*-alkane homologs.

Calcd. quantities:	*Three state model*	*Five state model*	*Experimental*
$(\langle r^2 \rangle_0 / nl^2)_\infty$	= 6.8 to 7.6	= 7.4	= 6.8
$d \ln (\langle r^2 \rangle_0)_\infty / dT$	= −1.15 (± 0.1) × 10⁻³ K⁻¹	= −1.15 (± 0.1) × 10⁻³ K⁻¹	= −1.15 (± 0.1) × 10⁻³ K⁻¹

S 007

polyethylene, polymethylene, PE

Jernigan, R. L.; Flory, P. J. *J. Chem. Phys.* **1969**, *50*, 4165, 4178.

Bond Length [pm]	Valence Angles [°]	Torsion Angles [°]	ξ (for 413 K)	ξ_o	E_ξ [kJ·mol^{-1}]		
C-C : 153	C-C-C : 112	t : 180	$\sigma = 0.54$	1	2.1		
		g$^+$: 60	$\omega = 0.087$	1	8.4	$U = \begin{bmatrix} 1 & \sigma & \sigma \\ 1 & \sigma & \sigma\omega \\ 1 & \sigma\omega & \sigma \end{bmatrix}$	rows and columns: t, g$^+$, g$^-$
		g$^-$: −60					

Comments:
Although the potentials affecting the rotation about a skeletal bond in a chain molecule such as PE usually depend only on rotations of immediately adjoining bonds, the interdependence of rotational states may be transmitted over greater distances. In the case of PE or of POM the effective range of correlation is only four or five bonds. This is established by calculating *a priori* probabilities for rotational states of a bond as a function of its location relative to the chain termini and of the total chain length.

Calcd. quantities: $<r^2>_o / nl^2$ (for $n \to \infty$: 6.87) $<s^2>_o / nl^2$

$d (\ln <r^2>_o) / dT$ $<r^4>_o / <r^2>_o^2$ $<r^6>_o / <r^2>_o^3$ $<r^8>_o / <r^2>_o^4$

S 008

polyethylene, polymethylene, PE

Liberman, M. H.; Abe, Y.; Flory, P. J. *Macromolecules* **1972**, *5*, 550.

Bond Length [pm]	Valence Angles [°]	Torsion Angles [°]	ξ (for 413 K)	ξ_0	E_ξ [kJ · mol^{-1}]		
C-C : 153	C-C-C : 112	t : 180	σ = 0.35 - 0.65	1	3.6 - 1.5		
	H-C-H : 109.5	g$^+$: 60 *or* 70	ω = 0 - 0.30	1	∞ - 4.1	$U = \begin{bmatrix} 1 & \sigma & \sigma \\ 1 & \sigma & \sigma\omega \\ 1 & \sigma\omega & \sigma \end{bmatrix}$	rows and columns: t, g$^+$, g$^-$
	H-C-C : 109.5	g$^-$: -60 *or* -70					

Comments:
Obtained from comparing computation with expected values of the strain birefringence of PE and for PDMS networks, unswollen and swollen with diluents, over the temperature ranges 388 - 493 K and 288 - 363 K, respectively.

Calcd. quantities: $<r^2>_0 / nl^2$ $d (\ln <r^2>_0) / dT$ $\Delta\alpha$ $d (\ln \Delta\alpha) / dT$

polyethylene, polymethylene, PE

Flory, P. J. *J. Polym. Sci.; Polym. Phys. Ed.* **1973**, *11*, 621.

Bond Length [pm]	Valence Angles [°]	Torsion Angles [°]	ξ (for 300 K)	ξ_0	E_ξ [kJ · mol^{-1}]
		t : 180	σ = 0.43	1	2.1
		t*⁺ : 140	σ^* = 0.025	1	9.2
		t*⁻ : −140	ω^* = 0.43	1	2.1
		g*⁺ : 100	$\beta^* \approx \sigma^*$		
		g*⁻ : −100			
		g⁺ : 60 *or* 65			
		g⁻ : −60 *or* −65			

$$U = \begin{bmatrix} 1 & \beta^* & \sigma^* & \sigma & \sigma & \sigma^* & \beta^* \\ 1 & \beta^* & \sigma^* & \sigma & \sigma & \sigma^* & \beta^* \\ 1 & \beta^* & \sigma^* & \sigma & \sigma\omega^* & \sigma^* & \beta^* \\ 1 & \beta^* & \sigma^* & \sigma & 0 & \sigma^*\omega^* & \beta^* \\ 1 & \beta^* & \sigma^*\omega^* & 0 & \sigma & \sigma^* & \beta^* \\ 1 & \beta^* & \sigma^* & \sigma\omega^* & \sigma & \sigma^* & \beta^* \\ 1 & \beta^* & \sigma^* & \sigma & \sigma & \sigma^* & \beta^* \end{bmatrix}$$

rows and columns: t, t*⁺, g*⁺, g⁺, g⁻, g*⁻, t*⁻

Comments:
The role of nonstaggered conformations in PE and PP is discussed in some detail. In incorporating such conformations into RIS treatments, it is essential to choose rotational states as to assure equitable sampling of configuration space. Tacit identification of rotational states with minima in the conformation energy surface, a common practice, may lead to serious errors. The significance and limitations of conformational energy calculations are discussed.

polyethylene, polymethylene, PE

Mattice, W. L.; Santiago, G. *Macromolecules* **1980**, *13*, 1560.

Bond Length [pm]	Valence Angles [°]	Torsion Angles [°]	ξ (for 300 K)	ξ_0	E_ξ [kJ·mol^{-1}]	
C-C : 153	C-C-C : 112	t : 180	$\sigma = 0.432$	1	2.1	*Unperturbed chain:*
		g$^+$: 60	$\omega = 0.034$	1	8.4	
		g$^-$: -60	$\tau = 1$	1	0 a)	
			$\psi = 1$	1	0 a)	

Unperturbed chain:

$$U = \begin{bmatrix} \tau & \sigma & \sigma \\ 1 & \sigma\psi & \sigma\omega \\ 1 & \sigma\omega & \sigma\psi \end{bmatrix} \text{ or } \begin{bmatrix} \tau & 2\sigma \\ 1 & \sigma(\psi+\omega) \end{bmatrix}$$

rows and columns: t, g$^+$, g$^-$
or:
rows and columns: t, g$^\pm$

Perturbed chain:

Connection between τ_i and ψ_i:

$$\tau_i = \sigma(\psi_i + \omega) + 2(p_t - p_{g\pm})\{2\sigma[1-(p_t - p_{g\pm})^2]^{-1}\}^{1/2}$$

$$= 1 + [1-(1-b/K)(n+1-2i)^2(n-3)^{-2}] K n^{1/5}$$

$$U_i = \begin{bmatrix} \tau & \sigma & \sigma \\ 1 & \sigma\psi & \sigma\omega \\ 1 & \sigma\omega & \sigma\psi \end{bmatrix}_i$$

rows and columns: t, g$^+$, g$^-$

---

a) For unperturbed chains.

Comments:
Matrix methods are adapted so that they reproduce several properties of chains perturbed by long-range interactions. This objective is achieved through a modification in the significance of certain elements in the statistical weight matrix. Matrices used can be of the same dimensions as those used to successfully treat the unperturbed chain. The model yields perturbed chains with the following characteristics: (1) Bond length and bond angles are the same as those for the unperturbed chain; (2) $(\alpha_s^5 - \alpha_s^3)/n^{1/2}$ reaches a nonzero asymptotic limit at large n; (3) large expansions are achieved without alteration in the *a priori* probability for the *trans* placement in a long chain. (4) The effect of the perturbation on the *i*th bond increases as n increases; (5) Long-range interactions exert perturbations preferentially in the middle of the chain; (6) Perturbations are independent of the direction selected for indexing bonds in the chain. The magnitude of the perturbation for a polymer of specified n in a particular polymer-solvent system depends on an adjustable parameter denoted by K.

Calcd. quantities: Expansion factors $\alpha_r^2 = <r^2>/<r^2>_0$ and $\alpha_s^2 = <s^2>/<s^2>_0$ are calculated for different values of τ and ψ.

α,ω-dibromo-n-alkanes

Khanarian, G.; Tonelli, A. E. *J. Chem. Phys.* **1981**, *75*, 5031.

Bond Length	Valence Angles	Torsion Angles	ξ	ξ_0	E_ξ				
[pm]	[°]	[°]	(for 298 K)		[kJ · mol^{-1}]				
C-C : 154		t : 180	σ = 0.43	1	2.1				
	g$^\pm$: ±60		ω = 0.034	1	8.4	$U_2 = \begin{bmatrix} 1 & \sigma' & \sigma' \\ 1 & \sigma' & \sigma' \\ 1 & \sigma' & \sigma' \end{bmatrix}$	$U_3 = \begin{bmatrix} 1 & \sigma & \sigma \\ 1 & \sigma & 0 \\ 1 & 0 & \sigma \end{bmatrix}$	$U = \begin{bmatrix} 1 & \sigma & \sigma \\ 1 & \sigma & \sigma\omega \\ 1 & \sigma\omega & \sigma \end{bmatrix}$	$U_{n-1} = \begin{bmatrix} 1 & \sigma' & \sigma' \\ 1 & \sigma' & 0 \\ 1 & 0 & \sigma' \end{bmatrix}$
			σ' = 1 (or ±0.7)	1	0 (or ±0.84)				
		Except for 2, n-2: g$^\pm$: ±80 to ±100						rows and columns: t, g$^+$, g$^-$	

Comments:
Based on comparison of calculation and experiment for the molar Kerr constants $_mK$ and the dipole moments squared $\langle\mu^2\rangle$ of the α,ω-dibromoalkanes (x ≤ 17) in cyclohexane.

Calcd. quantities: $_mK$ ≈ 145 × 10^{-12} cm^7 s.c.$^{-2}$ mol^{-1} (*Exptl.*: 157 × 10^{-12} cm^7 s.c.$^{-2}$ mol^{-1})

$\langle\mu^2\rangle$ ≈ 7.7 × 10^{-36} s.c.^{2} cm^2 (*Exptl.*: 7.70 × 10^{-36} s.c.^{2} cm^2)

S 012

polyethylene, polymethylene, PE

Abe, A. *Polymer J.* **1982**, *14*, 427.

Bond Length [pm]	Valence Angles [°]	Torsion Angles [°]	ξ (for 413 K)	ξ_o	E_ξ [kJ · mol^{-1}]		
C-C : 153	C-C-C : 112	t : 180	$\sigma = 0.46$	0.9	2.34		
			$\omega = 0.103$	1.1	8.12		
		g$^+$: 63.5	Alternatively:			$U = \begin{bmatrix} 1 & \sigma & \sigma \\ 1 & \sigma & \sigma\omega \\ 1 & \sigma\omega & \sigma \end{bmatrix}$	rows and columns: t, g$^+$, g$^-$
		g$^-$: −63.5	$\sigma = 0.51$ or 0.52	1	2.30 or 2.26		
			$\omega = 0.068$ or 0.084	1	9.21 or 8.50		

Comments:
Comment on a paper of *Oyama* and *Shiokawa* [Oyama, T.; Shiokawa, K. *Polym. J.* **1981**, *13*, 1145] who calculated bond conformation and unperturbed dimensions for the PE chain by using a direct integration method.

Calcd. quantities:

C_∞ = 7.84   (*Alternatively:* 7.65 or 7.51)

$d(\ln C_\infty)/dT$ = −1.23 × 10^{-3} K^{-1} (*Alternatively:* −1.06 × 10^{-3} K^{-1} or −1.10 × 10^{-3} K^{-1})

S 013

α,ω-dibromo-n-alkanes

Abe, A.; Furuya, H.; Toriumi, H. *Macromolecules* **1984**, *17*, 684.

Bond Length [pm]	Valence Angles [°]	Torsion Angles [°]	ξ (for 298 K)	ξ_0	E_ξ [kJ·mol^{-1}]
C-C : 153	C-C-C : 112	t : 180	σ = 0.43	1	2.1
C-Br : 194	Br-C-C : 112	g$^+$: 67.5	σ' = 1	1	0
		g$^-$: −67.5	$\omega = \omega' = 0.034$	1	8.4

$$U_2 = \begin{bmatrix} 1 & \sigma' & \sigma' \\ \sigma' & \sigma' & \sigma' \\ \sigma' & \sigma' & \sigma' \end{bmatrix} \quad U_3 = \begin{bmatrix} 1 & \sigma & \sigma \\ 1 & \sigma & \sigma\omega' \\ 1 & \sigma\omega' & \sigma \end{bmatrix} \quad U = \begin{bmatrix} 1 & \sigma & \sigma \\ 1 & \sigma & \sigma\omega \\ 1 & \sigma\omega & \sigma \end{bmatrix} \quad U_{n-1} = \begin{bmatrix} 1 & \sigma' & \sigma' \\ 1 & \sigma' & \sigma'\omega' \\ 1 & \sigma'\omega' & \sigma' \end{bmatrix}$$

rows and columns: t, g$^+$, g$^-$

Comments:
Studies are extended to include α,ω-dihydroperfluoroalkanes.

Calcd. quantities: Distribution curves $P(\theta)$ of the angle θ defined by the two terminal bond vectors; Order parameter S

S 014 (see also S 015 and S 016)

polyethylene, polymethylene, PE

Mattice, W. L. *Comput. Polym. Sci.* **1991**, *1*, 173.

Model A:

Bond Length [pm]	Valence Angles [°]	Torsion Angles [°]	ξ (for 413 K)	ξ_o	E_ξ [kJ · mol^{-1}]		
C-C : 153	C-C-C : 112	t : 180	$\sigma = 0.38$	1	3.3		
		g$^+$: 67.5	$\omega = 0.26$	1	12.5	$U = \begin{bmatrix} 1 & \sigma & \sigma \\ 1 & \sigma\psi & \sigma\omega \\ 1 & \sigma\omega & \sigma\psi \end{bmatrix}$	rows and columns: t, g$^+$, g$^-$
		g$^-$: −67.5	$\psi = 1$	1	0		

S 015 (see also S 014 and S 016)

t : 1

$g^+ : \sigma$

$g^- : \sigma$

ω

ψ

κ

Model B:

Bond Length [pm]	Valence Angles [°]	Torsion Angles [°]	ξ (for 413 K)	ξ_o	E_ξ [kJ · mol^{-1}]		
C-C : 153	C-C-C : 112	t : 180	$\sigma = 0.54$	1	2.09		
		$g^\pm : \pm 67.5$	$\omega = 0.16$	1	6.28		
		g (in $g^\pm g^\pm$) : 60 to 62	$\psi = 1.22$	1	− 0.67		
		t (in $g^\pm tg^\mp$) : 188	$\kappa = 0.86$	1	0.50		

$$U = \begin{bmatrix} 1 & 1 & 1 & 0 & 0 & 0 & 0 & 0 & 0 \\ 0 & 0 & 0 & \sigma & \sigma & \sigma & 0 & 0 & 0 \\ 0 & 0 & 0 & 0 & 0 & 0 & \sigma & \sigma & \sigma \\ 1 & 1 & \kappa & 0 & 0 & 0 & 0 & 0 & 0 \\ 0 & 0 & 0 & \sigma\psi & \sigma\psi & \sigma\psi & 0 & 0 & 0 \\ 0 & 0 & 0 & 0 & 0 & 0 & \sigma\omega & \sigma\omega & \sigma\omega \\ 1 & \kappa & 1 & 0 & 0 & 0 & 0 & 0 & 0 \\ 0 & 0 & 0 & \sigma\omega & \sigma\omega & \sigma\omega & 0 & 0 & 0 \\ 0 & 0 & 0 & 0 & 0 & 0 & \sigma\psi & \sigma\psi & \sigma\psi \end{bmatrix}$$

rows and columns:

tt, tg$^+$, tg$^-$, g$^+$t, g$^+$g$^+$,

g$^+$g$^-$, g$^-$t, g$^-$g$^+$, g$^-$g$^-$

S 016 (see also S 014 and S 015)

Model C:

Bond Length [pm]	Valence Angles [°]	Torsion Angles [°]	ξ (for 413 K)	ξ_0	E_ξ [kJ · mol^{-1}]		
C-C : 153	C-C-C : 112	t : 180	σ = 0.54	1	2.1		
		$g^\pm$: ± 67.5	ω = 0.16	1	6.3	$U = \begin{bmatrix} 1 & \sigma & \sigma \\ 1 & \sigma\psi & \sigma\omega \\ 1 & \sigma\omega & \sigma\psi \end{bmatrix}$	rows and columns: t, g^+, g^-
		g (in $g^\pm g^\pm$): 61.0	ψ = 1.22	1	−0.67		

Comments:

An investigation of the intramolecular interactions in n-alkanes by Tsuzuki et al. [Tsuzuki, S.; Schäfer, L.; Gotō, H.; Jemmis, E. D.; Hosoya, H.; Siam, K.; Tanabe, K.; Òsawa, E. J. Am. Chem. Soc. **1991**, *119*, 4665] provides evidence for a second-order interaction and a third-order interaction that are not included in the customary RIS model for unperturbed PE. Although the interactions individually are weak, it is suggested that they might have significant consequences when they accumulate in a polymer. Here, these interactions are incorporated into the RIS treatment for PE, by an appropriate expansion in the dimensions of the statistical weight matrix. Their implications for the unperturbed dimensions, and its temperature coefficient, are assessed. The most important new parameter is found to be the value of the dihedral angle at *gauche* placements that have at least one nearest neighbor that is a *gauche* placement of the same sign. This dihedral angle is displaced by about 6-7° from the dihedral angle for isolated *gauche* placements. This refinement can be accomodated in the calculations without expansion in the dimensions of U. The third-order interaction, which cannot be incorporated without an expansion in the dimensions of U, is of lesser importance than the second-order interaction.

Calcd. quantities:		Model A	Model B	Model C	Exptl.
$<r^2>_0 / nl^2$		= 7.08	= 6.44	= 6.7	= 6.7 (±0.3)
$d (\ln <r^2>_0) / dT$ [× 10^{-3} K^{-1}]		= −1.12	≈ −1.12	= −1.10	= −1.1 (±0.1)

S 017

n-hexane

Photinos, D. J.; Poliks, B. J.; Samulski, E. T.; Terzis, A. F.; Toriumi, H. *Mol. Phys.* **1991**, *72*, 333.

Bond Length [pm]	Valence Angles [°]	Torsion Angles [°]	ξ (for 413 K)	ξ_0	E_ξ [kJ·mol^{-1}]			
C-C : 153.3	C-C-C : 112.5	t : 180	$\sigma = 0.36$	1	3.54			
C-D : 110.0	in methylene: D-C-D : 109.0	g$^+$: 65.4	$\omega = 0.16$	1	6.28	$U = \begin{bmatrix} 1 & \sigma & \sigma \\ 1 & \sigma & \sigma\omega \\ 1 & \sigma\omega & \sigma \end{bmatrix}$	rows and columns: t, g$^+$, g$^-$	
	in terminal CD$_3$: D-C-D : 109.47	g$^-$: −65.4						

Comments:
The three-state RIS model of conformer statistics is used to analyze the 16 independent dipole coupling constants measured in a proton NMR study of *n*-hexane in a nematic liquid crystal solvent. The orientational ordering of the *n*-hexane molecule is treated in the context of the modular formulation of the potential of mean torque. This formulation gives an accurate description of alkane solute orientational order and conformer probabilities in the nematic solvent. Consequently, substantially more accurate calculated diplar couplings are obtained, and this is achieved without the need to resort to unconventionally high values of the *trans–gauche* energy difference E(g) in the RIS model.

Calcd. quantities: Dipolar coupling constants for the protons of *n*-hexane in the uniaxial phase, D_{zz}^{ij}.

polyethylene, polymethylene, PE

Raucci, R.; Vacatello, M. *Makromol. Chem., Theory Simul.* **1993**, *2*, 875.

Bond Length [pm]	Valence Angles [°]	Torsion Angles [°]	ξ (for 300 K)	ξ_0	E_ξ [kJ·mol^{-1}]
C-C : 153	C-C-C : 111	t : 180	σ = 0.69	0.982	1.17
C-H : 110	C-C-H : 109.5	g$^+$: 68	ψ = 1.03	0.942	−0.29
	H-C-H : 107.9	g$^-$: −68	ω = 0.236	0.942	4.61

$$U = \begin{bmatrix} 1 & \sigma & \sigma \\ 1 & \sigma\psi & \sigma\omega \\ 1 & \sigma\omega & \sigma\psi \end{bmatrix}$$

rows and columns: t, g$^+$, g$^-$

Comments:
The conformational distribution of unperturbed PE is studied utilizing the Monte-Carlo approach. The results are in excellent agreement with experiments on the average dimensions of PE chains, as well as with the molecular scattering function obtained in mixtures of deuterated and hydrogenous PE at 400 K. Maps of the free energy as a function of two consecutive torsional angles confirm that an approximate description of the conformational distribution of PE can be given in terms of three rotational isomeric states. The location of the states and the corresponding energy and entropy parameters, which can be separately evaluated from the Monte-Carlo results, are compared with literature results obtained by internal energy calculations for butane and pentane.

*Further calculations on **polyethylene** chains:*

S 019   Hayman, H. J. G.; Eliezer, I. *J. Chem. Phys.* **1958**, *28*, 890.

The theory of hindered rotation previously used for calculating the mean square length of polymer molecules [Smyth, C. P.; Walls, W. S. *J. Chem. Phys.* **1933**, *1*, 200; Ketelaar, J. A. A.; van Meurs, N. *Rec. Trav. Chim.* **1957**, *76*, 437, 495] is combined with the *Smith*, *Ree*, *Magee*, and *Eyring* theory of the inductive effect [*J. Am. Chem. Soc.* **1951**, *73*, 2263] to obtain a two-parameter expression for the dipole moment of α,ω-dibromoparaffins. The values which must be assigned to these parameters to give agreement with experiment indicate an intermediate degree of freedom of internal rotation in these molecules.

S 020   Lifson, S. *J. Chem. Phys.* **1959**, *30*, 964.

The internal rotations around the skeletal bonds of PE are hindered due to interaction between the neighboring hydrogens. Since second and higher orders of interactions are not negligible, the internal rotations are interdependent. The statistics of such interdependent rotations is developed and applied to obtain the configurational partition function and the mean-square end-to-end distance of the macromolecule.

S 021   Hayman, H. J. G.; Eliezer, I. *J. Chem. Phys.* **1961**, *35*, 644.

The dipole moments of the eight α,ω-dibromoparaffins from dibromopropane to dibromodecane together with those of *n*-propyl and *n*-butyl bromide are determined. The results obtained are in good quantitative agreement with the theory developed previously (**S 019**) on the assumption that the flexibility of these molecules is due to independent restricted rotations about the various C—C bonds. The data obtained are interpreted in terms of the picture of *gauche-trans* rotational isomerism on the assumption that the energy of a *gauche* isomer is 1.67 (± 0.50) kJ · mol^{-1} more than that of the corresponding *trans* isomer, except for the rotations of the CH_2Br groups where the energy difference is somewhat less, 1.42 (± 0.59) kJ · mol^{-1}.

S 022   Ciferri, A.; Hoeve, C. A. J.; Flory, P. J. *J. Am. Chem. Soc.* **1961**, *83*, 1015.

Stress-temperature coefficients are determined for cross-linked networks of PE and polyisobutylene elongated in the amorphous state. Interpretation of the indicated temperature coefficient of $<r^2>_o$ for PE according to the three-fold potential model for rotation around the C—C bonds is consistent with an energy difference of 2.1 kJ · mol^{-1} between *gauche* and *trans* states. The small temperature coefficient for isobutylene is due to steric interactions affecting bond rotations.

S 023 Nagai, K.; Ishikawa, T. *J. Chem. Phys.* **1962**, *37*, 496.

The mean-square end-to-end distance of very long PE in solutions is calculated under the assumptions that (*i*) the internal rotational angle ϕ may take only three discrete values (t, g$^+$, g$^-$), (*ii*) the *gauche*$^+$ and *gauche*$^-$ conformations are less stable than the *trans* conformation by a statistical weight σ (ca. 0.36 at 413 K), (*iii*) the g$^+$g$^-$ and g$^-$g$^+$ conformations for two consecutive bonds are inaccessible because of steric hinderances, and (*iv*) all interactions among the rotations about more than two consecutive bonds are negligible. Molecular averages such as the fraction of bonds being in the *trans* conformation are also calculated. Typical numerical calculations are carried out by setting the bond angle equal to the tetrahedral angle and $\phi_t = 180°$ and $|\phi_g| = 60°$. Detailed discussions are given on the validity of various assumptions made in defining the present physical model.

Calcd. quantities:	$<r^2>_o / nl^2$	=	8.0 ($\pm$ 0.2)	(Exptl.: 6.9)
	$d (\ln <r^2>_o) / dT$	=	-1.16×10^{-3} K^{-1}	(Exptl.: -1.16 ($\pm$ 0.10) $\times 10^{-3}$ K^{-1})

S 024 Flory, P. J. *Proc. Nat. Acad. Sci.* **1964**, *51*, 1060.

Previous treatments of the mean-square end-to-end distance and the dipole moment of chain molecules conforming to the RIS model [Birshtein, T. M.; Ptitsyn, O. B. *J. Tech. Phys., USSR* **1959**, *29*, 1048; Birshtein, T. M. *High Molec. Cmpds., USSR* **1959**, *1*, 798, 1086; Lifson, S. *J. Chem. Phys.* **1959**, *30*, 964; Nagai, K. *J. Chem. Phys.* **1959**, *31*, 1169; Hoeve, C. A. *J. J. Chem. Phys.* **1960**, *32*, 888] are restricted to chains (a) which are very long, and (b) in which identical structural units repeat with unerring regularity. Both of these restrictions are removed without approximation by the procedure presented. Numerical calculations are simultaneously simplified.

S 025 Nagai, K. *J. Chem. Phys.* **1964**, *40*, 2818.

By assuming the additivity of bond polarizabilities, a formal expression is derived for the difference Δγ in principal polarizabilities of a real polymer chain having constant end-to-end distance *r*. An expression in a matrix form is derived for Δγ of the PE chain.

S 026 Scott, R. A.; Scheraga, H. A. *J. Chem. Phys.* **1966**, *44*, 3054.

The conformational potential energy of the normal hydrocarbon molecules is considered to be a function of the dihedral angles of internal rotation about the C—C single bonds, the bond length and valence angles being held fixed. The energy equation contains terms of two kinds, a sum over torsional or internal-rotation terms, one for each C—C bond, which are attributed to exchange interactions of electrons in bonds adjacent to the bond about which internal rotation accurs, and a sum over all the pairwise nonbonded interactions between all of the atoms in the molecule. The calculations performed demonstrate that the traditional procedure of assuming that each bond in a hydrocarbon molecule or a polyethylene molecule can exist in three rotational isomeric states (t, g$^+$, g$^-$) is an oversimplification and that more minima occur than is predicted by the traditional procedure. The implications of these results for formulating statistical-mechanical theories of the normal hydrocarbons and of PE are discussed.

S 027 Jernigan, R. L.; Flory, P. J. *J. Chem. Phys.* **1967**, *47*, 1999.

The anisotropy of the polarizability is described by the tensor invariant γ^2, which determines the depolarization ratio. This invariant γ^2 is averaged over all configurations of the chain molecule treated in the RIS approximation. Interdependence of rotations about neighboring bonds is taken into account.
Calculations are carried out using $T = 298$ K, t :180, g$^\pm$: ± 60, θ : 112, $E_\sigma = 2.1$ kJ · mol^{-1} (σ = 0.43), $E_\omega = 10.5$ kJ · mol^{-1} (ω = 0.015).

Calcd. quantities:	$<\gamma^2>/n$ =	212 (to 249) $\times 10^{-50}$ cm^6	(for $n = 15$)	(Exptl.: 180 $\times 10^{-50}$ cm^6 (for $n = 15$)
		267 $\times 10^{-50}$ cm^6	(for $n \to \infty$)	

S 028   Nagai, K. *J. Chem. Phys.* **1967**, *47*, 2052.

Expressions for the optical anisotropy $\Delta\Gamma$ of Kuhn´s random link (an equivalent to the stress-optical coefficient) of stereo-irregular and multirepeat polymers are derived on the basis of the additivity principle of bond polarizabilities and the RIS approximation for rotations about skeletal bonds. Expressions for the unperturbed mean-square end-to-end distance $<r^2>_o$, which are required in the calculation of $\Delta\Gamma$, are also obtained.

S 029   Nagai, K. *J. Chem. Phys.* **1968**, *48*, 5646.

A method is developed for calculating even moments $<r^{2k}>$ of the end-to-end distance r of polymeric chains, on the basis of the RIS approximation for rotations about skeletal bonds. Expressions are obtained in a form which is applicable in principle to arbitrary k, but practical applications are limited by a tremendous increase in the order of the matrices to be treated, with increasing k. An application is made to the PE chain by using the familiar three-state model. Approximate values of the distribution function $W_n (r)$ of the end-to-end vector r, $W_n (0)$, and $<r^{-1}>$, are calculated from these even moments.

S 030   Nagai, K. *J. Chem. Phys.* **1968**, *49*, 4212.

The stress-optical coefficient of PE networks is calculated, and results are compared with experimental data. Observed temperature coefficients of $\Delta\Gamma$ and the optical anisotropy for unswollen samples are much larger than those calculated using acceptable values of E(g), the energy of the *gauche* conformation, relative to that of *trans*. It is concluded that observed temperature coefficients should include some contributions other than those implied in the theory, *i.e.*, those arising from the conformational change with temperature.

S 031   Flory, P. J.; Abe, Y. *Macromolecules* **1969**, *2*, 335.

A general theory of dichroism induced by strain in polymeric networks is developed by adaptation of methods developed earlier for treating strain birefringence. It is generally applicable to dichroic bands associated with any specified conformation involving sequences of one or more consecutive bonds. The transition dipole moment is introduced in the local framework of the skeletal bonds associated therewith. Possible differences in transition moments for various conformations and repeat units are taken into account. Numerical calculations for PE chains show *gauche* bonds, rather than *trans*, to be more favorably oriented with respect to the chain vector r.

S 032   Tonelli, A. E. *J. Chem. Phys.* **1970**, *53*, 4339.

The mean-square end-to-end distance and its temperature coefficient, the change in the intramolecular conformational entropy on melting, and the optical anisotropy, as manifested by the strain birefringence, and its temperature coefficient are calculated in the RIS approximation for unperturbed linear PE chains containing short stretches of skeletal bonds exclusively in the *all trans* or planar zigzag conformation. The results are compared with the corresponding quantities calculated for a PE chain free of such conformational constraints and with experimental data taken from the literature, obtained on undiluted bulk PE. This comparison shows that the optical properties of PE are more sensitive to the local segmental order, as defined in this study, than are its dimensions and their temperature coefficient or the change in intramolecular conformational entropy on melting.

Calcd. quantities: $<r^2>_o$ $d (\ln <r^2>_o) / dT$ $(S_u)_v$ $\Delta\alpha$ $d (\ln\Delta\alpha) / dT$

S 033   Abe, Y.; Flory, P. J. *J. Chem. Phys.* **1970**, *52*, 2814.

A general theory is presented on the effect of elongation of a polymer chain on the apportionment of its bonds and bond sequences among various rotational isomeric states. Numerical calculations are presented for PE chains and for syndiotactic vinyl polymers.

S 034 Flory, P. J.; Abe, Y. *J. Chem. Phys.* **1971**, *54*, 1351.

The higher even moments $<r^{2p}>_0$ of the end-to-end vector *r* of a chain molecule are formulated according to a straightforward procedure which is free of complications for large values of p. It is generally applicable also to averages of other configuration-dependent quantities, and simplifies their treatment. The condensation of self-direct matrix products derived by *Nagai* [Nagai, K. *J. Chem. Phys.* **1963**, *38*, 924; Nagai, K.; Ishikawa, T. *J. Chem. Phys.* **1966**, *45*, 3128] for reducing the complexity of computation of higher moments is achieved directly through an appropriate transformation, which is defined for such a product of any degree. Major reductions of the orders of the statistical weight matrix and associated expressions permissible for symmetric chains are achieved succinctly by similar matrix transformations. It is pointed out that further condensation can be realized through deletion of superfluous rows and columns from the generator matrices (**G**, **G** ⊗ **G**, etc.) for the moments.

S 035 Heatley, F. *Polymer* **1972**, *13*, 218.

Calculations of the characteristic ratio and its temperature dependence for PE and isotactic PP have been performed using a RIS model that takes account of non-staggered conformations and the interdependence of the rotational potentials in sequences of four chain bonds. The experimental values are shown to be reproducible satisfactorily by a set of energy parameters consistent with the similarity between steric interactions in the two polymers.

S 036 Patterson, G. D.; Flory, P. J. *J. Chem. Soc.; Farad. Trans.* 2 **1972**, *68*, 1098.

A procedure is developed whereby the mean-square optical anisotropy $<\gamma^2>$ associated with isolated solute molecules can be obtained from measurements of the depolarized Rayleigh scattering by their solutions in optically isotropic and geometrically symmetric solvents. Values of $<\gamma^2>$ for the *n*-alkanes obtained by extrapolating the present results to infinite dilution are considerably smaller than those reported by other workers. The dependence of $<\gamma^2>$ on the chain length is well reproduced by calculations carried out using the valence optical scheme and RIS theory, *gauche* states being assigned an energy of 2.10 (± 0.42) kJ · mol^{-1} relative to *trans* in conformity with other evidence.

S 037 Jernigan, R. L. in *Dielectric Properties of Polymers*, Karasy, F. E., ed., Plenum, New York, **1972**, p 99.

A dynamic rotational isomeric state description is presented of α,ω-dibromo-*n*-alkanes, including their high frequency dielectric dispersion.

S 038 Fixman, M.; Alben, R. *J. Chem. Phys.* **1973**, *58*, 1553.

The probability distribution *P(r)* of the end-to-end distance is studied for the RIS model of polymer chains. A Monte-Carlo investigation provided reliable numerical data for *P(r)*, which was then compared with results from two related analytical studies.

S 039 Fixman, M. *J. Chem. Phys.* **1973**, *58*, 1559.

The probability distribution of the end-to-end vector for the RIS model of polymer chains is discussed in terms of the characteristic function. For PE the characteristic function calculated from the RIS model is found to be in good agreement with the much simpler worm model.

S 040   Patterson, G. D. *Macromolecules* **1974**, *7*, 220.

The theory of Rayleigh light scattering by independent isotropic systems composed of anisotropic units developed by *Nagai* [Nagai, K. *J. Polym. J.* **1972**, *3*, 67] is extended according to the RIS model. Calculations for the PE chain indicate that for molecules in the random-coil state in dilute solution the terms arising from the anisotropy of the units should be small in comparison to those arising from the scalar mean polarizability of the molecules.

S 041   Flory, P. J.; Yoon, D. Y. *J. Chem. Phys.* **1975**, *61*, 5358, 5366.

Moments and distribution functions for PE chains of finite length are calculated: $<r^2>_0$ $<r^4>_0/<r^2>^2$ $<r^6>_0/<r^2>^3$ $a \equiv <r>$ $<g>$ $W_a(\rho)$ $W(r)$

S 042   Tonelli, A. E. *Macromolecules* **1976**, *9*, 863.

Some comments on the liquid state conformations of *n*-alkanes are made. Calculated are the average probability $<P_g>$ and number $<N_g>$ of *gauche* bond rotations assumed by internal bonds, and the probability P_g(central) that the central bond in each of the even members (C_6, C_8,...,C_{22}) is in a *gauche* rotational state.

S 043   Gény, F.; Monnerie, L. *Macromolecules* **1977**, *10*, 1003.

Within the frame of a program on simulation of the Brownian motion of chain molecules, the conformational static and dynamic properties of a model of PE are studied. In the present paper the same properties are systematically derived by using the RIS theory. As expected, there is good agreement for static properties such as conformational averages and chain dimensions. In addition the local mobility of the chain is favorably compared by the aid of the two approaches.

S 044   Baram, A.; Gelbart, W. M. *J. Chem. Phys.* **1977**, *66*, 617.

The equilibrium structure is considered of flexible polymer chains within the RIS model. This model is solved by an irreducible tensor method which is somewhat different from, and simpler than, the approach of *Flory* and others. The results are used to compute the light scattering intensities from dilute solutions of flexible polymer chains, and the angle dependence is found to be negligible.

S 045   Baram, A. *J. Chem. Phys.* **1977**, *66*, 3128.

An approximate analytical solution is presented for the RIS model of flexible polymer configurations. The method is applied to calculate $<r^2_n>$ and $<D^2_{00}(\Omega_n)>$.

S 046   Baram, A.; Gelbart, W. M. *J. Chem. Phys.* **1977**, *66*, 4666.

In an earlier paper (**S 044**), anomalies in the angular moments $<P_J(\cos\theta_n)>$ of the monomer-monomer distribution function for flexible polymers are established. It is shown here how these anomalies arise from the tetrahedral symmetry of the three-state RIS model and how they disappear in the continuum limit of torsional conformations. It is concluded that the eighth and higher radial moments contain spurious contributions when calculated within the usual three-state model.

S 047   Feigin, R. I.; Napper, D. H. *J. Coll. Interf. Sci.* **1979**, *71*, 117.

The simulation of polymer chains attached at one end to an inert interface ("tails") is described, using a combination of Monte-Carlo techniques and the RIS scheme. Four types of chains that exhibit different structural and geometrical characteristics are examined: PE, POM, POE, and PDMS. The overall effects on the conformation of the attached chains arising from the presence of the interface are qualitatively similar. The chains are significantly extended, as measured by the mean square displacement, in the direction normal to the interface, but essentially unaffected parallel to the interface. A comparison is made of intramolecular excluded volume with excluded volume arising from an impenetrable interface. It is shown that the two types of excluded volume exhibit qualitatively different effects on the conformation.

S 048   Winnik, M. A.; Rigby, D.; Stepto, R. F. T.; Lemaire, B. *Macromolecules* **1980**, *13*, 699.

Various properties of hydrocarbon chains are calculated with the use of three different variants of the RIS model: lattice chains, unperturbed chains, and off-lattice chains with excluded volume. The properties examined are $<r^2>$, $<s^2>$, and $<r^6>$. It is found that lower and inverse moments of r are more sensitive to excluded volume effects than $<r^2>$.

S 049   Cook, R.; Moon, M. *Macromolecules* **1980**, *13*, 1537.

The effect of bond-rotational angle fluctuations on the equilibrium properties of a macromolecule is considered by introducing rotational-angle flexibility into the configurational statistics methods. For $\delta_\phi = 15°$, the decrease in C_∞ is only about 5.5 % for $\sigma = 0.54$ and $\omega = 0.088$ but becomes markedly larger as σ decreases, rising to 13.1% for $\sigma = 0.10$. Experimental estimates for PE of C_∞ and its temperature coefficient at 413 K are 6.7 (± 0.3) and -1.1 (± 0.1) $\times 10^{-3}$ K^{-1}, respectively. Using the best estimate of the *gauche* rotational angle of 67.5° and bond angles of 112°, one finds for the strict RIS model that E(g) = 1.348 kJ · mol^{-1} ($\sigma = 0.68$) and E(g$^+$g$^-$) = 5.903 kJ · mol^{-1} ($\omega = 0.18$) best fit these values. Where flexibility is added with $\delta_\phi = 15°$, the best fits occur for E(g) = 1.758 kJ · mol^{-1} ($\sigma = 0.60$) and E(g$^+$g$^-$) = 6.175 kJ · mol^{-1} ($\omega = 0.17$). The development of distribution functions for flexible chains by Monte-Carlo generation is also discussed.

S 050   Mansfield, M. L. *Macromolecules* **1981**, *14*, 1822.

Investigations are presented concerning the relationship between the RIS and wormlike chain models.

S 051   Oyama, T.; Shiokawa, K. *Polymer J.* **1981**, *13*, 1145.

Evaluation of the unperturbed dimension by using potential energy surface: the end-to-end distance and averages of bond conformations are calculated without using the RIS approximation. An integral equation, the kernel of which corresponds to statistical weights of rotational angles, is solved numerically for a model PE chain. The authors conclude that the RIS approximation is very simple and useful for the calculation of $<r^2>$, but is not suitable for calculating the averages of bond conformations.

S 052   Mattice, W. L.; Napper, D. H. *Macromolecules* **1981**, *14*, 1066.

PE chains in which the chain atom at one end is attached to an impenetrable interface are studied by using an RIS model which accurately reproduces configuration-dependent physical properties of the unperturbed chain in free solution. Properties investigated are the mean-square radius of gyration, asymmetry of the distribution of the chain atoms, and the probability of observing a particular rotational state at bonds near the site of attachment to the impenetrable interface. The limiting behaviour of infinitely long chains is characterized by studying finite chains with up to 1000 bonds.

S 053   Mattice, W. L. *Macromolecules* **1981**, *14*, 1485.

The generator matrix treatment of simple chains with excluded volume described earlier (**S 010**) properly reproduces the known chain length dependence of the mean-square dimensions in the limit of infinite chains. The purpose of this paper is to compare the behaviour of finite generator matrix chains with that of Monte-Carlo chains in which atoms participating in long-range interactions behave as hard spheres. The model for the unperturbed chain is that developed by *Flory et al.* for PE (**S 027**).

S 054   Mattice, W. L. *Macromolecules* **1981**, *14*, 1491.

Expansion of subchains in a linear PE chain with excluded volume is evaluated by two methods: (1) Monte-Carlo chains with methylene groups participating in long-range interactions behave as impenetrable spheres with a diameter of 300 pm, and (2) generator matrix calculations in which expansion is produced without any effect on the probability of a *trans* placement in an infinitely long chain.

S 055   Mattice, W. L. *Macromolecules* **1982**, *15*, 579.

Angular scattering functions, $P(\mu)$, are computed for subchains located in the middle and at the end of a PE chain. The RIS model developed by *Flory et al.* (**S 004** - **S 006**) is used for the unperturbed chain. Chain expansion is introduced using a matrix treatment which satisfactorily reproduces several configuration-dependent properties of macromolecules perturbed by long-range interactions.

S 056   Guttman, C. M.; DiMarzio, E. A. *Macromolecules* **1982**, *15*, 525.

A Monte-Carlo simulation of a PE-like polymer chain between two plates is performed. This continuum treatment augments previous analytical lattice treatments of completely flexible chains between plates [DiMarzio, E. A.; Guttman, C. M. *Polymer* **1981**, *21*, 733; Guttman, C. M.; DiMarzio, E. A.; Hoffman, J. D. *Polymer* **1981**, *22*, 1466]. The Monte-Carlo results show that the simple concept of statistical length appropriate to unconfined bulk polymer is also appropriate to chain portions residing in the amorphous regions of lamellar semicrystalline polymer. Thus, the "gambler's ruin" method, with the statistical length of the polymer used as the fundamental step length, is a valid method to obtain quantitative estimates of quantities such as length of loops, length of ties, and fraction of loops or ties for moderately stiff polymers. Previous estimates of the amount of chain folding in PE are thus shown to retain their validity for the more realistic isomeric state model.

S 057   Gil, V. M. S.; Varandas, A. J. C.; Murrell, J. N. *Can. J. Chem.* **1983**, *61*, 163.

A critical appraisal is made of the validity of the RIS approximation in studies of rotational isomerism, especially as far as the determination of energy differences for the stable conformers is concerned. By using simple model potential energy functions for internal rotation appropriate to some ethane derivatives, a comparison is made between the thermodynamic parameters that can be extracted from such potential functions and those obtained by applying the RIS method to continuum averaged values of conformation dependent properties. The effects due to the form of this conformation dependence, to the features of the potential energy curve, and to the temperature are discussed. The errors can be very large for realistic situations, and variable with temperature and with the property being studied; cases for which the errors are small are usually accidental.

S 058   Mansfield, M. L. *Macromolecules* **1983**, *16*, 1863.

Methods for introducing independent fluctuations about the bond angles and bond rotational angles of RIS models are discussed. The given examples indicate that as a rule, the greater the tendency for a chain to favor a helix (or specifically, to occur in repeated rotational states), the less likely it will be that the chain can be accurately modeled by the standard RIS approximation. They also indicate that models of high persistence length polymers should be viewed with at least some suspicion, due to the high sensitivity of such models to perturbations.

S 059   Kajiwara, K.; Burchard, W. *Macromolecules* **1984**, *17*, 2669.

Monte-Carlo simulations are applied to estimate the characteristic ratios and ρ parameters from the RIS models for PE, POM, polybutadiene, and polyisoprene. Here the ρ parameter is defined as the ratio of the radius of gyration to the hydrodynamic radius. The ρ parameters of these real chains in the unperturbed state show only a slight dependence on the microconformation in the limit of large molecular weights and are found close to 1.504, which is the value for an idealized Gaussian chain. The estimated ρ parameters of the real chains appear to be correlated to the chain stiffness and increase with the characteristic ratios.

S 060   Mattice, W. L. *J. Phys. Chem.* **1984**, *88*, 6492.

Two approaches are employed for evaluation of the expansion of realistic RIS models of infinite PE chains. Simulations are used for chains of 100-750 bonds in which atoms participating in long-range interactions behave as hard spheres. An approximate generator matrix method permits extension of the study to longer chains. The ratio of expansion factors for the mean-square end-to-end distance and mean-square radius of gyration approaches a limit that is significantly smaller than that estimated from several earlier studies of lattice and off-lattice chains with hard-sphere interactions. The present limit for $<r^2>/<s^2>$ is closer to Debye´s limit for $<r^2>_0/<s^2>_0$ than limits estimated previously by using lattice chains.

S 061   Mattice, W. L.; Carpenter, D. K. *Macromolecules* **1984**, *17*, 625.

Response of the mean square dipole moment, $<\mu^2>$, to excluded volume is evaluated for several chains via Monte-Carlo methods. The chains differ in the manner in which dipolar moment vectors are attached to the local coordinate systems for the skeletal bonds. In the unperturbed state, configurational statistics are those specified by the usual RIS model for linear PE chains. Excluded volume is introduced by requiring chain atoms participating in long-range interactions to behave as hard spheres.

S 062   Mark, J. E.; Curro, J. G. *J. Chem. Phys.* **1984**, *81*, 6408.

Distribution functions for very short *n*-alkane chains are calculated. The results obtained implicate the discrete nature of the RIS approximation as the origin of the multimodal nature of the short-chain distribution, and suggest that the best representation of such results would be a simple unimodal curve averaging out all of the minima and retaining only the most prominent maximum.

S 063   Henyey, F. S.; Rabin, Y. *J. Chem. Phys.* **1985**, *82*, 4362.

A simple RIS model of polymers in elongational flows is developed and used to analyze the coil stretching and chain retraction as a function of polymer and flow parameters. The results are in agreement with available experimental data on dilute polymer solutions in strong elongation flows.

S 064   Mattice, W. L.; Carpenter, D. K.; Barkley, M. D.; Kestner, N. R. *Macromolecules* **1985**, *18*, 2236.

Investigation of the multivariate Gaussian distribution and the dipole moments of perturbed chains; expansion factors for perturbed chains.

S 065   Janik, B.; Samulski, E. T.; Toriumi, H. *J. Phys. Chem.* **1987**, *91*, 1842.

^{2}H NMR spectra for homologous series of perdeuterated *n*-alkanes solubilized in nematic solvents are reported. These flexible solutes acquiesce to the uniaxial environment of the solvent and thereby reflect the nature of the nematic mean field. Quantitative simulations of the quadrupolar splittings exhibited by the alkanes are carried out using a parametrized potential of mean torque in conjunction with an ensemble average over alkane conformers. Two parametrizations are selected in order to gauge the relative importance of attractive (dispersion) forces and repulsive (excluded volume) forces. A detailed examination of the resulting angular dependence of the potential is shown for hexane along with a critical evaluation of the RIS approximation itself. The findings suggest that while the latter approximation is adequate, a more elaborate specification of the orientational potential of mean torque for solutes is required — one that explicitly and rigorously couples attractive and repulsive intermolecular interactions.

S 066   Bahar, I.; Erman, B. *Macromolecules* **1987**, *20*, 1368.

The internal dynamics of a short sequence in a chain is studied according to the dynamic RIS scheme. Conformational transitions with dynamic pair correlations are considered. Resistance to dynamic rearrangements resulting from environmental effects and constraints operating at the ends of a sequence are incorporated into the calculation scheme. Calculations for a short sequence in a PE chain show that pair correlations do not significantly affect the orientational relaxation of a vector affixed to a bond in the sequence. Contributions from constraints, on the other hand, are dominant and slow down the orientational motions.

S 067   Perico, A. *J. Chem. Phys.* **1988**, *88*, 3996.

The viscoelastic theory of *Perico* and *Guenza* [Perico, A.; Guenza, M. *J. Chem. Phys.* **1985**, *83*, 3103; Perico, A.; Guenza, M. *J. Chem. Phys.* **1986**, *84*, 510] for the time correlation function probed in different experiments on segment relaxation is compared with the master equation approach of *Helfand* [Hall, C. K.; Helfand, E. *J. Chem. Phys.* **1982**, *77*, 3275] and applied to detailed RIS models in the optimized Rouse-Zimm dynamic approximation. In the free draining limit, for the vector autocorrelation function of a segment in the central position on an ideally flexible chain, a zero order modified Bessel function form is obtained as previously found for Helfand's conformational time correlation function. The correlation time for the alignment memory function is explicitly calculated for PE, isotactic PP, isotactic PS, and PDMS.

S 068   Bahar, I.; Erman, B.; Monnerie, L. *Macromolecules* **1989**, *22*, 431.

The dynamic RIS model is used to calculate the conformational and first and second orientational autocorrelation functions for PE. Various sequence lengths and directions in the chain are considered.

S 069   Bahar, I. *J. Chem. Phys.* **1989**, *91*, 6525.

The stochastic process of conformational transitions between isomeric states in polymer chains is considered. In analogy with the conventional treatments of chain statistics where equilibrium configurations are assigned statistical weights based on near neighbor intramolecular potentials, stochastic weights are defined for the configurational transitions undergone by chains of pairwise interdependent bonds. A matrix multiplication scheme is devised to determine the fraction of bonds or segments that undergo specific isomeric transitions in a given time interval.

S 070   Noid, D. W.; Sumpter, B. G.; Wunderlich, B. *Macromolecules* **1990**, *23*, 664.

Using a realistic model for PE, the molecular dynamics technique is used to simulate atomic motion in a crystal. The calculations reveal conformational disorder above a critical temperature. The customarily assumed RIS model is found to be a poor description of the crystal at elevated temperature.

S 071 Pannikottu, A.; Mattice, W. L. *Macromolecules* **1990**, *23*, 867.

K_e, the intramolecular excimer equilibrium constant under conditions where the dynamics or rotational isomeric transitions is suppressed, is examined on the basis of the theory of macrocyclization [Flory, P. J.; Suter, U. W.; Mutter, M. *J. Am. Chem. Soc.* **1976**, *98*, 5733, 5740, 5745]. The conformational averages of the required moments and polynomials are calculated using their exact matrix generation method. The calculations can rationalize the appearance of a maximum near $m = 15$ in the I_E/I_M that have been reported for pyrene-$(CH_2)_m$-pyrene. In order to achieve this agreement, it is imperative that the theoretical analysis incorporate the probability of a proper angular correlation of the two pyrene ring systems when the separation of their centers is 0.35 nm.

S 072 Bahar, I.; Mattice, W. L. *Macromolecules* **1990**, *23*, 2719.

A matrix formulation of the time-dependent transition partition function is combined with a generator matrix formalism to permit rapid and accurate calculation of the first and second orientation autocorrelation functions for a chain molecule.

S 073 Rodríguez, A. L.; Vega, C.; Freire, J. J.; Lago, S. *Mol. Phys.* **1991**, *73*, 691.

The second virial coefficients of a number of *n*-alkanes (from butane to hexadecane) are evaluated by using the RIS model. It is shown that a simple site-site potential model is able to reproduce the second virial coefficient of several *n*-alkanes in a wide range of temperature.

S 074 Sasanuma, Y.; Abe, A. *Polymer J.* **1991**, *23*, 117.

Conformational anisotropy of *n*-alkane chains incorporated in a nematic environment is investigated. Experimental values of proton-proton dipolar couplings of *n*-hexane and ^{2}H NMR quadrupolar splittings of a series of *n*-C_nH_{2n+2} are analyzed. The treatment proceeds as follows: 1. All possible configurations are enumerated for a free molecule within the framework of the RIS approximation. 2. For each conformer, the molecular axis is defined along the "longest" principal axis of inertia. 3. Conformational statistical weight factors assigned to the individual bond rotations are adjusted according to the simplex method so as to reproduce the observations. The convergence of iteration is monitored by the reliability factor. The agreement between theory and experiment is found to be satisfactory in all examples. The bond conformations derived from the observed data are compared with those estimated for the free state. *n*-Alkane chains are found to be highly anisotropic in the nematic media.

S 075 Rubio, A. M.; Freire, J. J.; Horta, A.; de Piérola, I. F. *Macromolecules* **1991**, *24*, 5167.

The end-to-end distribution of short polymer molecules (represented by a RIS model that includes long-range interactions through a hard-sphere potential) is calculated by means of a Monte-Carlo method. The model predictions are contrasted with experimental data of the equilibrium cyclization constants.

S 076 McCoy, J. D.; Honnell, K. G.; Curro, J. G.; Schweizer, K. S.; Honeycutt, J. D. *Macromolecules* **1992**, *25*, 4905.

The RIS model is usually considered to be an excellent description of the single-chain structure of polymer chains, the manifestation of which is the correlation function $\omega(r)$. The evaluation of $\omega(r)$ from the RIS model requires laborious statistical averages, and, as a consequence, various approximations of $\omega(r)$ are of importance. In the present paper, an approximation is presented which is accurate on all length scales.

S 077

polytetrafluoroethylene, PTFE

Bates, T. W.; Stockmayer, W. H. *J. Chem. Phys.* **1966**, *45*, 2321.

Bond Length [pm]	Valence Angles [°]	Torsion Angles [°]	ξ (for 298 K)	ξ_0	E_ξ [kJ·mol^{-1}]
C-C : 153	C-C-C : 116	t : 180	σ = 0.16	1	4.6 (± 1.3)
		g$^+$: 65 *or* 60	σ' ≈ 2.0	1	– 1.7 [a]
		g$^-$: – 65 *or* – 60	β = 0.63 [b]	1	1.1

$$U = \begin{bmatrix} 1 & \sigma & \sigma \\ 1 & \sigma & 0 \\ 1 & 0 & \sigma \end{bmatrix} \quad U_3 = \begin{bmatrix} 1 & \sigma & \sigma \\ 1 & \sigma & \sigma\beta \\ 1 & \sigma\beta & \sigma \end{bmatrix} \quad U_n = \begin{bmatrix} 1 & \sigma' & \sigma' \\ 1 & \sigma' & \sigma'\beta \\ 1 & \sigma'\beta & \sigma' \end{bmatrix}$$

$$U_2 = \begin{bmatrix} 1 & \sigma' & \sigma' \end{bmatrix}$$

rows and columns: t, g$^+$, g$^-$

[a] There is a misprint in the original paper ($E(\sigma')$ = + 1.7 kJ·mol^{-1}).

[b] The value of β is calculated from $\sigma\sigma'\beta$ = 0.2 which is given in the paper.

Calcd. quantities: $\langle\mu^2\rangle$

S 078 (see also S 079)

polytetrafluoroethylene, PTFE

Bates, T. W.; Stockmayer, W. H. *Macromolecules* **1968**, *1*, 12, 17.

Model A:

Bond Length [pm]	Valence Angles [°]	Torsion Angles [°]	ξ (for 298 K)	ξ_0	E_ξ [kJ · mol^{-1}]
C - C : 153	C-C-C : 116	t : 180	$\sigma \approx 0.13$	1	5.0 (± 0.8)
C - F : 109		g$^+$: 65	$\sigma' \approx 2.0$ (± 0.1)	1	– 1.72 (± 0.01)
		g$^-$: – 65	$\beta \approx 0.8$ a)	1	≈ 0.9

$$U = \begin{bmatrix} 1 & \sigma & \sigma \\ 1 & \sigma & 0 \\ 1 & 0 & \sigma \end{bmatrix} \quad U_2 = \begin{bmatrix} 1 & \sigma' & \sigma' \end{bmatrix} \quad U_3 = \begin{bmatrix} 1 & \sigma & \sigma \\ 1 & \sigma & \sigma\beta \\ 1 & \sigma\beta & \sigma \end{bmatrix} \quad U_n = \begin{bmatrix} 1 & \sigma' & \sigma' \\ 1 & \sigma' & \sigma'\beta \\ 1 & \sigma'\beta & \sigma' \end{bmatrix}$$

rows and columns: t, g$^+$, g$^-$

a) The parameter β is chosen so that $\sigma\sigma'\beta$ is between 0.3 and 0.1; usually, $\sigma\sigma'\beta$ = 0.2 is used for which the value of β is given.

Calcd. quantities: $\langle \mu^2 \rangle^{1/2}$ $d(\ln \langle \mu^2 \rangle)/dT$

S 079 (see also S 078)

Model B:

Bond Length [pm]	Valence Angles [°]	Torsion Angles [°]	ξ (for 298 K)	ξ_0	E_ξ [kJ·mol^{-1}]
C - C : 153	C-C-C : 116	t_+ : 165	$\sigma = 0.09$	1	5.9 (± 1.7)
C - F : 109		t_- : −165	$\sigma' = 2.0$	1	−1.7 (± 0.4)
		g^+ : 60	$\omega = 0.16$	1	4.6 (± 2.9)
		g^- : −60	$\beta \approx 1.1$ [a]	1	≈ -0.2

$$U = \begin{bmatrix} 1 & \sigma & 0 & \omega \\ 1 & \sigma & 0 & 0 \\ 0 & 0 & \sigma & 1 \\ \omega & 0 & \sigma & 1 \end{bmatrix} \quad U_3 = \begin{bmatrix} 1 & \sigma & 0 & \omega \\ 1 & \sigma & \sigma\beta & 1 \\ 1 & \sigma\beta & \sigma & 1 \\ \omega & 0 & \sigma & 1 \end{bmatrix} \quad U_n = \begin{bmatrix} 1 & \sigma' & \sigma' & \omega \\ 1 & \sigma' & \sigma'\beta & 0 \\ 0 & \sigma'\beta & \sigma' & 1 \\ \omega & \sigma' & \sigma' & 1 \end{bmatrix}$$

$$U_2 = \begin{bmatrix} 1 & \sigma' & \sigma' & 1 \end{bmatrix}$$

rows and columns: t_+, g^+, g^-, t_-

[a] The parameter β is chosen so that $\sigma\sigma'\beta$ is between 0.5 and 0.1; usually, $\sigma\sigma'\beta = 0.2$ is used.

Calcd. quantities: $\langle \mu^2 \rangle^{1/2}$ $d(\ln \langle \mu^2 \rangle)/dT$

S 080 (see also S 081)

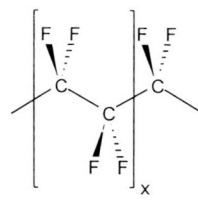

polytetrafluoroethylene, PTFE

Smith, G. D.; Jaffe, R. L.; Yoon, D. Y. *Macromolecules* **1994**, *27*, 3166.

Model A:

Bond Length [pm]	Valence Angles [°]	Torsion Angles [°]	ξ a) (for 600 K)	ξ_0	E_ξ [kJ·mol^{-1}]
C-C : 153	C-C-C : 116	t_+ : 165	$\sigma = 0.61$	1	2.5
		t_- : –165	$\sigma' = 0.19$	1	8.4
		g^+_+ : 56	$\psi = 0.61$	1	2.5
		g^+_- : 96	$\omega = 0.43$	1	4.2
		g^-_+ : –96			
		g^-_- : –56			

$$U = \begin{bmatrix} 1 & 0 & \sigma & 0 & \sigma' & 0 \\ 0 & 1 & 0 & \sigma' & 0 & \sigma \\ 1 & 0 & \sigma\psi & 0 & \sigma'\omega & 0 \\ 0 & 1 & 0 & \sigma' & 0 & \sigma\omega \\ 1 & 0 & \sigma\omega & 0 & \sigma' & 0 \\ 0 & 1 & 0 & \sigma'\omega & 0 & \sigma\psi \end{bmatrix}$$

rows and columns: $t_+, t_-, g^+_+, g^+_-, g^-_+, g^-_-$

a) The statistical weight matrix has only four parameters: two first-order statistical weights for the $g^\pm_\pm$ (σ) and $g^\mp_\pm$ (σ') states, and two second-order statistical weights for the $g^\pm_\pm g^\pm_\pm$ (ψ) and $g^\pm_\pm g^\mp_\pm$ (ω) states.

Comments:
Conformational characteristics of PTFE chains are studied in detail, based upon ab initio electronic structure calculations on perfluorobutane, perfluoropentane, and perfluorohexane. The found conformational characteristics are fully represented by a six-state RIS model. This six-state model, with no adjustment of the geometric or energy parameters as determined from the *ab initio* calculations, predicts the unperturbed chain dimensions, and the fraction of *gauche* bonds as a function of temperature, in good agreement with available experimental values.

Calcd. quantities: $\langle r^2 \rangle_0 / nl^2 = 9.8$ (Exptl.: 8 (± 2.5)) $d(\ln \langle r^2 \rangle)/dT = -1.2 \times 10^{-3}$ K^{-1}
gauche probability $P_{g\pm}$

S 081   (see also S 080)

Model B:

Bond Length [pm]	Valence Angles [°]	Torsion Angles [°]	ξ (for 600 K)	ξ_0	E_ξ [kJ·mol^{-1}]		
C - C : 153	C-C-C : 116	t_+ : 165	$\sigma = 0.61$	1	2.5		
C - F : 109		t_- : -165	$\psi = 0.61$	1	2.5	$U = \begin{bmatrix} 1 & 0 & \sigma & 0 \\ 0 & 1 & 0 & \sigma \\ 1 & 0 & \sigma\psi & \sigma\omega' \\ 0 & 1 & \sigma\omega' & \sigma\psi \end{bmatrix}$	rows and columns: t_+, t_-, g^+, g^-
		g^+ : 56	$\omega' = 0.13$	1	10.0		
		g^- : -56					

Comments:

The above six-state model is reduced to a four-state model where the g^+_- and g^-_+ states combine with the g^+_+ and g^-_- states, respectively. Here, the g^+ and g^- states are assumed to have the geometry of the g^+_+ and g^-_- conformation, respectively, and the $E(\sigma)$ and $E(\psi)$ values are the same as above. Conformations of the type $g^+_+g^-_+$ are represented as g^+g^-, requiring a new second-order parameter ω'.

Calcd. quantities: $<r^2>_0 / nl^2$ = 9.4   (Exptl.: 8 (±2.5)) $d (\ln <r^2>) / dT = -1.5 \times 10^{-3}$ K^{-1} *gauche* probability $P_{g\pm}$

Further calculations on **halogenated n-alkane chains**:

S 082   Abe, A.; Furuya, H.; Toriumi, H. *Macromolecules* **1984**, *17*, 684.

The dipole moments of α,ω-dihaloalkanes and α,ω-dihydroperfluoroalkanes are analyzed within the framework of the RIS scheme. Parameters required for the analysis are mostly taken from the previous studies by *Leonard et al.* (**S 002**) and *Bates et al.* (**S 077 - S 081**). The three state and the four state model are applied.

Calcd. quantities: $<\mu^2>/2m^2$ Distribution curves for the angle ψ, defined by the two terminal C—H bond vectors
$P(\psi), <\psi>, <\cos^2\psi>$ Order parameter $S = \frac{1}{2} (3<\cos^2\psi> - 1)$

S 083   Rosi-Schwartz, B.; Mitchell, G. R. *Polymer* **1994**, *35*, 3139.

A new methodology is presented that couples neutron diffraction experiments over a wide Q range with single chain modelling in order to explore, in a quantitative manner, the intrachain organization of non-crystalline polymers. The method is successfully applied to the study of molten PTFE. From analysis of the experimental data a model is derived with C—C and C—F bond length of 158 and 136 pm, respectively, a backbone valence angle of 110° and a torsional angle distribution which is characterized by four isomeric states, namely a split *trans* state at ±162°, and two *gauche* states at ±68°.

catena-poly[sulfur] and *catena*-poly[selenium]

Semlyen, J. A. *Trans. Farad. Soc.* **1967**, *63*, 743, 2342; *ibid.* **1968**, *64*, 1396.

Bond Length [pm]	Valence Angles [°]	Torsion Angles [°]	ξ (for 433 K)	ξ_o	E_ξ [a) [kJ·mol^{-1}]		
S-S : 206 (±2)	S-S-S : 106 (±2)	ϕ^+ : 90 (±10) ϕ^- : −90 (±10)	$\omega \approx 1.4$ (1.2 to 1.7)	1	≈ -1.1	$U = \begin{bmatrix} 1 & \omega \\ \omega & 1 \end{bmatrix}$	rows and columns: ϕ^+, ϕ^-
Se-Se : 234 (±2)	Se-Se-Se : 104 (±2)	ϕ^+ : 90 (±10) ϕ^- : −90 (±10)	$\omega' \approx 1.6$ (1.6 to 2.4)	1	≈ -1.7	$U = \begin{bmatrix} 1 & \omega' \\ \omega' & 1 \end{bmatrix}$	rows and columns: ϕ^+, ϕ^-

a) E_ξ represents the energy difference between the states + − (or − +) and + + (or − −).

Calcd. quantities: $\langle r^2 \rangle_o / nl^2$: Calculated values (*from* 1 *to* 2) depend on the choice of ω (*or* ω'), θ, and ϕ.

S 085

catena-poly[sulfur] and *catena*-poly[selenium]

Mark, J. E.; Curro, J. G. *J. Chem. Phys.* **1984**, *80*, 5262.

Bond Length [pm]	Valence Angles [°]	Torsion Angles [°]	ξ (for 298 K)	ξ_0	E_ξ [kJ · mol^{-1}]		
S-S : 206	S-S-S : 106	ϕ^+ : 90	$\sigma(S) = 1.5$	1	-1.005	$U = \begin{bmatrix} 1 & \sigma \\ \sigma & 1 \end{bmatrix}$	rows and columns: ϕ^+, ϕ^-
Se-Se : 234	Se-Se-Se : 104	ϕ^- : -90	$\sigma(Se) = 2.0$	1	-1.717		

Comments:
Distribution functions for the end-to-end separation of polymeric sulfur and selenium are obtained from Monte-Carlo simulations which take into account the chains' geometric characteristics and conformational preferences. Comparisons with the corresponding information on PE demonstrate the remarkable equilibrium flexibility or compactness of these two molecules. Use of the S and Se distribution functions in the three-chain model for rubberlike elasticity in the affine limit gives elastomeric properties very close to those of non-Gaussian networks, even though their distribution functions appear to be significantly non-Gaussian.

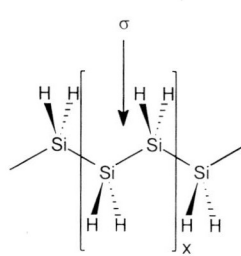

polysilane

Welsh, W. J.; DeBolt, L.; Mark, J. E. *Macromolecules* **1986**, *19*, 2978.

Bond Length [pm]	Valence Angles [°]	Torsion Angles [°]		ξ [b) (for 298 K)	ξ_0	E_ξ [c) [kJ·mol^{-1}]		
		For NR / FR: a)		For NR / FR: a)		For NR / FR: a)		
Si-Si : 234	Si-Si-Si : 109.4	t (in tt) :	180 / 180	σ = 1.6 / 1.6	1.0	−1.2 / −1.2	$U = \begin{bmatrix} 1 & \sigma & \sigma \\ 1 & \sigma\psi & \sigma\omega \\ 1 & \sigma\omega & \sigma\psi \end{bmatrix}$	rows and columns: t, g$^+$, g$^-$
Si-H : 148	Si-Si-H : 110.0	g (in tg) :	55 / 58.6	ψ = 1.5 / 2.0	1.0	−1.0 / −1.7		
		g (in gg) :	55 / 54.7	ω = 0.52 / 1.0	1.0	1.6 / 0.0		(see: d))
		g$^-$ (in g$^+$g$^-$): −60 / −68.8						

a) Techniques used to calculate conformational energies are: *NR* for "no relaxation", *PR* for "partial relaxation", and *FR* for "full relaxation".
b) In the case of the *NR* and *PR* calculations, the statistical weights σ, ψ, and ω are determined from the respective values of z derived from the potential energy maps. Values of the statistical weight parameters determined in this manner take explicit consideration of the relative size of the domains for each state, as denoted by the so-called "entropy factor" ξ_0. In the case of the *FR* calculations, for which the absence of potential energy maps precludes computation of z values, the statistical weight parameters are given as simple Boltzmann factors ($\xi_0 = 1$).
c) Calculated for $\xi_0 = 1$ throughout.
d) In the original paper, the matrix $U = \begin{bmatrix} 1 & \sigma\psi & \sigma \\ 1 & \sigma\psi & \sigma\omega \\ 1 & \sigma\omega & \sigma\psi \end{bmatrix}$ is given.

Comments:
The given model uses results of conformational energy calculations by Damewood, Jr., J. R.; West, R. *Macromolecules* **1985**, *18*, 159.

Calcd. quantities: $\langle r^2 \rangle_0 / nl^2$ = 4.1 (for *NR*)
= 3.9 (for *FR*)

S 087

polydimethylsilylene

Welsh, W. J.; DeBolt, L.; Mark, J. E. *Macromolecules* **1986**, *19*, 2978.

Bond Length [pm]	Valence Angles [°]	Torsion Angles [°]	ξ [b)] (for 298 K)	ξ_0	E_ξ [c)] [kJ · mol^{-1}]		
Si-Si : 235	Si-Si-Si : 115.4	For *NR* / *PR* / *FR* : [a)]	For *NR* / *PR* / *FR* : [a)]				
Si-C : 187	Si-Si-C : 108.5	t (in tt) :180 / 180 / 180	σ = 0.27 / 0.82 / 1.2	1.0	3.2 / 0.5 / −0.5	$U = \begin{bmatrix} 1 & \sigma & \sigma \\ 1 & \sigma\psi & \sigma\omega \\ 1 & \sigma\omega & \sigma\psi \end{bmatrix}$	rows and columns: t, g$^+$, g$^-$
C-H : 110	Si-C-H : 110.0	g (in tg) : 55 / 82 / 59.7	ψ = 0.00 / 0.56 / 3.8	1.0	∞ / 1.4 / −3.3		
		g (in gg) : 55 / 91 / 54.7	ω = 0.00 / 0.00 / 0.0	1.0	∞ / ∞ / ∞		(see: [d)])
		g$^-$ (in g$^+$g$^-$): −65 / -- / −72.3					

[a)] Techniques used to calculate conformational energies are: *NR* for *"no relaxation"*, *PR* for *"partial relaxation"*, and *FR* for *"full relaxation"*.

[b)] In the case of the *NR* and *PR* calculations, the statistical weights σ, ψ, and ω are determined from the respective values of *z* derived from the potential energy maps. Values of the statistical weight parameters determined in this manner take explicit consideration of the relative size of the domains for each state, as denoted by the so-called "entropy factor" ξ_0. In the case of the *FR* calculations, for which the absence of potential energy maps precludes computation of *z* values, the statistical weight parameters are given as simple Boltzmann factors (ξ_0 = 1).

[c)] Calculated for ξ_0 = 1 throughout.

[d)] In the original paper, the matrix $U = \begin{bmatrix} 1 & \sigma\psi & \sigma \\ 1 & \sigma\psi & \sigma\omega \\ 1 & \sigma\omega & \sigma\psi \end{bmatrix}$ is given.

Comments:
The given model uses results of conformational energy calculations by Damewood, Jr., J. R.; West, R. *Macromolecules* **1985**, *18*, 159.

Calcd. quantities: $<r^2>_0 / nl^2$

= 15.0 (for *NR*)
= 13.2 (for *PR*)
= 12.5 (for *FR*)

![Structure of polyisobutylene repeat unit with methyl substituents, showing bonds labeled a and b]

polyisobutylene, poly(1,1-dimethylethylene), PIB

Allegra, G.; Benedetti, E.; Pedone, C. *Macromolecules* **1970**, *3*, 727.

Bond Length [pm]	Valence Angles [°]	Torsion Angles [°]	ξ (for 297 K)	E_ξ [kJ·mol^{-1}]			
C-C : 154	C-CH$_2$-C : 124	t : 180	ε [a] = 0.03 to 0.001	E = 8.4 - 16.7			
C-H : 110	CH$_2$-C-CH$_2$: 109.5	g$^+$: 60	η [b] = 6.5 to 31.0	E$_1$ = 4.2 - 8.4	$U_a = \begin{bmatrix} 2 & \eta & \eta \\ \eta & \varepsilon & 0 \\ \eta & 0 & \varepsilon \end{bmatrix}$	$U_b = \begin{bmatrix} 1 & 1 & 1 \\ 1 & 1 & 1 \\ 1 & 1 & 1 \end{bmatrix}$	rows and columns: t, g$^+$, g$^-$
		g$^-$: -60					

[a] $\varepsilon = \exp(-E/RT)$
[b] $\eta = 1 + \exp(-E_1/RT)$

Comments:
The conformational analysis of PIB is performed, with explicit allowance for elastic bending of the chain C-C-C bond angles.

Calcd. quantities: $<r^2>_0 / 2nl^2$ = 6.4 to 6.75   (*Exptl.*: 6.6)
                        $d(\ln <r^2>_0)/dT$ = -0.1 (to -0.4) × 10^{-3} K^{-1} (*Exptl.*: -0.27 (to -0.28) × 10^{-3} K^{-1})

polyisobutylene, poly(1,1-dimethylethylene), PIB

Boyd, R. H.; Breitling, S. M. *Macromolecules* 1972, 5, 1.

Bond Length [pm]	Valence Angles [°]	Torsion Angles [°]	ξ (for 298 K)	ξ_0	E_ξ [kJ · mol^{-1}]	
C-C : 153	C-CH$_2$-C : 123	t$_+$: 195	ω = 2.3	1	−2.1	
	CH$_2$-C-CH$_2$: 110	t$_-$: 165	β = 0.18	1	4.2	An 18 × 18 statistical weight matrix for the bond sequences $(i\text{-}1, i)/(i\text{-}2, i\text{-}1)$ is given in the original paper.
		g$^+_+$: 74	δ = 0.36	1	2.5	
		g$^+_-$: 53	α = 0.72	1	0.8	
		g$^-_+$: −53	γ = 0.11	1	5.4	
		g$^-_-$: −74				

Comments:
Energy diagrams for bond rotation in 2,2,4,4-tetramethylpentane, 2,2,4,4,6,6-hexamethylheptane, and 2,2,4,4,6,6,8,8-octamethylnonane are generated in a completely *a priori* manner. A relatively simple conformational model gives a good representation of the conformations calculated, and permits a statistical mechanical calculation of the characteristic ratio.

Calcd. quantities:

$\langle r^2 \rangle_0 / nl^2 = 4.59$ (*Exptl.*: 6.6)

$d(\ln \langle r^2 \rangle_0)/dT = -0.12 \times 10^{-3}$ K^{-1} (*Exptl.*: −0.20 (± 0.20) × 10^{-3} K^{-1})

S 090

[Structural diagram of polyisobutylene showing repeat unit with bonds labeled a and b, methyl groups (H₃C, CH₃) on quaternary carbons and H, H on methylene carbons, repeating x times]

polyisobutylene, poly(1,1-dimethylethylene), PIB

Liberman, M. H.; DeBolt, L. C.; Flory, P. J. *J. Polym. Sci., Polym. Phys. Ed.* **1974**, *12*, 187.

Bond Length [pm]	Valence Angles [°]	Torsion Angles [°]	ξ (for 298 K)	ξ_0	E_ξ [kJ · mol^{-1}]			
C-C : 153	C-CH$_2$-C : 122	t : 180	σ = 0.91 to 0.72	1	0.2 to 0.8			
	CH$_2$-C-CH$_2$:109.5	g$^+$: 60 to 70	ψ = 0.08 to 0.03	1	6.3 to 8.4	$U_a = \begin{bmatrix} 1 & \sigma & \sigma \\ 1 & \sigma\psi & 0 \\ 1 & 0 & \sigma\psi \end{bmatrix}$	$U_b = \begin{bmatrix} 1 & \sigma & \sigma \\ 1 & \sigma & \sigma \\ 1 & \sigma & \sigma \end{bmatrix}$	rows and columns: t, g$^+$, g$^-$
		g$^-$: −60 to −70						

Comments:
Experimental results on the characteristic ratio, $<r^2>_0/nl^2$ = 6.6, and on the temperature coefficient, $d (\ln <r^2>_0) / dT$ = −0.28 · 10^{-3} K^{-1}, are reproduced within limits of ±0.2 and ± 0.15 × 10^{-3}, respectively. The effect of a 10° change in $\phi(g)$ may be compensated by adjustment of either ψ or σ. On the other hand, both $\Delta\alpha$ and its temperature coefficient are much greater than calculated from RIS theory assuming additivity of bond polarizabilities. The disparity (more than tenfold for $\Delta\alpha$) cannot be relieved by any rational adjustment of the structural parameters.

S 091 (see also S 092)

polyisobutylene, poly(1,1-dimethylethylene), PIB

Suter, U. W.; Saiz, E.; Flory, P. J. *Macromolecules* **1983**, *16*, 1317.

Bond Length [pm]	Valence Angles [°]	Torsion Angles [°]	ξ (for 300 K)	ξ_0	E_ξ [kJ·mol^{-1}]
C-C : 153	C-CH$_2$-C : 123	t_+ : 165	α = 0.034	0.5	6.7 (± 0.8)
C-H : 110	CH$_2$-C-CH$_2$: 109	t_- : -165	β = 0.22	0.6	2.5 (± 1.3)
		g^+_+ : 50	β' = 0.00052	0.6	17.6 (± 0.8)
		g^+_- : 75	γ = 0.000042	0.5	23.4 (± 1.7)
		g^-_+ : -75	δ = 0.000018	0.6	26.0 (± 0.8)
		g^-_- : -50	ε = 3.77	0.7	-4.2 (± 1.3)
			ζ = 0.024	0.7	8.4 (± 1.3)
			ξ = 0.00044	1.0	19.3 (± 2.9)
			ρ = 0.031	0.9	8.4 (± 2.1)
			ψ = 0.48	1.1	2.1 (± 1.3)

$$U_a = \begin{bmatrix} \gamma & \delta & 1 & \varepsilon & \zeta & \xi \\ \delta & \gamma & \xi & \zeta & \varepsilon & 1 \\ 1 & \xi & 0 & \rho & 0 & 0 \\ \varepsilon & \zeta & \rho & \psi & 0 & 0 \\ \zeta & \varepsilon & 0 & 0 & \psi & \rho \\ \xi & 1 & 0 & 0 & \rho & 0 \end{bmatrix} \qquad U_b = \begin{bmatrix} 1 & 0 & \beta & 0 & \beta' & 0 \\ 0 & 1 & 0 & \beta' & 0 & \beta \\ \beta & 0 & \beta^2 & 0 & \alpha\beta' & 0 \\ 0 & \beta' & 0 & (\beta')^2 & 0 & \alpha\beta' \\ \beta' & 0 & \alpha\beta' & 0 & (\beta')^2 & 0 \\ 0 & \beta & 0 & \alpha\beta' & 0 & \beta^2 \end{bmatrix} \text{(see: a))}$$

$\alpha = \sigma\omega / \omega'$
$\beta = \sigma\omega' / \omega''$
$\beta' = \sigma'\omega' / \omega''$

rows and columns: $t_+, t_-, g^+_+, g^+_-, g^-_+, g^-_-$

a) U_a contains second-order parameters only.

S 092   (see also S 091)

polyisobutylene, poly(1,1-dimethylethylene), PIB

Suter, U. W.; Saiz, E.; Flory, P. J. *Macromolecules* **1983**, *16*, 1317.
DeBolt, L. C.; Suter, U. W. *Macromolecules* **1987**, *20*, 1424.

Bond Length [pm]	Valence Angles [°]	Torsion Angles [°]		ξ (for 300 K)	ξ_0	E_ξ [kJ·mol^{-1}]
C-C : 153	C-CH$_2$-C : 124	t$_+$	: 155	ξ = 0.007	1.0	12.47
C-H : 110	CH$_2$-C-CH$_2$: 110	t$_-$	: -155	γ < 10^{-4}		
		g$^+_+$	: 60	δ < 10^{-4}		
		g$^-_-$	: -60			

$$U_a = \begin{bmatrix} \gamma & \delta & 1 & \xi \\ \delta & \gamma & \xi & 1 \\ 1 & \xi & 0 & 0 \\ \xi & 1 & 0 & 0 \end{bmatrix} \quad U_b = \begin{bmatrix} 1 & 0 & \beta & 0 \\ 0 & 1 & 0 & \beta \\ \beta & 0 & \beta^2 & 0 \\ 0 & \beta & 0 & \beta^2 \end{bmatrix} \equiv diag\,(1,1,\beta,\beta) \begin{bmatrix} 1 & 0 & 1 & 0 \\ 0 & 1 & 0 & 1 \\ 1 & 0 & 1 & 0 \\ 0 & 1 & 0 & 1 \end{bmatrix} diag\,(1,1,\beta,\beta)$$

After renormalization, and $\gamma = \delta = 0$:

$$U_a = \begin{bmatrix} 0 & 0 & 1 & \xi \\ 0 & 0 & \xi & 1 \\ 1 & \xi & 0 & 0 \\ \xi & 1 & 0 & 0 \end{bmatrix} \quad U_b = \begin{bmatrix} 1 & 0 & 1 & 0 \\ 0 & 1 & 0 & 1 \\ 1 & 0 & 1 & 0 \\ 0 & 1 & 0 & 1 \end{bmatrix}$$

rows and columns: t$_+$, t$_-$, g$^+_+$, g$^-_-$

$\beta = \sigma\omega'/\omega''$

Comments:
Based on calculations on structures CH$_3$-[C(CH$_3$)$_2$-CH$_2$]$_x$-H with x = 2, 3, and 4, and on a hexad of structure C-CH$_2$-[C(CH$_3$)$_2$-CH$_2$]$_6$-C.

Calcd. quantities: $<r^2>_0 / nl^2$ = 6.6 (± 0.1)   (Exptl.: 6.6 - 6.9)
   $d\ln(<r^2>_0)/dT$

polyisobutylene, poly(1,1-dimethylethylene), PIB

Vacatello, M.; Yoon, D. Y. *Macromolecules* **1992**, *25*, 2502.

Bond Length [pm]	Valence Angles [°]	Torsion Angles [°]	Diad (P_d) and interdiad (P_i) matrices of the *a priori* probabilities for two-bond sequences (for 400 K)	
C-C : 153	C-CH$_2$-C : 126	t_+ : 165		
C-H : 110	CH$_2$-C-CH$_2$: 110	t_- : -165		
		g^+_+ : 50	$P_d = \begin{bmatrix} 0.056 & 0.012 & 0.072 & 0.029 & 0.054 & 0.004 \\ 0.012 & 0.056 & 0.004 & 0.054 & 0.029 & 0.072 \\ 0.072 & 0.004 & 0.036 & 0.007 & 0.019 & 0.001 \\ 0.029 & 0.054 & 0.007 & 0.015 & 0.011 & 0.019 \\ 0.054 & 0.029 & 0.019 & 0.011 & 0.015 & 0.007 \\ 0.004 & 0.072 & 0.001 & 0.019 & 0.007 & 0.036 \end{bmatrix}$	$P_i = \begin{bmatrix} 0.003 & 0.002 & 0.120 & 0.075 & 0.019 & 0.015 \\ 0.002 & 0.003 & 0.015 & 0.019 & 0.075 & 0.120 \\ 0.120 & 0.015 & 0.000 & 0.001 & 0.000 & 0.000 \\ 0.075 & 0.019 & 0.001 & 0.033 & 0.000 & 0.000 \\ 0.019 & 0.075 & 0.000 & 0.000 & 0.033 & 0.001 \\ 0.015 & 0.120 & 0.000 & 0.000 & 0.001 & 0.000 \end{bmatrix}$
		g^+_- : 75		
		g^-_+ : -75		
		g^-_- : -50	rows and columns: $t_+, t_-, g^+_+, g^+_-, g^-_+, g^-_-$	

Comments:
The Monte-Carlo method is utilized to investigate the conformational distribution in the central section of a PIB decamer at various temperatures. It is checked that a six-state RIS model based on the two matrices P_d and P_i constitutes a description of the conformational distribution in PIB. The Monte-Carlo results are in excellent agreement with the experimental data on the average dimensions of PIB chains, as well as with the molecular scattering functions of this polymer in solution and in bulk.

Calcd. quantities:

$<r^2>_0 / nl^2$ = 6.7 (± 0.2) (*Exptl.*: 6.6 - 6.9)

$d (\ln <r^2>_0) / dT$ = - (0.1 to 0.3) × 10^{-3} K^{-1} (*Exptl.*: - 0.2 (± 0.2) × 10^{-3} K^{-1})

polyisobutylene, poly(1,1-dimethylethylene), PIB

Vacatello, M.; Yoon, D. Y. *Macromolecules* **1992**, *25*, 2502.

Bond Length [pm]	Valence Angles [°]	Torsion Angles [°]	Diad (P_d) and interdiad (P_i) matrices of the *a priori* probabilities for two-bond sequences (for 400 K)		
C-C : 153	C-CH$_2$-C : 126	t$_+$: 165			
C-H : 110	CH$_2$-C-CH$_2$: 110	t$_-$: -165	$P_d = \begin{bmatrix} 0.068 & 0.000 & 0.101 & 0.058 \\ 0.000 & 0.068 & 0.058 & 0.101 \\ 0.101 & 0.058 & 0.065 & 0.050 \\ 0.058 & 0.101 & 0.050 & 0.065 \end{bmatrix}$	$P_i = \begin{bmatrix} 0.000 & 0.000 & 0.195 & 0.034 \\ 0.000 & 0.000 & 0.034 & 0.195 \\ 0.195 & 0.034 & 0.038 & 0.000 \\ 0.034 & 0.195 & 0.000 & 0.038 \end{bmatrix}$	rows and columns: t$_+$, t$_-$, g$^+$, g$^-$
		g$^+$: 68			
		g$^-$: -68			

Comments:
In spite of the fact that a six-state model is a natural choice for describing the conformational distribution in PIB chains, a simpler model can be obtained by labeling all the bond conformations in the domain [60,180°] and all those in the domain [−60, −180°] with the symbols g$^+$ and g$^-$, respectively. This leads to a four state model formally analogous to the older ones but leaves the exact location of the *gauche* states undefined.

Calcd. quantities: $<r^2>_0 / nl^2$ = 6.7 (± 0.2)   (*Exptl.*: 6.6 to 6.9)

$d (\ln <r^2>_0) / dT$ = −0.1 (*to* −0.3) × 10^{-3} K^{-1} (*Exptl.*: −0.2 (± 0.2) × 10^{-3} K^{-1})

*Further calculations on **polyisobutylene** chains:*

S 095   Hoeve, C. A. J. *J. Chem. Phys.* **1960**, *32*, 888.

Methods are given for the calculation of polymer chain dimensions unperturbed by excluded volume effects but dependent on restricted rotation around the chain bonds as a result of interactions between substituents on neighboring chain atoms. The method developed by *Lifson* [Lifson, S. *J. Chem. Phys.* **1958**, *29*, 80, and *J. Chem. Phys.* **1959**, *30*, 964] to account for interactions between any two consecutive rotations in chains of the type (CR_2) are extended to chains of the type (CH_2-CR_2) and isotactic chains of the type (CH_2-CHR). As an illustration the chain dimensions of PIB are calculated.

Calcd. quantities: $\langle r_o^2 \rangle / 2nl^2$ = 3.31   (*Exptl.*: ~ 3.8)
(for 320 K) $d(\ln \langle r_o^2 \rangle) / d(\ln T)$ = −0.06   (*Exptl.*: −0.06)

S 096   Cho, D.; Neuburger, N. A.; Mattice, W. L. *Macromolecules* **1992**, *25*, 322.

A molecular dynamics trajectory is computed for methyl-terminated PIB at 400 K. Several time-dependent properties (mean-square end-to-end distance, averaged bond angles, and the number and locations of rotational isomeric states) deduced from the trajectory are in reasonable agreement with the results of earlier experiments and earlier theoretical investigations of the static properties of this polymer.

S 097   Born, R.; Spiess, H. W.; Kutzelnigg, W.; Fleischer, U.; Schindler, M. *Macromolecules* **1994**, *27*, 1500.

^{13}C NMR chemical shifts are calculated on an *ab initio* level for the central carbons of a tetramer model molecule, employing the IGLO method. Remarkable agreement between experimental and simulated spectra is obtained for PIB using the conformational statistics as obtained by *Vacatello* and *Yoon* (**S 094**). The observed experimental spread of ≈ 20 ppm for the CH_2 resonance is quantitatively reproduced in the calculation as is the γ-*gauche* effect. Correlations of the chemical shift with specific geometrical aspects as C−C bond length are established.

S 098

poly(vinylidene chloride), poly(1,1-dichloroethylene), PVDC

Matsuo, K.; Stockmayer, W. H. *Macromolecules* **1975**, *8*, 660.

Bond Length [pm]	Valence Angles [°]	Torsion Angles [°]	ξ (for 298 K)	E_ξ [kJ·mol^{-1}]			
C-Cl : 176	Cl-C-Cl : 112	t : 180	ε [a] = 0.033	E = 8.4			
C-C : 154	C-CH$_2$-C : 123	g$^+$: 60	η [b] = 1.09	E_1 = 5.9	$U_a = \begin{bmatrix} 2 & \eta & \eta \\ \eta & \varepsilon & 0 \\ \eta & 0 & \varepsilon \end{bmatrix}$	$U_b = \begin{bmatrix} 1 & \alpha & \alpha \\ \alpha & \beta & \gamma \\ \alpha & \gamma & \beta \end{bmatrix}$	rows and columns: t, g$^+$, g$^-$
	CH$_2$-C-CH$_2$: 112	g$^-$: -60	$\beta > \alpha > 1 > \gamma$				

[a] $\varepsilon = \exp(-E/RT)$
[b] $\eta = 1 + \exp(-E_1/RT)$

Comments:
The matrix U_a is identical in form with that in PIB, and is thus conceived to be steric in origin. The elements of the matrix U_b reflect differences in electrostatic energy.

Calcd. quantities: $<r^2>_0 / nl^2$ = 6.7 *to* 11.5 (*Exptl.*: 8 (± 1))

S 099

poly(vinylidene chloride), poly(1,1-dichloroethylene), PVDC

Boyd, R. H.; Kesner, L. *J. Polym. Sci., Polym. Phys. Ed.* **1981**, *19*, 393.

Bond Length [pm]	Valence Angles [°]	Torsion Angles [°]	ξ (for 298 K)	ξ_0	E_ξ [a)] [kJ·mol^{-1}]
C-CH$_2$-C : 121		t_+ : 195	α = 2.32	1	−2.09
C-CCl$_2$-C : 114		t_- : 165	α' = 0.057	1	7.11
		g^+_+ : 90	β = 1	1	0
		g^+_- : 52	γ = 2.41	1	−2.18
		g^-_+ : −52	γ' = 0.24	1	3.56
		g^-_- : −90	δ = 0.66	1	1.04

$$U_a = \begin{bmatrix} \beta & \beta & 1 & \delta & 1 & 1 \\ \beta & \beta & 1 & 1 & \delta & 1 \\ 1 & 1 & \gamma' & \gamma & 0 & 0 \\ \delta & 1 & \gamma & 0 & 0 & 0 \\ 1 & \delta & 0 & 0 & 0 & \gamma \\ 1 & 1 & 0 & 0 & \gamma & \gamma' \end{bmatrix} \qquad U_b = \begin{bmatrix} 1 & 0 & \alpha' & 0 & \alpha & 0 \\ 0 & 1 & 0 & \alpha & 0 & \alpha' \\ \alpha' & 0 & \alpha'^2 & 0 & \alpha'\alpha & 0 \\ 0 & \alpha & 0 & \alpha^2 & 0 & \alpha\alpha' \\ \alpha & 0 & \alpha\alpha' & 0 & \alpha^2 & 0 \\ 0 & \alpha' & 0 & \alpha\alpha' & 0 & \alpha'^2 \end{bmatrix}$$

rows and columns: t_+, t_-, g^+_+, g^+_-, g^-_+, g^-_-

a) Parameter set III of the original paper.

Comments:
A polarization model for representing polar bond effects in conformational energy calculations is applied to PVDC. The geometries and conformational energies of a number of conformers of hexachloroheptane are calculated. The geometries are found to be similar to the hydrocarbon analog PIB in that steric crowding results in the usual t, g$^+$, g$^-$ states being split into + or − distortions of the torsional angles away from the traditional values. The same statistical model along with energy parameters previously calculated also gives agreement with experiment for the characteristic ratio of PIB.

Calcd. quantities:
$\langle r^2 \rangle_0 / nl^2$ = 7.70 (298 K)   (Exptl.: 8 (± 1), 298 K)
$d \ln(\langle r^2 \rangle_0) / dT$ = −0.44 × 10^{-4} K^{-1}
$\langle \mu^2 \rangle_0 / nm^2$ = 0.89 (333 K)   (Exptl.: ~ 0.86 , 333 K)

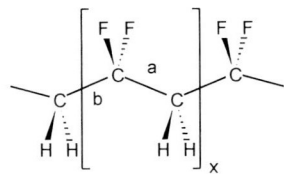

poly(vinylidene fluoride), poly(1,1-difluoroethylene), PVDF

Carballeira, L.; Pereiras, A. J.; Rios, M. A. *Macromolecules* **1990**, *23*, 1309.

Bond Length [pm]	Valence Angles [°]	Torsion Angles [°]		ξ (for 298 K)	ξ_o	E_ξ [kJ · mol^{-1}]
	C-CH$_2$-C : 116	t_+	: 195	α = 2.75	1	−2.51
	CH$_2$-C-CH$_2$: 113	t_-	: 175	α' = 0.014	1	10.65
		g^+_+	: 65	β = 1	1	0
		g^+_-	: 55	δ = 0.33	1	2.78
		g^-_+	: −55	γ = 14.88	1	−6.69
		g^-_-	: −65	γ' = 0.21	1	3.81

$$U_a = \begin{bmatrix} \beta & \beta & 1 & \delta & 1 & 1 \\ \beta & \beta & 1 & 1 & \delta & 1 \\ 1 & 1 & \gamma' & \gamma & 0 & 0 \\ \delta & 1 & \gamma & 0 & 0 & 0 \\ 1 & \delta & 0 & 0 & 0 & \gamma \\ 1 & 1 & 0 & 0 & \gamma & \gamma' \end{bmatrix} \qquad U_b = \begin{bmatrix} 1 & 0 & \alpha' & 0 & \alpha & 0 \\ 0 & 1 & 0 & \alpha & 0 & \alpha' \\ \alpha' & 0 & \alpha'^2 & 0 & \alpha'\alpha & 0 \\ 0 & \alpha & 0 & \alpha^2 & 0 & \alpha\alpha' \\ \alpha & 0 & \alpha\alpha' & 0 & \alpha^2 & 0 \\ 0 & \alpha' & 0 & \alpha\alpha' & 0 & \alpha'^2 \end{bmatrix}$$

rows and columns: t_+, t_-, g^+_+, g^+_-, g^-_+, g^-_-

Comments:
The conformational properties of trimer molecules modeling PVDB (**S 100**) and PVDF are analyzed by the molecular mechanics method of *Boyd* and *Kesner* [*J. Chem. Phys.* **1980**, *72*, 2179], which takes into account both steric and electrostatic energy. Total conformational energies are used to calculate a set of intramolecular interaction energies that, by means of the RIS model, allowed estimation of the characteristic ratios and dipole moment ratios of PVDB and PVDF under unperturbed conditions.

Calcd. quantities: $<r^2>_o / nl^2$ = 5.7 (at 298 K) , 5.5 (at 463 K)   (Exptl.: 5.6 (at 463 K))
  $<\mu^2>_o / nm^2$ = 0.84 (at 333 K)

poly(vinylidene bromide), poly(1,1-dibromoethylene), PVDB

Carballeira, L.; Pereiras, A. J.; Rios, M. A. *Macromolecules* **1990**, *23*, 1309.

Bond Length [pm]	Valence Angles [°]	Torsion Angles [°]	ξ (for 298 K)	ξ_0	E_ξ [kJ·mol^{-1}]
	C-CH$_2$-C : 125	t_+ : 200	α = 5.81	1	-4.36
	CH$_2$-C-CH$_2$: 110	t_- : 165	α' = 0.16	1	4.54
		g^+_+ : 80	β = 1	1	1
		g^+_- : 40	δ = 0.035	1	8.29
		g^-_+ : -45	γ = 0.53	1	1.57
		g^-_- : -80	γ' = 0.19	1	4.06

$$U_a = \begin{bmatrix} \beta & \beta & 1 & \delta & 1 & 1 \\ \beta & \beta & 1 & 1 & \delta & 1 \\ 1 & 1 & \gamma' & \gamma & 0 & 0 \\ \delta & 1 & \gamma & 0 & 0 & 0 \\ 1 & \delta & 0 & 0 & 0 & \gamma \\ 1 & 1 & 0 & 0 & \gamma & \gamma' \end{bmatrix} \qquad U_b = \begin{bmatrix} 1 & 0 & \alpha' & 0 & \alpha & 0 \\ 0 & 1 & 0 & \alpha & 0 & \alpha' \\ \alpha' & 0 & \alpha'^2 & 0 & \alpha'\alpha & 0 \\ 0 & \alpha & 0 & \alpha^2 & 0 & \alpha\alpha' \\ \alpha & 0 & \alpha\alpha' & 0 & \alpha^2 & 0 \\ 0 & \alpha' & 0 & \alpha\alpha' & 0 & \alpha'^2 \end{bmatrix}$$

rows and columns: $t_+, t_-, g^+_+, g^+_-, g^-_+, g^-_-$

Comments:
The conformational properties of trimer molecules modeling PVDB and PVDF (**S 100**) are analyzed by the molecular mechanics method of *Boyd* and *Kesner*. [*J. Chem. Phys.* **1980**, *72*, 2179], which takes into account both steric and electrostatic energy. Total conformational energies are used to calculate a set of intramolecular interaction energies that by means of the RIS model allowed estimation of the characteristic ratios and dipole moment ratios of PVDB and PVDF under unperturbed conditions.

Calcd. quantities: $\langle r^2 \rangle_0 / nl^2$ = 8.2 (at 298 K)
$\langle \mu^2 \rangle_0 / nm^2$ = 0.74 (at 333 K)

*Further calculations on **poly(vinylidene halogenide)** chains:*

S 102   Tonelli, A. E. *Macromolecules* **1976**, *9*, 547.

Approximate conformational energy estimates are utilized to evaluate the RIS model of PVDF. Occasional (0-20%) head to head:tail to tail (H-H:T-T) addition of monomer units in a random fashion is accounted for in the calculation of these conformational properties. In general it is found that the calculated conformational properties are relatively insensitive to the amount of H-H:T-T addition assumed, but are instead markedly dependent upon the value of the dielectric constant (ε) selected to mediate the electrostatic interactions encountered along a PVDF chain.

Calcd. quantities: $<r^2>_0 / nl^2$ = 5.06 (for 463 K) ; 4.93 (for 483 K)   (*Exptl.:* 5.6 ($\pm$ 0.3) at 463 K)
(ranges correspond to $d(\ln <r^2>_0)/dT$ = -1.26×10^{-3} K^{-1}
0 *to* 20% HH:TT addition) $<\mu^2>_0 / nm^2$ = 0.92 *to* 0.76 (for 483 K)
                         $d(\ln <\mu^2>_0)/dT$ = -1.00 (*to* -0.49) $\times 10^{-3}$ K^{-1}

Conformational entropies S_a (amorphous chains) and S_c (crystalline chains)

S 103

polyoxymethylene, POM

Flory, P. J.; Mark, J. E. *Makromol. Chem.* **1964**, *75*, 11.

Bond Length [pm]	Valence Angles [°]	Torsion Angles [°]	ξ (for 473.2 K)	ξ_0	E_ξ [kJ·mol^{-1}] [a]			
C-O : 143	C-O-C : 110	t : 180	σ = 8.5	1	−8.4			
	O-C-O : 110	g$^+$: 60 or 80	σ' = 8.5	1	−8.4	$U_a = \begin{bmatrix} 1 & \sigma & \sigma \\ 1 & \sigma' & 0 \\ 1 & 0 & \sigma' \end{bmatrix}$	$U_b = \begin{bmatrix} 1 & \sigma & \sigma \\ 1 & \sigma & \delta \\ 1 & \delta & \sigma \end{bmatrix}$	rows and columns: t, g$^+$, g$^-$
		g$^-$: −60 or −80	δ = 0	1	∞			

[a] Given parameters lead to the *Calcd.* quantities given below; other combinations of parameters are provided, too.

Comments:
The POM chain is discussed in terms of *trans* and *gauche* rotational states and is compared with PE and PDMS. Examination of distances between non-bonded atoms and groups in various conformations suggest dispersion interactions between O and CH$_2$ as being primarily responsible for preference for the *gauche* states; coulombic interactions also favor the *gauche* conformation, but should be comparatively small.

Calcd. quantities:

$\langle r^2 \rangle_0 / nl^2 = 8.7$

$d(\ln \langle r^2 \rangle_0)/dT = -2.6 \times 10^{-3} \text{ K}^{-1}$

polyoxymethylene; POM

Jernigan, R. L.; Flory, P. J. *J. Chem. Phys.* **1969**, *50*, 4165, 4178.

Bond Length [pm]	Valence Angles [°]	Torsion Angles [°]	ξ (for 473 K)	ξ_0	E_ξ [kJ·mol^{-1}]			
C-O : 143	C-O-C: 110	t : 180	σ = 12	1	−9.8	$U_a = \begin{bmatrix} 1 & \sigma & \sigma \\ 1 & \sigma & \sigma\omega \\ 1 & \sigma\omega & \sigma \end{bmatrix}$	$U_b = \begin{bmatrix} 1 & \sigma & \sigma \\ 1 & \sigma & 0 \\ 1 & 0 & \sigma \end{bmatrix}$	rows and columns t, g$^+$, g$^-$
	C-O-C: 110	g$^+$: 60	ω = 0.05	1	11.8			
		g$^-$: −60						

Comments:
Numerical calculations are presented for the PE chains and for the POM chains. The former displays a preference for *trans* conformations among its skeletal bonds; the latter manifests a strong preference for the *gauche* conformation.

Calcd. quantities: *A priori* singlet and pair probabilities for infinite chains
A priori rotational state probabilities averaged over all internal bonds

polyoxymethylene, POM

Abe, A.; Mark, J. E. *J. Am. Chem. Soc.* **1976**, *98*, 6468.

Bond Length [pm]	Valence Angles [°]	Torsion Angles [°]	ξ (for 298 K)	ξ_0	E_ξ [kJ·mol^{-1}]			
C-O : 142	C-O-C : 112	t : 180	σ = 10.8	1	−5.9			
C-H : 110	O-C-O : 112	g$^+$: 65	ω = 0.079	1	6.3	$U_a = \begin{bmatrix} 1 & \sigma & \sigma \\ 1 & \sigma & \sigma\omega \\ 1 & \sigma\omega & \sigma \end{bmatrix}$	$U_b = \begin{bmatrix} 1 & \sigma & \sigma \\ 1 & \sigma & \sigma\omega' \\ 1 & \sigma\omega' & \sigma \end{bmatrix}$	rows and columns: t, g$^+$, g$^-$
		g$^-$: −65	ω' = 0	1	∞			

Comments:
Conformational energies of the first four members (y = 1-4) of the polyoxide series CH$_3$O{(CH$_2$)$_y$-O-}$_x$-CH$_3$ are calculated using semiempirical potential energy functions.

Calcd. quantities: [a] $<r^2>_0 / nl^2$ = 9.5 (298 K)   (Exptl.: 10 (±2) (298 K))
                             $d\ln(<r^2>_0) / dT$ = −6.6 × 10^{-3} K^{-1} (Exptl.: "negative and large")
                             $<\mu^2> / nm^2$ = 0.3 (473 K)   (Exptl.: 0.1 for x = 1; 0.29 for x = 2 (333 K); 0.3 for x → ∞ (473 K))

[a] x = 1 : dimethoxy methane, x = 2 : dioxymethylene dimethyl ether.

polyoxymethylene, POM

Kajiwara, K.; Burchard, W. *Macromolecules* **1984**, *17*, 2669.

Bond Length [pm]	Valence Angles [°]	Torsion Angles [°]	ξ (for 413 K)	ξ_0	E_ξ [kJ · mol^{-1}]
C-O : 143	C-O-C : 110	t : 180	$\sigma = 12$	1	-8.5
C-H : 110	O-C-O : 110	g$^+$: 60	$\omega = 0.05$	1	10.3
		g$^-$: -60			

$$U_a = \begin{bmatrix} 1 & \sigma & \sigma \\ 1 & \sigma & \sigma\omega \\ 1 & \sigma\omega & \sigma \end{bmatrix} \quad U_b = \begin{bmatrix} 1 & \sigma & \sigma \\ 1 & \sigma & 0 \\ 1 & 0 & \sigma \end{bmatrix}$$

rows and columns: t, g$^+$, g$^-$

Comments:
Monte-Carlo simulations are applied to estimate the characteristic ratios and ρ parameters (which is defined as the ratio of the radius of gyration to the hydrodynamic radius) from the RIS models for PE, POM, polybutadiene, and polyisoprene.

The bond pair probabilities are (400 K):
(denoted conditional probabilities in the reference)

$$P_a = \begin{bmatrix} 0.0089 & 0.0567 & 0.0567 \\ 0.0567 & 0.3636 & 0.0182 \\ 0.0567 & 0.0182 & 0.3636 \end{bmatrix} \quad P_b = \begin{bmatrix} 0.0087 & 0.0568 & 0.0568 \\ 0.0568 & 0.3820 & 0.0000 \\ 0.0568 & 0.0000 & 0.3820 \end{bmatrix}$$

Calcd. quantities: (400 K)

$<r^2>_0 / nl^2 = 8.16$

$\rho_\infty = 1.60$

polyoxymethylene, POM

Curro, J. G.; Schweizer, K. S.; Adolf, D.; Mark, J. E. *Macromolecules* **1986**, *19*, 1739.

Bond Length [pm]	Valence Angles [°]	Torsion Angles [°]	ξ (for 300 K)	ξ_o	E_ξ [kJ·mol^{-1}]			
C-O : 143	C-O-C : 110	t : 180	σ = 10.6	1	−5.9	$U_a = \begin{bmatrix} 1 & \sigma & \sigma \\ 1 & \sigma & \sigma\omega_a \\ 1 & \sigma\omega_a & \sigma \end{bmatrix}$	$U_b = \begin{bmatrix} 1 & \sigma & \sigma \\ 1 & \sigma & \sigma\omega_b \\ 1 & \sigma\omega_b & \sigma \end{bmatrix}$	rows and columns: t, g$^+$, g$^-$
	O-C-O : 110	g$^+$: 65	ω_a = 0.08	1	6.3			
		g$^-$: −65	ω_b = 0	1	∞			

Comments:
In POM, the *gauche* state has a significantly lower energy than the *trans* state. This energy difference, which is thought to be due to attractive interactions between CH$_2$ and O groups, is estimated to be −5.9 kJ·mol^{-1} on the basis of semiempirical conformational energy calculations and dipole moment measurements. The preference for *gauche* configurations, coupled with severe repulsive interactions for alternate *gauche* bonds of opposite sign, gives POM a strong tendency to form helical sequences. Monte-Carlo calculations are performed on POM chains using the RIS model. The end-to-end distribution function was found to exhibit bimodal behavior, characteristic of a helix/coil coexistence, over a range of temperature and chain length. The first-order *gauche* distribution (P_g) and the second-order *gauche* pair distribution (P_{gg}) does not show any bimodality. The sequence length distribution, however, does show pronounced bimodality. The transition-like behavior is found to become sharper and shifts to lower temperature with a logarithmic dependence of the chain length. Thus the helix/coil coexistence behavior is a finite chain effect, with the transition temperature approaching 0 K for the infinite system. As expected, an external force on the ends of the chains is found to shift the coexistence temperature to higher temperatures. These results can be understood by analogy with the one-dimensional Ising model.

Calcd. quantities: End-to-end distribution function $W(r)$

*Further calculations on **polyoxymethylene** chains:*

S 108   Allegra, G.; Calligaris, M.; Randaccio, L. *Macromolecules* **1973**, *6*, 390.

For POM, a matrix algorithm for the statistical mechanical treatment of an unperturbed -A-B-A-B- polymer chain with energy correlation between first-neighboring skeletal rotations is described. The results of the unperturbed dimensions are in satisfactory agreement with experimental data. In addition, if the same energy data are used, the results are rather close to those obtained by the RIS scheme usually adopted. The RIS scheme is shown to be also adequate for the calculation of the average intramolecular conformational energy, if the torsional oscillation about skeletal bonds is taken into account in the harmonic approximation.

S 109   Abe, A. *J. Am. Chem. Soc.* **1976**, *98*, 6477.

Conformational energies of 2-methoxytetrahydropyran, a model compound for methyl aldopyranoside, are computed using semiempirical energy expressions. The free-energy difference between the equatorial and axial conformers, and the dipole moment for an equilibrium mixture are calculated, the extra stabilization energy ΔE associated with the *gauche* C-O-C-O arrangements being treated as a variable. The observed values are both satisfactorily reproduced at a value of $\Delta E = 4.6$ kJ · mol^{-1}, thus leading to the axial conformer. The stabilization energy of the same magnitude is estimated for the *gauche* conformation about a C-O bond in POM chain. This strongly suggest that the anomalous preference for *gauche* conformations found in POM is related in its origin to the "anomeric effect" known in carbohydrate chemistry.

S 110   Miyasaka, T.; Kinai, Y.; Imamura, Y. *Makromol. Chem.* **1981**, *182*, 3533.

The conformational energies of the lower members of POM, 2,4-dioxapentane and 2,4,6-trioxaheptane are estimated by the empirical force field method. The *gauche* states of the internal rotation around the skeletal C-O bonds are successfully predicted to be of lower energies in both molecules. In order to calculate the unperturbed dimension and dipole moment of POM, RIS approximations are made by using the results obtained from the force field calculations on 2,4,6-trioxaheptane. Although these parameters are significantly different from those estimated earlier, they reproduce the observed values fairly well.

	Bond Length C-O [pm]	Valence Angle COC [°]	Valence Angle OCO [°]	Torsion Angle ϕ_t [°]	Torsion Angle ϕ_t [°]	E_σ	E_ϕ	E_ω	$E_{\phi'}$	$E_{\omega'}$
								[kJ · mol^{-1}]		
Set I	141.1	112.10	118.42	173.5	± 63.2	−14.14	3.01	14.73	6.69	23.14
Set II	141.3	112.26	111.14	176.5	± 67.7	−12.89	0.71	8.20	6.44	18.62
Abe [a]	142	112.0	112.0	180.0	± 65.0	−5.86	0.00	6.28	0.00	∞

[a] see model **S 105**

S 111   Curro, J. G.; Mark, J. E. *J. Chem. Phys.* **1985**, *82*, 3820.

Monte-Carlo simulations based on the RIS models are used to generate distribution functions for the end-to-end separation of polyoxide chains having the repeat unit -$(CH_2)_m$—O-.

S 112   Bahar, I.; Mattice, W. L. *Macromolecules* **1991**, *24*, 877.

Computation, using the dynamic RIS model, of the relaxation times for POM helices. The bimodal distribution of relaxation times is rationalized with a simple model.

polydimethylsilmethylene, polydimethylsilaethylene, PDMSM

Ko, J. H.; Mark, J. E. *Macromolecules* **1975**, *8*, 869, 874.

Bond Length [pm]	Valence Angles [°]	Torsion Angles [°]	ξ (for 333 K)	ξ_0	E_ξ [kJ·mol^{-1}]			
Si-C : 190	CH$_2$-Si-CH$_2$: 109.5	t : 180	ω = 0.75	1	0.80			
	Si-CH$_2$-Si : 115	g$^+$: 60				$U_a = \begin{bmatrix} 1 & 1 & 1 \\ 1 & 1 & \omega \\ 1 & \omega & 1 \end{bmatrix}$	$U_b = \begin{bmatrix} 1 & 1 & 1 \\ 1 & 1 & 0 \\ 1 & 0 & 1 \end{bmatrix}$	rows and columns: t, g$^+$, g$^-$
		g$^-$: −60						

Comments:
The lack of any strong conformational preference is the origin of the relatively small value of the unperturbed dimensions and of the insensitivity of both the unperturbed dimensions and dipole moments to changes in temperature.

Calcd. quantities:

$\langle r^2 \rangle_0 / nl^2$ = ~ 4.2   (*Exptl.*: 5.32 (± 0.01))

$d \ln(\langle r^2 \rangle_0) / dT$ = ~ −0.30 × 10^{-3} K^{-1} (*Exptl.*: 0.2 × 10^{-3} K^{-1})

$\langle \mu^2 \rangle_0 / nm^2$ = ~ 0.44   (*Exptl.*: 0.39 (± 0.03))

$d \ln(\langle \mu^2 \rangle_0) / dT$ = ~ −0.13 × 10^{-3} K^{-1} (*Exptl.*: 0.0)

S 114

polydimethylsilmethylene, polydimethylsilaethylene, PDMSM

Sundararajan, P. R. *Comput. Polym. Sci.* **1991**, *1*, 18.

Bond Length [pm]	Valence Angles [°]	Torsion Angles [°]	ξ	ξ_o	E_ξ [kJ · mol^{-1}]
Si-C : 187	C-Si-C : 109.5	t : 170 *or* 180			
	Si-C-Si : 121	g$^+$: 80			
		g$^-$: −80			

$$U_a = \begin{bmatrix} 1 & 1 & 1 \\ 1 & 1 & 1 \\ 1 & 1 & 1 \end{bmatrix} \quad U_b = \begin{bmatrix} 1 & 1 & 1 \\ 1 & 1 & 0 \\ 1 & 0 & 1 \end{bmatrix}$$

rows and columns: t, g$^+$, g$^-$

Comments:
The energy calculations on the diad of PDMSM show that all the allowed rotational isomeric states are of equal energy. Although the minima are displaced from perfect staggering, the energy surface in the vicinity of the minima is shallow. The conformational map shows that near-free but heterogeneous rotations of the diad segments are possible. This finding can be used in the interpretation of molecular mobility of PDMSM, if such studies are undertaken. Excellent agreement between the experimental and calculated values of $<r^2>_o / nl^2$ and $<\mu^2> / nm^2$ is achieved by using torsion angles corresponding to non-staggered *gauche* rotational states. A brief comparison of the conformational features of this chain and PIB is presented.

Calcd. quantities: $<r^2>_o / nl^2$ = 5.5 (*Exptl.*: 5.32)
 $<\mu^2>_o / nm^2$ = 0.39 (*Exptl.*: 0.39 (± 0.03))

Further calculations on **polydimethylsilmethylene** *chains:*

S 115 Llorente, M. A.; Mark, J. E. *J. Polym. Sci., Polym. Phys. Ed.* **1983**, *21*, 1173.

Elastomeric networks are prepared from PDMSM, and the birefringence in elongation is found to be qualitatively similar to that of PDMS, in that there is no evidence for strain-induced crystallization. However, the values of the optical-configuration parameter $\Delta\alpha$ are considerably larger, and both $\Delta\alpha$ and its temperature coefficient are essentially the same as those of PIB. Results obtained from RIS theory considerably underestimate both $\Delta\alpha$ and its temperature coefficient for PDMSM, as they do for PIB. Although the origin of the discrepancy is not necessarily the same for both polymers, the results on PDMSM suggest that the discrepancy for PIB is not due to the severe steric congestion known to be present in this polymer.

polydimethylsiloxane, PDMS

Flory, P. J.; Crescenzi, V.; Mark, J. E. *J. Am. Chem. Soc.* **1964**, *86*, 146.

Bond Length [pm]	Valence Angles [°]	Torsion Angles [°]	ξ (for 343 K)	ξ_0	E_ξ [kJ·mol^{-1}]			
Si-O : 164	O-Si-O : 110	t : 180	σ = 0.286	1	3.6	$U_a = \begin{bmatrix} 1 & \sigma & \sigma \\ 1 & \sigma' & 0 \\ 1 & 0 & \sigma' \end{bmatrix}$	$U_b = \begin{bmatrix} 1 & \sigma & \sigma \\ 1 & \sigma & \delta \\ 1 & \delta & \sigma \end{bmatrix}$	rows and columns: t, g$^+$, g$^-$
Si-C : 190	Si-O-Si : 143	g$^+$: 60	σ' = 0.286	1	3.6			
		g$^-$: -60	$\delta \approx$ 0.06	1	8.0			

Comments:
Owing to the inequality of the bond angles, a decrease in the *trans* population increases $<r^2>_0/nl^2$. Hence, the positive temperature coefficient for this ratio denotes a lower energy for the *trans* state. This energy is attributed to favorable interaction between CH$_3$ pairs separated by 380 pm in the planar (*trans*) conformation. The larger value observed for the ratio in a less polar medium is in the direction predicted for enhanced electrostatic interaction within a chain of partially ionic Si—O bonds.

Calcd. quantities:

$<r^2>_0 / nl^2$ = 6.3 (343 K)   (*Exptl.*: 6.3 to 7.7, depending on the solvent)

$d \ln(<r^2>_0) / dT$ = 0.67 × 10^{-3} K^{-1} (*Exptl.*: 0.75 (± 0.06) × 10^{-3} K^{-1})

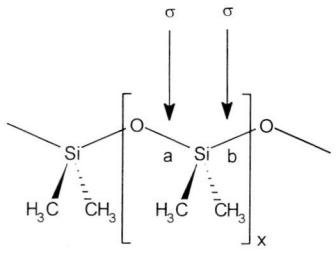

polydimethylsiloxane, PDMS

Mark, J. E. *J. Chem. Phys.* **1968**, *49*, 1398.

Bond Length [pm]	Valence Angles [°]	Torsion Angles [°]	ξ (for 298 K)	ξ_0	E_ξ [kJ · mol^{-1}]			
Si-O : 164	O-Si-O : 110	t : 180	$\sigma = 0.23$	1	3.6	$U_a = \begin{bmatrix} 1 & \sigma & \sigma \\ 1 & \sigma & \sigma\omega \\ 1 & \sigma\omega & \sigma \end{bmatrix}$	$U_b = \begin{bmatrix} 1 & \sigma & \sigma \\ 1 & \sigma\psi & 0 \\ 1 & 0 & \sigma\psi \end{bmatrix}$	rows and columns: t, g$^+$, g$^-$
Si-C : 190	Si-O-Si : 143	g$^+$: 60	$\omega = 0.17$	1	4.4			
		g$^-$: -60	$\psi = 1$	1	0			

Comments:
A RIS model with neighbor dependence is used to calculate mean-square dipole moments and their temperature coefficients for PDMS chains over a wide range of molecular weight. Chain conformational energies required in the calculations are obtained from a previous analysis of the random-coil dimensions of PDMS chains in the limit of large *x* (**S 116**).

Calcd. quantities:

$<\mu^2> / xm^2 = 0.23 \; (x \to \infty)$

$d \ln(<\mu^2>) / dT = 0.71 \times 10^{-3} \; K^{-1} \; (x \to \infty)$

polydimethylsiloxane, PDMS

Beevers, M. S.; Semlyen, J. A. *Polymer* **1972**, *13*, 385.

Bond Length [pm]	Valence Angles [°]	Torsion Angles [°]	ξ a) (for 383 K)	ξ_0	E_ξ [kJ · mol^{-1}]
Si-O : 164	O-Si-O : 110	τ : 180	σ = 0.327	1	3.56
Si-C : 190	Si-O-Si : 143	g^+ : 60	δ = 0.082	1	7.96
		g^- : −60			

$$U_a = \begin{bmatrix} 1 & \sigma & \sigma \\ 1 & \sigma & 0 \\ 1 & 0 & \sigma \end{bmatrix} \quad U_b = \begin{bmatrix} 1 & \sigma & \sigma \\ 1 & \sigma & \delta \\ 1 & \delta & \sigma \end{bmatrix}$$

rows and columns: t, g^+, g^-

a) $\omega = \delta/\sigma = 0.251$

Comments:
Theoretical molar cyclization equilibrium constants, K_x, for small, unstrained cyclics (H$_2$SiO)$_x$ (x = 4-8) and [(CH$_3$)$_2$SiO]$_x$ (x = 4-9) in undiluted polydihydrogensiloxane and PDMS equilibrates are calculated using the *Jacobson-Stockmayer* theory, without assuming that the corresponding chain molecules obey Gaussian statistics. The RIS model gives theoretical molar cyclization equilibrium constants in excellent agreement with the experimental values.

Calcd. quantities: Molar cyclization equilibrium constants, K_x

polydimethylsiloxane, PDMS

Liberman, M. H.; Abe, Y.; Flory, P. J. *Macromolecules* **1972**, *5*, 550.

Bond Length [pm]	Valence Angles [°]	Torsion Angles [°]	ξ (for 343 K)	ξ_0	E_ξ [kJ · mol^{-1}]			
Si-O : 164	Si-O-Si : 143	t : 180	$\sigma = 0.29$	1	3.56			
Si-C : 190	O-Si-O : 110	g$^+$: 60	$\omega = 0.20$	1	4.40	$U_a = \begin{bmatrix} 1 & \sigma & \sigma \\ 1 & \sigma\psi & 0 \\ 1 & 0 & \sigma\psi \end{bmatrix}$	$U_b = \begin{bmatrix} 1 & \sigma & \sigma \\ 1 & \sigma & \sigma\omega \\ 1 & \sigma\omega & \sigma \end{bmatrix}$	rows and columns: t, g$^+$, g$^-$
	C-Si-C : 112	g$^-$: −60	$\psi = 1.0$	1	0			
	O-Si-C : 108							

Comments:
The strain birefringence of PDMS networks is investigated over the temperature range 288-363 K. Temperature coefficients of the optical-configuration parameter, $\Delta\alpha$, are determined experimentally. The value of $\Delta\alpha$ is markedly reduced by swelling with decalin, with cyclohexane, and especially with CCl$_4$. The observed (positive) temperature coefficient considerably exceeds theoretical predictions, and would be at variance with the supposition of order in the amorphous polymer. The vanishingly small optical anisotropy of PDMS casts doubt on the significance of the discrepancy with theory.

Calcd. quantities: Optical anisotropy-configuration parameter, $\Delta\alpha$ $d \ln \Delta\alpha / dT$

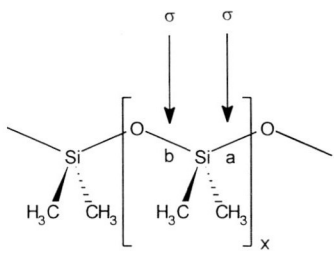

polydimethylsiloxane, PDMS

Flory, P. J.; Chang, V. W. C. *Macromolecules* **1976**, *9*, 33.

Bond Length [pm]	Valence Angles [°]	Torsion Angles [°]	ξ (for 383 K)	ξ_0	E_ξ [kJ · mol^{-1}]			
Si-O : 164	O-Si-O : 110	t : 180	σ = 0.327	1	3.56			
Si-C : 190	Si-O-Si : 143	g$^+$: 60	ω = 0.236	1	4.61	$U_a = \begin{bmatrix} 1 & \sigma & \sigma \\ 1 & \sigma & 0 \\ 1 & 0 & \sigma \end{bmatrix}$	$U_b = \begin{bmatrix} 1 & \sigma & \sigma \\ 1 & \sigma & \sigma\omega \\ 1 & \sigma\omega & \sigma \end{bmatrix}$	rows and columns: t, g$^+$, g$^-$
		g$^-$: −60						

Comments:

The persistence vector $\mathbf{a} \equiv \langle \mathbf{r} \rangle$ and the "center of gravity" vector $\langle \mathbf{g} \rangle$ are calculated for PDMS chains, both vectors being expressed in a reference frame with x axis along the initial Si−O bond, and y axis in the plane defined by this bond and the following one. The respective vectors converge with increase in chain length to the limiting persistence $\mathbf{a}_\infty$ of magnitude 735 pm and direction virtually coincident with the x axis. Cartesian tensors up to the sixth rank formed from the displacement vector $\rho = \mathbf{r} - \mathbf{a}$, where $\mathbf{r}$ is the end-to-end vector for the chain of n bonds, are evaluated as the average over all configurations for n = 2-100.

Calcd. quantities: Persistence vector $\mathbf{a} \equiv \langle \mathbf{r} \rangle$:

$|\mathbf{a}_\infty|$ = 735 and 840 pm[a)]

$\langle r^2 \rangle_0 / nl^2$ = 6.43

Density distribution functions $W_a(\rho)$

[a)] Depending on the choice of the first bond (silicon-to-oxygen or oxygen-to-silicon).

S 121

polydimethylsiloxane, PDMS

Mattice, W. L. *Macromolecules* **1978**, *11*, 517.

Bond Length [pm]	Valence Angles [°]	Torsion Angles [°]	ξ (for 298 K)	ξ_0	E_ξ [kJ · mol^{-1}]			
Si - O : 164	O-Si-O : 109.5	t : 180	$\sigma = 0.238$	1	3.56	$U_a = \begin{bmatrix} 1 & \sigma & \sigma \\ 1 & \sigma & 0 \\ 1 & 0 & \sigma \end{bmatrix}$	$U_b = \begin{bmatrix} 1 & \sigma & \sigma \\ 1 & \sigma & \sigma\omega \\ 1 & \sigma\omega & \sigma \end{bmatrix}$	rows and columns: t, g$^+$, g$^-$
Si - C : 190	Si-O-Si : 143	g$^+$: 60	$\omega = 0.156$	1	4.61			
		g$^-$: −60						

Comments:
Unperturbed dimensions and dipole moments of polydialkylsiloxanes are investigated using RIS theory. Polymers are treated as branched molecules in which each silicon atom constitutes a tetrafunctional branch point. All significant first- and second-order interactions are included in the configuration partition function. Higher order interactions not suppressed by second-order interactions are also evaluated and accounted for in the statistical weights used.

Calcd. quantities: $(\langle r^2 \rangle_0 / nl^2)_\infty$ $(\langle \mu^2 \rangle_0 / nm^2)_\infty$

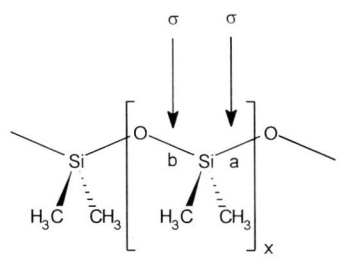
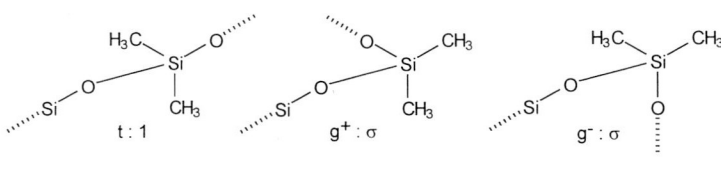

polydimethylsiloxane, PDMS

DeBolt, L. C.; Mark, J. E. *Macromolecules* **1987**, *20*, 2369.

Bond Length [pm]	Valence Angles [°]	Torsion Angles [°]	ξ (for 300 K)	ξ_o	E_ξ [kJ · mol^{-1}]
			$\sigma = 0.240$	1	3.56
			$\omega = 0.158$	1	4.61

$$U' \equiv \begin{bmatrix} 1 & \sigma & \sigma \\ 1 & \sigma & 0 \\ 1 & 0 & \sigma \end{bmatrix} = \begin{bmatrix} 1 & 0.240 & 0.240 \\ 1 & 0.240 & 0 \\ 1 & 0 & 0.240 \end{bmatrix}$$

The *a priori* probability matrix **P** for a single bond and matrices **P'** and **P''** representing bond pairs are: a)

$$\mathbf{P} = \begin{bmatrix} 0.734 & 0.132 & 0.132 \end{bmatrix}$$

$$\mathbf{P'} = \begin{bmatrix} 0.517 & 0.109 & 0.109 \\ 0.109 & 0.023 & 0.000 \\ 0.109 & 0.000 & 0.023 \end{bmatrix}$$

$$\mathbf{P''} = \begin{bmatrix} 0.522 & 0.107 & 0.107 \\ 0.107 & 0.022 & 0.004 \\ 0.107 & 0.004 & 0.022 \end{bmatrix}$$

$$U'' \equiv \begin{bmatrix} 1 & \sigma & \sigma \\ 1 & \sigma & \sigma\omega \\ 1 & \sigma\omega & \sigma \end{bmatrix} = \begin{bmatrix} 1 & 0.240 & 0.240 \\ 1 & 0.240 & 0.038 \\ 1 & 0.038 & 0.240 \end{bmatrix}$$

Conditional probability matrices **Q** are obtained as follows:

$$\mathbf{Q'} = \begin{bmatrix} 0.703 & 0.148 & 0.148 \\ 0.826 & 0.174 & 0.000 \\ 0.826 & 0.000 & 0.174 \end{bmatrix}$$

$$\mathbf{Q''} = \begin{bmatrix} 0.710 & 0.145 & 0.145 \\ 0.808 & 0.166 & 0.026 \\ 0.808 & 0.026 & 0.166 \end{bmatrix}$$

rows and columns: t, g$^+$, g$^-$

a) It is assumed that the *a priori* conformational probability of a bond pair is independent of its location within the chain.

Comments:
Two theoretical models are developed to model the trapping of PDMS rings present at the time of network formation by the end-linking of PDMS chains. Monte-Carlo methods are first used to generate representative samples of cyclics having degrees of polymerization DP of 20, 40, 75, 120, and 200. Criteria are developed to determine whether a particular ring would be topologically trapped by one or more network chains. The more realistic model yields values for the percent trapped that are in excellent agreement with experiment over the entire DP range.

polydimethylsiloxane, PDMS

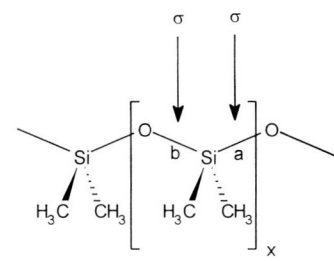

Bahar, I.; Zuniga, I.; Dodge, R.; Mattice, W. L. *Macromolecules* **1991**, *24*, 2986, 2993.

Bond Length [pm]	Valence Angles [°]	Torsion Angles [°]	ξ (for 343 K)	ξ_0	E_ξ [kJ·mol^{-1}]
Si-O : 163	O-Si-O : 110	t : 180	$\sigma = 0.64$	1	1.26
Si-C : 190	Si-O-Si : 143 *or* 150	g$^+$: 60	$\tau = 0.60$	1	1.47
C-H : 110	O-Si-C : 109.5	g$^-$: -60	$\tau' = 1.16$	1	-0.42
	C-Si-C : 109.5		$\omega = 0.41$	1	2.51
	H-Si-C : 109.5		$\omega' = 0.60$	1	1.47
	C-O-Si : 122.0		$\psi = 0.69$	1	1.05
			$\psi' = 0.74$	1	0.84

$$U_a = \begin{bmatrix} \tau & \sigma & \sigma \\ 1 & \sigma\psi & \sigma\omega \\ 1 & \sigma\omega & \sigma\psi \end{bmatrix} \quad U_b = \begin{bmatrix} \tau' & \sigma & \sigma \\ 1 & \sigma\psi' & \sigma\omega' \\ 1 & \sigma\omega' & \sigma\psi' \end{bmatrix}$$

rows and columns: t, g$^+$, g$^-$

Comments:
The probability distribution of isomeric conformations in PDMS is investigated by both conformational energy considerations and by molecular dynamics simulations. A comparatively smooth distribution of isomeric states is obtained from both approaches. A new RIS treatment, compatible with the molecular mechanics and dynamics considerations, is introduced for describing the conformational statistics of PDMS.

Calcd. quantities:

$<r^2>_0 / nl^2$	= 6.43 (343 K)	(*Exptl.*: 6.3)
$d \ln(<r^2>_0) / dT$	= -2.34×10^{-4} K^{-1}	(*Exptl.*: 0.7×10^{-3} K^{-1})
$<\mu^2>_0 / nm^2$	= 0.25	(*Exptl.*: 0.33 (298 K))
$d \ln(<\mu^2>) / dT$	= "negative"	(*Exptl.*: "negative")
Cyclization equilibrium constant K_x		

polydimethylsiloxane, PDMS

Besbes, S.; Cermelli, I.; Bokobza, L.; Monnerie, L.; Bahar, I.; Erman, B.; Herz, J. *Macromolecules* **1992**, *25*, 1949.

Bond Length [pm]	Valence Angles [°]	Torsion Angles [°]	ξ a) (for 300 K)	ξ_0	E_ξ a) [kJ·mol^{-1}]
Si-O : 164	O-Si-O : 143	t : 180	$\sigma = 0.60 / 0.60$	1	1.26 / 1.26
			$\tau = 0 / 0.55$	1	∞ / 1.47
	Si-O-Si : 110	g^+ : 60	$\tau' = 0.17 / 0.85$	1	4.40 / 0.42
			$\omega = 0 / 0.37$	1	∞ / 2.51
		g^- : −60	$\omega' = 0 / 0.55$	1	∞ / 1.47
			$\psi = 0 / 0.66$	1	∞ / 1.05
			$\psi' = 0 / 0.71$	1	∞ / 0.84

$$U_a = \begin{bmatrix} \tau & \sigma & \sigma \\ 1 & \sigma\psi & \sigma\omega \\ 1 & \sigma\omega & \sigma\psi \end{bmatrix} \qquad U_b = \begin{bmatrix} \tau' & \sigma & \sigma \\ 1 & \sigma\psi' & \sigma\omega' \\ 1 & \sigma\omega' & \sigma\psi' \end{bmatrix}$$

rows and columns: t, g^+, g^-

a) Two different sets of data are considered for the intramolecular energies associated with the statistical weight parameters. The former, introduced by *Crescenzi*, *Mark*, and *Flory* (**S 116**), and the latter from a work of *Bahar* and *Mattice* (**S 123**). The present results are not significantly altered with the choice of the energy parameters. Nevertheless, the second set of energy parameters is found to favor more *gauche* placements compared to the previous model, leading to a more uniform distribution of the isomeric states.

Comments:
Segmental orientation in model networks of PDMS in uniaxial tension is measured by infrared dichroism. Measurements are made for four tetrafunctional end-linked networks. Results of experiments are compared with predictions of calculations based in (i) the widely used Kuhn expression and (ii) the RIS formalism. The Kuhn expression is found to considerably overestimate the segmental orientation. The RIS approach leads to values of segmental orientation that fall between predictions of the affine and phantom network models. This indicates that the nematic-like intermolecular contributions to orientation are not significant.

*Further calculations on **polydimethylsiloxane** chains:*

S 125   Flory, P. J.; Semlyen, J. A. *J. Am. Chem. Soc.* **1966**, *88*, 3209.

Mean-square end-to-end dimensions, $<r^2>_0/xl^2$, and macrocyclization equilibrium constants, K_x, are calculated of unperturbed PDMS chains, as a function of chain length. The calculations are carried out by the method of *Flory* and *Jernigan* [*J. Chem. Phys.* **1965**, *42*, 3509; *Proc. Natl. Acad. Sci. U. S.* **1964**, *51*, 1060], the statistical weights used in these calculations ($\sigma = 0.326$ and $\delta = 0.08$) are taken from **S 116**.

S 126   Semlyen, J. A.; Wright, P. V. *Polymer* **1969**, *10*, 543.

The molar cyclization equilibrium constants, K_x, of PDMS are measured. Using the *Jacobson* and *Stockmayer* equilibrium theory of macrocyclization, the dimensions of PDMS chains with 40-80 chemical bonds in the bulk polymer at 383 K are deduced. Dilution effects in the PDMS systems are contrasted with predictions of the *Jacobson-Stockmayer* theory, and the experimental molar cyclization equilibrium constants of the smallest siloxane rings are discussed in terms of the statistical properties of the corresponding oligomeric chains using the RIS model of PDMS of *Flory, Crescenzi,* and *Mark* [**S 116**].

S 127   Sutton, C.; Mark, J. E. *J. Chem. Phys.* **1971**, *54*, 5011.

Dielectric constants are determined for pure liquid dimethylsiloxane oligomers. Mean-square dipole moments, calculated from the Onsager equation, are in good agreement with predicted values based on the RIS model (**S 117**) with neighbor dependence and chain conformational energies obtained in an independent analysis of the random-coil dimensions of such chains. In addition, the observed temperature coefficients of $<\mu^2>$ are in qualitative agreement with calculated results.

S 128   Liao, S. C.; Mark, J. E. *J. Chem. Phys.* **1973**, *59*, 3825.

Dielectric constants are determined for PDMS chains in the thermodynamically good solvent cyclohexane. Unfortunately, comparison of the experimental values of the dipole moment ratio with those predicted from RIS theory is complicated by pronounced specific solvent effects and comparison of experimental and theoretical values of the temperature coefficient is also difficult because of the very small magnitude of this coefficient.

S 129   DeBolt, L. C.; Mark, J. E. *J. Polym. Sci.* **1988**, *26*, 989.

Evidence of only a low barrier to inversion in Si—O—Si sequences could be important with regard to the interpretation of the statistical properties of silicone polymers. The effects are estimated for the temperature coefficients of the unperturbed dimensions, dipole moments, and the optical anisotropy for PDMS.

S 130   Grigoras, S. *Polym. Prepr. (Am. Chem. Soc., Div. Polym. Chem.)* **1991**, *32(3)*, 720.

A method is presented for the direct computation of statistical weights from conformational analysis for the RIS analysis of PDMS.

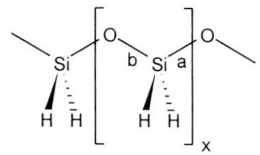

polydihydrogensiloxane

Beevers, M. S.; Semlyen, J. A. *Polymer* **1972**, *13*, 385.

Bond Length [pm]	Valence Angles [°]	Torsion Angles [°]	ξ	ξ_0	E_ξ [kJ·mol^{-1}]			
Si-O : 164	O-Si-O : 110	t : 180						
Si-H : 148	Si-O-Si : 143	g$^+$: 60				$U_a = \begin{bmatrix} 1 & 1 & 1 \\ 1 & 1 & 1 \\ 1 & 1 & 1 \end{bmatrix}$	$U_b = \begin{bmatrix} 1 & 1 & 1 \\ 1 & 1 & 1 \\ 1 & 1 & 1 \end{bmatrix}$	rows and columns t, g$^+$, g$^-$
		g$^-$: -60						

Comments:
Theoretical molar cyclization equilibrium constants, K_x, for small, unstrained cyclics (H$_2$SiO)$_x$ (x = 4-8) and [(CH$_3$)$_2$SiO]$_x$ (x = 4-9) in undiluted polydihydrogensiloxane and PDMS equilibrates are calculated using the *Jacobson-Stockmayer* theory, without assuming that the corresponding chain molecules obey Gaussian statistics. The RIS model gives theoretical molar cyclization equilibrium constants in excellent agreement with the experimental values.

Calcd. quantities: $\langle r^2 \rangle_0 / xl^2 = 3.3$ Molar cyclization equilibrium constants, K_x

S 132 and S 133

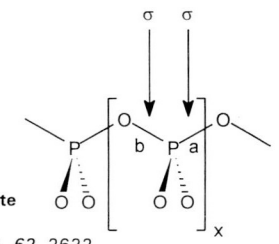

catena-poly[(dioxophosphorus)-μ-oxo], polyphosphate

Semlyen, J. A.; Flory, P. J. *Trans. Farad. Soc.* **1966**, *62*, 2622.

Bond Length [pm]	Valence Angles [°]	Torsion Angles [°]	ξ a) (for 298 K)	ξ_0	E_ξ [kJ·mol^{-1}]			

S 132 Model A:

Main chain:	Main chain:	t : 180	$0.7 < \sigma < 1.0$			$U_a = \begin{bmatrix} 1 & \sigma & \sigma \\ 1 & 0 & 0 \\ 1 & 0 & 0 \end{bmatrix}$	$U_b = \begin{bmatrix} 1 & \sigma & \sigma \\ 1 & \sigma\psi & 0 \\ 1 & 0 & \sigma\psi \end{bmatrix}$	rows and columns: t, g$^+$, g$^-$
P-O : 162	P-O-P : 130	g$^+$: 60	$0.001 < \omega < 0.1$					
	O-P-O : 101.5	g$^-$: −60	Most of the calculations were carried out with:					
Pendant:	Pendant:		$\sigma = 1$	1	0			(see: a))
P-O : 148	O-P-O : 121.0		$\omega = 0$ or 0.1	1	∞ or 5.7			

S 133 Model B: (The inherent torsional potential of the P−O bond is ignored).

see: above	see: above	t : 180	$\tau \approx \sigma$			$U_a = \begin{bmatrix} 1 & \tau \\ \tau & 1 \end{bmatrix}$	$U_b = \begin{bmatrix} 1 & \tau \\ 1 & 0 \end{bmatrix}$	rows and columns: t, c
		c : 0						

a) ω (= $\sigma\psi$) is used in the original paper.

Comments:
Rotation about one bond of the chain, while all its neighbours are *trans*, appears to be subject to little hinderance by either electrostatic or steric interactions. Simultaneous large rotations about pairs of adjoining bonds, irrespective of their relative signs, are made unlikely by both types of interactions.

Calcd. quantities: $<r^2>_0 / nl^2$ = ~ 7.0 (Exptl.: 6.6 - 7.2, depending on the solvent)
 $d \ln(<r^2>_0) / dT$ = ~ −0.02 × 10^{-3} K^{-1}

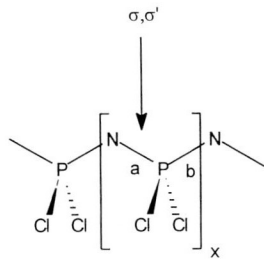

catena-poly[(dichlorophosphorus)-*μ*-nitrido], polydichlorophosphazene, PDCP

Saiz, E. *J. Polym. Sci.; Polym. Phys. Ed.* **1987**, *25*, 1565.

Bond Length [pm]	Valence Angles [°]	Torsion Angles [°]	ξ (for 298 K)	ξ_0	E_ξ [kJ · mol^{-1}]			
P-N : 152	N-P-N : 118	t : 180	σ = 11.3	1	−6.0 (± 0.3)			
P-Cl : 199	P-N-P : 130	g$^+$: 45	σ' = 14.9	1	−6.7 (± 0.1)	$U_a = \begin{bmatrix} 1 & \sigma & \sigma & \sigma \\ \sigma & \sigma' & \sigma' & \sigma' \\ \sigma & \sigma' & 0 & \sigma' \\ \sigma & \sigma' & \sigma' & \sigma' \end{bmatrix}$	$U_b = \begin{bmatrix} 1 & \omega & \omega & \omega \\ \omega & 0 & 0 & 0 \\ \omega & 0 & 0 & 0 \\ \omega & 0 & 0 & 0 \end{bmatrix}$	rows and columns: t, g$^+$, c, g$^-$
Cl-P-Cl : 102		c : 0	ω = 5.4	1	−4.2			
		g$^-$: −45						

Comments:
A theoretical analysis of the conformational energies of PDCP is presented. The results indicate that the bond pair P−N−P possesses a considerable conformational freedom, whereas the bond pair N−P−N is relatively rigid. This difference explains the low glass transition temperatures and large end-to-end distances measured for polyphosphazenes. All the calculated magnitudes are extremely sensitive to the energy E(σ) that controls the statistical weight of the conformations tg, tc, tg$^-$, gt, ct, and g$^-$t, relative to tt for the bond pair P−N−P. A qualitative explanation for this sensitivity is discussed.

Calcd. quantities:

$\langle r^2 \rangle_0 / nl^2$ = 13.5

$d \ln(\langle r^2 \rangle_0) / dT$ = −3.0 × 10^{-3} K^{-1}

$\langle \mu^2 \rangle_0 / nm^2$ = 0.35

$d \ln(\langle \mu^2 \rangle_0) / dT$ = −3.4 × 10^{-3} K^{-1}

*Further calculations on **phosphorus containing** chains:*

S 135   Allcock, H. R.; Allen, R. W.; Meister, J. J. *Macromolecules* **1976**, *9*, 950.

The polydihalophosphazenes are examined by conformational analysis using nonbonding intramolecular interactions based on a 6-12 Lennard-Jones potential and a Coulombic term. The results provide an insight into the reasons for the low glass transition temperatures, the high chain flexibilities, and the conformational preferences of these molecules. Minimum energy conformations are discussed.

S 136   Allen, R. W.; Allcock, H. R. *Macromolecules* **1976**, *9*, 956.

A number of polyorganophosphazanes are studied by conformational analysis techniques with the use of intramolecular nonbonding 6-12-Lennard-Jones and Coulombic potentials.

polyoxyethylene, poly(ethylene oxide), POE, PEO, PM$_2$O

Mark, J. E.; Flory, P. J. *J. Am. Chem. Soc.* **1965**, *87*, 1415.

Bond Length [pm]	Valence Angles [°]	Torsion Angles [°]	ξ a) (for 333 K)	ξ_0	E_ξ [kJ·mol^{-1}]
C-C : 153	C-O-C : 110	t : 180	σ = 0.144	1	5.3
C-O : 143	C-C-O : 110	g$^+$: 60	σ' = 1.80	1	−1.6
	C-C-H : 110	g$^-$: −60	β = 0.70	1	1.0
			$\alpha = \gamma = 1$	1	0

$$U_a = \begin{bmatrix} 1 & \sigma & \sigma \\ 1 & \sigma\alpha & \sigma\beta \\ 1 & \sigma\beta & \sigma\alpha \end{bmatrix} \quad U_b = \begin{bmatrix} 1 & \sigma & \sigma \\ 1 & \sigma\gamma & 0 \\ 1 & 0 & \sigma\gamma \end{bmatrix} \quad U_c = \begin{bmatrix} 1 & \sigma' & \sigma' \\ 1 & \sigma'\alpha & \sigma'\beta \\ 1 & \sigma'\beta & \sigma'\alpha \end{bmatrix}$$

rows and columns: t, g$^+$, g$^-$

a) Other sets of acceptable parameters for β, σ, and σ' are given in the original paper.

Comments:
The energy for *gauche* rotational states about CH$_2$−O and O−CH$_2$ bonds appears to exceed that for the *trans* state owing to interactions, primarily steric, between adjoining methylene groups. The reverse holds for the CH$_2$−CH$_2$ bond according to comparison of observed and calculated results. A lower energy for the *gauche* rotational state compared to the *trans* for this bond is attributable to a favourable dispersion interaction between the adjoining O atoms. The dominant effect of increasing temperature is to increase the *trans* population about CH$_2$−CH$_2$ bonds, and this is the main factor causing $d(\ln \langle r^2 \rangle_0)/dT$ to be positive. An unambiguous assignment of the more important statistical weights for the various rotational states is thus achieved by comparison of calculated and experimental values of both $\langle r^2 \rangle_0 / nl^2$ and $d(\ln \langle r^2 \rangle_0)/dT$.

Calcd. quantities:
$\langle r^2 \rangle_0 / nl^2$ = ~ 4.1 (*Exptl.*: 4.1 (±0.4))
$d(\ln \langle r^2 \rangle_0)/dT$ = ~ 0.2 (*Exptl.*: 0.23 (±0.02) × 10^{-3} K^{-1})

polyoxyethylene, poly(ethylene oxide), POE, PEO, PM$_2$O

Mark, J. E.; Flory, P. J. *J. Am. Chem. Soc.* **1966**, *88*, 3702.

Bond Length [pm]	Valence Angles [°]	Torsion Angles [°]	ξ (for 298 K)	ξ_0	E_ξ [kJ·mol^{-1}]
C-C : 153	C-O-C : 110	t : 180	σ = 0.220	1	3.77
C-O : 143	C-C-O : 110	g$^+$: 60	σ' = 2.07	1	−1.80
	C-C-H : 110	g$^-$: −60	ω = 0.566	1	1.42

For R = C$_2$H$_5$: $U_1 = U_a$, $U_2 = U_b$, $U_3 = U_c$, $U_{n-1} = U_a$, $U_n = U_b$

For R = H:

$$U_a = \begin{bmatrix} 1 & \sigma & \sigma \\ 1 & \sigma & \sigma\omega \\ 1 & \sigma\omega & \sigma \end{bmatrix} \quad U_b = \begin{bmatrix} 1 & \sigma & \sigma \\ 1 & \sigma & 0 \\ 1 & 0 & \sigma \end{bmatrix} \quad U_c = \begin{bmatrix} 1 & \sigma' & \sigma' \\ 1 & \sigma' & \sigma'\omega \\ 1 & \sigma'\omega & \sigma' \end{bmatrix}$$

$$U_2 = U_{n-1} = \begin{bmatrix} 1 & 1 & 1 \\ 1 & 1 & 1 \\ 1 & 1 & 1 \end{bmatrix} \quad U_3 = \begin{bmatrix} 1 & \sigma' & \sigma' \\ 1 & \sigma' & \sigma' \\ 1 & \sigma' & \sigma' \end{bmatrix}$$

rows and columns: t, g$^+$, g$^-$

Comments:
Mean-square dipole moments are calculated for POE, using a RIS model with neighbor dependence. In a previous study of POE (**S 137**), the mean extension of the chain and its temperature coefficient were used to obtain statistical weights for rotational states about skeletal bonds. These statistical weights, with minor modification, are shown to give dipole moments in good agreement with measured values over a wide range of molecular weight. A comparatively large, positive temperature coefficient of the dipole moment is predicted for POE.

Calcd. quantities:

$\langle r^2 \rangle_0 / nl^2$ = 4.0 (Exptl.: 4.1 (± 0.4))

$d(\ln \langle r^2 \rangle_0)/dT$ = 0.12 × 10^{-3} K^{-1} (for 333 K) (Exptl.: 0.23 (±0.02) × 10^{-3} K^{-1})

$\langle \mu^2 \rangle_0 / nm^2$ = 0.58 (for x = 120) (Exptl.: 0.58)

$d(\ln \langle \mu^2 \rangle_0)/dT$ = 2.51 × 10^{-3} K^{-1} (for x = 120) (Exptl.: 2.0 × 10^{-3} K^{-1} (for x = 6, R = C$_2$H$_5$))

polyoxyethylene, poly(ethylene oxide), POE, PEO, PM$_2$O

Patterson, G. D.; Flory, P. J. *J. Chem. Soc.; Farad. Trans. 2* **1972**, *68*, 1111.

Bond Length [pm]	Valence Angles [°]	Torsion Angles [°]	ξ (for 298 K)	ξ_0	E_ξ [kJ·mol^{-1}]
C-C : 153	C-O-C : 110	t : 180	$\sigma = 0.22$	1	3.8 (± 0.8)
C-O : 143	C-C-O : 110	g$^+$: 60	$\sigma' = 1.53$	1	−1.05 (± 0.63)
	C-C-H : 110	g$^-$: −60	$\omega = 0.26$	1	3.3 (± 2.1)

$$U_a = \begin{bmatrix} 1 & \sigma & \sigma \\ 1 & \sigma & \sigma\omega \\ 1 & \sigma\omega & \sigma \end{bmatrix} \quad U_b = \begin{bmatrix} 1 & \sigma & \sigma \\ 1 & \sigma & 0 \\ 1 & 0 & \sigma \end{bmatrix} \quad U_c = \begin{bmatrix} 1 & \sigma' & \sigma' \\ 1 & \sigma' & \sigma'\omega \\ 1 & \sigma'\omega & \sigma' \end{bmatrix}$$

rows and columns: t, g$^+$, g$^-$

Comments:
The mean-square optical anisotropies $\langle\gamma^2\rangle$ of four oligomers of polyoxyethylene glycol dimethyl ether are determined from measurements of depolarized Rayleight scattering. They are interpreted using the valence optical scheme and RIS theory.

Calcd. quantities:

$\langle r^2\rangle_0 / nl^2$ = 4.8 (*Exptl.*: 4.8)

$d(\ln \langle r^2\rangle_0) / dT$ = 0.18 × 10^{-3} K^{-1} (*Exptl.*: 0.23 (± 0.02) × 10^{-3} K^{-1})

$\langle\gamma^2\rangle$ = 80 (± 10) pm^6 (*Exptl.*: 82 (± 10) pm^6)

polyoxyethylene, poly(ethylene oxide), POE, PEO, PM$_2$O

Abe, A.; Mark, J. E. *J. Am. Chem. Soc.* **1976**, *98*, 6468.

Bond Length [pm]	Valence Angles [°]	Torsion Angles [°]	ξ (for 303 K)	ξ_0	E_ξ [kJ · mol^{-1}]
C-H : 110	C-O-C : 111.5	t : 180	σ = 2.0 to 2.3	1	−1.7 to −2.1
C-O : 143	C-C-O : 111.5	g$^+$: 70	σ' = 0.22	1	3.8
C-C : 153		g$^-$: −70	ω = 0.51	1	1.7

$$U_a = \begin{bmatrix} 1 & \sigma & \sigma \\ 1 & \sigma & \sigma\omega \\ 1 & \sigma\omega & \sigma \end{bmatrix} \quad U_b = \begin{bmatrix} 1 & \sigma & \sigma \\ 1 & \sigma & 0 \\ 1 & 0 & \sigma \end{bmatrix} \quad U_c = \begin{bmatrix} 1 & \sigma' & \sigma' \\ 1 & \sigma' & \sigma'/\omega \\ 1 & \sigma'/\omega & \sigma' \end{bmatrix}$$

rows and columns: t, g$^+$, g$^-$

Comments:
Conformational energies of the first four members (y = 1-4) of the polyoxide series CH$_3$O[(CH$_2$)$_y$-O-]$_x$-CH$_3$ are calculated using semiempirical potential energy functions. The semiempirical methods do not in general successfully predict conformational energies in this class of chain molecules. The present study implicates the intramolecular interactions involving oxygen atoms as the origin of this disagreement between theory and experiment and gives a quantitative estimate of the magnitude of the discrepancy for each of the interactions thus involved. The energy differences thus established should be included, as corrections, in conformational energy calculations on such molecules.

Calcd. quantities: [a,b]

$\langle r^2 \rangle_0 / nl^2$	= 5.0	(303 K)	(*Exptl.*: 5.2)
$d(\ln \langle r^2 \rangle_0)/dT$	= 1.4 × 10^{-3} K^{-1}	(333 K)	(*Exptl.*: 0.23 (±0.02) × 10^{-3} K^{-1})
$\langle \mu^2 \rangle_0 / nm^2$	= 0.49	(293 K)	(*Exptl.*: 0.51 (± 0.02))
$d(\ln \langle \mu^2 \rangle_0)/dT$	= 2.9 × 10^{-3} K^{-1}	(313 K)	(*Exptl.*: 2.6 × 10^{-3} K^{-1})

[a] For $E(\sigma)$ = −2.1 kJ · mol^{-1}
[b] The respective values are also given for some oligomers CH$_3$-O-(CH$_2$-CH$_2$-O)$_x$-CH$_3$

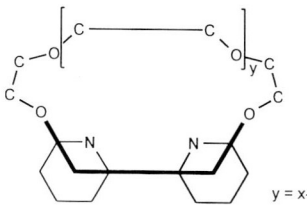

polyoxyethylene, poly(ethylene oxide), POE, PEO, PM$_2$O

Mattice, W. L.; Newkome, G. R. *J. Am. Chem. Soc.* **1979**, *101*, 4477.

Bond Length [pm]	Valence Angles [°]	Torsion Angles [°]	ξ (for 298 K)	ξ_o	E_ξ [kJ·mol^{-1}]
C-C : 153	C-C-O : 110	t : 180 g$^+$: 60	σ = 0.22	1	3.768 (± 0.293)
C-O : 143	C-O-C : 110	g$^-$: -60	σ' = 2.07	1	-1.800 (± 0.293)
C^{ar}-O : 136	$\theta_1 = \theta_n$: 117.6	For 2 and n: ϕ^+ : 90	ω = 0.55	1	1.465 (± 0.837)
l_1 = 445		ϕ^- : -90	σ'' = 0.03	1	8.374
$l_2 = l_n$ = 269		For 3 and n: t : 180 g$^+$: 60 g$^-$: -60			

rows and/or columns: t, g$^+$, g$^-$ or ϕ^+, ϕ^-

$$U_{C-C} = \begin{bmatrix} 1 & \sigma' & \sigma' \\ 1 & \sigma' & \sigma'\omega \\ 1 & \sigma'\omega & \sigma' \end{bmatrix} \quad U_{C-O} = \begin{bmatrix} 1 & \sigma & \sigma \\ 1 & \sigma & \sigma\omega \\ 1 & \sigma\omega & \sigma \end{bmatrix} \quad U_{O-C} = \begin{bmatrix} 1 & \sigma & \sigma \\ 1 & \sigma & 0 \\ 1 & 0 & \sigma \end{bmatrix}$$

$$U_3 = \begin{bmatrix} 1 & \sigma'' & \sigma'' \\ 1 & \sigma'' & \sigma'' \end{bmatrix} \quad U_{n-1} = \begin{bmatrix} 1 & \sigma'' & \sigma'' \\ 1 & \sigma'' & \sigma''\omega \\ 1 & \sigma''\omega & \sigma'' \end{bmatrix} \quad U_n = \begin{bmatrix} 1 & 1 \\ 1 & 1 \\ 1 & 1 \end{bmatrix}$$

$U_1 = U_2$ = row (1,1) U_{n+1} = col (1,1)

Comments:
A study of the conformational characteristics of cyclic POE oligomers (x-mers) is undertaken to attempt an identification of the molecular origin of the temperature-dependent NMR which is reported for certain cyclic molecules prepared from 2,2'-bipyridyl and polyoxyethylene oligomers. The POE chain in an acyclic analogue is assumed to behave according to the RIS model developed by *Mark* and *Flory* (**S 137**, **S 138**). Monte-Carlo calculations are performed using *a priory* and conditional probabilities deduced from the RIS model. Calculations predict that cyclization is possible for x = 2, barely possible for x = 3, and most readily achieved with x = 5-7, in reasonable accord with experiment. Many POE chain conformations are consistent with cyclization when x = 6. Cyclization is achieved with little change in probabilities for occupancy of *trans* and *gauche* states. According to these calculations, the temperature-dependent NMR is a consequence of thermal alteration in the distribution of POE chain conformations consistent with cyclization.

polyoxyethylene, poly(ethylene oxide), POE, PEO, PM_2O

Khanarian, G.; Tonelli, A. E. *Macromolecules* **1982**, *15*, 145.

Bond Length [pm]	Valence Angles [°]	Torsion Angles [°]	ξ (for 298 K)	ξ_o	E_ξ [kJ·mol^{-1}]
C-H : 110	C-O-C : 111.5	t : 180	$\sigma' = 0.22 - 0.11$ 1		3.8 to 5.4
C-O : 143	C-C-O : 111.5	g^+ : 60 or 70	$\sigma = 1.99$ 1		-1.7
C-C : 153		g^- : -60 or -70	$\omega = 1$ or 0.65 1		0 or 1.05

$$U_a = \begin{bmatrix} 1 & \sigma' & \sigma' \\ 1 & \sigma' & \sigma'\omega \\ 1 & \sigma'\omega & \sigma' \end{bmatrix} \quad U_b = \begin{bmatrix} 1 & \sigma' & \sigma' \\ 1 & \sigma' & 0 \\ 1 & 0 & \sigma' \end{bmatrix} \quad U_c = \begin{bmatrix} 1 & \sigma & \sigma \\ 1 & \sigma & \sigma\omega \\ 1 & \sigma\omega & \sigma \end{bmatrix} \quad \text{(see: }^{a)}\text{)}$$

rows and columns: t, g^+, g^-

a) The matrices, **U**, for the terminal bonds *1, 2, 3, n-1,* and *n* are given by *Mark* and *Flory* (**S 137, S 138, S 140**).

Comments:
Molar Kerr constants $_mK$ and dipole moments squared $<\mu^2>$ of poly(oxyethylene glycol)s (POEG) and poly(oxyethylene dimethyl ether)s (POEDE) are reported in the isotropically polarizable solvents carbon tetrachloride, cyclohexane, and dioxane. Data for $_mK/x$ for POEG appear to reach an asymptotical value. Calculations of $_mK/x$ and $<\mu^2>/x$ based on the RIS model show good agreement with the experimental results.

Calcd. quantities:
(for R = H)
$_mK/x = 1.15 \times 10^{12}$ cm^7 s.c.$^{-2}$ mol^{-1} ($x = 320$)   (*Exptl.*: 1.15×10^{12} cm^7 s.c.$^{-2}$ mol^{-1}, $x = 320$)
$<\mu^2>/x = \sim 1.2 \times 10^{-36}$ s.c.^{2}cm^2 ($x = 320$)   (*Exptl.*: 1.12×10^{-36} s.c.^{2}cm^2, $x = 154$)

polyoxyethylene, poly(ethylene oxide), POE, PEO, PM$_2$O

Miyasaka, T.; Yoshida, T.; Imamura, Y. *Makromol. Chem.* **1983**, *184*, 1285.

Bond Length [pm]	Valence Angles [°]	Torsion Angles [°]	ξ (for 303 K)	ξ_0	E_ξ [kJ · mol^{-1}] a)
C-O : 141	C-O-C : 113.6	For C-O bonds:	$\sigma = 0.26$	1	3.35
		t : 179.9			
C-C : 153	C-C-O : 109.8	g$^+$: 75	$\sigma' = 0.22$	1	3.85
		g$^-$: −75			
			$\omega = 8.26$	1	−5.32
		For C-C bonds:			
		t : 174.3	$\omega' = 0.12$	1	5.28
		g$^+$: 67			
		g$^-$: −67	$\phi = 0.42$	1	2.18
			$\phi' = 0.50$	1	1.76

$$U_a = \begin{bmatrix} 1 & \sigma & \sigma \\ 1 & \sigma\phi & \sigma\omega \\ 1 & \sigma\omega & \sigma\phi \end{bmatrix} \quad U_b = \begin{bmatrix} 1 & \sigma & \sigma \\ 1 & \sigma\phi' & \sigma\omega' \\ 1 & \sigma\omega' & \sigma\phi' \end{bmatrix} \quad U_c = \begin{bmatrix} 1 & \sigma' & \sigma' \\ 1 & \sigma'\phi & \sigma'\omega \\ 1 & \sigma'\omega & \sigma'\phi \end{bmatrix}$$

rows and columns: t, g$^+$, g$^-$

a) Given values are those from the parameter *set I* of the original paper.

Comments:
Conformational energies of various oligooxyethylene isomers are calculated by the empirical force field method, and statistical mechanics calculations of the chain dimensions and the dipole moments are carried out.

Calcd. quantities:

$\langle r^2 \rangle_0 / nl^2$	= 3.4	(*Exptl.*: 4.1 to 4.8)
$d(\ln \langle r^2 \rangle_0)/dT$	= 0.09 × 10^{-3} K^{-1}	(*Exptl.*: 0.23 (± 0.02) × 10^{-3} K^{-1})
$\langle \mu^2 \rangle_0 / nm^2$	= 0.34	(*Exptl.*: 0.63)
$d(\ln \langle \mu^2 \rangle_0)/dT$	= 1.6 × 10^{-3} K^{-1}	(*Exptl.*: 2.6 × 10^{-3} K^{-1})

polyoxyethylene, poly(ethylene oxide), POE, PEO, PM$_2$O

Tasaki, K.; Abe, A. *Polymer J.* **1985**, *17*, 641; Abe, A.; Tasaki, K.; Mark, J. E. *Polymer J.* **1985**, *17*, 883.

Bond Length [pm]	Valence Angles [°]	Torsion Angles [°]	ξ (for 298 K)	ξ$_0$	E$_ξ$ [kJ·mol^{-1}]
C-C : 153	C-C-O : 111.5	t : 180	σ = 2.31	0.99	−2.1
C-O : 143	C-O-C : 111.5	For C-C bonds: g$^+$: 68 g$^-$: −68	ρ = 0.16 ω = 0.36	0.61 0.88	3.3 2.2
		For C-O bonds: g$^+$: 80 g$^-$: −80			

$$U_a = \begin{bmatrix} 1 & \rho & \rho \\ 1 & \rho & \rho\omega \\ 1 & \rho\omega & \rho \end{bmatrix} \quad U_b = \begin{bmatrix} 1 & \rho & \rho \\ 1 & \rho & 0 \\ 1 & 0 & \rho \end{bmatrix} \quad U_c = \begin{bmatrix} 1 & \sigma & \sigma \\ 1 & \sigma & \sigma\omega \\ 1 & \sigma\omega & \sigma \end{bmatrix}$$

rows and columns: t, g$^+$, g$^-$

Comments:
Conformational energies of the POE chain are determined by the ^{1}H and ^{13}C NMR studies on 1,2-dimethoxyethane and POE. The unperturbed chain dimensions $C_n = \langle r^2 \rangle_0/nl^2$, the dipole moment $\langle \mu^2 \rangle/nm^2$, and the molar Kerr constant $\langle _mK \rangle/x$ are calculated as a function of the RIS parameters such as the bond angles, rotational angles, and conformational energies. The value of $C_n = 5.2$ determined in organic solvents is found to be favorably reproduced by these calculations. For the C—C bond, conformational energy E(σ) is found to vary over a range from −2.1 to −5.0 kJ·mol^{-1}, depending on the solvent system.

Calcd. quantities:

$\langle r^2 \rangle_0 / nl^2$	= 5.2	(Exptl.: 5.2)
$d(\ln \langle r^2 \rangle_0)/dT$	= 0.74 × 10^{-3} K^{-1}	(Exptl.: 0.2 (± 0.2) × 10^{-3} K^{-1})
$\langle \mu^2 \rangle_0 / nm^2$	= 0.35	(Exptl.: 0.48 to 0.53)
$d(\ln \langle \mu^2 \rangle_0)/dT$	= 3.0 × 10^{-3} K^{-1}	(Exptl.: 2.6 × 10^{-3} K^{-1})
$\langle _mK \rangle / x$	= −2.9 × 10^{-12} cm^5 s.c.$^{-2}$ mol^{-1}	(Exptl.: −2.0 × 10^{-12} cm^5 s.c.$^{-2}$ mol^{-1})

Temperature dependence of the vicinal NMR coupling constants is also analyzed on the basis of the RIS scheme.

polyoxyethylene, poly(ethylene oxide), POE, PEO, PM$_2$O; α,ω-diiodo oligomers, $x = 1 - 4$

Abe, A.; Tasaki, K. *Macromolecules* **1986**, *19*, 2647.

Bond Length [pm]	Valence Angles [°]	Torsion Angles [°]	ξ (for 293 K)	ξ_0	E_ξ [kJ·mol^{-1}]
C-I : 218	C-C-I : 109.0	t : 180	$\eta = 1.48$	1.00	−0.96
C-C : 154	C-C-O : 111.5	For C-C bonds: g$^+$: 60	$\rho = 0.15$	0.61	3.35
C-O : 143	C-O-C : 111.5	g$^-$: −60	$\sigma = 2.33$	0.99	−2.09
		For C-O bonds: g$^+$: 80 g$^-$: −80	$\omega = 0.35$	0.88	2.22

$$U_a = \begin{bmatrix} 1 & \rho & \rho \\ 1 & \rho & \rho\omega \\ 1 & \rho\omega & \rho \end{bmatrix} \quad U_b = \begin{bmatrix} 1 & \rho & \rho \\ 1 & \rho & 0 \\ 1 & 0 & \rho \end{bmatrix} \quad U_c = \begin{bmatrix} 1 & \sigma & \sigma \\ 1 & \sigma & \sigma\omega \\ 1 & \sigma\omega & \sigma \end{bmatrix} \quad U_4 = \begin{bmatrix} 1 & \eta & \eta \\ 1 & \eta & 0 \\ 1 & 0 & \eta \end{bmatrix}$$

$U_2 = \text{diag}[1, \eta, \eta] \qquad U_3 = U_b \qquad$ rows and columns: t, g$^+$, g$^-$

Calcd. quantities:	End-to-end distances $\langle r \rangle$, X-ray scattering functions.

1,2-dimethoxyethane, DME

Inomata, K.; Abe, A. *J. Phys. Chem.* **1992**, *96*, 7934.

Bond Length [pm]	Valence Angles [°]	Torsion Angles [°]		ξ (for 293 K)	ξ_0	E_ξ [kJ·mol^{-1}]
C-C : 154	C-C-O : 111.5	*For bond a:*		ρ = 0.21	1	3.77
		t (in tt) :	180			
C-O : 143	C-O-C : 111.5	t (in tg$^+$):	-179.5	σ = 2.36	1	-2.09
		g$^+$ (in g$^+$t):	80			
		g$^+$ (in g^{++}g):	78	ω = 0.58	1	1.34
		g$^-$ (in g$^-$g$^+$):	-89			
		For bond b:				
		t (in tt):	180			
		g (in tg$^+$):	62.5			
		t (in g$^+$t):	177			
		g (in g$^+$g$^+$):	57.5			
		g (in g$^-$g$^+$):	71			

$$U_a = \begin{bmatrix} 1 & \rho & \rho \\ 1 & \rho & \rho \\ 1 & \rho & \rho \end{bmatrix} \quad U_b = \begin{bmatrix} 1 & \sigma & \sigma \\ 1 & \sigma & \sigma\omega \\ 1 & \sigma\omega & \sigma \end{bmatrix} \quad U_c = \begin{bmatrix} 1 & \rho & \rho \\ 1 & \rho & \rho\omega \\ 1 & \rho\omega & \rho \end{bmatrix}$$

rows and columns: t, g$^+$, g$^-$

Comments:
Conformational characteristics of DME in the gas phase are studied by the NMR method. Observed ^{1}H-^{1}H and ^{13}C-^{1}H NMR vicinal coupling constants are compared with those determined in nonpolar solvents. The values observed in the gas phase and in solution do not exhibit any appreciable discontinuity at the transition. RIS simulations of these vicinal coupling constants yield conformational energies given above. In these treatments, the neighbor-dependent character of the bond rotation is rigorously taken into account.

Calcd. quantities: Conformer populations, $^3J_{HH}$ and $^3J_{CH}$ vicinal NMR coupling constants

S 147

polyoxyethylene, poly(ethylene oxide), POE, PEO, PM$_2$O

Smith, G. D.; Yoon, D. Y.; Jaffe, R. L. *Macromolecules* **1993**, *26*, 5213.

Bond Length [pm]	Valence Angles [°]	Torsion Angles [°]	ξ (for 298 K)	ξ$_0$	E$_\xi$ [kJ·mol^{-1}]
C-C : 152	C-C-O : 111	t : 180	σ = 0.85	1	0.4
C-O : 140	C-O-C : 115	For C-C bonds:	ρ = 0.09	1	5.9
		g$^+$: 75			
		g$^-$: -75	ω = 8.84	1	-5.4
		For C-O bonds:	ω' = 0.11	1	5.4
		g$^+$: 80			
		g$^-$: -80	τ = 8.84	1	-5.4

Comments:
A three-state RIS model is derived for POE, based upon ab initio electronic structure analyses of model molecules DME and DEE. It is demonstrated that the low energy of the tg$^\pm$g$^\mp$ conformation of DME, resulting from strong O···H attractions, as indicated by the ab initio studies, necessitates the inclusion of third-order interactions in the RIS model. This is realized by adopting 9 × 9 statistical weight matrices.

DME : 1,2-dimethoxy ethane;   *DEE* : 1,2-diethoxy ethane

$$U_b = \begin{bmatrix} 1 & \rho & \rho & 0 & 0 & 0 & 0 & 0 & 0 \\ 0 & 0 & 0 & 1 & \rho & \rho\omega' & 0 & 0 & 0 \\ 0 & 0 & 0 & 0 & 0 & 1 & \rho\omega' & \rho \\ 1 & \rho & \rho & 0 & 0 & 0 & 0 & 0 & 0 \\ 0 & 0 & 0 & 1 & \rho & \rho\omega' & 0 & 0 & 0 \\ 0 & 0 & 0 & 0 & 0 & 1 & 0 & \rho \\ 1 & \rho & \rho & 0 & 0 & 0 & 0 & 0 & 0 \\ 0 & 0 & 0 & 1 & \rho & 0 & 0 & 0 & 0 \\ 0 & 0 & 0 & 0 & 0 & 1 & \rho\omega' & \rho \end{bmatrix}$$

$$U_c = \begin{bmatrix} 1 & \sigma & \sigma & 0 & 0 & 0 & 0 & 0 & 0 \\ 0 & 0 & 0 & 1 & \sigma & \sigma\omega & 0 & 0 & 0 \\ 0 & 0 & 0 & 0 & 0 & 1 & \sigma\omega & \sigma \\ 1 & \sigma & \sigma & 0 & 0 & 0 & 0 & 0 & 0 \\ 0 & 0 & 0 & 1 & \sigma & \sigma\omega & 0 & 0 & 0 \\ 0 & 0 & 0 & 0 & 0 & 1 & 0 & \sigma \\ 1 & \sigma & \sigma & 0 & 0 & 0 & 0 & 0 & 0 \\ 0 & 0 & 0 & 1 & \sigma & 0 & 0 & 0 & 0 \\ 0 & 0 & 0 & 0 & 0 & 1 & \sigma\omega & \sigma \end{bmatrix}$$

rows *(i-1 / i)* and columns *(i-2 / i-1)*:
tt, tg$^+$, tg$^-$, g$^+$t, g$^+$g$^+$, g$^+$g$^-$, g$^-$t, g$^-$g$^+$, g$^-$g$^-$

$$U_a = \begin{bmatrix} 1 & \rho & \rho & 0 & 0 & 0 & 0 & 0 & 0 \\ 0 & 0 & 0 & 1 & \rho & \rho\omega & 0 & 0 & 0 \\ 0 & 0 & 0 & 0 & 0 & 1 & \rho\omega & \rho \\ 1 & \rho & \rho & 0 & 0 & 0 & 0 & 0 & 0 \\ 0 & 0 & 0 & 1 & \rho\tau & \rho\omega & 0 & 0 & 0 \\ 0 & 0 & 0 & 0 & 0 & 1 & 0 & \rho \\ 1 & \rho & \rho & 0 & 0 & 0 & 0 & 0 & 0 \\ 0 & 0 & 0 & 1 & \rho & 0 & 0 & 0 & 0 \\ 0 & 0 & 0 & 0 & 0 & 1 & \rho\omega & \rho\tau \end{bmatrix}$$

Calcd. quantities: [a]

$<r^2>_0 / nl^2$	= 4.5 (/ 5.0)	(Exptl.: 4.0 - 5.6)	
$d (\ln <r^2>_0) / dT$	= 0.08 (/ 0.11) × 10^{-3} K^{-1}	(Exptl.: 0.2 × 10^{-3} K^{-1})	
$<\mu^2>_0 / n\mu_0^2$	= 0.29 (/0.27)	(Exptl.: 0.27, 0.30)	
$d (\ln <\mu^2>_0 / dT$	= 2.9 (/2.4) × 10^{-3} K^{-1}	(Exptl.: 2.6 × 10^{-3} K^{-1})	

[a]   Given values correspond to the third order model without/with prefactor weighting: ξ$_0$ = 1 (/ 0.7) for g$^\pm$g$^\mp$ of C-O-C-C-O conformation.

*Further calculations on **polyoxyethylene** chains:*

S 148   Mark, J. E. *J. Chem. Phys.* **1977**, *67*, 3300.

RIS theory is used to calculate values of the configurational partition function z and entropy S for polyoxide chains having the repeat unit $-(CH_2)_y-O-$. The values of z are at least qualitatively useful in interpreting the effect of the number y of the methylene groups in the repeat unit on the characteristic ratio of the unperturbed dimensions relative to the number of skeletal bonds and the average square of their length. The configurational entropy S, however, shows a very poor correlation with either z or $<r^2>_0/nl^2$.

S 149   Kelly, K. M.; Patterson, G. D.; Tonelli, A. E. *Macromolecules* **1977**, *10*, 859.

Molar Kerr constants, $_mK$, are obtained for the first four oligomers of POE, $CH_3-(O-CH_2-CH_2)_x-O-CH_3$, and for a higher molecular weight POE ($x = 91$) terminated with hydroxyl groups, from electric birefringence measurements. In addition, the dipole moments of each of the PEO's are measured. The results are compared with average $<_mK>$ and mean square dipole moments, $<\mu^2>$, calculated for POE from its RIS model through adoption of the valence optical scheme, which assumes bond polarizability tensor additivity. Agreement between measured and calculated $_mK$ is poor except for the sign of $_mK$ and its dependence upon chain length. The calculated and measured $<\mu^2>$ are in good agreement including the dependence of $<\mu^2>$ upon chain length.

S 150   Mattice, W. L. *Macromolecules* **1979**, *12*, 944.

Cyclization of polyoxyethylene to form the macrocycle 3*x*-crown-*x* is studied for even *x* from 4 to 20 units using Monte-Carlo methods. POE is assumed to behave in accord with the model developed by *Mark* and *Flory* (**S 137, S 138**). The fraction of acyclic chains satisfying criteria for cyclozation reaches a maximum of 1.2×10^{-3} for $x = 6$ and falls of to 0.1×10^{-3} when x is 20. The implication that 18-crown-6 is the most easily formed macrocycle is in harmony with experiment. Conditional probabilities and *a priori* probabilities are evaluated at 298 K; conditional probabilities, in their matrix representations, are:

$$Q_{C-C} = \begin{bmatrix} 0.2050 & 0.3975 & 0.3975 \\ 0.2491 & 0.4832 & 0.2676 \\ 0.2491 & 0.2676 & 0.4832 \end{bmatrix} \quad Q_{C-O} = \begin{bmatrix} 0.7247 & 0.1376 & 0.1376 \\ 0.7721 & 0.1466 & 0.0812 \\ 0.7721 & 0.0812 & 0.1466 \end{bmatrix} \quad Q_{O-C} = \begin{bmatrix} 0.7352 & 0.1324 & 0.1324 \\ 0.8474 & 0.1526 & 0.0000 \\ 0.8474 & 0.0000 & 0.1526 \end{bmatrix}$$

S 151   Mattice, W. L.; Newkome, G. R. *J. Am. Chem. Soc.* **1982**, *104*, 5942.

A Monte-Carlo study is conducted of the formation of benzo-crown ethers from an acyclic precursor. The major features of the computed cyclization constants are in harmony with trends observed in kinetic constants reported.

S 152 Abe, A.; Tasaki, K. *J. Mol. Struc.* **1986**, *145*, 309.

MM2 calculations are carried out for a series of glycol ethers. For 1,2-dimethoxyethane, the observed energy of the *gauche* state (E(σ) = -2.1 kJ·mol^{-1}) for the O-C—C-O bond is found to be approximately reproduced by the calculation when the dipole-dipole interaction term is suppressed by adopting a relatively large dielectric constant. The *gauche* energy of 1,2-dimethoxy-2-methylpropane is calculated to be E(γ) = -3.77 kJ·mol^{-1}, which is at variance with the observed value of 2.1 kJ·mol^{-1}. In 1,2-dimethoxypropane, the two *gauche* forms around the O-C—C-O bond are non-equivalent, and the stability of the conformers varies in the order $g_\alpha > t > g_\beta$. These experimental observations are not reproduced by the calculation. The MM2 program failed to provide an adequate explanation for the *gauche* oxygen effect.

S 153 Bahar, I.; Erman, B.; Monnerie, L. *Polymer* **1988**, *29*, 349.

The dynamic RIS model of polymer chains is applied to the interpretation of nuclear magnetic relaxation measurements of local chain dynamics. According to the proposed model, the relaxation times T_{1C} and T_{1H} may be related to the chemical structure of a specific polymer.

S 154 Bahar, I.; Erman, B.; Monnerie, L. *Macromolecules* **1989**, *22*, 2396.

The dynamic RIS model developed for investigating local chain dynamics is further improved and applied to POE. A set of eigenvalues characterizes the dynamic behaviour of a given segment of *N* motional bonds, with ν isomeric states available to each bond. The rates of transitions between isomeric states are assumed to be inversely proportional to solvent viscosity. Predictions are in satisfactory agreement with the isotropic correlation times and spin-lattice relaxation times from ^{13}C and ^{1}H NMR experiments for POE.

S 155 Abe, A.; Inomata, K. *J. Mol. Struc.* **1991**, *245*, 399.

The conformational energies derived from gas phase NMR of 1,2-dimethoxyethane can be most reliably compared with the results of theoretical calculations where a single molecule is treated. Adoption of the J_t and J_g values obtained leads to the energy difference E(σ) = -1.26 kJ·mol^{-1} between the *gauche* and the *trans* state. This result is consistent with E(σ) = 2.1 kJ·mol^{-1} obtained from measurements in cyclohexane.

S 156 Müller-Plathe, F.; van Gunsteren, W. F. *Macromolecules* **1994**, *27*, 6040.

Ab initio quantum-chemical calculations are reported at the level of second-order many-body perturbation theory aimed at the equilibrium between the all-*trans* (ttt) and the *trans-gauche-trans* (tgt) conformations of dimethoxyethane. It is concluded that the *gauche* effect in dimethoxyethane and by analogy POE is mainly due to the presence of a polarizable environment and not to some intrinsic conformational preference.

poly(trimethylene oxide), POM$_3$

Abe, A.; Mark, J. E. *J. Am. Chem. Soc.* **1976**, *98*, 6468.

Bond Length [pm]	Valence Angles [°]	Torsion Angles [°]	ξ (for 293 K)	ξ_0	E_ξ [kJ·mol^{-1}] [a]
C-H : 110	C-O-C : 111.5	t : 180	σ = 2.00	1	-1.7
C-O : 143	C-C-O : 111.5	For C-O bonds: g$^+$: 70	σ' = 0.21	1	3.8
C-C : 153	C-C-C : 111.5	g$^-$: -70	ω = 0.36	1	2.5
		For C-C bonds: g$^+$: 60 g$^-$: -60			

$$U_a = \begin{bmatrix} 1 & \sigma' & \sigma' \\ 1 & \sigma' & 0 \\ 1 & 0 & \sigma' \end{bmatrix} \quad U_b = \begin{bmatrix} 1 & \sigma & \sigma \\ 1 & \sigma & 0 \\ 1 & 0 & \sigma \end{bmatrix} \quad U_c = \begin{bmatrix} 1 & \sigma & \sigma \\ 1 & \sigma & \sigma\omega \\ 1 & \sigma\omega & \sigma \end{bmatrix} \quad U_d = \begin{bmatrix} 1 & \sigma' & \sigma' \\ 1 & \sigma' & 0 \\ 1 & 0 & \sigma' \end{bmatrix}$$

rows and columns: t, g$^+$, g$^-$

[a] Given values correspond to the parameter *set II* of the original paper.

Comments:
The semiempirical methods do not in general successfully predict conformational energies in this class of chain molecules, as was pointed out in several earlier but more limited comparisons of polymers of this type. The present study implicates the intramolecular interactions involving oxygen atoms as the origin of this disagreement between theory and experiment and gives a quantitative estimate of the magnitude of the discrepancy for each of the interactions thus involved.

Calcd. quantities: (for x = 129)

$<r^2>_0 / nl^2$	= 4.3	(for 303 K)	(Exptl.: 3.7 ± 0.1)
$d (\ln <r^2>_0) / dT$	= 0.32 × 10^{-3} K^{-1}	(for 333 K)	(Exptl.: 0.08 (± 0.08) × 10^{-3} K^{-1})
$<\mu^2> / nm^2$	= 0.38	(for 293 K)	(Exptl.: 0.41 ± 0.01)
$d (\ln <\mu^2>) / dT$	= 2.1 × 10^{-3} K^{-1}	(for 293 K)	(Exptl.: 1.5 × 10^{-3} K^{-1})

1,3-dimethoxypropane, 1,3-DMP

Inomata, K.; Phataralaoha, N.; Abe, A. *Comput. Polym. Sci.* **1991**, *1*, 126.

Bond Length [pm]	Valence Angles [°]	Torsion Angles [°]	ξ (for 298 K)	ξ_0	E_ξ [kJ·mol^{-1}]
C-H : 110	C-O-C : 111.5	t : 180	ρ = 0.18	1	4.2
C-O : 143	O-C-C : 111.5	For C-O bonds:	σ = 2.33	1	-2.1
		g$^+$: 80			
C-C : 153	C-C-C : 111.5	g$^-$: -80	ω = 0.25	1	3.4
		For C-C bonds:			
		g$^+$: 62			
		g$^-$: -62			

$$U_2 = \begin{bmatrix} 1 & \rho & \rho \\ 1 & \rho & \rho \\ 1 & \rho & \rho \end{bmatrix} \quad U_3 = \begin{bmatrix} 1 & \sigma & \sigma \\ 1 & \sigma & 0 \\ 1 & 0 & \sigma \end{bmatrix} \quad U_4 = \begin{bmatrix} 1 & \sigma & \sigma \\ 1 & \sigma & \sigma\omega \\ 1 & \sigma\omega & \sigma \end{bmatrix} \quad U_5 = \begin{bmatrix} 1 & \rho & \rho \\ 1 & \rho & 0 \\ 1 & 0 & \rho \end{bmatrix}$$

rows and columns: t, g$^+$, g$^-$

Comments:
The conformation of α,ω-dimethoxyalkanes such as CH$_3$O(CH$_2$)$_y$OCH$_3$ (y = 3,4) and CH$_3$OCH$_2$C(CH$_3$)$_2$CH$_2$OCH$_3$ are studied by the NMR method. Conformational energies of the internal C—O and C—C bonds are estimated from the observed vicinal coupling constants. Molecular mechanics calculations are used as a supplemental tool to elucidate the characteristic feature of the potential energy surface. Values of the conformational energy for the rotation around O-C—C-C are found to be slightly more negative than those calculated. The differences (ca. 1.3 kJ·mol^{-1}) between the calculated and the observed energies are small as compared with those (4 to 8 kJ·mol^{-1}) encountered in lower homologs such as CH$_3$O(CH$_2$)$_y$OCH$_3$ (y = 1,2). The conformational energy of the central C-C—C-C bond of 1,4-dimethoxybutane is found to be positive (1.3 kJ·mol^{-1}). The results of the present analysis are reasonably consistent with those previously derived from the statistical analysis of conformation-dependent properties of polymers. Following the previous treatment, the characteristic ratios of the dimension and the dipole moment of polymers are calculated, and the results are compared with relevant experimental data.

Calcd. quantities: Vicinal $^3J_{HH}$ and $^3J_{CH}$ NMR coupling constants

S 159

poly(trimethylene oxide), POM$_3$

Inomata, K.; Phataralaoha, N.; Abe, A. *Comput. Polym. Sci.* **1991**, *1*, 126.

Bond Length [pm]	Valence Angles [°]	Torsion Angles [°]	ξ (for 298 K)	ξ_0	E_ξ [kJ·mol^{-1}]
C-H : 110	C-O-C : 111.5	t : 180	σ = 2.33	1	-2.1
C-O : 143	O-C-C : 111.5	For C-O bonds: g$^+$: 80 g$^-$: -80	σ' = 0.18	1	4.2
C-C : 153	C-C-C : 111.5		ω = 0.50	1	1.7
		For C-C bonds: g$^+$: 62 g$^-$: -62			

$$U_a = U_d = \begin{bmatrix} 1 & \sigma' & \sigma' \\ 1 & \sigma' & 0 \\ 1 & 0 & \sigma' \end{bmatrix} \quad U_b = \begin{bmatrix} 1 & \sigma & \sigma \\ 1 & \sigma & 0 \\ 1 & 0 & \sigma \end{bmatrix} \quad U_c = \begin{bmatrix} 1 & \sigma & \sigma \\ 1 & \sigma & \sigma\omega \\ 1 & \sigma\omega & \sigma \end{bmatrix}$$

rows and columns: t, g$^+$, g$^-$

Comments:
The conformation of α,ω-dimethoxyalkanes such as CH$_3$O(CH$_2$)$_y$OCH$_3$ (y = 3,4) and CH$_3$OCH$_2$C(CH$_3$)$_2$CH$_2$OCH$_3$ are studied by the NMR method. Conformational energies of the internal C—O and C—C bonds are estimated from the observed vicinal coupling constants. Molecular mechanics calculations are used as a supplemental tool to eludicate the characteristic feature of the potential energy surface. Values of the conformational energy for the rotation around O-C—C-C are found to be slightly more negative than those calculated. The differences (ca. 1.3 kJ·mol^{-1}) between the calculated and the observed energies are small as compared with those (4 to 8 kJ·mol^{-1}) encountered in lower homologs such as CH$_3$O(CH$_2$)$_y$OCH$_3$ (y = 1,2). The conformational energy of the central C-C—C-C bond of 1,4-dimethoxybutane is found to be positive (1.3 kJ·mol^{-1}). The results of the present analysis are reasonably consistent with those previously derived from the statistical analysis of conformation-dependent properties of polymers. Following the previous treatment, the characteristic ratios of the dimension and the dipole moment of polymers are calculated, and the results are compared with relevant experimental data.

Calcd. quantities:

$<r^2>_0 / nl^2$	= 4.6	(*Exptl.*: 3.9)
$d(\ln <r^2>_0)/dT$	= 0.38 × 10^{-3} K^{-1}	(*Exptl.*: 0.08 × 10^{-3} K^{-1})
$<\mu^2> / nm^2$	= 0.33	(*Exptl.*: 0.42)
$d(\ln <\mu^2>)/dT$	= 2.3 × 10^{-3} K^{-1}	(*Exptl.*: 1.8 × 10^{-3} K^{-1})

poly(tetramethylene oxide), POM$_4$

Mark, J. E. *J. Am. Chem. Soc.* **1966**, *88*, 3708.

Bond Length [pm]	Valence Angles [°]	Torsion Angles [°]	ξ (for 293 K)	ξ_0	E_ξ [kJ·mol^{-1}]
C-H : 110	C-O-C : 110	t : 180	σ = 0.215	1	3.74
C-O : 143	O-C-C : 110	g$^+$: 60	σ' = 1 to 2	1	0 to −1.69
C-C : 153	C-C-C : 110	g$^-$: −60	σ'' = 0.426	1	2.08
			ω = 0.563	1	1.40

$$U_a = U_b = \begin{bmatrix} 1 & \sigma & \sigma \\ 1 & \sigma & 0 \\ 1 & 0 & \sigma \end{bmatrix} \quad U_c = \begin{bmatrix} 1 & \sigma' & \sigma' \\ 1 & \sigma' & 0 \\ 1 & 0 & \sigma' \end{bmatrix} \quad U_d = \begin{bmatrix} 1 & \sigma'' & \sigma'' \\ 1 & \sigma'' & \sigma''\omega \\ 1 & \sigma''\omega & \sigma'' \end{bmatrix} \quad U_e = \begin{bmatrix} 1 & \sigma' & \sigma' \\ 1 & \sigma' & \sigma'\omega \\ 1 & \sigma'\omega & \sigma' \end{bmatrix}$$

rows and columns: t, g$^+$, g$^-$

Comments:
Interactions between nonbonded atoms and groups in this chain are found to be either identical with or similar to interactions arising in PE, POM, and POE. Statistical weights obtained in the analysis of the dimensions and dipole moments of these chains are thus applicable to the present investigation of the poly(tetramethylene oxide) chain. The calculated dipole moments are in good agreement with preliminary published results obtained on the undiluted, amorphous polymer.

Calcd. quantities:

$\langle r^2 \rangle_0 / nl^2$ = 5.3 to 4.6

$d(\ln \langle r^2 \rangle_0)/dT$ = −1.3 (to −1.2) × 10^{-3} K^{-1}

$\langle \mu^2 \rangle / nm^2$ = 0.53 to 0.64

$d(\ln \langle \mu^2 \rangle)/dT$ = 1.0 (to 1.5) × 10^{-3} K^{-1}

poly(tetramethylene oxide), POM$_4$

Abe, A.; Mark, J. E. J. Am. Chem. Soc. **1976**, 98, 6468.

Bond Length [pm]	Valence Angles [°]	Torsion Angles [°]	ξ (for 293 K)	ξ_0	E_ξ [kJ·mol^{-1}] [a]
C-H : 110	C-O-C : 111.5	t : 180	$\sigma = 1.30$	1	$-0.63 (\pm 0.2)$
C-O : 143	C-C-O : 111.5	For CC-CC: g$^+$: 67.5	$\sigma' = 0.42$	1	2.1
C-C : 153	C-C-C : 111.5	g$^-$: -67.5	$\sigma'' = 0.13$	1	5.0
		For C-O bonds: g$^+$: 80 g$^-$: -80	$\omega = 0.36$	1	2.5
		For CC-CO: g$^+$: 60 g$^-$: -60			

$$U_a = U_e = \begin{bmatrix} 1 & \sigma'' & \sigma'' \\ 1 & \sigma'' & 0 \\ 1 & 0 & \sigma'' \end{bmatrix} \quad U_b = \begin{bmatrix} 1 & \sigma & \sigma \\ 1 & \sigma & 0 \\ 1 & 0 & \sigma \end{bmatrix} \quad U_c = \begin{bmatrix} 1 & \sigma' & \sigma' \\ 1 & \sigma' & \sigma'\omega \\ 1 & \sigma'\omega & \sigma' \end{bmatrix} \quad U_d = \begin{bmatrix} 1 & \sigma & \sigma \\ 1 & \sigma & \sigma\omega \\ 1 & \sigma\omega & \sigma \end{bmatrix}$$

rows and columns: t, g$^+$, g$^-$

[a] Given values correspond to the parameter *set I* of the original paper.

Comments:
The semiempirical methods do not in general successfully predict conformational energies in this class of chain molecules, as was pointed out in several earlier but more limited comparisons of polymers of this type. The present study implicates the intramolecular interactions involving oxygen atoms as the origin of this disagreement between theory and experiment and gives a quantitative estimate of the magnitude of the discrepancy for each of the interactions thus involved.

Calcd. quantities: (for x = 97)

$<r^2>_0 / nl^2$	$= 6.4 \pm 0.1$	(for 303 K)	(Exptl.: 6.1 ± 0.1)
$d (\ln <r^2>_0) / dT$	$= -1.0 \times 10^{-3}$ K^{-1}	(for 333 K; set I, but: $E(\sigma) = -0.8$ kJ·mol^{-1})	(Exptl.: -1.33×10^{-3} K^{-1})
$<\mu^2> / nm^2$	$= 0.48 \pm 0.01$	(for 293 K)	(Exptl.: 0.50)
$d (\ln <\mu^2>) / dT$	$= 1.6 \times 10^{-3}$ K^{-1}	(for 313 K; set I, but: $E(\sigma) = -0.8$ kJ·mol^{-1})	(Exptl.: 2.7 (or 1.9) $\times 10^{-3}$ K^{-1})

1,4-dimethoxybutane, 1,4-BBM

Inomata, K.; Phataralaoha, N.; Abe, A. *Comput. Polym. Sci.* **1991**, *1*, 126.

Bond Length [pm]	Valence Angles [°]	Torsion Angles [°]	ξ (for 298 K)	ξ_0	E_ξ [kJ·mol^{-1}]
C-H : 110	C-O-C : 111.5	t : 180	σ_1 = 1.38	1	−0.8
C-O : 143	O-C-C : 111.5	For C-O bonds: g^+ : 80	σ_2 = 0.59	1	1.3
C-C : 153	C-C-C : 111.5	g^- : −80	ρ = 0.18	1	4.2
		For C_α-C_β bonds: g^+ : 63 g^- : −63	ω = 0.45	1	2.0
		For C_β-C_γ bonds: g^+ : 72 g^- : −72			

$$U_2 = U_6 = \begin{bmatrix} 1 & \rho & \rho \\ 1 & \rho & 0 \\ 1 & 0 & \rho \end{bmatrix} \quad U_3 = \begin{bmatrix} 1 & \sigma_1 & \sigma_1 \\ 1 & \sigma_1 & 0 \\ 1 & 0 & \sigma_1 \end{bmatrix} \quad U_4 = \begin{bmatrix} 1 & \sigma_2 & \sigma_2 \\ 1 & \sigma_2 & \sigma_2\omega \\ 1 & \sigma_2\omega & \sigma_2 \end{bmatrix} \quad U_5 = \begin{bmatrix} 1 & \sigma_1 & \sigma_1 \\ 1 & \sigma_1 & \sigma_1\omega \\ 1 & \sigma_1\omega & \sigma_1 \end{bmatrix}$$

rows and columns: t, g^+, g^-

Comments:
The conformation of α,ω-dimethoxyalkanes such as $CH_3O(CH_2)_yOCH_3$ (y = 3,4) and $CH_3OCH_2C(CH_3)_2CH_2OCH_3$ are studied by the NMR method. Conformational energies of the internal C−O and C−C bonds are estimated from the observed vicinal coupling constants. Molecular mechanics calculations are used as a supplemental tool to elucidate the characteristic feature of the potential energy surface. Values of the conformational energy for the rotation around O-C−C-C are found to be slightly more negative than those calculated. The differences (ca. 1.3 kJ·mol^{-1}) between the calculated and the observed energies are small as compared with those (4 to 8 kJ·mol^{-1}) encountered in lower homologs such as $CH_3O(CH_2)_yOCH_3$ (y = 1,2). The conformational energy of the central C-C−C-C bond of 1,4-dimethoxybutane is found to be positive (1.3 kJ·mol^{-1}). The results of the present analysis are reasonably consistent with those previously derived from the statistical analysis of conformation-dependent properties of polymers. Following the previous treatment, the characteristic ratios of the dimension and the dipole moment of polymers are calculated, and the results are compared with relevant experimental data.

Calcd. quantities: Vicinal $^3J_{HH}$ and $^3J_{CH}$ NMR coupling constants

S 163

poly(tetramethylene oxide), POM$_4$

Inomata, K.; Phataralaoha, N.; Abe, A. *Comput. Polym. Sci.* **1991**, *1*, 126.

Bond Length [pm]	Valence Angles [°]	Torsion Angles [°]	ξ (for 298 K)	ξ_0	E_ξ [kJ·mol^{-1}]
C-H : 110	C-O-C : 111.5	t : 180	σ = 1.38	1	−0.8
C-O : 143	O-C-C : 111.5	*For C-O bonds:* g$^+$: 80 g$^-$: −80	σ' = 0.18	1	4.2
C-C : 153	C-C-C : 111.5		σ'' = 0.59	1	1.3
		For C$_\alpha$-C$_\beta$ bonds: g$^+$: 63 g$^-$: −63	ω = 0.45	1	1.97
		For C$_\beta$-C$_\gamma$ bonds: g$^+$: 72 g$^-$: −72			

$$U_a = U_e = \begin{bmatrix} 1 & \sigma' & \sigma' \\ 1 & \sigma' & 0 \\ 1 & 0 & \sigma' \end{bmatrix} \quad U_b = \begin{bmatrix} 1 & \sigma & \sigma \\ 1 & \sigma & 0 \\ 1 & 0 & \sigma \end{bmatrix} \quad U_c = \begin{bmatrix} 1 & \sigma'' & \sigma'' \\ 1 & \sigma'' & \sigma''\omega \\ 1 & \sigma''\omega & \sigma'' \end{bmatrix} \quad U_d = \begin{bmatrix} 1 & \sigma & \sigma \\ 1 & \sigma & \sigma\omega \\ 1 & \sigma\omega & \sigma \end{bmatrix}$$

rows and columns: t, g$^+$, g$^-$

Comments:

The conformation of α,ω-dimethoxyalkanes such as CH$_3$O(CH$_2$)$_y$OCH$_3$ (y = 3,4) and CH$_3$OCH$_2$C(CH$_3$)$_2$CH$_2$OCH$_3$ are studied by the NMR method. Conformational energies of the internal C—O and C—C bonds are estimated from the observed vicinal coupling constants. Molecular mechanics calculations are used as a supplemental tool to elucidate the characteristic feature of the potential energy surface. Values of the conformational energy for the rotation around O-C—C-C are found to be slightly more negative than those calculated. The differences (ca. 1.3 kJ·mol^{-1}) between the calculated and the observed energies are small as compared with those (4 to 8 kJ·mol^{-1}) encountered in lower homologs such as CH$_3$O(CH$_2$)$_y$OCH$_3$ (y = 1,2). The conformational energy of the central C-C—C-C bond of 1,4-dimethoxybutane is found to be positive (1.3 kJ·mol^{-1}). The results of the present analysis are reasonably consistent with those previously derived from the statistical analysis of conformation-dependent properties of polymers. Following the previous treatment, the characteristic ratios of the dimension and the dipole moment of polymers are calculated, and the results are compared with relevant experimental data.

Calcd. quantities:

$<r^2>_0 / nl^2$	= 5.9	(*Exptl.*: 5.4 to 6.2)
$d(\ln <r^2>_0) / dT$	= −0.71 × 10^{-3} K^{-1}	(*Exptl.*: −1.3 × 10^{-3} K^{-1})
$<\mu^2> / nm^2$	= 0.54	(*Exptl.*: 0.50 to 0.52)
$d(\ln <\mu^2>) / dT$	= 0.76 × 10^{-3} K^{-1}	(*Exptl.*: 1.8 (*to* 2.7) × 10^{-3} K^{-1})

poly(hexamethylene oxide), POM$_6$

Riande, E. *J. Polym. Sci., Polym. Phys. Ed.* **1976**, *14*, 2231.

Bond Length [pm]	Valence Angles [°]	Torsion Angles [°]	ξ (for 308 K)	ξ_0	E_ξ [kJ·mol^{-1}]
C-C : 153	C-O-C : *111.5*	t : 180	σ = 0.44	1	2.10
C-O : 143	C-C-O : *111.5*	g$^+$: 60	σ' = 1.39	1	−0.84
O-H : 96	C-C-C : *111.5*	g$^-$: −60	σ'' = 0.22	1	3.88
			ω = 0.56	1	1.48

$$U_a = U_b = \begin{bmatrix} 1 & \sigma'' & \sigma'' \\ 1 & \sigma'' & 0 \\ 1 & 0 & \sigma'' \end{bmatrix}$$

rows and columns: t, g$^+$, g$^-$

$$U_c = \begin{bmatrix} 1 & \sigma' & \sigma' \\ 1 & \sigma' & 0 \\ 1 & 0 & \sigma' \end{bmatrix} \quad U_d = \begin{bmatrix} 1 & \sigma & \sigma \\ 1 & \sigma & \sigma\omega \\ 1 & \sigma\omega & \sigma \end{bmatrix} \quad U_g = \begin{bmatrix} 1 & \sigma' & \sigma' \\ 1 & \sigma' & \sigma'\omega \\ 1 & \sigma'\omega & \sigma' \end{bmatrix} \quad U_e = U_f = \begin{bmatrix} 1 & \sigma & \sigma \\ 1 & \sigma & 0 \\ 1 & 0 & \sigma \end{bmatrix}$$

Comments:
The mean-square dipole moments of POE and POM$_6$ are determined from dielectric constant measurements on dilute solutions in benzene. The values obtained are in good agreement with those predicted using the RIS models for these chains. In addition, the unperturbed dimensions of POM$_6$ are calculated as a function of molecular weight using the RIS theory.

Calcd. quantities: (for x = 75)

$<r^2>_0 / nl^2$	= 5.09	(*Exptl.*: 5.27)
$d (\ln <r^2>_0) / dT$	= −0.88 × 10^{-3} K^{-1}	(*Exptl.*: ~ −1.0 × 10^{-3} K^{-1})
$<\mu^2> / nm^2$	= 0.594	(*Exptl.*: 0.636)
$d (\ln <\mu^2>) / dT$	= 0.6 × 10^{-3} K^{-1}	(*Exptl.*: 1.8 × 10^{-3} K^{-1})

poly(decamethylene oxide), POM$_{10}$

Riande, E. *Makromol. Chem.* **1977**, *178*, 2001.

Bond Length [pm]	Valence Angles [°]	Torsion Angles [°]	ξ (for 308 K)	ξ_0	E_ξ [kJ·mol^{-1}]
C-C : 153	C-O-C : 111.5	t : 180	σ = 0.44	1	2.10
C-O : 143	C-C-O : 111.5	g$^+$: 60	σ' = 1.39 (or 1)	1	−0.84 (or 0)
O-H : 96	C-C-C : 111.5	g$^-$: −60	σ'' = 0.22	1	3.88
			ω = 0.56	1	1.48

$$U_a = U_b = \begin{bmatrix} 1 & \sigma'' & \sigma'' \\ 1 & \sigma'' & 0 \\ 1 & 0 & \sigma'' \end{bmatrix}$$

rows and columns: t, g$^+$, g$^-$

$$U_e = \ldots = U_j = \begin{bmatrix} 1 & \sigma & \sigma \\ 1 & \sigma & 0 \\ 1 & 0 & \sigma \end{bmatrix} \quad U_c = \begin{bmatrix} 1 & \sigma' & \sigma' \\ 1 & \sigma' & 0 \\ 1 & 0 & \sigma' \end{bmatrix} \quad U_d = \begin{bmatrix} 1 & \sigma & \sigma \\ 1 & \sigma & \sigma\omega \\ 1 & \sigma\omega & \sigma \end{bmatrix} \quad U_k = \begin{bmatrix} 1 & \sigma' & \sigma' \\ 1 & \sigma' & \sigma'\omega \\ 1 & \sigma'\omega & \sigma' \end{bmatrix}$$

Comments:
Dielectric constants are determined for POM$_4$ and POM$_{10}$ chains in the thermodynamically good solvent benzene. The data for the former polymer indicate that the dipole moment ratio, $<\mu^2>/nm^2$, is independent of the chain length as has been predicted for chains of such structural features. The value of this ratio for α-hydro-ω-hydroxypoly(oxydecamethylene) chains is in the same range as that for POE, POM$_4$ and POM$_6$ and is in fair agreement with that predicted by the RIS theory.

Calcd. quantities: (for x = 75)

$<r^2>_0/nl^2$	= 5.72	(Exptl.: 5.88)
$d(\ln <r^2>_0)/dT$	= −0.9 × 10^{-3} K^{-1} (for σ' = 1)	(Exptl.: ~ −1.0 × 10^{-3} K^{-1})
$<\mu^2>/nm^2$	= 0.630	(Exptl.: 0.642)
$d(\ln <\mu^2>)/dT$	= 0.2 × 10^{-3} K^{-1} (for x = 13)	(Exptl.: 1.8 × 10^{-3} K^{-1})

poly(1,3-dioxolane)

Andrews, J. M.; Semlyen, J. A. *Polymer* **1972**, *13*, 142.

Bond Length [pm]	Valence Angles [°]	Torsion Angles [°]	ξ (for 333 K)	ξ_0	E_ξ [kJ·mol^{-1}]
C-C : 153	C-O-C : 111.5	t : 180	σ = 0.26	1	3.7
			σ' = 1.9	1	−1.8
C-O : 143	C-C-O : 111.5	g$^+$: 60	σ'' = 9.7	1	−6.3
	C-C-C : 111.5	g$^-$: −60	ψ = 1	1	0
			ψ' = 1	1	0
			ψ'' = 1	1	0
			ω = 0	1	∞
			ω' = 0.6	1	1.4
			ω'' = 0	1	∞

$$U_a = \begin{bmatrix} 1 & \sigma & \sigma \\ 1 & \sigma\psi & \sigma\omega \\ 1 & \sigma\omega & \sigma\psi \end{bmatrix} \quad U_b = \begin{bmatrix} 1 & \sigma' & \sigma' \\ 1 & \sigma'\psi' & \sigma'\omega' \\ 1 & \sigma'\omega' & \sigma'\psi' \end{bmatrix} \quad U_c = \begin{bmatrix} 1 & \sigma & \sigma \\ 1 & \sigma\psi' & \sigma\omega' \\ 1 & \sigma\omega' & \sigma\psi' \end{bmatrix}$$

$$U_d = \begin{bmatrix} 1 & \sigma'' & \sigma'' \\ 1 & \sigma''\psi & \sigma''\omega \\ 1 & \sigma''\omega & \sigma''\psi \end{bmatrix} \quad U_e = \begin{bmatrix} 1 & \sigma'' & \sigma'' \\ 1 & \sigma''\psi'' & \sigma''\omega'' \\ 1 & \sigma''\omega'' & \sigma''\psi'' \end{bmatrix}$$

rows and columns: t, g$^+$, g$^-$

Comments:
Cyclic oligomers with x = 2-9 are found to be present in poly(1,3-dioxolane) samples prepared by monomer-polymer-equilibrations using boron trifluoride diethyl etherate as catalyst. The molecular cyclization equilibrium constants K_x are measured and the values are in agreement with those calculated by the Jacobson-Stockmayer theory, using an RIS model to describe the statistical conformations of the corresponding chains and assuming that the chains obey Gaussian statistics.

Calcd. quantities: K_x

poly(1,3-dioxolane)

Riande, E.; Mark, J. E. *Macromolecules* **1978**, *11*, 956.

Bond Length [pm]	Valence Angles [°]	Torsion Angles [°]	ξ (for 298 K)	ξ_0	E_ξ [kJ·mol^{-1}]
C-C : 153	C-O-C : 111.5	t : 180	$\sigma = 7.52$	1	−5.0
C-O : 143	C-C-O : 111.5	g$^+$: 60	$\sigma' = 0.22$	1	3.8
C-C : 153	C-C-C : 111.5	g$^-$: −60	$\sigma'' = 1.99$	1	−1.7
			$\omega = 0.50$	1	1.7

$$U_a = \begin{bmatrix} 1 & \sigma & \sigma \\ 1 & \sigma & 0 \\ 1 & 0 & \sigma \end{bmatrix} \quad U_b = \begin{bmatrix} 1 & \sigma' & \sigma' \\ 1 & \sigma' & \sigma'\omega \\ 1 & \sigma'\omega & \sigma' \end{bmatrix} \quad U_c = \begin{bmatrix} 1 & \sigma'' & \sigma'' \\ 1 & \sigma'' & \sigma''\omega \\ 1 & \sigma''\omega & \sigma'' \end{bmatrix}$$

$$U_d = \begin{bmatrix} 1 & \sigma' & \sigma' \\ 1 & \sigma' & \sigma'\omega \\ 1 & \sigma'\omega & \sigma' \end{bmatrix} \quad U_e = \begin{bmatrix} 1 & \sigma & \sigma \\ 1 & \sigma & \sigma\omega \\ 1 & \sigma\omega & \sigma \end{bmatrix}$$

rows and columns: t, g$^+$, g$^-$

Comments:

Mean-square dipole moments of poly(1,3-dioxolane) chains are determined as a function of temperature by means of dielectric constant measurements in the thermodynamically good solvent benzene. The experimental results are found to be in good agreement with theoretical results based on an RIS model in which the required conformational energies of the chains are obtained from previous configurational analyses of POM and POE. The model assumes perfect alteration of oxymethylene units and oxyethylene units along the chain, rather than a more irregular distribution. The good agreement found between theory and experiment, therefore, is consistent with this regularly alternating structure. The present analysis also indicates that poly(1,3-dioxolane), and a number of other polyformals, should have a high degree of conformational randomness.

Calcd. quantities:

$\langle \mu^2 \rangle / nm^2 \approx 0.17$ (Exptl.: 0.17)

$d(\ln \langle \mu^2 \rangle)/dT = 6.1 \times 10^{-3}$ K^{-1} (Exptl.: 6.0×10^{-3} K^{-1})

*Further calculations on **poly(A_xB_y)** chains:*

S 168 Riande, E.; Mark, J. E. *Polymer* **1979**, *20*, 1188.

Samples of poly(1,3-dioxocane) [-$CH_2O(CH_2)_5O$-] and poly(1,3-dioxonane) [-$CH_2O(CH_2)_6O$-] are prepared, and fractions of both polymers are studied in solution by means of dielectric constant measurements from 20 to 60°C. Mean-square dipole moments thus obtained are compared with theoretical results based on the RIS models of the two chains. Good agreement is obtained.

polythioethylene, poly(ethylene sulfide), PTE, PES, PM$_2$S

Abe, A. *Macromolecules* **1980**, *13*, 546.

Bond Length [pm]	Valence Angles [°]	Torsion Angles [°]	ξ (for 298 K)	ξ$_0$	E$_\xi$ [kJ·mol^{-1}]
C-C : 153	C-S-C : 100	t : 180	σ = 0.43	1	2.1
C-S : 181.5	C-C-S : 114	g$^+$: 70	σ' = 1.18	1	-0.4
C-H : 110	C-C-H : 110	g$^-$: -70	ω' = 0.16	1	4.6
			ω'' = 0.50	1	1.7

$$U_a = \begin{bmatrix} 1 & \sigma' & \sigma' \\ 1 & \sigma' & \sigma'\omega' \\ 1 & \sigma'\omega' & \sigma' \end{bmatrix} \quad U_b = \begin{bmatrix} 1 & \sigma' & \sigma' \\ 1 & \sigma' & \sigma'\omega'' \\ 1 & \sigma'\omega'' & \sigma' \end{bmatrix} \quad U_c = \begin{bmatrix} 1 & \sigma & \sigma \\ 1 & \sigma & \sigma\omega' \\ 1 & \sigma\omega' & \sigma \end{bmatrix}$$

rows and columns: t, g$^+$, g$^-$

Comments:
The configurational characteristics of PTE are estimated by calculation based on the information acquired through the analysis of the polythiopropylene chain [Abe, A. *Macromolecules* **1980**, *13*, 541]. Results suggest that the polymer chain is quite flexible. The flexibility of the chain estimated in terms of the number of allowed conformations for a monomeric residue may be arranged in the order, PTE > PTP > POE > POP. The results of the present analysis are consistent with the view presented by *Takahashi, Tadokoro*, and *Chatani* [*J. Macromol. Sci.-Phys.* **1968**, *2*, 361] who demonstrated that the strong (intermolecular) dipole-dipole interactions play a key role in enhancing the enthalpy of fusion and thus the melting point of the polymer.

Calcd. quantities: (for x = 200)

$\langle r^2 \rangle_0 / nl^2$ = 4.2

$d(\ln \langle r^2 \rangle_0)/dT$ = -0.33 × 10^{-3} K^{-1}

$\langle \mu^2 \rangle_0 / nm^2$ = 0.42

$d(\ln \langle \mu^2 \rangle_0)/dT$ = 2.8 × 10^{-3} K^{-1}

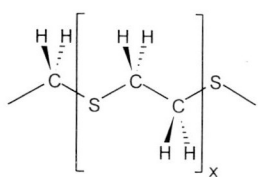

polythioethylene, poly(ethylene sulfide), PTE, PES, PM$_2$S

Mattice, W. L. *J. Am. Chem. Soc.* **1980**, *102*, 2242.

Bond Length [pm]	Valence Angles [°]	Torsion Angles [°]	(for 300 K)
C - C : 153	C-S-C : 102	t : 180	
C - S : 182	C-C-S : 114	g$^+$: 60	
C - H : 110		g$^-$: −60	

$$U_{C-C} = U_{C-S} = \begin{bmatrix} 1.00 & 1.00 & 1.00 \\ 1.00 & 0.71 & 0.17 \\ 1.00 & 0.17 & 0.71 \end{bmatrix} \quad U_{C-S} = \begin{bmatrix} 1.00 & 1.00 & 1.00 \\ 1.00 & 0.33 & 0.07 \\ 1.00 & 0.07 & 0.33 \end{bmatrix} \quad \text{rows and columns: t, g}^+\text{, g}^-$$

Conditional probabilities, given in matrix form, are:

$$Q_{C-C} = \begin{bmatrix} 0.414 & 0.293 & 0.293 \\ 0.616 & 0.310 & 0.074 \\ 0.616 & 0.074 & 0.310 \end{bmatrix} \quad Q_{C-S} = \begin{bmatrix} 0.480 & 0.260 & 0.260 \\ 0.676 & 0.260 & 0.062 \\ 0.676 & 0.072 & 0.260 \end{bmatrix} \quad Q_{S-C} = \begin{bmatrix} 0.427 & 0.287 & 0.287 \\ 0.788 & 0.175 & 0.037 \\ 0.788 & 0.037 & 0.175 \end{bmatrix}$$

Comments:
Cyclization of an acyclic polythioethylene chain to form $(CH_2CH_2S)_x$ is studied by Monte-Carlo methods for even x ranging from 4 to 24. Unperturbed acyclic chains are assumed to behave in accord with a RIS treatment which incorporates first- and second-order interactions. The most readily formed macrocycle is that with $x = 6$, in harmony with results obtained in the POE series. However, cyclization at all x considered is found to be more difficult for polythioethylene than for the equivalent polyoxyethylene.

*Further calculations on **polythioethylene** chains:*

S 171　　　　　Bhaumik, D.; Mark, J. E. *Macromolecules* **1981**, *14*, 162.

Semiempirical potential energy functions are used to characterize interchain interactions in PTE and in POE, the primary purpose being to eludicate the very high melting point of PES (216°C) relative to that of PEO (68°C). In the case of the PEO chain, the partial charges on the atoms could be calculated by the CNDO/2 method. The charges thus obtained show only a slight dependence on the conformation, thus supporting the assumption of conformation-independent charges usually made in conformation analyses. The total interchain interactions, significantly attractive in both polymers, are much larger in PTE than in POE, which is consistent with the interpretation of its unusually high melting point in terms of its enthalpy of fusion. The difference in intermolecular attractions is primarily due to van der Waals interactions, rather than to Coulombic (dipolar) effects.

poly(trimethylene sulfide), PM$_3$S

Guzmán, J.; Riande, E.; Welsh, W. J.; Mark, J. E. *Makromol. Chem.* **1982**, *183*, 2573.

Bond Length [pm]	Valence Angles [°]	Torsion Angles [°]	ξ (for 298 K)	ξ_0	E_ξ [kJ·mol^{-1}]
C-C : 153	C-S-C : 100	t : 180	σ = 0.83	1	0.46
C-S : 181.5	C-C-S : 114	For bonds a,d: g$^+$: 70 g$^-$: -70	σ' = 0.86	1	0.38
C-H : 110	C-C-H : 110	For bonds b,c: g$^+$: 67 g$^-$: -67			

$$U_a = U_d = \begin{bmatrix} 1 & \sigma & \sigma \\ 1 & \sigma & 0 \\ 1 & 0 & \sigma \end{bmatrix} \qquad U_b = U_c = \begin{bmatrix} 1 & \sigma' & \sigma' \\ 1 & \sigma' & 0 \\ 1 & 0 & \sigma' \end{bmatrix}$$

rows and columns: t, g$^+$, g$^-$

Comments:
Using conformational energies derived from semiempirical potential energy functions, a three-rotational-state model gives values of the dipole moment in good agreement with experiment. Theoretical and experimental values of the temperature coefficient of $<\mu^2>_0$ are in disagreement, however, and a five-rotational-state model failes to remove this discrepancy.

Calcd. quantities:

	Three State Scheme	Five State Scheme	
$<\mu^2>$ / nm^2	= 0.691	= 0.63	(*Exptl.*: 0.61)
$d (\ln <\mu^2>) / dT$	= -0.08 × 10^{-3} K^{-1}	= -0.2 × 10^{-3} K^{-1}	(*Exptl.*: 1.3 × 10^{-3} K^{-1})

S 173 (see also S 172)

poly(trimethylene sulfide), PM$_3$S

Guzmán, J.; Riande, E.; Welsh, W. J.; Mark, J. E. *Makromol. Chem.* **1982**, *183*, 2573.

Bond Length [pm]	Valence Angles [°]	Torsion Angles [°]		ξ (for 298 K)	ξ_0	E_ξ [kJ·mol^{-1}]	
C-C : 153	C-S-C : 100	t : 180		σ_1^* = 0.034	1	8.4	(for bonds a or d)
				σ_2^* = 0.014	1	10.5	(for bonds b or c)
C-S : 181.5	C-C-S : 114	*For bond pairs da:*					
		g$^{*\pm}$ g$^{\mp}$: ± 100 , ∓ 70	→	ω^* = 1.44	1	− 0.9	
C-H : 110	C-C-H : 110	g$^{\pm}$ g$^{*\mp}$: ± 70 , ∓ 100	→	ω^* = 1.44	1	− 0.9	
		For bond pairs ab:					
		g$^{*\pm}$ g$^{\mp}$: ± 100 , ∓ 65	→	ω^* = 1.22	1	− 0.5	
		g$^{\pm}$ g$^{*\mp}$: ± 65 , ∓ 100	→	ω^* = 0	1	> 25	
		For bond pairs bc:					
		g$^{*\pm}$ g$^{\mp}$: ± 100 , ∓ 65	→	ω^* = 0.021	1	9.6	
		g$^{\pm}$ g$^{*\mp}$: ± 65 , ∓ 100	→	ω^* = 0.021	1	9.6	
		For bond pairs cd:					
		g$^{*\pm}$ g$^{\mp}$: ± 100 , ∓ 70	→	ω^* = 1.69	1	− 1.3	
		g$^{\pm}$ g$^{*\mp}$: ± 70 , ∓ 100	→	ω^* = 0	1	> 25	
				σ = 0.82	1	0.5	
				σ' = 0.85	1	0.4	

$$U = \begin{bmatrix} 1 & \sigma^* & \sigma & \sigma & \sigma^* \\ 1 & \sigma^* & \sigma & \sigma\omega^* & 0 \\ 1 & \sigma^* & \sigma & 0 & \sigma^*\omega^* \\ 1 & \sigma^*\omega^* & 0 & \sigma & \sigma^* \\ 1 & 0 & \sigma\omega^* & \sigma & \sigma^* \end{bmatrix}$$

rows and columns: t, g^{*+}, g^{+}, g^{-}, g^{*-}

Comments:
Using conformational energies derived from semiempirical potential energy functions, a three-rotational-state model gives values of the dipole moment in good agreement with experiment. Theoretical and experimental values of the temperature coefficient of $<\mu^2>_0$ are in disagreement, however, and a five-rotational-state model failes to remove this discrepancy.

Calcd. quantities:

		Three State Scheme	*Five State Scheme*	
	$<\mu^2> / nm^2$	= 0.691	= 0.63	(*Exptl.*: 0.61)
	$d (\ln <\mu^2>) / dT$	= − 0.08 × 10^{-3} K^{-1}	= − 0.2 × 10^{-3} K^{-1}	(*Exptl.*: 1.3 × 10^{-3} K^{-1})

poly(pentamethylene sulfide), PM$_5$S

Riande, E.; Guzmán, J.; Welsh, W. J.; Mark, J. E. *Makromol. Chem.* **1982**, *183*, 2555.

Bond Length [pm]	Valence Angles [°]	Torsion Angles [°]	ξ (for 298 K)	ξ_0	E_ξ [kJ·mol^{-1}]
C-C : 153	C-S-C : 100	t : 180	σ = 0.43	1	2.1
C-S : 181.5	C-C-S : 114	For bonds a,f: g^+ : 70 g^- : -70	σ' = 0.55	1	1.5
C-H : 110	C-C-C : 111.5		σ'' = 0.82	1	0.5
	C-C-H : 110	For bonds b,e: g^+ : 65 g^- : -65			
		For bonds c,d: g^+ : 68 g^- : -68			

$$U_a = U_f = \begin{bmatrix} 1 & \sigma'' & \sigma'' \\ 1 & \sigma'' & 0 \\ 1 & 0 & \sigma'' \end{bmatrix} \quad U_b = U_e = \begin{bmatrix} 1 & \sigma' & \sigma' \\ 1 & \sigma' & 0 \\ 1 & 0 & \sigma' \end{bmatrix} \quad U_c = U_d = \begin{bmatrix} 1 & \sigma & \sigma \\ 1 & \sigma & 0 \\ 1 & 0 & \sigma \end{bmatrix}$$

rows and columns: t, g^+, g^-

Calcd. quantities:
$\langle\mu^2\rangle / nm^2$ = 0.745   (Exptl.: 0.757)
$d(\ln\langle\mu^2\rangle)/dT$ = 0.09 × 10^{-3} K^{-1} (Exptl.: 1.09 × 10^{-3} K^{-1})

S 175

poly(1,3-dithiocane), PDTC

Welsh, W. J.; Mark, J. E.; Guzmán, J.; Riande, E. *Makromol. Chem.* **1982**, *183*, 2565.

Bond Length [pm]	Valence Angles [°]	Torsion Angles [°]	ξ (for 298 K)	ξ_0	E_ξ [kJ·mol^{-1}]	
C-C : 153	C-S-C : 100	t : 180	σ = 0.36	1	2.5	
C-S : 181.5	C-C-S : 114	For bonds a-c,h: g^+ : 70	σ' = 0.43	1	2.1	$U_a = U_b = \begin{bmatrix} 1 & \sigma'' & \sigma'' \\ 1 & \sigma'' & 0 \\ 1 & 0 & \sigma'' \end{bmatrix}$
C-H : 110	C-C-C : 110	g^- : −70	σ'' = 7.52	1	−5.0	
	S-C-S : 114	For bonds d,g: g^+ : 65 g^- : −65	σ''' = 0.82	1	0.5	
			$\omega \approx 0$	1	> 25	$U_c = U_h = \begin{bmatrix} 1 & \sigma''' & \sigma''' \\ 1 & \sigma''' & 0 \\ 1 & 0 & \sigma''' \end{bmatrix}$ $U_d = U_g = \begin{bmatrix} 1 & \sigma & \sigma \\ 1 & \sigma & 0 \\ 1 & 0 & \sigma \end{bmatrix}$ $U_e = U_f = \begin{bmatrix} 1 & \sigma' & \sigma' \\ 1 & \sigma' & 0 \\ 1 & 0 & \sigma' \end{bmatrix}$
		For bonds e,f: g^+ : 68 g^- : −68				rows and columns: t, g^+, g^-

Comments:
In this study, the dipole moments at 298 K and the corresponding temperature coefficient of PDTC are calculated in the RIS approximation. The results are compared to the values determined by experiment. The results indicate that an extra stabilization energy of about 3.8 kJ · mol^{-1} must be added to these *gauche* states relative to *trans* state. This provides evidence of a large attractive sulfur *gauche* effect in the polysulfides, about equal in magnitude to that found for the analogous POM.

Calcd. quantities:

$\langle \mu^2 \rangle / nm^2$ = 0.26   (*Exptl.*: 0.265)

$d(\ln \langle \mu^2 \rangle)/dT$ = 4.63 × 10^{-3} K^{-1} (*Exptl.*: 4.9 × 10^{-3} K^{-1})

alternating copolymers of ethylene sulfide and pentamethylene sulfide, PXS

Riande, E.; Guzmán, J. *Macromolecules* **1981**, *14*, 1234.

Bond Length [pm]	Valence Angles [°]	Torsion Angles [°]	ξ (for 303 K)	ξ_0	E_ξ [kJ·mol^{-1}]
C-C : 153	C-S-C : 100	t : 180	σ = 0.43	1	2.1
C-S : 181.5	C-C-S : 114	For CC-SC and CC-CS:	σ' = 0.60	1	1.3
C-H : 110	C-C-C : 110	g$^+$: 70	σ'' = 1.17	1	−0.4
		g$^-$: −70			
			ω' = 0.06	1	7.1
		For CC-CC:			
		g$^+$: 60	ω'' = 0.51	1	1.7
		g$^-$: −60			
			σ_p = 0.40	1	2.3

$$U_a = U_d = \begin{bmatrix} 1 & \sigma' & \sigma' \\ 1 & \sigma' & 0 \\ 1 & 0 & \sigma' \end{bmatrix} \quad U_b = U_c = \begin{bmatrix} 1 & \sigma & \sigma \\ 1 & \sigma & 0 \\ 1 & 0 & \sigma \end{bmatrix} \quad U_e = \begin{bmatrix} 1 & \sigma'' & \sigma'' \\ 1 & \sigma'' & 0 \\ 1 & 0 & \sigma'' \end{bmatrix}$$

$$U_f = U_i = \begin{bmatrix} 1 & \sigma'' & \sigma'' \\ 1 & \sigma'' & \sigma''\omega' \\ 1 & \sigma''\omega' & \sigma'' \end{bmatrix} \quad U_g = \begin{bmatrix} 1 & \sigma_p & \sigma_p \\ 1 & \sigma_p & \sigma_p\omega' \\ 1 & \sigma_p\omega' & \sigma_p \end{bmatrix} \quad U_h = \begin{bmatrix} 1 & \sigma'' & \sigma'' \\ 1 & \sigma'' & \sigma''\omega' \\ 1 & \sigma''\omega' & \sigma'' \end{bmatrix}$$

rows and columns: t, g$^+$, g$^-$

Comments:
The conformational energy $E(\sigma_p)$ associated with *gauche* states about CH_2-CH_2 bonds in poly(ethylene sulfide) (PSE) is estimated from the RIS analysis of experimental dipole moments and their temperature coefficients corresponding to the alternating copolymer of pentamethylene sulfide and ethylene sulfide (PXS) as well as to 1,2-bis(butylthio)ethane.

Calcd. quantities:

	$<\mu^2> / nm^2$	= 0.604	(Exptl.: 0.635)
	$d(\ln <\mu^2>)/dT$	= 0.94 × 10^{-3} K^{-1}	(Exptl.: 1.7 × 10^{-3} K^{-1})

Further calculations on ***poly($A_xB_yC_z$)*** *chains:*

S 177   Tasaki, K.; Sasanuma, Y.; Ando, I.; Abe, A. *Bull. Chem. Soc. Jpn.* **1984**, *57*, 2391.

The energy difference between the two rotational isomers of isopropyl methyl ether is elucidated from the observed vicinal ^{13}C-^{1}H coupling constant. The conventional Gutowsky method based on the simple RIS model tends to give an underestimate, especially when the energy difference is large. A more elaborate treatment, which takes account of the overall profile of the torsional potential energy courve, is attempted.

1,2-dimethoxy-2-methylpropane, DMMP

Kato, K.; Araki, K.; Abe, A. Polymer J. **1981**, *13*, 1055.

Bond Length [pm]	Valence Angles [°]	Torsion Angles [°]	ξ (for 298 K)	ξ₀	E_ξ [kJ·mol⁻¹]			
C-C : 153	θ_1 : 111.5	t : 180	σ = 0.43	1	2.1 (1.7 to 3.3)			
C-O : 143	θ_2 : 112	g^+ : 60	σ' = 0.13 to 0.11	1	5.0 to 5.4	$U_2 = \begin{bmatrix} 1 & \sigma' & \sigma' \\ 1 & \sigma' & \sigma' \\ 1 & \sigma' & \sigma' \end{bmatrix}$	$U_3 = \begin{bmatrix} 1 & \sigma & \sigma \\ 0 & 0 & \sigma\omega \\ 0 & \sigma\omega & 0 \end{bmatrix}$	$U_4 = \begin{bmatrix} 1 & 1 & 1 \\ 1 & 1 & \omega' \\ 1 & \omega' & 1 \end{bmatrix}$
C-H : 110	θ_3 : 109.5	g^- : −60	ω = 0.72 to 0.59	1	0.8 to 1.3			
	θ_4 : 118		ω' = 0.22	1	3.8			rows and columns: t, g^+, g^-

Calcd. quantities: $\langle\mu^2\rangle / nm^2$ = 0.58, 0.56, or 0.52 (depending on E(σ))   (*Exptl.*: 0.58 (benzene), 0.56 (CCl₄), 0.52 (cyclohexane))

(for n = 5) $d(\ln\langle\mu^2\rangle)/dT$ = 0.5 × 10⁻³ K⁻¹ (σ = 0.43)   (*Exptl.*: 3.5 (±3.8) × 10⁻³ K⁻¹ (benzene); 4.4 (± 4.8) × 10⁻³ K⁻¹ (cyclohexane))

poly(oxy-1,1-dimethylethylene), PODME

Kato, K.; Araki, K.; Abe, A. *Polymer J.* **1981**, *13*, 1055;
Ando, I.; Sato, K.; Kato, K.; Abe, A. *ibid.*, 1063;
Abe, A.; Ando, I.; Kato, K.; Uematsu, I. *ibid.*, 1069.

Bond Length [pm]	Valence Angles [°]	Torsion Angles [°]	ξ (for 298 K)	ξ_0	E_ξ [kJ·mol^{-1}]
C-C : 153	θ_1 : 111.5	t : 180	$\sigma = 0.43$	1	2.1
C-O : 143	C-O-C : 118	g$^+$: 60	$\omega' = 0.22$	1	3.8
C-H : 110	O-CH$_2$-C : 112	g$^-$: −60			
	CH$_2$-C-O : 109.5				

$$U_a = \begin{bmatrix} 1 & \sigma & \sigma \end{bmatrix} \quad U_b = \begin{bmatrix} 1 & 1 & 1 \\ 1 & 1 & \omega' \\ 1 & \omega' & 1 \end{bmatrix} \quad U_c = \begin{bmatrix} 1 \\ 1 \\ 1 \end{bmatrix}$$

rows and columns: t, g$^+$, g$^-$

Comments:
The configurational characteristics of PODME are established by calculations based on the RIS model. By using an appropriate set of conformational energy parameters, the characteristic ratio and the dipole moment ratio are evaluated.

Calcd. quantities: (for $x = 200$)

$\langle r^2 \rangle_0 / nl^2 = 6.1$

$d(\ln \langle r^2 \rangle_0)/dT = -0.87 \times 10^{-3}$ K^{-1}

$\langle \mu^2 \rangle / nm^2 = 0.48$

$d(\ln \langle \mu^2 \rangle)/dT = 0.81 \times 10^{-3}$ K^{-1}

1,3-dimethoxy-2,2-dimethylpropane, DDP

Inomata, K.; Phataralaoha, N.; Abe, A. *Comput. Polym. Sci.* **1991**, *1*, 126.

Bond Length [pm]	Valence Angles [°]	Torsion Angles [°]	ξ (for 298 K)	ξ_0	E_ξ [kJ·mol^{-1}]
C-H : 110	C-O-C : 111.5	t : 180	σ = 2.74	1	−2.5
C-O : 143	O-C-C : 112.0	For C-C bonds: g^+ : 60	ω = 0.17	1	4.4
C-C : 153	C-C-C : 109.5	g^- : −60	ρ = 0	1	∞

$$U_2 = \begin{bmatrix} 1 & \rho & \rho \end{bmatrix} \quad U_3 = \begin{bmatrix} 1 & \sigma & \sigma \\ 1 & \sigma & 0 \\ 1 & 0 & \sigma \end{bmatrix} \quad U_4 = \begin{bmatrix} 1 & \sigma & \sigma \\ 1 & \sigma & \sigma\omega \\ 1 & \sigma\omega & \sigma \end{bmatrix} \quad U_5 = \begin{bmatrix} 1 & \rho & \rho \\ 1 & \rho & 0 \\ 1 & 0 & \rho \end{bmatrix}$$

rows and columns: t, g^+, g^-

Comments:
The conformation of α,ω-dimethoxyalkanes such as $CH_3O(CH_2)_yOCH_3$ (y = 3,4) and $CH_3OCH_2C(CH_3)_2CH_2OCH_3$ are studied by the NMR method. Conformational energies of the internal C−O and C−C bonds are estimated from the observed vicinal coupling constants. Molecular mechanics calculations are used as a supplemental tool to elucidate the characteristic feature of the potential energy surface. Values of the conformational energy for the rotation around O-C−C-C are found to be slightly more negative than those calculated. The differences (ca. 1.3 kJ·mol^{-1}) between the calculated and the observed energies are small as compared with those (4 to 8 kJ·mol^{-1}) encountered in lower homologs such as $CH_3O(CH_2)_yOCH_3$ (y = 1,2). The conformational energy of the central C-C−C-C bond of 1,4-dimethoxybutane is found to be positive (1.3 kJ·mol^{-1}). The results of the present analysis are reasonably consistent with those previously derived from the statistical analysis of conformation-dependent properties of polymers. Following the previous treatment, the characteristic ratios of the dimension and the dipole moment of polymers are calculated, and the results are compared with relevant experimental data. *Trans* states are strongly prefered at bonds 2 and 5 in DDP.

Calcd. quantities: Vicinal $^3J_{HH}$ and $^3J_{CH}$ NMR coupling constants

S 181

poly(3,3-dimethyl oxetane), PDO

Saiz, E.; Riande, E.; Guzmán, J.; de Abajo, J. *J. Chem. Phys.* **1980**, *73*, 958.

Bond Length [pm]	Valence Angles [°]	Torsion Angles [°]	σ (for 298 K)	σ_0	E_ξ [kJ·mol^{-1}]
C-C : 153	C-C-O : 110	t : 180	σ = 2.33	1	−2.1
C-O : 143	C-O-C : 110	g$^+$: 60	ω = 0.11	1	5.4
	C-C-C : 110	g$^-$: −60			

$$U_a = \begin{bmatrix} 1 & 0 & 0 \\ 0 & 0 & 0 \\ 0 & 0 & 0 \end{bmatrix} \quad U_b = \begin{bmatrix} 1 & \sigma & \sigma \\ 0 & 0 & 0 \\ 0 & 0 & 0 \end{bmatrix} \quad U_c = \begin{bmatrix} 1 & \sigma & \sigma \\ 1 & \sigma & \sigma\omega \\ 1 & \sigma\omega & \sigma \end{bmatrix} \quad U_d = \begin{bmatrix} 1 & 0 & 0 \\ 1 & 0 & 0 \\ 1 & 0 & 0 \end{bmatrix}$$

rows and columns: t, g$^+$, g$^-$

Comments:
Semiempirical potentials are used to compute conformational energies as function of rotations over the backbone bonds; the results of these calculations indicate that skeletal C−O bonds are always placed in *trans* conformation, whereas *gauche* states over C−C bonds have an energy of about 2 kJ·mol^{-1} lower than their *trans* conformation.

Calcd. quantities:

$\langle r^2 \rangle_0 / nl^2$ = 4.2

$d(\ln \langle r^2 \rangle_0)/dT$ = 0.5 × 10^{-3} K^{-1}

$\langle \mu^2 \rangle / nm^2$ = 0.25   (*Exptl.*: ≈ 0.25)

$d(\ln \langle \mu^2 \rangle)/dT$ = 2.7 × 10^{-3} K^{-1} (*Exptl.*: 4.4 × 10^{-3} K^{-1})

poly(3,3-dimethyloxetane), PDO

Garrido, L.; Riande, E.; Guzmán *J. Polym. Sci., Polym. Phys. Ed.* **1982**, *20*, 1805.

Bond Length [pm]	Valence Angles [°]	Torsion Angles [°]	ξ (for 303 K)	ξ_0	E_ξ [kJ·mol^{-1}]
C-O : 143	C-O-C : 110	t : 180	$\tau = 0$	1	∞
C-C : 153	C-C-C : 110	g$^+$: 60	$\sigma = 2.3$	1	-2.1
C-H : 110	C-C-O : 110	g$^-$: -60	$\omega = 0.12$	1	5.4
	C-C-H : 110				

$$U_a = U_d = \begin{bmatrix} 1 & \tau & \tau \\ 1 & \tau & \tau \\ 1 & \tau & \tau \end{bmatrix} \quad U_b = \begin{bmatrix} 1 & \sigma & \sigma \\ 1 & \sigma & \sigma \\ 1 & \sigma & \sigma \end{bmatrix} \quad U_c = \begin{bmatrix} 1 & \sigma & \sigma \\ 1 & \sigma & \sigma\omega \\ 1 & \sigma\omega & \sigma \end{bmatrix}$$

rows and columns: t, g$^+$, g$^-$

Calcd. quantities:
- $d(\ln \langle r^2 \rangle)/dT = 0.4 \times 10^{-3}$ K^{-1} (for 333 K) (*Exptl.*: 1.1×10^{-3} K^{-1} (for 333 K))
- $\langle \mu^2 \rangle / nm^2 = 0.25$ (*Exptl.*: 0.21)
- $d(\ln \langle \mu^2 \rangle)/dT = 2.7 \times 10^{-3}$ K^{-1} (*Exptl.*: 2.5×10^{-3} K^{-1})

S 183

poly((2,2-dimethyl)trimethylene oxide), PODM$_3$

Inomata, K.; Phataralaoha, N.; Abe, A. *Comput. Polym. Sci.* **1991**, *1*, 126.

Bond Length [pm]	Valence Angles [°]	Torsion Angles [°]	ξ (for 298 K)	ξ_σ	E_ξ [kJ·mol^{-1}]
C-H : 110	C-O-C : 111.5	t : 180	σ = 2.33 to 2.74	1	−2.1 to −2.5
C-O : 143	O-C-C : 112.0	For C-C bonds: g$^+$: 60	σ' = 0	1	∞
C-C : 153	C-C-C : 109.5	g$^-$: −60	ω = 0.169	1	4.40

$$U_a = U_d = \begin{bmatrix} 1 & 0 & 0 \\ 1 & 0 & 0 \\ 1 & 0 & 0 \end{bmatrix} \quad U_b = \begin{bmatrix} 1 & \sigma & \sigma \\ 1 & \sigma & 0 \\ 1 & 0 & \sigma \end{bmatrix} \quad U_c = \begin{bmatrix} 1 & \sigma & \sigma \\ 1 & \sigma & \sigma\omega \\ 1 & \sigma\omega & \sigma \end{bmatrix}$$

rows and columns: t, g$^+$, g$^-$

Comments:
The conformation of α,ω-dimethoxyalkanes such as CH$_3$O(CH$_2$)$_y$OCH$_3$ (y = 3,4) and CH$_3$OCH$_2$C(CH$_3$)$_2$CH$_2$OCH$_3$ are studied by the NMR method. Conformational energies of the internal C−O and C−C bonds are estimated from the observed vicinal coupling constants. Molecular mechanics calculations are used as a supplemental tool to elucidate the characteristic feature of the potential energy surface. Values of the conformational energy for the rotation around O-C−C-C are found to be slightly more negative than those calculated. The differences (ca. 1.3 kJ·mol^{-1}) between the calculated and the observed energies are small as compared with those (4 to 8 kJ·mol^{-1}) encountered in lower homologs such as CH$_3$O(CH$_2$)$_y$OCH$_3$ (y = 1,2). The conformational energy of the central C-C−C-C bond of 1,4-dimethoxybutane is found to be positive (1.3 kJ·mol^{-1}). The results of the present analysis are reasonably consistent with those previously derived from the statistical analysis of conformation-dependent properties of polymers. Following the previous treatment, the characteristic ratios of the dimension and the dipole moment of polymers are calculated, and the results are compared with relevant experimental data.

Calcd. quantities:

$<r^2>_0 / nl^2$	= 3.9	(*Exptl.*: 4.34)
$d(\ln <r^2>_0)/dT$	= 0.32 × 10^{-3} K^{-1}	(*Exptl.*: 1.1 × 10^{-3} K^{-1})
$<\mu^2> / nm^2$	= 0.23	(*Exptl.*: 0.21 to 0.26)
$d(\ln <\mu^2>)/dT$	= 2.6 × 10^{-3} K^{-1}	(*Exptl.*: 2.5 (to 4.4) × 10^{-3} K^{-1})

poly(3,3-dimethylthietane), PDS

Riande, E.; Guzmán, J.; Saiz, E.; de Abajo, J. *Macromolecules* **1981**, *14*, 608.

Bond Length [pm]	Valence Angles [°]	Torsion Angles [°]	ξ (for 303 K)	ξ_o	E_ξ [kJ·mol^{-1}]
C-S : 181.5	C-S-C : 100	For C-S bonds: t : 180 g$^-$: −105 For C-C bonds: t : 180 g$^+$: 60 g$^-$: −60	$\tau = 0.08$	1	6.3
C-H : 110	C-C-S : 114		$\omega = 0.08$	1	6.3
	C-C-H : 110				

$$U_a = U_d = \begin{bmatrix} 1 & \tau & \tau \\ 1 & \tau & \tau \\ 1 & \tau & \tau \end{bmatrix} \quad U_b = \begin{bmatrix} 1 & \sigma & \sigma \\ 1 & \sigma & \sigma \\ 1 & \sigma & \sigma \end{bmatrix} \quad U_c = \begin{bmatrix} 1 & \sigma & \sigma \\ 1 & \sigma & \sigma\omega \\ 1 & \sigma\omega & \sigma \end{bmatrix}$$

rows and columns: t, g$^+$, g$^-$

Calcd. quantities:

$\langle r^2 \rangle_o / nl^2 = 5.1$

$\langle \mu^2 \rangle / nm^2 = 0.57$ (Exptl.: 0.61)

$d(\ln \langle \mu^2 \rangle)/dT = 0.86 \times 10^{-3}$ K^{-1} (Exptl.: 0.8×10^{-3} K^{-1})

poly(thiodiethylene glycol)

Riande, E.; Guzmán, J. *Macromolecules* **1979**, *12*, 952.

Bond Length [pm]	Valence Angles [°]	Torsion Angles [°]	ξ (for 298 K)	ξ_0	E_ξ [kJ·mol^{-1}]
C-C : 153	C-S-C : 100	t : 180	$\sigma = 1.18$	1	-0.42
C-O : 143	S-C-C : 114	g$^+$: 60 or 65	$\sigma' = 0.47$ or 0.51	1	1.88 or 1.67
C-S : 181.5	O-C-C : 110	g$^-$: -60 or -65	$\sigma'' = 0.22$	1	3.77
	C-O-C : 110		$\omega = 0.60$	1	1.26
			$\omega' = 0.20$	1	3.99

$$U_a = \begin{bmatrix} 1 & \sigma & \sigma \\ 1 & \sigma & 0 \\ 1 & 0 & \sigma \end{bmatrix} \quad U_b = \begin{bmatrix} 1 & \sigma' & \sigma' \\ 1 & \sigma' & \sigma'\omega \\ 1 & \sigma'\omega & \sigma' \end{bmatrix} \quad U_c = \begin{bmatrix} 1 & \sigma'' & \sigma'' \\ 1 & \sigma'' & \sigma''\omega' \\ 1 & \sigma''\omega' & \sigma'' \end{bmatrix}$$

$$U_d = \begin{bmatrix} 1 & \sigma'' & \sigma'' \\ 1 & \sigma'' & 0 \\ 1 & 0 & \sigma'' \end{bmatrix} \quad U_e = \begin{bmatrix} 1 & \sigma' & \sigma' \\ 1 & \sigma' & \sigma'\omega' \\ 1 & \sigma'\omega' & \sigma' \end{bmatrix} \quad U_f = \begin{bmatrix} 1 & \sigma & \sigma \\ 1 & \sigma & \sigma\omega \\ 1 & \sigma\omega & \sigma \end{bmatrix}$$

rows and columns: t, g$^+$, g$^-$

Comments:

Mean-square dipole moments of poly(thiodiethylene gycol), an alternating copolymer of ethylene oxide and ethylene sulfide, are determined from dielectric constant measurements on dilute solutions of the polymer in benzene. Since the configuration-dependent properties of one of the parent homopolymers, PES, are unknown, because of its insolubility in ordinary solvents, the results are preferably compared with those of POE chains. It is found that the dipole moments of poly(thiodiethylene glycol) are somewhat larger than those of POE.

Calcd. quantities:

$\langle \mu^2 \rangle / nm^2 = 0.62$ (*Exptl.*: 0.62 ± 0.01)

$d(\ln \langle \mu^2 \rangle)/dT = 1.0 \times 10^{-3}$ K^{-1} (*Exptl.*: 1.6×10^{-3} K^{-1})

poly(1,3-dioxa-6-thiocane)

Riande, E.; Guzmán, J. *Macromolecules* **1979**, *12*, 1117.

Bond Length [pm]	Valence Angles [°]	Torsion Angles [°]	ξ (for 298 K)	ξ_0	E_ξ [kJ·mol^{-1}]
C-C : 153	C-C-S : 114	t : 180	$\sigma = 1.20$	1	-0.4
C-O : 143	C-C-O : 110	g$^+$: 60	$\sigma_p = 8$	1	-5.0
C-S : 181.5	C-O-C : 110	g$^-$: -60	$\sigma' = 0.59$	1	1.3
	C-S-C : 100		$\sigma'' = 0.22$	1	3.8
			$\omega = 0.56$	1	1.7
			$\omega' = 0.20$	1	4.2

$$U_a = \begin{bmatrix} 1 & \sigma'' & \sigma' \\ 1 & \sigma'' & 0 \\ 1 & 0 & \sigma'' \end{bmatrix} \quad U_b = \begin{bmatrix} 1 & \sigma' & \sigma' \\ 1 & \sigma' & \sigma'\omega \\ 1 & \sigma'\omega' & \sigma' \end{bmatrix} \quad U_c = \begin{bmatrix} 1 & \sigma & \sigma \\ 1 & \sigma & \sigma\omega \\ 1 & \sigma\omega & \sigma \end{bmatrix} \quad U_d = \begin{bmatrix} 1 & \sigma & \sigma \\ 1 & \sigma & 0 \\ 1 & 0 & \sigma \end{bmatrix}$$

$$U_e = \begin{bmatrix} 1 & \sigma' & \sigma' \\ 1 & \sigma' & \sigma'\omega \\ 1 & \sigma'\omega & \sigma' \end{bmatrix} \quad U_f = \begin{bmatrix} 1 & \sigma'' & \sigma'' \\ 1 & \sigma'' & \sigma''\omega' \\ 1 & \sigma''\omega' & \sigma'' \end{bmatrix} \quad U_g = U_h = \begin{bmatrix} 1 & \sigma_p & \sigma_p \\ 1 & \sigma_p & 0 \\ 1 & 0 & \sigma_p \end{bmatrix}$$

rows and columns: t, g$^+$, g$^-$

Comments:
Dielectric measurements are carried out on solutions of poly(1,3-dioxa-6-thiocane) in benzene over a range of 20-60°C. Conformational energies arising from first-order interactions between sulfur and oxygen atoms are obtained by analysis of the dipole moments in terms of the RIS theory of chain configurations. The present study indicates that intramolecular interactions involving S and O atoms have higher energy in *gauche* than in *trans* states, in agreement with the results found in previous studies on the configurational properties of poly(thiodiethylene) glycol.

Calcd. quantities:

$<\mu^2> / nm^2 = 0.42$ (*Exptl.*: 0.42 (± 0.01))

$d(\ln <\mu^2>)/dT = 0.7 \times 10^{-3} \, K^{-1}$ (*Exptl.*: $0.7 \times 10^{-3} \, K^{-1}$)

poly(1-oxa-3-thiacyclopentane), POTC

Riande, E.; Guzmán, J. *Macromolecules* **1981**, *14*, 1511.

Bond Length [pm]	Valence Angles [°]	Torsion Angles [°]	ξ	ξ_0	E_ξ [kJ · mol^{-1}]
C-S : 181.5	C-S-C : 100	t : 180	σ_1 = 0.56	1	1.5 (1.3 to 1.7)
C-C : 153	C-C-S : 114	For a,c,d,f,h,i: g^+ : 66	σ_2 = 1.18	1	−0.4
C-O : 143	S-C-S : 114	g^- : −66	σ' = 6.13	1	−4.6
	O-C-O : 112	For b,e,g,j: g^+ : 70	σ'' = 0.22	1	3.8
	C-C-O : 111.5	g^- : −70	σ = 7.23	1	−5.0
			ω = 0.51 $^{a)}$	1	1.7

$$U_a = \begin{bmatrix} 1 & \sigma_1 & \sigma_1 \\ 1 & \sigma_1 & 0 \\ 1 & 0 & \sigma_1 \end{bmatrix} \quad U_b = \begin{bmatrix} 1 & \sigma_2 & \sigma_2 \\ 1 & \sigma_2 & \sigma_2\omega \\ 1 & \sigma_2\omega & \sigma_2 \end{bmatrix} \quad U_c = U_d = \begin{bmatrix} 1 & \sigma' & \sigma' \\ 1 & \sigma' & 0 \\ 1 & 0 & \sigma' \end{bmatrix} \quad U_e = \begin{bmatrix} 1 & \sigma_2 & \sigma_2 \\ 1 & \sigma_2 & 0 \\ 1 & 0 & \sigma_2 \end{bmatrix}$$

$$U_f = \begin{bmatrix} 1 & \sigma_1 & \sigma_1 \\ 1 & \sigma_1 & \sigma_1\omega \\ 1 & \sigma_1\omega & \sigma_1 \end{bmatrix} \quad U_g = U_j = \begin{bmatrix} 1 & \sigma'' & \sigma'' \\ 1 & \sigma'' & 0 \\ 1 & 0 & \sigma'' \end{bmatrix} \quad U_h = U_i = \begin{bmatrix} 1 & \sigma & \sigma \\ 1 & \sigma & 0 \\ 1 & 0 & \sigma \end{bmatrix}$$

rows and columns: t, g^+, g^-

$^{a)}$ A value of ω = 0.61 (corresponding to $E(\omega)$ = 1.2 kJ · mol^{-1}) is also given in the original paper.

Comments:
Experimental results are found to be in very good agreement with theoretical results based on an RIS model which assumed perfect alternation of 1,3-dioxolane and 1,3-dithiolane, rather than a more irregular distribution which could conceivably occur in the type of ring-opening polymerization used to prepare the polymer.

Calcd. quantities: $\langle\mu^2\rangle / nm^2$ = 0.248 (Exptl.: 0.255)
(for x = 48) $d(\ln\langle\mu^2\rangle)/dT$ = 4.7 × 10^{-3} K^{-1} (Exptl.: 4.2 × 10^{-3} K^{-1})

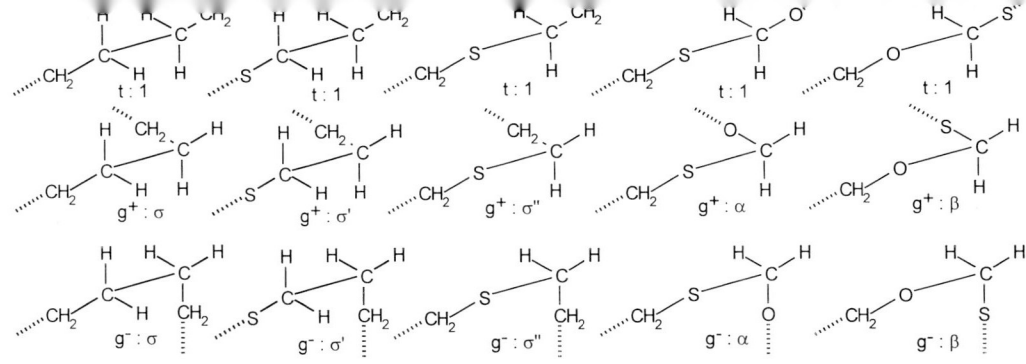

poly(3-oxa-1,5-dithiadecamethylene), PODTD

Riande, E.; Guzmán, J. *Macromolecules* **1986**, *19*, 2956.

Bond Length [pm]	Valence Angles [°]	Torsion Angles [°]	ξ (for 303 K)	ξ_o	E_ξ [kJ·mol^{-1}]				
C-C : 153	C-C-C : 111.5	t : 180 g$^+$: 60	$\sigma = 0.43$	1	2.1	$U_a = \begin{bmatrix} 1 & \sigma'' & \sigma'' \\ 1 & \sigma'' & 0 \\ 1 & 0 & \sigma'' \end{bmatrix}$	$U_b = \begin{bmatrix} 1 & \alpha & \alpha \\ 1 & \alpha & \alpha\omega \\ 1 & \alpha\omega & \alpha \end{bmatrix}$	$U_c = \begin{bmatrix} 1 & \beta & \beta \\ 1 & \beta & 0 \\ 1 & 0 & \beta \end{bmatrix}$	$U_d = \begin{bmatrix} 1 & \beta & \beta \\ 1 & \beta & 0 \\ 1 & 0 & \beta \end{bmatrix}$
C-S : 181.5	C-C-S : 114	g$^-$: -60	$\sigma' = 0.37$	1	2.5				
C-O : 143	C-S-C : 100	For C-O bonds: t : 180	$\sigma'' = 1.17$	1	-0.4				
C-H : 109	S-C-O : 114	g$^+$: 75 g$^-$: -75	$\alpha = 4.52$	1	-3.8	$U_e = \begin{bmatrix} 1 & \alpha & \alpha \\ 1 & \alpha & 0 \\ 1 & 0 & \alpha \end{bmatrix}$	$U_f = \begin{bmatrix} 1 & \sigma'' & \sigma'' \\ 1 & \sigma'' & \sigma''\omega \\ 1 & \sigma''\omega & \sigma'' \end{bmatrix}$	$U_g = \begin{bmatrix} 1 & \sigma' & \sigma' \\ 1 & \sigma' & 0 \\ 1 & 0 & \sigma' \end{bmatrix}$	$U_h = \begin{bmatrix} 1 & \sigma & \sigma \\ 1 & \sigma & 0 \\ 1 & 0 & \sigma \end{bmatrix}$
			$\beta = 4.52$	1	-3.8				
			$\omega = 0.37$	1	2.5	$U_i = \begin{bmatrix} 1 & \sigma & \sigma \\ 1 & \sigma & 0 \\ 1 & 0 & \sigma \end{bmatrix}$	$U_j = \begin{bmatrix} 1 & \sigma' & \sigma' \\ 1 & \sigma' & 0 \\ 1 & 0 & \sigma' \end{bmatrix}$	rows and columns: t, g$^+$, g$^-$	

Comments:
The critical analysis of the experimental results, carried out by statistical mechanical procedures, indicates that *gauche* states about the C—O and C—S bonds are strongly favored with respect to the alternative *trans* states. This analysis, in conjunction with calculations of the conformational energies using semiempirical potential energy functions, suggests that *gauche* states about these bonds present an extra stabilization energy which in the case of C—O bonds is larger than that observed in POM.

Calcd. quantities: $<\mu^2> / nm^2 = 0.340$ (Exptl.: 0.325)

 $d(\ln <\mu^2>)/dT = 1.65 \times 10^{-3} \text{ K}^{-1}$ (Exptl.: 2.2×10^{-3} K^{-1})

polyfluoromethylene, PFM

Tonelli, A. E. *Macromolecules* **1980**, *13*, 734.

Bond Length [pm]	Valence Angles [°]	Torsion Angles [°]	(see: a))		
C-C : 153	C-C-C : 117.5	t : 180			
C-F : 136	H-C-F : 106	g^+ : 60	$U(lll) = \begin{bmatrix} 0.045 & 0.037 & 0.237 \\ 0.283 & 0.083 & 0.165 \\ 0.052 & 0.005 & 0.097 \end{bmatrix}$	$U(ldl) = \begin{bmatrix} 0.291 & 0.134 & 0.123 \\ 0.174 & 0.039 & 0.009 \\ 0.170 & 0.021 & 0.039 \end{bmatrix}$	
C-H : 110		g^- : −60			
			$U(ldd) = \begin{bmatrix} 0.207 & 0.283 & 0.093 \\ 0.032 & 0.177 & 0.011 \\ 0.100 & 0.080 & 0.017 \end{bmatrix}$	$U(ldd) = \begin{bmatrix} 0.205 & 0.023 & 0.071 \\ 0.319 & 0.154 & 0.073 \\ 0.126 & 0.010 & 0.019 \end{bmatrix}$	rows and columns: t, g^+, g^-

a) Values given for 323 K; values for 273, 373, 425, and 473 K are given in the original paper as well.

Comments:
Conformational energy estimates are employed to determine the conformational characteristics of poly(vinyl fluoride) (PVF), polyfluoromethylene (PFM), and polytrifluoroethylene (PTF$_3$). Effects of stereoconfiguration and, in the case of PVF and PTF$_3$, the presence of head-to-head:tail-to-tail (HH:TT) defect structures are considered. The calculated results are compared to corresponding values found for poly(vinylidene fluoride), polytetrafluoroethylene, and polyethylene, and the equilibrium flexibilities of PVF, PFM, and PTF$_3$ are discussed on this basis.

Calcd. quantities: $<r^2>_0 / nl^2$ $d (\ln <r^2>_0) / dT$ $<\mu^2> / xm^2$ $d (\ln <\mu^2>) / dT$ Conformational entropy S_c

A 002

2,4-dichloro-*n*-pentane, DCP, and
2,4,6-trichloro-*n*-heptane, TCH

Flory, P. J.; Williams, A. D. *J. Am. Chem. Soc.* **1969**, *91*, 3118.

Bond Length [pm]	Valence Angles [°]	Torsion Angles [°]	ξ a) (for 343 K)	ξ_o	E_ξ [kJ·mol^{-1}]
C-C : 153	C-C-C : 112	t : 180	η = 3.5	1	−3.6
		g : 60	τ = 0.5	1	2.0
		$\bar{g}$: −60	ω = 0.05	1	8.5
			ω' = 0.10	1	6.6
			ω'' = 0.05	1	8.5

$$U_p = \begin{bmatrix} \eta & 1 & \tau \\ \eta & \omega & \tau \\ \eta & 1 & \tau\omega \end{bmatrix} \quad U_m = \begin{bmatrix} \eta\omega'' & 1 & \tau\omega' \\ \eta & \omega & \tau\omega' \\ \eta\omega' & \omega' & \tau\omega\omega'' \end{bmatrix} \quad U_r = \begin{bmatrix} \eta & \omega' & \tau\omega'' \\ \eta\omega' & 1 & \tau\omega \\ \eta\omega'' & \omega & \tau\omega'^2 \end{bmatrix}$$

rows and columns: t, g, $\bar{g}$

a) Other sets of statistical weight parameters are given as well in the original paper, e.g.: η = 4.0, τ = 0.5, ω = ω'' = 0.3, ω' = 0.07 (for 298 K).

Comments:
Stereochemical equilibration of DCP in DMSO at 343 K in the presence of LiCl yields a mixture containing 36.4 (±0.3) % of the *meso* isomer. The statistical weight parameters evaluated from this result are used for theoretical calculation of the proportions of various conformers in *meso* and *racemic* DCP, and also in the three diastereoisomers of TCH. Calculations for TCH are compared with estimates of others for NMR coupling constants. It is shown that the less-favoured conformations, often ignored, contribute appreciably to the conformer populations of the TCH isomer.

Calcd. quantities: Conformer fractions for *meso* and *racemic* diads of DCP and TCH.

poly(vinyl chloride), PVC

Mark, J. E. *J. Chem. Phys.* **1972**, *56*, 451.

Bond Length [pm]	Valence Angles [°]	Torsion Angles [°]	ξ (for 298 K)	ξ_0	E_ξ [kJ·mol^{-1}]
C-C : 153	C-C-C : 112	t : 180	η = 4.2	1	−3.6
		g$^+$: 60	τ = 0.45	1	2.0
		g$^-$: −60	ω = 0.032	1	8.5
			ω' = 0.071	1	6.6
			ω'' = 0.032	1	8.5

$$U_d = \begin{bmatrix} \eta & 1 & \tau \\ \eta & 1 & \tau\omega \\ \eta\omega & \omega & \tau \end{bmatrix} \quad U_l = \begin{bmatrix} \eta & \tau & 1 \\ \eta & \tau & \omega \\ \eta & \tau\omega & 1 \end{bmatrix} \quad U_{dl} = \begin{bmatrix} \eta & \omega' & \tau\omega'' \\ \eta\omega' & 1 & \tau\omega \\ \eta\omega'' & \omega & \tau\omega'^2 \end{bmatrix}$$

$$U_{ld} = \begin{bmatrix} \eta & \tau\omega'' & \omega' \\ \eta\omega'' & \tau\omega'^2 & \omega \\ \eta\omega' & \tau\omega & 1 \end{bmatrix} \quad U_{dd} = \begin{bmatrix} \eta\omega'' & \tau\omega' & 1 \\ \eta & \tau\omega' & \omega \\ \eta\omega' & \tau\omega\omega'' & \omega' \end{bmatrix} \quad U_{ll} = \begin{bmatrix} \eta\omega'' & 1 & \tau\omega' \\ \eta\omega' & \omega' & \tau\omega\omega'' \\ \eta & \omega & \tau\omega' \end{bmatrix}$$

rows and columns: t, g$^+$, g$^-$

Comments:

A RIS model with neighbor interactions is used to calculate mean-square unperturbed dimensions and dipole moments for vinyl chloride chains having degrees of polymerization ranging from x = 1 to 150 and stereochemical structures ranging from perfect syndiotacticity to perfect isotacticity. Conformational energies used in these calculations are those which have been established in the analysis based on the stereochemical equilibration of 2,4-dichloro-n-pentane by *Flory* and *Williams* (**A 002**).

Calcd. quantities: $<r^2>_0 / nl^2$ $d(\ln <r^2>_0)/dT$ $<\mu^2>/xm^2$ $d(\ln <\mu^2>)/dT$

A 004

poly(vinyl chloride), PVC

Flory, P. J.; Pickles, C. J. *J. Chem. Soc., Farad. Trans. 2* **1973**, *69*, 632.

Bond Length [pm]	Valence Angles [°]	Torsion Angles [°]	ξ (for 293 K)	ξ_o	E_ξ [kJ·mol^{-1}]
C-C : 153	C-C-C : 112	t : 180	η = 4.5	1	−3.7
		g : 60	τ = 0.5	1	1.7
		$\bar{g}$: −60	ω = 0.03	1	8.5
			ω' = 0.07	1	6.5
			ω'' = 0.03	1	8.5

$$U_p = \begin{bmatrix} \eta & 1 & \tau \\ \eta & \omega & \tau \\ \eta & 1 & \tau\omega \end{bmatrix} \qquad U_m = \begin{bmatrix} \eta\omega'' & 1 & \tau\omega' \\ \eta & \omega & \tau\omega' \\ \eta\omega' & \omega' & \tau\omega\omega'' \end{bmatrix} \qquad U_r = \begin{bmatrix} \eta & \omega' & \tau\omega'' \\ \eta\omega' & 1 & \tau\omega \\ \eta\omega'' & \omega & \tau\omega'^2 \end{bmatrix}$$

rows and columns: t, g, $\bar{g}$

Comments:
Stereochemical equilibration of 2,4,6-trichloro-*n*-heptane (TCH) in DMSO at 70°C in the presence of LiCl yields a mixture containing 11.1 ±02 % of the isotactic isomer and 42.7 ± 0.2 % of the syndiotactic isomer. These results are interpreted according to the theory of stereochemical equilibrium. The statistical weight parameters thus evaluated are used for the theoretical calculation of the proportions of various conformers in the three diastereomers of TCH. These calculations are compared with estimates from NMR coupling constants. It is confirmed that the less-favoured conformations contribute appreciably to the conformer populations in the isomers of TCH, owing to the small size of the chlorine substituent. Stereochemical equilibria and conformer populations calculated for the PVC from the same parameters show significant departures from those of the oligomers. The calculated average length of sequences of the preferred conformation (tgtg, etc.) for isotactic PVC is much smaller than for an isotactic vinyl polymer having a larger substituent.

Calcd. quantities: Conformer populations of the diad placements in stereoregular PVC chains.

poly(vinyl chloride), PVC

Blasco Cantera, F.; Riande, E.; Almendro, J. P.; Saiz, E. *Macromolecules* **1981**, *14*, 138.

Bond Length [pm]	Valence Angles [°]	Torsion Angles [°]	ξ (for 298 K)	ξ_o	E_ξ [kJ · mol^{-1}]
C-C : 153	C-C-C : 112	t : 180	η = 4.2	1	−3.56
C-H : 110	others : 109.5	g : 60	τ = 0.45	1	1.98
C-Cl : 185		$\bar{g}$: −60	ω = 0.032	1	8.53
			ω' = 0.071	1	6.55
			ω'' = 0.032	1	8.53

$$U_p = \begin{bmatrix} 1 & 1 & 1 \\ 1 & 1 & \omega \\ 1 & \omega & 1 \end{bmatrix} \quad U_m = \begin{bmatrix} \eta^2\omega'' & \eta\tau\omega' & \eta \\ \eta & \tau\omega' & \omega \\ \eta\tau\omega' & \tau^2\omega'' & \tau\omega' \end{bmatrix} \quad U_r = \begin{bmatrix} \eta^2 & \eta\omega' & \eta\tau\omega'' \\ \eta\omega' & 1 & \tau\omega \\ \eta\tau\omega'' & \tau\omega & \tau^2\omega/2 \end{bmatrix}$$

rows and columns: t, g, $\bar{g}$

Comments:
The mean-square dipole moment of two fractions of PVC is measured in dioxane solution at different temperatures. The experimental values of $<\mu^2>$ increase with increasing syndiotacticity and show negative temperature coefficient. Earlier theoretical calculations carried out assuming that the orientation of the vector dipole moment m corresponding to the repeat unit coincides with that of the C−Cl bond dipole moment give values of $<\mu^2>$ significantly higher than the experimental results. The present analysis shows, however, that if the contribution to m from the vector dipole associated with the C− bond of the HCCl group is taken into account, the theoretical calculations reproduce the experimental results.

Calcd. quantities: $<\mu^2> / xm^2$ $d (\ln <\mu^2>) / dT$

A 006

poly(vinyl chloride), PVC

Boyd, R. H.; Kesner, L. J. Polym. Sci.; Polym. Phys. Ed. **1981**, 19, 375.

Bond Length [pm]	Valence Angles [°]	Torsion Angles [°]	ξ a,b,c) (for 298 K)	ξ_0	E_ξ [kJ·mol^{-1}]
C-C : 153	C-CH$_2$-C : 115	t : 180	η = 1.63 / 1.91	1	−1.21 / −1.60
	CH$_2$-C-CH$_2$: 113	g$^+$: 60	τ = 0.82 / 0.60	1	0.50 / 1.26
		g$^-$: −60	ω = 0.014/0.023	1	10.60 / 9.39
			ω' = 0.069/0.066	1	6.63 / 6.75
			ω'' = 0.021/0.052	1	9.58 / 7.34
			ω_p = 0.45 / 0.32	1	1.98 / 2.86

$$U_I = \begin{bmatrix} \eta & \tau & 1 \\ \eta & \tau & \omega \\ \eta & \tau\omega & 1 \end{bmatrix} \quad U_{II} = \begin{bmatrix} \eta\omega'' & 1 & \tau\omega' \\ \eta\omega' & \omega' & 0 \\ \eta & \omega\omega_p & \tau\omega' \end{bmatrix} \quad U_{Id} = \begin{bmatrix} \eta & \tau\omega'' & \omega' \\ \eta\omega'' & 0 & \omega \\ \eta\omega' & \tau\omega & \omega_p \end{bmatrix}$$

rows and columns: t, g$^+$, g$^-$

a) τ is called sk in the original paper.
b) The interaction energies are calculated for: conformational energies / conformational + solvation energies.
c) The interaction $E(\omega_p)$ is largely electrostatic in origin.

Comments:
Inductive effects on dipole moments and the effects of intervening atoms on electrostatic interaction energies are represented by polarizability centers in conjunction with bond centered dipoles. Solvation energies are estimated by means of a continuum dipole-quadrupole electrostatic model. Calculated energies of a number of conformations of *meso* and *racemic* 2,4-dichloropentane and the *iso*, *syndio*, and *hetero* forms of 2,4,6-trichloroheptane give satisfactory representations of isomer and conformer populations. Electrostatic effects are found to be quite important.

Calcd. quantities: $<r^2>_0 / nl^2$ = 11.0 (± 0.4)   (Calculated for the parameters including solvation energies with replication probability = 0.43)
$<\mu^2> / xm^2$ = 0.72 (± 0.10)

poly(vinyl chloride), PVC

Khanarian, G.; Schilling, F. C.; Cais, R. E.; Tonelli, A. E. *Macromolecules* **1983**, *16*, 278.

Bond Length [pm]	Valence Angles [°]	Torsion Angles [a)] [°]	ξ (for 298 K)	ξ_0	E_ξ [kJ·mol^{-1}]			
C-C : 153	C-C-C : 112	t : 180 − Δφ	η = 4.2 to 2.0	1	−3.56 to −1.67			
		g : 60 + Δφ	τ = 0.43	1	2.1	$U_p = \begin{bmatrix} \eta & 1 & \tau \\ \eta & \omega & \tau \\ \eta & 1 & \tau\omega \end{bmatrix}$	$U_m = \begin{bmatrix} \eta\omega'' & 1 & \tau\omega' \\ \eta & \omega & \tau\omega' \\ \eta\omega' & \omega' & \tau\omega\omega'' \end{bmatrix}$	$U_r = \begin{bmatrix} \eta & \omega' & \tau\omega'' \\ \eta\omega' & 1 & \tau\omega \\ \eta\omega'' & \omega & \tau\omega'^2 \end{bmatrix}$
		$\bar{g}$: −60	ω = 0.034	1	8.4			
			ω' = 0.014	1	10.5			
						rows and columns: t, g, $\bar{g}$		
			ω'' = 0.006	1	12.6			

a) Δφ = 3° (0 to 5°)

Calcd. quantities: $\langle r^2 \rangle_0 / nl^2$ $d(\ln \langle r^2 \rangle_0)/dT$ $_mK/x$ $\langle \mu^2 \rangle / x$ $\langle \gamma^2 \rangle / x$

A 008

poly(vinyl chloride), PVC

Wang, S.; Mark, J. E. *Comput. Polym. Sci.* **1991**, *1*, 188.

Bond Length [pm]	Valence Angles [a)] [°]	Torsion Angles [°]		ξ (for 298 K)	ξ_0	E_ξ [kJ·mol^{-1}]	(see: a,b))
C-C : 153	CH$_2$-C-CH$_2$: 112	t	: 180	η = 4.2	1	-3.56	
C-Cl : 176	C-CH$_2$-C : 112	g$^+$	: 60	τ = 0.45	1	1.98	
		g$^-$	: -60	$\sigma \approx 1$	1	≈ 0	
				ω = 0.032	1	8.53	
				ω' = 0.071	1	6.55	
				ω'' = 0.032	1	8.53	

t : η g$^+$: 1   g$^-$: τ g (for SSU, USS): σ

$$U_{UUS(l)} = \begin{bmatrix} \eta & 1 & \tau \\ \eta\omega' & 1 & \tau\omega \\ \eta & \omega & \tau\omega' \end{bmatrix} \quad U_{SUU(l)} = \begin{bmatrix} 1 & 1 & \omega' \\ 1 & \omega' & \omega \\ 1 & \omega & 1 \end{bmatrix} \quad U_{SSU(II)} = \begin{bmatrix} \eta & 1 & \tau\omega' \\ \eta & \omega' & \tau\omega \\ \eta & \omega & \tau \end{bmatrix}$$

$$U_{SSU(II)} = \begin{bmatrix} \eta^2\sigma & \sigma & \tau\eta \\ \eta^2\sigma\omega' & \sigma & \tau\eta\omega \\ \eta^2\sigma & \sigma\omega & \tau\eta\omega' \end{bmatrix} \quad U_{SSU(ld)} = \begin{bmatrix} \eta & \tau & \omega' \\ \eta & \tau\omega' & \omega \\ \eta & \tau\omega & 1 \end{bmatrix} \quad U_{USS(dl)} = \begin{bmatrix} \eta^2 & \tau\sigma & \tau\sigma \\ \eta^2\omega' & \tau\sigma & \tau\sigma\omega \\ \eta^2 & \tau\sigma\omega & \tau\sigma\omega' \end{bmatrix}$$

rows and columns: t, g$^+$, g$^-$

a) Subscript U corresponds to a CH$_2$ group (unsubstituted), subscript S to a CHCl group (substituted).
b) The matrices $U_{USU(l)}$, $U_{SUS(dd)}$, and $U_{SUS(dl)}$ for the H-T structure are the same as those described in (**A 003**).

Comments:
RIS theory is used to calculate the unperturbed dimensions for chains of PVDF and PVC for the entire range of fraction of head-to-head (H-H) monomer placements. The Monte-Carlo method is used to generate representative chain sequences with specified fractions of head-to-head and head-to-tail (H-T) components and, in the case of PVC, with the desired stereochemical structures of the sequences. The characteristic ratios of the PVDF chains are quite insensitive to H-H placements, presumably because of the relatively small difference between the sizes of H atoms and F atoms. In the case of the PVC, the syndiotactic form show the greatest sensitivity to H-H placements, with the limiting cases of the arrangements that are all H-H and all H-T giving larger values of the characteristic ratio than the more random arrangements.

Calcd. quantities: $<r^2>_0 / nl^2$ (depending on the probability of *isotactic* placement and H-T placement).

A 009

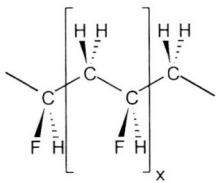

poly(vinyl fluoride), PVF

Tonelli, A. E. *Macromolecules* **1980**, *13*, 734.

Bond Length [pm]	Valence Angles [°]	Torsion Angles [°]			
C-C : 153	C-C-C : 112	t : 180			
C-F : 136	H-C-F : 106	g^+ : 60			
C-H : 110	H-C-H : 109.5	g^- : -60			

(see: a))

$$U_{FHF} (dd, HT) = \begin{bmatrix} 0.151 & 0.115 & 0.188 \\ 0.188 & 0.074 & 0.077 \\ 0.118 & 0.016 & 0.073 \end{bmatrix}$$

$$U_{FHF} (dl, HT) = \begin{bmatrix} 0.301 & 0.122 & 0.096 \\ 0.125 & 0.114 & 0.053 \\ 0.105 & 0.046 & 0.038 \end{bmatrix}$$

$$U_{HFH} (d, HT) = \begin{bmatrix} 0.276 & 0.148 & 0.165 \\ 0.162 & 0.055 & 0.014 \\ 0.122 & 0.008 & 0.050 \end{bmatrix}$$

$$U_{FFH} (dl, HH:TT) = \begin{bmatrix} 0.184 & 0.118 & 0.100 \\ 0.144 & 0.125 & 0.023 \\ 0.188 & 0.061 & 0.057 \end{bmatrix}$$

$$U_{FFH} (dd, HH:TT) = \begin{bmatrix} 0.245 & 0.074 & 0.217 \\ 0.115 & 0.065 & 0.048 \\ 0.152 & 0.019 & 0.065 \end{bmatrix}$$

$$U_{FHH} (d, HH:TT) = \begin{bmatrix} 0.245 & 0.118 & 0.143 \\ 0.140 & 0.081 & 0.033 \\ 0.117 & 0.033 & 0.090 \end{bmatrix}$$

rows and columns: t, g^+, g^-

a) Values given for 323 K; values for 273, 373, 425, and 473 K are given in the original paper as well.

Comments:
Conformational energy estimates are employed to determine the conformational characteristics of poly(vinyl fluoride) (PVF), polyfluoromethylene (PFM), and polytrifluoroethylene (PTF$_3$). Effects of stereoconfiguration and, in the case of PVF and PTF$_3$, the presence of head-to-head:tail-to-tail (HH:TT) defect structures are considered. The calculated results are compared to corresponding values found for poly(vinylidene fluoride), polytetrafluoroethylene, and polyethylene, and the equilibrium flexibilities of PVF, PFM, and PTF$_3$ are discussed on this basis.

Calcd. quantities: $<r^2>_0 / nl^2$ $d (\ln <r^2>_0) / dT$ $<\mu^2> / xm^2$ $d (\ln <\mu^2>) / dT$ Conformational entropy S_c

A 010

poly(vinyl fluoride), PVF

Carballeira, L.; Pereiras, A. J.; Rios, M. A. *Macromolecules* **1989**, *22*, 2668.

Bond Length [pm]	Valence Angles [a)] [°]	Torsion Angles [°]	ξ [b,c] (for 323 K)	ξ_o	E_ξ [kJ·mol^{-1}]
C-C : 153	CH$_2$-C-CH$_2$: 113	t : 180	τ = 1.26	1	−0.61
	C-CH$_2$-C : 114	g$^+$: 60	η = 2.60	1	−2.57
		g$^-$: −60	ω_p = 0.81	1	0.56
			ω = 0.068	1	7.22
			ω' = 0.59	1	1.42
			ω'' = 0.088	1	6.54

$$U_I = \begin{bmatrix} \eta & \tau & 1 \\ \eta & \tau & \omega \\ \eta & \tau\omega & 1 \end{bmatrix} \quad U_{II} = \begin{bmatrix} \eta\omega'' & 1 & \tau\omega' \\ \eta\omega' & \omega' & 0 \\ \eta & \omega\omega_p & \tau\omega' \end{bmatrix} \quad U_{Id} = \begin{bmatrix} \eta & \tau\omega'' & \omega' \\ \eta\omega'' & 0 & \omega \\ \eta\omega' & \tau\omega & \omega_p \end{bmatrix}$$

rows and columns: t, g$^+$, g$^-$

a) For both angles: ± 2°.
b) τ = sk in the original paper.
c) Calculated for a replication probability of 0.46.

Comments:
The conformational properties of molecules modeling PVB and PVF dimers and trimers are analyzed using a molecular mechanics method which takes into account both steric and electronic energies. The geometric and energy results show that the steric effects of changing the halide are greater than the electrostatic effects. Total conformational energies are used to determine a set of intramolecular interaction energies that by means of the RIS model allow estimation of the characteristic ratio and dipole moment ratio of PVB and PVF. The results agree satisfactorily with available experimental values.

Calcd. quantities: $\langle r^2 \rangle_o / nl^2$ = 5.0 $\quad$ $\langle \mu^2 \rangle / xm^2$ = 0.49

A 011

[Structure diagram of polytrifluoroethylene repeat unit]

polytrifluoroethylene, PTF$_3$

Tonelli, A. E. *Macromolecules* **1980**, *13*, 734.

Bond Length [pm]	Valence Angles [°]	Torsion Angles [°]	
C-C : 153	C-C-C : 116	t : 180	(see: a))
C-F : 136	H-C-F : 106	g$^+$: 60	
C-H : 110	F-C-F : 106	g$^-$: −60	

$$U_{HFH}(dd,HT) = \begin{bmatrix} 0.333 & 0.253 & 0.064 \\ 0.064 & 0.015 & 0.000 \\ 0.253 & 0.003 & 0.015 \end{bmatrix}$$

$$U_{HFH}(dl,HT) = \begin{bmatrix} 0.539 & 0.099 & 0.082 \\ 0.099 & 0.002 & 0.002 \\ 0.082 & 0.001 & 0.094 \end{bmatrix}$$

$$U_{FHF}(d,HT) = \begin{bmatrix} 0.151 & 0.147 & 0.205 \\ 0.201 & 0.076 & 0.030 \\ 0.083 & 0.005 & 0.102 \end{bmatrix}$$

$$U_{HHF}(dl,HH{:}TT) = \begin{bmatrix} 0.246 & 0.274 & 0.027 \\ 0.107 & 0.015 & 0.000 \\ 0.275 & 0.007 & 0.050 \end{bmatrix}$$

$$U_{HHF}(dd,HH{:}TT) = \begin{bmatrix} 0.357 & 0.117 & 0.126 \\ 0.225 & 0.010 & 0.004 \\ 0.097 & 0.000 & 0.065 \end{bmatrix}$$

$$U_{HHF}(d,HH{:}TT) = \begin{bmatrix} 0.382 & 0.164 & 0.080 \\ 0.132 & 0.014 & 0.003 \\ 0.175 & 0.005 & 0.046 \end{bmatrix}$$

rows and columns: t, g$^+$, g$^-$

a) Values given for 323 K; values for 273, 373, 425, and 473 K are given in the original paper as well.

Comments:
Conformational energy estimates are employed to determine the conformational characteristics of poly(vinyl fluoride) (PVF), polyfluoromethylene (PFM), and polytrifluoroethylene (PTF$_3$). Effects of stereoconfiguration and, in the case of PVF and PTF$_3$, the presence of head-to-head:tail-to-tail (HH:TT) defect structures are considered. The calculated results are compared to corresponding values found for poly(vinylidene fluoride), polytetrafluoroethylene, and polyethylene, and the equilibrium flexibilities of PVF, PFM, and PTF$_3$ are discussed on this basis.

Calcd. quantities: $<r^2>_0 / nl^2$ $d(\ln <r^2>_0)/dT$ $<\mu^2>/xm^2$ $d(\ln <\mu^2>)/dT$ Conformational entropy S_c

A 012

poly(vinyl bromide), PVB

Saiz, E.; Riande, E.; Delgado, M. P.; Barrales-Rienda, J. M. *Macromolecules* **1982**, *15*, 1152.

Bond Length [pm]	Valence Angles [°]	Torsion Angles [°]	ξ (for 303 K)	ξ_0	E_ξ [kJ·mol^{-1}]
C-C : 153	CH$_2$-C-CH$_2$: 112	t : 180	η = 1.3	1	−0.63
C-H : 110	C-CH$_2$-C : 114	g : 60	τ = 0.43	1	2.1
C-Br : 195	others : 109.5	$\bar{g}$: −60	ω = 0.015	1	10.5
			ω' = 0.036	1	8.4
			ω'' = 0.0025	1	15.1

$$U_l = \begin{bmatrix} 1 & 1 & 1 \\ 1 & 1 & \omega \\ 1 & \omega & 1 \end{bmatrix} \quad U_m = \begin{bmatrix} \eta^2\omega'' & \eta\tau\omega' & \eta \\ \eta & \tau\omega' & \omega \\ \eta\tau\omega' & \tau^2\omega\omega'' & \tau\omega' \end{bmatrix} \quad U_r = \begin{bmatrix} \eta^2 & \eta\omega' & \eta\tau\omega'' \\ \eta\omega' & 1 & \tau\omega \\ \eta\tau\omega'' & \tau\omega & \tau^2\omega'^2 \end{bmatrix}$$

rows and columns: t, g, $\bar{g}$

Comments:
Samples of PVB are prepared, having an isotactic content of 46 %. Dielectric measurements are performed in solutions of both samples in dioxane and 1-methylnaphthalene at several temperatures, and dipole ratios $D_x = \langle\mu^2\rangle/xm^2$ are found to be 0.52 and 0.45, respectively. No noticeable dependence of D_x with molecular weight is found; the variation of D_x with temperature is too small to allow an accurate determination of its temperature coefficient. An RIS model is derived and used to calculate dipole and characteristic ($C_n = \langle r^2\rangle_0/nl^2$) ratios.

Calcd. quantities: $\langle r^2 \rangle_0 / nl^2$ $\quad$ d (ln $\langle r^2 \rangle_0$) / dT $\quad$ $\langle\mu^2\rangle / xm^2$ $\quad$ d (ln $\langle\mu^2\rangle$) / dT

poly(vinyl bromide), PVB

Carballeira, L.; Pereiras, A. J.; Rios, M. A. *Macromolecules* **1989**, *22*, 2668.

Bond Length [pm]	Valence Angles [a] [°]	Torsion Angles [°]	ξ [b,c] (for 303 K)	ξ_0	E_ξ [kJ·mol^{-1}]
C-C : 153	CH$_2$-C-CH$_2$: 113	t : 180	τ = 1.50	1	-1.02
	C-CH$_2$-C : 116	g$^+$: 60	η = 1.66	1	-1.27
		g$^-$: -60	ω_p = 0.73	1	0.78
			ω = 0.013	1	10.96
			ω' = 0.039	1	8.17
			ω'' = 0.004	1	13.84

$$U_I = \begin{bmatrix} \eta & \tau & 1 \\ \eta & \tau & \omega \\ \eta & \tau\omega & 1 \end{bmatrix} \quad U_{II} = \begin{bmatrix} \eta\omega'' & 1 & \tau\omega' \\ \eta\omega' & \omega' & 0 \\ \eta & \omega\omega_p & \tau\omega' \end{bmatrix} \quad U_{Id} = \begin{bmatrix} \eta & \tau\omega'' & \omega' \\ \eta\omega'' & 0 & \omega \\ \eta\omega' & \tau\omega & \omega_p \end{bmatrix}$$

rows and columns: t, g$^+$, g$^-$

[a] For both angles: ± 2°.
[b] τ = sk in the original paper.
[c] Calculated for a replication probability of 0.46.

Comments:
The conformational properties of molecules modeling PVB and PVF dimers and trimers are analyzed using a molecular mechanics method which takes into account both steric and electronic energies. The geometric and energy results show that the steric effects of changing the halide are greater than the electrostatic effects. Total conformational energies are used to determine a set of intramolecular interaction energies that by means of the RIS model allow estimation of the characteristic ratio and dipole moment ratio of PVB and PVF. The results agree satisfactorily with available experimental values.

Calcd. quantities: $<r^2>_0 / nl^2$ = 6.0 $<\mu^2> / xm^2$ = 0.30

*Further calculations on **poly(vinyl halogenide)** chains:*

A 014   Carlson, C. W.; Flory, P. J. *J. Chem. Soc.; Farad. Trans. II* **1977**, 1505.

A procedure is demonstrated for separating the intrinsic molecular component of the depolarized Rayleigh scattering from the transient, collision-induced contribution. Conformational averages $<\gamma^2>$ for $ClCH_2-CH_2Cl$ and for *meso* and *racemic* $CH_3-CHCl-CH_2-CHCl-CH_3$ calculated by RIS analysis using $\Gamma_{CCl} = 1.5$ Å are in good agreement with values measured in CCl_4.

A 015   Tonelli, A. E.; Khanarian, G.; Cais, R. E. *Macromolecules* **1985**, *18*, 2324.

Two different RIS models are used to calculate dipole moments and molar Kerr constants for PVB oligomers. The obtained results are compared with experimental results.

A 016   Mattice, W. L.; Lloyd, A. C. *Macromolecules* **1986**, *19*, 2250.

Dipole moments are calculated of perturbed poly(vinyl chloride), poly(vinyl bromide) and poly(*p*-chlorostyrene). Recent work (*Macromolecules* **1984**, *17*, 625; **1985**, *18*, 2236) has shown that the mean square dipole moment, $<\mu^2>$, of model chains with a finite number of bonds, n, may depend on excluded volume even if $<\mathbf{r} \cdot \boldsymbol{\mu}>_o = 0$.
The present work demonstrates that (1) conclusions reached earlier for model chains also apply to realistic chains, (2) effects seen with finite chains may survive in extremely long chains, (3) the limit at large n for $(\alpha_\mu^2 - 1)/(\alpha_r^2 - 1)$ provides little information about the effect of excluded volume on the dipole moment of infinitely long chains, and (4) an alternative relationship between α_μ^2 and α_r^2 may provide useful information on the relationship of $<\mu^2>$ and $<\mu^2>_o$ for long chains.

A 017   Mark, J. E.; DeBolt, L. C.; Curro, J. G. *Macromolecules* **1986**, *19*, 491.

Effects of stereochemical structure on distribution functions are determined for short polypropylene and poly(vinyl chloride) chains.

A 018   Mattice, W. L.; Saiz, E. *J. Polym. Sci.: Polym. Phys.* **1986**, *24*, 2669.

The influence of the chain expansion produced by excluded volume on the mean-square optical anisotropy is studied in six types of polymers [PE, PVC, PVB, PS, poly(*p*-chlorostyrene), poly(*p*-bromostyrene]. RIS models are used for the configuration statistics of the unperturbed chains. The mean-square optical anisotropy of PE is found to be insensitive to excluded volume. The mean-square optical anisotropy of the five other polymers, on the other hand, is sensitive to the imposition of the excluded volume if the stereochemical composition is exclusively *racemic*. Much smaller effects are seen in *meso* chains and in chains with Bernoullian statistics and an equal probability for *meso* and *racemic* diads.

A 019   Mattice, W. L. *Macromolecules* **1988**, *21*, 3320.

Generator matrix methods are used to compute $<r^2\mu^2>_o$, $<r^2>_o$, and $<\mu^2>_o$ for PVB, PVC, and PS chains as function of the stereochemical composition. Simulations that permit introduction of excluded volume show that $<\mu^2>$ for all three chains is insensitive to $<r^2>/<r^2>_o$ unless the stereochemical composition is predominantly *racemic*. The response of $<\mu^2>$ to chain expansion is more dramatic in *racemic* PVC than in the other two polymers.

A 020

polypropylene, PP

Abe, Y.; Tonelli, A. E.; Flory, P. J. *Macromolecules* **1970**, *3*, 294, 303.

Bond Length [pm]	Valence Angles [a) [°]	Torsion Angles [a) [°]	ξ (for 481 K)	ξ_0	E_ξ [kJ·mol^{-1}]
C-C : 153	C-CH$_2$-C : 112	t : 180 – Δφ	η = 1.0	1	0
	CH$_2$-C-CH$_2$: 112	g : 60 + Δφ	τ = 0.5	1	2.8
	C-C*-H : 106.8	$\bar{g}$: – 60	ω = 0 to 0.05	1	∞ to 12
	H-C-H : 109	τ* = 1.0	1	0	

$$U_p = \begin{bmatrix} \eta\tau^* & 1 & \tau \\ \eta & \omega & \tau \\ \eta & 1 & \tau\omega \end{bmatrix} \quad U_m = \begin{bmatrix} \eta\omega & 1 & \tau\omega \\ \eta & \omega & \tau\omega \\ \eta\omega & \omega & \tau\omega^2 \end{bmatrix} \quad U_r = \begin{bmatrix} \eta & \omega & \tau\omega \\ \eta\omega & 1 & \tau\omega \\ \eta\omega & \omega & \tau\omega^2 \end{bmatrix}$$

rows and columns: t, g, $\bar{g}$

a) Δφ = 0, 10 or 20°.

Comments:
The theory of strain birefringence is elaborated in terms of the RIS model as applied to vinyl polymer chains. Additivity of the polarizability tensors for constituent groups is assumed. Stress-birefringence coefficients are calculated for PP and for PS. Statistical weight parameters which affect the incidences of various rotational states are varied over ranges consistent with other evidence. The effects of these variations are explored in detail for *isotactic* and *syndiotactic* chains.

Calcd. quantities: Strain-birefringence coefficient, Γ_2 Depolarized anisotropy per unit, $\langle\gamma^2\rangle / x$

A 021

polypropylene, PP

Tonelli, A. E. *Macromolecules* **1972**, *5*, 563.

Bond Length [pm]	Valence Angles [°]	Torsion Angles [°]	ξ (for 481 K)	ξ_0	E_ξ [kJ · mol^{-1}]			
C-C : 153	C-CH$_2$-C : 112	t : 180	η = 1.0	1	0			
	CH$_2$-C-CH$_2$: 112	g : 60	τ = 0.59 to 0.073	1	2.1 to 10.5	$U_p = \begin{bmatrix} \eta & 1 & \tau \\ \eta & 1 & \tau\omega \\ \eta & \omega & \tau \end{bmatrix}$	$U_m = \begin{bmatrix} \eta\omega & \tau\omega & 1 \\ \eta & \tau\omega & \omega \\ \eta\omega & 0 & \omega \end{bmatrix}$	rows and columns: t, g, $\bar{g}$
		$\bar{g}$: −60	ω = 0.12 to 0.043	1	8.4 to 12.6			

Comments:
The increase in the conformational or intramolecular entropy accompanying the fusion process is calculated for *isotactic* PP in the RIS approximation. When the calculated change in the intramolecular entropy is compared to the experimental entropy of fusion, after correcting for the entropy resulting from the volume expansion of melting, good agreement is found, providing the RIS parameters τ and ω are both small.

Calcd. quantities: Conformational contribution to the entropy and energy of fusion of *isotactic* PP.

A 022

polypropylene, PP

Boyd, R. H.; Breitling, S. M. *Macromolecules* **1972**, *5*, 279.

Bond Length [pm]	Valence Angles [°]	Torsion Angles [°]		ξ a) (for 418 K)	ξ_0	E_ξ [kJ · mol^{-1}]
C-C : 153	C-C-C : 112	t_-	: 135	τ = 0.49	1	2.5
		t	: 180	σ = 0.61	1	1.7
		t_+	: 225	ω = 0.21	1	5.4
		g^+	: 60			
		g^+_+	: 100			
		g^-_-	: -100			
		g^-	: -60			

rows and columns: $t_-, t, t_+, g^+, g^+_+, g^-_-, g^-$
or: t, g^+, g^-

(see: b))

$$U_I = \begin{bmatrix} \tau\omega & \tau\omega & \tau\omega & \tau\omega & \tau\omega & \tau\omega & \tau\omega \\ \tau & \tau & \tau & \tau & \tau & \tau & \tau \\ \tau\omega & \tau\omega & \tau\omega & \tau\omega & \tau\omega & \tau\omega & \tau\omega \\ \sigma & \sigma & \sigma & \sigma & \sigma & 0 & 0 \\ \sigma\omega & \sigma\omega & \sigma\omega & \sigma\omega & \sigma\omega & 0 & 0 \\ \tau\sigma\omega & \tau\sigma\omega & \tau\sigma\omega & 0 & 0 & \tau\sigma\omega & \tau\sigma\omega \\ \tau\sigma & \tau\sigma & \tau\sigma & 0 & 0 & \tau\sigma & \tau\sigma \end{bmatrix} \text{ or } \begin{bmatrix} \tau & \tau & \tau \\ \sigma & \sigma & 0 \\ \tau\sigma & 0 & \tau\sigma \end{bmatrix}$$

$$U_{II} = \begin{bmatrix} 0 & 0 & 0 & 0 & 0 & 0 & 0 \\ \tau & 0 & 0 & \tau & 0 & 0 & 0 \\ 0 & \tau\omega & 0 & 0 & 0 & 0 & \tau\omega \\ \sigma\tau & 0 & 0 & 0 & \sigma\tau & 0 & 0 \\ 0 & 0 & 0 & 0 & 0 & 0 & 0 \\ 0 & 0 & 0 & \sigma\omega & 0 & 0 & \sigma\omega \\ 0 & \sigma & 0 & 0 & \sigma & 0 & 0 \end{bmatrix} \text{ or } \begin{bmatrix} \tau\omega & \tau & \tau\omega \\ \sigma\tau\omega & \sigma\tau\omega & 0 \\ \sigma & \sigma\omega & \sigma\omega \end{bmatrix}$$

$$U_{dl} = \begin{bmatrix} 0 & 0 & 0 & 0 & 0 & 0 & 0 \\ 0 & \tau & 0 & 0 & 0 & \tau & 0 \\ 0 & 0 & 0 & \tau\omega & 0 & 0 & \tau\omega \\ 0 & 0 & \sigma\tau & 0 & 0 & \sigma\tau & 0 \\ 0 & 0 & 0 & 0 & 0 & 0 & 0 \\ 0 & \sigma\omega & 0 & \sigma\omega & 0 & 0 & 0 \\ 0 & 0 & \sigma & 0 & 0 & 0 & 0 \end{bmatrix}$$

a) τ = SK, σ = g in the original paper.
b) Dropping the distinction between distorted and undistorted states but not regarding any state as degenerate, the 3 × 3 matrices are obtained.

Comments:
Conformational energy calculations are carried out on 2,4,6-trimethylheptane as a model for PP using energy minimization with all internal degrees of freedom allowed to participate.

Calcd. quantities: $<r^2>_0 / nl^2$ $d (\ln <r^2>_0) / dT$ [for values of p_r = 1 (perfectly *isotactic*) to p_r = 0 (perfectly *syndiotactic*)]

A 023

polypropylene, PP

Biskup, U.; Cantow, H.-J. *Macromolecules* **1972**, *5*, 546.

Bond Length [pm]	Valence Angles [°]	Torsion Angles [a)] [°]	ξ (for 403 K)	ξ_o	E_ξ [kJ·mol^{-1}]
C-C : 153	C-C-C : 112	t : 180 – $\Delta\phi$	$\eta = 1$	1	0
		g : 60 + $\Delta\phi$	$\tau = 0.6$	1	1.7
		$\bar{g}$: –60	$\omega = 0.1356$	1	6.7
			$\omega' = 0.0565$	1	9.6
			$\omega'' = 0.0565$	1	9.6

$$U_p = \begin{bmatrix} \eta & 1 & \tau \\ \eta & \omega & \tau \\ \eta & 1 & \tau\omega \end{bmatrix} \quad U_i = \begin{bmatrix} \eta\omega'' & 1 & \tau\omega' \\ \eta & \omega & \tau\omega' \\ \eta\omega & \omega' & \tau\omega\omega'' \end{bmatrix} \quad U_s = \begin{bmatrix} \eta & \omega' & \tau\omega'' \\ \eta\omega' & 1 & \tau\omega \\ \eta\omega'' & \omega & \tau\omega'^2 \end{bmatrix}$$

rows and columns: t, g, $\bar{g}$

a) $\Delta\phi = 0$ or $5°$.

Comments:
The model parameters are varied systematically within reasonable limits to fit the experimental results. The minimum positions of the rotational bond angles probably do not deviate more than 5° from planar *trans* and from symetrically staggered *gauche*, respectively. Entropy contributions to the free energies of the rotational isomers are discussed with respect to the influence on the temperature coefficient.

Calcd. quantities: $<r^2>_o / nl^2$ $d(\ln <r^2>_o)/dT$ (evaluated as a function of tacticity)

A 024

polypropylene, PP

Heatley, F. *Polymer* **1972**, *13*, 218.

Conformers (with ω interactions shown): t:1, t':1, g':1, g:1, $\bar{g}:\tau$

Valence Angles [°]	Torsion Angles [a)] [°]		ξ (for 403 K)	ξ_0	E_ξ [kJ·mol^{-1}]
C-C-C : 111	t :	$180 - \Delta\phi$	$\tau = 0.1$ to 0.5	1	5.2 ($\pm$ 2.0)
H-C-H : 109	t' :	140	$\omega = 0.1$	1	8.2 ($\pm$ 0.4)
	g :	$60 + \Delta\phi$			
	g' :	100			
Bond length [pm]	$\bar{g}$:	-60			
C-C : 150					
C-H : 109					

$$U_{mm} = \begin{bmatrix}
0 & 0 & 0 & 0 & 1 & \omega & \omega & \tau\omega & \tau\omega & \tau\omega & 0 & 0 & 1 \\
0 & 0 & \omega & 0 & 1 & \omega & \omega & \tau\omega & \tau\omega & \tau\omega & 0 & 0 & 1 \\
\omega & \omega & 1 & \tau\omega & 1 & \omega & \omega & \tau\omega & \tau\omega & \tau\omega & 0 & 0 & 1 \\
\omega & \omega & 1 & \tau\omega & 1 & \omega & \omega & \tau\omega & 0 & 0 & 0 & 0 & 1 \\
\omega^2 & 0 & 0 & \tau\omega^2 & 1 & \omega & \omega & \tau\omega & \tau\omega & \tau\omega & 0 & \omega & 1 \\
\omega & \omega & 1 & \tau\omega & 1 & \omega & \omega & \tau\omega & \tau\omega & \tau\omega & 0 & 0 & 1 \\
\omega & \omega & 1 & \tau\omega & 1 & \omega & \omega & \tau\omega & \tau\omega & \tau\omega & 0 & 0 & 1 \\
\omega & \omega & 1 & \tau\omega & 1 & \omega & \omega & \tau\omega & 0 & 0 & 0 & 0 & 1 \\
0 & 0 & \omega & 0 & 1 & \omega & \omega & \tau\omega & \tau\omega & \tau\omega & 0 & 0 & 1 \\
\omega & \omega & 1 & \tau\omega & 1 & \omega & \omega & \tau\omega & \tau\omega & \tau\omega & 0 & 0 & 1 \\
\omega & \omega & 1 & \tau\omega & 1 & \omega & \omega & \tau\omega & 0 & 0 & 0 & 0 & 1 \\
\omega & \omega & 1 & \tau\omega & 1 & \omega & \omega & \tau\omega & \tau\omega & \tau\omega & 0 & 0 & 1 \\
0 & 0 & \omega & 0 & 0 & 0 & 0 & 0 & 0 & 0 & 0 & 0 & 0
\end{bmatrix}$$

$$U_{rr} = \begin{bmatrix}
\omega & \omega & 1 & \tau\omega & 0 & 0 & 0 & 0 & \tau\omega & \tau\omega & 0 & 0 & 0 & 0 \\
\omega & \omega & 1 & \tau\omega & \omega & 0 & 0 & 0 & \tau\omega & \tau\omega & 0 & 0 & \omega & 0 \\
\omega & \omega & 1 & \tau\omega & 1 & \omega & \omega & \tau\omega & \tau\omega & \tau\omega & 0 & 0 & 1 & 0 \\
\omega & \omega & 1 & \tau\omega & 1 & \omega & \omega & \tau\omega & 0 & 0 & 0 & 0 & 1 & 0 \\
\omega & \omega & 1 & \tau\omega & 0 & \omega^2 & 0 & \tau\omega^2 & \tau\omega & \tau\omega & 0 & \omega & 0 & \omega \\
\omega & \omega & 1 & \tau\omega & 1 & \omega & \omega & \tau\omega & \tau\omega & \tau\omega & 0 & 0 & 1 & 0 \\
\omega & \omega & 1 & \tau\omega & 1 & \omega & \omega & \tau\omega & \tau\omega & \tau\omega & 0 & 0 & 1 & 0 \\
\omega & \omega & 1 & \tau\omega & 1 & \omega & \omega & \tau\omega & 0 & 0 & 0 & 0 & 1 & 0 \\
\omega & \omega & 1 & \tau\omega & \omega & 0 & 0 & 0 & \tau\omega & \tau\omega & 0 & 0 & \omega & 0 \\
\omega & \omega & 1 & \tau\omega & 1 & \omega & \omega & \tau\omega & \tau\omega & \tau\omega & 0 & 0 & 1 & 0 \\
\omega & \omega & 1 & \tau\omega & 1 & \omega & \omega & \tau\omega & 0 & 0 & 0 & 0 & 1 & 0 \\
\omega & \omega & 1 & \tau\omega & 0 & \omega^2 & 0 & \tau\omega^2 & \tau\omega & \tau\omega & 0 & 0 & 0 & \omega \\
0 & 0 & 0 & 0 & \omega & 0 & 0 & 0 & 0 & 0 & 0 & 0 & \omega & 0 \\
0 & 0 & 0 & 0 & \omega & 0 & 0 & 0 & 0 & 0 & 0 & 0 & \omega & 0
\end{bmatrix}$$

For U_{mm}: rows and columns: g'g, gg', gt, g'$\bar{g}$, tg, t't, tt', t'$\bar{g}$, $\bar{g}g'$, gt', $\bar{g}\bar{g}$, g't, tg'

For U_{rr}: rows and columns: t'g, tg', tt, t'$\bar{g}$, gg, g't, gt', g'$\bar{g}$, $\bar{g}g'$, $\bar{g}t'$, $\bar{g}\bar{g}$, g'g, gg', g'g'

(see: [b])

[a)] $\Delta\phi = 6 \ (\pm \ 2)°$

[b)] Statistical weight matrices for bond pairs; the 14 × 13 matrix U_{rm} for an *isotactic* diad preceeded by a *syndiotactic* is the same as U_{mm} except that a row 14 identical to row 13 is added, row 12 is replaced by a row identical to row 5, and the rows are labelled as in U_{rr}. The 13 × 14 matrix U_{mr} is the same as U_{rr} except that row 14 is deleted, row 12 is replaced by a row identical to row 3 and the rows are labelled as in U_{mm}.

Calcd. quantities: $<r^2>_0 / nl^2$ $d \ (\ln <r^2>_0) / dT$

A 025

polypropylene, PP

Flory, P. J. J. Polym. Sci.; Polym. Phys. Ed. **1973**, *11*, 621.

Bond Length [pm]	Valence Angles [°]	Torsion Angles [°]		ξ (for 300 K)
		t : 180		σ^*
		t* : 140		ω^*
		g* : 100		
		g : 60		

$$U_p = \begin{bmatrix} 1 & \sigma^* & \sigma^* & 1 \\ 1 & \sigma^* & \sigma^* & 1 \\ 1 & \sigma^* & 0 & \omega^* \\ 1 & \sigma^* & \sigma^*\omega^* & 0 \end{bmatrix}$$

$$U_m = \begin{bmatrix} 0 & \sigma^*\omega^* & \sigma^* & 1 \\ \omega^* & 0 & \sigma^* & 1 \\ 1 & \sigma^* & 0 & \omega^* \\ 1 & \sigma^* & \sigma^*\omega^* & 0 \end{bmatrix}$$

rows and columns: t, t*, g*, g

Comments:
The role of nonstaggered conformations in PE and PP is discussed in some detail. In incorporating such conformations into RIS treatments, it is essential to so choose rotational states as to assure equitable sampling of configuration space. Tacit identification of rotational states with minima in the conformation energy surface, a common practice, may lead to serious errors. The significance and limitations of conformational energy calculations are discussed.

A 026

**polypropylene, PP,
2,4,6,8-tetramethylnonane and
2,4,6,8,10-pentamethylundecane**

Suter, U. W.; Pucci, S.; Pino, P. *J. Am. Chem. Soc.* **1975**, *97*, 1018.

Bond Length [pm]	Valence Angles [°]	Torsion Angles [°]	ξ (for 413 K)	ξ_o	E_ξ [kJ·mol^{-1}]			
C-C : 153	C-CH$_2$-C : 112	t : 180	η = 0.92	1	0.29 (± 0.04)			
C-H : 110	CH$_2$-C-CH$_2$: 112	g$^+$: 60	ω = 0.10	1	8.0 (± 0.4)	$U_d = \begin{bmatrix} \eta & 1 & \tau \\ \eta & 1 & \tau\omega \\ \eta & \omega & \tau \end{bmatrix}$	$U_{dd} = \begin{bmatrix} \eta\omega & \tau\omega & 1 \\ \eta & \tau\omega & \omega \\ \eta\omega & \tau\omega^2 & \omega \end{bmatrix}$	$U_{dl} = \begin{bmatrix} \eta & \omega & \tau\omega \\ \eta\omega & 1 & \tau\omega \\ \eta\omega & \omega & \tau\omega^2 \end{bmatrix}$
		g$^-$: –60	τ = 0.33	1	3.8 (± 1.7)			

rows and columns: t, g$^+$, g$^-$

Comments:
The epimerization equilibria for the diastereomers of 2,4,6,8-tetramethylnonane and 2,4,6,8,10-pentamethylundecane are determined and the results are interpreted in terms of a RIS model. The results yield correct values for the optical activity and are consistent with conformational energies calculated from experimental values of the unperturbed dimensions of PP.

Calcd. quantities: Conformational equilibrium of diastereomers, optical activities and their temperature coefficients.

A 027

polypropylene, PP

Suter, U. W.; Flory, P. J. *Macromolecules* **1975**, *8*, 765.

Bond Length [pm]	Valence Angles [°]	Torsion Angles [°]	ξ a,b) (for 413 K)	ξ_o	E_ξ [kJ·mol^{-1}]
C-C : 153	C-CH$_2$-C : 112	t : 165	η = 0.90	1.0	0.25
C-H : 110	CH$_2$-C-CH$_2$: 112	t* : 130	τ = 0.22	0.4	2.1
	C-CH-H : 109	g* : 110	ω^* = 0.13	0.9	6.7
	C-CH$_2$-H : 110	g : 75			
		$\bar{g}$: -65			

$$U_p = \begin{bmatrix} 1 & 1 & 1 & 1 & 1 \\ 1 & 1 & 1 & 1 & 1 \\ 1 & 1 & 0 & 0 & 1 \\ 1 & 1 & 0 & 0 & 1 \\ 1 & 1 & 1 & 1 & 0 \end{bmatrix}$$

rows and columns: t, t*, g*, g, $\bar{g}$

$$U_m = \begin{bmatrix} 0 & \eta\omega^* & 0 & \eta & 0 \\ \eta\omega^* & 0 & 0 & 0 & \tau\omega^* \\ 0 & 0 & 0 & \omega^* & \tau\omega^* \\ \eta & 0 & \omega^* & 0 & 0 \\ 0 & \tau\omega^* & \tau\omega^* & 0 & 0 \end{bmatrix} \quad U_r = \begin{bmatrix} \eta^2 & 0 & \eta\omega^* & 0 & 0 \\ 0 & 0 & 0 & \omega^* & \tau\omega^* \\ \eta\omega^* & 0 & 0 & 0 & \tau\omega^* \\ 0 & \omega^* & 0 & 1 & 0 \\ 0 & \tau\omega^* & \tau\omega^* & 0 & 0 \end{bmatrix}$$

a) $\omega^* = \eta^* \omega$
b) Four other parameter sets are discussed as well.

Comments:
The intramolecular energy of the chain segment of PP is computed as a function of its conformation, interactions between every pair of atoms being included. Contributions from methyl group rotations and from the skeletal conformation are separable in good approximation. At each conformation, the two methyl groups of the segment are therefore assigned the rotations that minimize the energy. Ten accessible energy domains (minima) are clearly delineated for the *meso* diad, and ten for the *racemic* diad. Boltzmann averages over the rotation angles for the diad pair yield mean conformations for each domain. These are well represented by combinations of five states for each bond. In contrast to the predictions of the three state model used heretofore, C_∞ is predicted to be greatest for *syndiotactic* PP, and to decrease monotonically as the proportion of *meso* dyads increases, C_∞ for the *isotactic* chain actually being somewhat smaller than for the *atactic* chain.

Calcd. quantities: $<r^2>_o / nl^2$ $d (\ln <r^2>_o) / dT$

A 028

polypropylene, PP

HT-placement (dd)

TT-placement (dd)

HH-placement (dd)

(For HT) t : η   g : 1 $g^* : \tau$

(For TT) t : α   g : β $g^* : \alpha$

(For HH) t : 1   g : σ $g^* : \sigma$

Asakura, T.; Ando, I.; Nishioka, A. *Makromol. Chem.* **1976**, *177*, 1493.

Bond Length [pm]	Valence Angles [°]	Torsion Angles [°]		ξ (for 413 K)	ξ_0	E_ξ [kJ·mol^{-1}]
C-C : 153	C-C-C : 112	t	: 180	η = 1.0	1	0
		g^+	: 60	ω = 0.1	1	5.7
		g^-	: -60	τ = 0.5	1	1.7
				σ = 0.5	1	1.7
				α = 0.6	1	1.3
				β = 1.0	1	0

$$U^{HT}_{a(d)} = \begin{bmatrix} \eta & 1 & \tau \\ \eta & 1 & \tau\omega \\ \eta & \omega & \tau \end{bmatrix} \quad U^{HT}_{a(l)} = \begin{bmatrix} \eta & \tau & 1 \\ \eta & \tau & \omega \\ \eta & \tau\omega & 1 \end{bmatrix} \quad U^{HT}_{b(dd)} = \begin{bmatrix} \eta\omega & \tau\omega & 1 \\ \eta & \tau\omega & \omega \\ \eta\omega & \tau\omega^2 & \omega \end{bmatrix} \quad U^{HT}_{b(dl)} = \begin{bmatrix} \eta & \omega & \tau\omega \\ \eta\omega & 1 & \tau\omega \\ \eta\omega & \omega & \tau\omega^2 \end{bmatrix}$$

$$U^{HT}_{b(ld)} = \begin{bmatrix} \eta & \tau\omega & \omega \\ \eta\omega & \tau\omega^2 & \omega \\ \eta\omega & \tau\omega & 1 \end{bmatrix} \quad U^{HT}_{b(ll)} = \begin{bmatrix} \eta\omega & 1 & \tau\omega \\ \eta\omega & \omega & \tau\omega^2 \\ \eta & \omega & \tau\omega \end{bmatrix} \quad U^{TT}_{a(dd)} = \begin{bmatrix} \beta & \alpha & \alpha \\ \beta & \alpha\omega & \alpha\omega \\ \beta\omega & \alpha\omega & \alpha \end{bmatrix} \quad U^{TT}_{b(dd)} = \begin{bmatrix} \eta & \omega & \tau \\ \eta & 1 & \tau\omega \\ \eta & \omega & \tau\omega \end{bmatrix}$$

$$U^{TT}_{a(dl)} = \begin{bmatrix} \alpha & \beta & \alpha \\ \alpha\omega & \beta & \alpha\omega \\ \alpha & \beta\omega & \alpha\omega \end{bmatrix} \quad U^{TT}_{b(dl)} = \begin{bmatrix} \eta & \tau\omega & 1 \\ \eta & \tau & \omega \\ \eta & \tau\omega & \omega \end{bmatrix} \quad U^{TT}_{a(ld)} = \begin{bmatrix} \alpha & \alpha & \beta \\ \alpha & \alpha\omega & \beta\omega \\ \alpha\omega & \alpha\omega & \beta \end{bmatrix} \quad U^{TT}_{b(ld)} = \begin{bmatrix} \eta & 1 & \tau\omega \\ \eta & \omega & \tau\omega \\ \eta\omega & \omega & \tau \end{bmatrix} \quad U^{TT}_{a(ll)} = \begin{bmatrix} \beta & \alpha & \alpha \\ \beta\omega & \alpha & \alpha\omega \\ \beta & \alpha\omega & \alpha\omega \end{bmatrix} \quad U^{TT}_{b(ll)} = \begin{bmatrix} \eta & \tau & \omega \\ \eta & \tau\omega & \omega \\ \eta & \tau\omega & 1 \end{bmatrix} \quad U^{HH}_{a(dd)} = \begin{bmatrix} 1 & \sigma\omega & \sigma \\ 1 & \sigma & \sigma\omega \\ 1 & \sigma\omega & \sigma\omega \end{bmatrix} \quad U^{HH}_{b(dd)} = \begin{bmatrix} \eta & \tau & 1 \\ \eta & \tau\omega & \omega \\ \eta\omega & \tau\omega & 1 \end{bmatrix}$$

$$U^{HH}_{a(dl)} = \begin{bmatrix} 1 & \sigma\omega & \sigma \\ 1 & \sigma & \sigma\omega \\ 1 & \sigma\omega & \sigma\omega \end{bmatrix} \quad U^{HH}_{b(dl)} = \begin{bmatrix} \eta & 1 & \tau \\ \eta\omega & 1 & \tau\omega \\ \eta & \omega & \tau\omega \end{bmatrix} \quad U^{HH}_{a(ld)} = \begin{bmatrix} 1 & \sigma & \sigma\omega \\ 1 & \sigma\omega & \sigma\omega \\ 1 & \sigma\omega & \sigma \end{bmatrix} \quad U^{HH}_{b(ld)} = \begin{bmatrix} \eta & \tau & 1 \\ \eta & \tau\omega & \omega \\ \eta\omega & \tau\omega & 1 \end{bmatrix} \quad U^{HH}_{a(ll)} = \begin{bmatrix} 1 & \sigma & \sigma\omega \\ 1 & \sigma\omega & \sigma\omega \\ 1 & \sigma\omega & \sigma \end{bmatrix} \quad U^{HH}_{b(ll)} = \begin{bmatrix} \eta & 1 & \tau \\ \eta\omega & 1 & \tau\omega \\ \eta & \omega & \tau\omega \end{bmatrix}$$

rows and columns: t, g^+, g^-

Comments:
The characteristic ratio of PP is calculated taking into account the chemical inversions such as head-to-head (HH) and tail-to-tail (TT) units in the chain and compared with the values calculated previously as a function of the tacticity.

A 029

polypropylene, PP

Asakura, T.; Ando, I.; Nishioka, A. *Makromol. Chem.* **1976**, *177*, 523.

Bond Length [pm]	Valence Angles [°]	Torsion Angles [°]	ξ (for 413 K)	ξ_o	E_ξ [kJ·mol^{-1}]			
C-C : 154	C-C-C : 109.28	t : 180	η = 1.0	1	0	$U_d = \begin{bmatrix} \eta & 1 & \tau \\ \eta & \omega & \tau \\ \eta & 1 & \tau\omega \end{bmatrix}$	$U_{dd} = \begin{bmatrix} \eta\omega & 1 & \tau\omega \\ \eta & \omega & \tau\omega \\ \eta\omega & \omega & \tau\omega^2 \end{bmatrix}$	$U_{dl} = \begin{bmatrix} \eta & \omega & \tau\omega \\ \eta\omega & 1 & \tau\omega \\ \eta\omega & \omega & \tau\omega^2 \end{bmatrix}$
C-H : 110	C-C-H : 109.28	g^+ : 60	ω = 0 to 0.1	1	∞ to 7.9			
		g^- : -60	τ = 0.5	1	2.4			

rows and columns: t, g^+, g^-

Comments:
The proton NMR shifts of diad, triad, and tetrad protons in an isolated chain of PP are calculated, taking into account its configurations and conformations using *Pople's* approximation. The agreement between the observed and calculated results, however, is insufficient with respect to the large difference between *syn* and *anti* protons and the order of tetrad protons. Here, *a priori* probabilities of specified conformations necessary to the estimation of the chemical shifts are calculated using *Flory's* matrix method.

Calcd. quantities: Conformational probabilities P_{conf}.   Proton NMR shifts.

A 030

polypropylene, PP

Alfonso, G. C.; Yan, D.; Zhou, Z. *Polymer* **1993**, *34*, 2830.

Bond Length [pm]	Valence Angles [°]	Torsion Angles [°]	ξ (for 403 K)	ξ_0	E_ξ [kJ · mol^{-1}]
C-C : 153	C-C-C : 112	t : 175	η = 0.9	1	0.35
		g : 65	τ = 0.6	1	1.7
		$\bar{g}$: –60	ω = 0.0932	1	8.0
			ω' = 0.0565	1	9.6
			ω'' = 0.0565	1	9.6

$$U_p = \begin{bmatrix} \eta & 1 & \tau \\ \eta & \omega & \tau \\ \eta & 1 & \tau\omega \end{bmatrix} \quad U_i = \begin{bmatrix} \eta\omega'' & 1 & \tau\omega' \\ \eta & \omega & \tau\omega' \\ \eta\omega' & \omega' & \tau\omega\omega'' \end{bmatrix} \quad U_s = \begin{bmatrix} \eta & \omega' & \tau\omega'' \\ \eta\omega' & 1 & \tau\omega \\ \eta\omega'' & \omega & \tau\omega'^2 \end{bmatrix}$$

rows and columns: t, g, $\bar{g}$

Comments:
The configurational-conformational characteristics of PP are discussed by considering every polymer chain as constituted by the periodic repetition of a sequence of monomeric units in a given configuration. Calculations are presented for the special case in which *meso* and *racemic* diads are distributed according to Bernoullian statistics. Numerical results show that the characteristic ratio of atactic PP reaches an asymptotic value of 5.34 when the size of the periodic sequence corresponds to six monomeric units. The temperature coefficient is calculated to be – 1.34 × 10^{-3} K^{-1}, in good agreement with experimental data.

Calcd. quantities: $<r^2>_0 / nl^2$ $\quad$ $d (\ln <r^2>_0) / dT$ $\quad$ $<r_g^2> / nl^2$

*Further calculations on **polypropylene** chains:*

A 031 Flory, P. J., Fujiwara, Y. *Macromolecules* **1969**, *2*, 327.

The environment of a methylenic proton in a vinyl polymer is affected in major degree by the occurrence of the conformation gt for the pair of main-chain bonds preceding the methylene group, and similarly by the tg conformation of the following bond pair. These conformations are suggested as being of overriding importance in determining the NMR chemical shift of the methylenic protons in PP in which the substituent CH_3 resembles CH_2. Probabilities of these conformations are calculated for the several kinds of tetrads situated either in stereoregular chains or in atactic chains generated for the various diad compositions.

A 032 Heatley, F.; Salovey, R.; Bovey, F. A. *Macromolecules* **1969**, *2*, 619.

The proton NMR spectrum of a highly isotactic sample of PP is examined. The polymer is shown to contain 2% *racemic* diads occurring randomly at junctions of isotactic sequences of opposite configurations. The mean-square end-to-end distance of this polymer is measured under ϑ conditions. Comparison of the value obtained with theoretical predicitons of *Flory, Mark*, and *Abe* [*J. Am. Chem. Soc.* **1966**, *88*, 639] permit an approximate measurement of the strength of the steric interactions within the PP chain.

A 033 Flory, P. J. *Macromolecules* **1970**, *3*, 613.

The chemical shifts observed in the proton NMR spectra of stereoirregular PP compare favorably with deductions from theoretical calculations. Compelling arguments are presented against assignment of the unresolved resonance for the *mrm* tetrad of the peak identified with the *rrr* tetrad. The intensities for various tetrads in the spectra of the stereoirregular polymers reveal a pronounced tendency favoring perturbation of sequences of diads of the same kind (*m* or *r*).

A 034 Allegra, G.; Calligaris, M.; Randaccio, L.; Moraglio, G. *Macromolecules* **1973**, *6*, 397.

An algorithm for the statistical mechanical treatment of -A-B-A-B- chains with neighbor interactions is applied to *isotactic* PP. It allows accounting for all possible skeletal rotations (ASR scheme), in contrast with the usual method based on the RIS approximation. The unperturbed mean-square end-to-end distance as well as its temperature coefficient are calculated using two distinct sets of non-bonded interaction parameters. Within the ASR scheme the results for either set may be put in good agreement with the experimental data by adopting a reasonable value of the van der Waals radius of the methyl group. Assumption of steric defects is therefore unnecessary. The RIS scheme is not easily applicable using the same energy maps because of the strong correlation between neighboring rotation angles, which makes the choice of the rotational isomers subject to some uncertainty. The chain conformational energy at various temperatures is also calculated: the intramolecular contribution, which is about 20% of the heat of melting, has approximately the same value whichever method is applied.

A 035 Provasoli, A.; Ferro, D. R. *Macromolecules* **1977**, *10*, 874.

It is investigated whether the stereochemical ^{13}C NMR chemical shifts in the resonance peaks can be ascribed to differences in the conformations in the various stereoisomers. The authors follow *Boyd* and *Breitling* (**A 022**) in their statistical treatment of the PP chain, with the exception that here conformational sequences are not excluded of the type XG/G'Y for two adjacent diads unless XG or G'Y imply another ω (syn-axial) interaction within either diad. Therefore this treatment, which is more rigorous but consistent with *Boyd* and *Breitling's* energy calculations, requires statistical weights which are functions of three adjacent torsional angles.

A 036   Tonelli, A. E. *Macromolecules* **1978**, *11*, 565.

The *Suter-Flory* RIS model of PP (**A 027**) is employed to calculate the ^{13}C NMR chemical shifts expected at the 9-C$^\alpha$ and the CH$_3$ carbons and at the 8- and 10-CH$_2$ carbons in the various stereoisomers of the PP model compound 3,5,7,9,11,13,15-heptamethylheptadecane. Differences in the chemical shifts of the same carbon atom in the various stereoisomers are assumed to be attributable solely to stereo-sequence dependent differences in the probability that the given carbon atom is involved in three-bond *gauche* or γ interactions with other carbon atoms. The *Suter-Flory* model provides an accurate description of the conformational characteristics of PP which permits a detailed understanding of its ^{13}C NMR spectrum. On the other hand, the failure of *Provasoli* and *Ferro's* calculations [*Macromolecules* **1977**, *10*, 874] is directely attributed to the inadequacies of the *Boyd* and *Breitling* three-state RIS model of PP (**A 022**).

A 037   Ferro, D. R.; Zambelli, A.; Provasoli, A.; Locatelli, P.; Rigamonti, E. *Macromolecules* **1980**, *13*, 179.

An improved semiempirical method for calculating the ^{13}C NMR chemical shifts of methyl and methylene carbons is presented. The mixture of diastereomers of 2,4,6,8,10,12-hexamethyltridecane, ^{13}C enriched on carbon C(7), is prepared, and the observed spectrum is reported, together with the assignment of each tetrad resonance of carbon C(7). The conformational origin of the stereochemical shifts of PP is qualitatively discussed. It is found that the quantitative agreement between calculated and observed chemical shifts for both the CH$_2$ and CH$_3$ carbons in PP model compounds is improved by taking into account the effect of distorsions of the tetrahedral angles on the γ shielding parameter. The value of γ is found to be considerably smaller for the CH$_2$ carbon than for CH$_3$, suggesting caution in transferring such best-fitting parameters from one carbon to another. The *Boyd-Breitling* model of the PP chain (**A 022**), as used in the present calculations, leads to theoretical CH$_3$ and CH$_2$ chemical shifts in good agreement with experiment and also reproduces thermodynamical data satisfactorily. Equally accurate results are obtained with the statistical model of *Suter* and *Flory* (**A 027**), when the effects of angular distortions on the shielding are properly taken into account.

A 038   Schilling, F. C.; Tonelli, A. E. *Macromolecules* **1980**, *13*, 270.

^{13}C NMR spectra are recorded for a low molecular weight *atactic* PP dissolved in a variety of solvents over a broad temperature range (293 - 393 K). Comparison of chemical shifts calculated via the γ effect method with the observed resonances, whose relative chemical shifts are solvent independent, permits their assignment to most of the methyl heptad, methylene hexad, and methine pentad stereosequences. Agreement between observed and calculated chemical shifts requires γ effects, *i.e.*, upfield chemical shifts produced by a *gauche* arrangement of carbon atoms separated by three bonds, of ca. – 5 ppm for the methyl and methine carbons and ca. – 4 ppm for the methylene carbons.

A 039   Ferro, D. R.; Ragazzi, M. *Macromolecules* **1981**, *14*, 1830.

Proton NMR coupling constants are calculated as Boltzmann averages over a discrete number of states, in the frame of the RIS theory. Different models are used for these calculations.

A 040   Theodorou, D. N.; Suter, U. W. *Macromolecules* **1985**, *18*, 1206.

Large numbers of conformations of unperturbed PP chains are generated in Monte-Carlo experiments, based on a RIS scheme. The average instantaneous shape in the system of principal axes of gyration is evaluated. Several new shape measures are introduced to characterize the shape anisotropy, asphericity, and acylindricity. Significant differences are found between short- and medium-length chains of different tacticity, while for long chains all shape measures converge to common limits.

A 041   Mark, J. E.; DeBolt, L. C.; Curro, J. G. *Macromolecules* **1986**, *19*, 491.

Effects of stereochemical structure on distribution functions for short PP and PVC chains are examined.

polyethylethylene, poly(n-butene-1)

Biskup, U.; Cantow, H.-J. *Makromol. Chem.* **1973**, *168*, 315.

Bond Length [pm]	Valence Angles [°]	Torsion Angles [°]	ξ a) (for 338 K)	ξ₀	E_ξ [kJ·mol⁻¹]
C-C : 153	C-CH₂-C : 112	t : 170	η = 1	1	0
C-H : 110	CH₂-C-CH₂: 112	g : 70	τ = 0.5	1	2.0
		ḡ : −60	τ* = 0.85	1	0.46
			ω = 0.059	1	8.0
			ω' = 0.024	1	10.5
			ω'' = 0.024	1	10.5

$$U_p = \begin{bmatrix} \eta\tau^* & 1 & \tau \\ \eta & \omega & \tau \\ \eta & 1 & \tau\omega \end{bmatrix} \quad U_i = \begin{bmatrix} \eta\omega'' & 1 & \tau\omega' \\ \eta & \omega & \tau\omega' \\ \eta\omega' & \omega' & \tau\omega\omega'' \end{bmatrix} \quad U_s = \begin{bmatrix} \eta & \omega' & \tau\omega'' \\ \eta\omega' & 1 & \tau\omega \\ \eta\omega'' & \omega & \tau\omega'^2 \end{bmatrix}$$

rows and columns: t, g, ḡ

a) For two consecutive angles φ in the *trans* position, considerable interactions may occur between side chains R and the main chain. These interactions are taken into account by the parameter τ*.

Comments:
The unperturbed dimensions $<r^2>_0/nl^2$ and their temperature coefficients are evaluated for polypropylethylene, polyethylethylene, and polystyrene with the RIS model. The calculated values of $<r^2>_0/nl^2$ for *atactic* and *isotactic* chains are in good agreement with the experimental data reported in the literature. The values of the model parameters required for good agreement change in a meaningful way with the length of the side chain. The measured temperature coefficients, however, are described satisfactorally by the model for atactic polypropylethylene and polyethylethylene only.

Calcd. quantities: $<r^2>_0/nl^2$ $d(\ln <r^2>_0)/dT$

A 043

polypropylethylene, poly(n-pentene-1)

Biskup, U.; Cantow, H.-J. *Makromol. Chem.* **1973**, *168*, 315.

Bond Length [pm]	Valence Angles [°]	Torsion Angles [°]	ξ a) (for 338 K)	ξ_0	E_ξ [kJ·mol^{-1}]
C-C : 153	C-CH$_2$-C : 112	t : 165	η = 1	1	0
C-H : 110	CH$_2$-C-CH$_2$: 112	g : 75	τ = 0.3	1	3.4
		$\bar{g}$: −60	τ* = 0.7	1	1.0
			ω = 0.033	1	9.6
			ω' = 0.011	1	12.6
			ω'' = 0.011	1	12.6

$$U_p = \begin{bmatrix} \eta\tau^* & 1 & \tau \\ \eta & \omega & \tau \\ \eta & 1 & \tau\omega \end{bmatrix} \quad U_i = \begin{bmatrix} \eta\omega'' & 1 & \tau\omega' \\ \eta & \omega & \tau\omega' \\ \eta\omega' & \omega' & \tau\omega\omega'' \end{bmatrix} \quad U_s = \begin{bmatrix} \eta & \omega' & \tau\omega'' \\ \eta\omega' & 1 & \tau\omega \\ \eta\omega'' & \omega & \tau\omega/2 \end{bmatrix}$$

rows and columns: t, g, $\bar{g}$

a) For two consecutive angles φ in the *trans* position, considerable interactions may occur between side chains R and the main chain. These interactions are taken into account by the parameter τ*.

Comments:
The unperturbed dimensions $<r^2>_0/nl^2$ and their temperature coefficients are evaluated for polypropylethylene, polyethylethylene, and polystyrene with the RIS model. The calculated values of $<r^2>_0/nl^2$ for *atactic* and *isotactic* chains are in good agreement with the experimental data reported in the literature. The values of the model parameters required for good agreement change in a meaningful way with the length of the side chain. The measured temperature coefficients, however, are described satisfactorily by the model for atactic polypropylethylene and polyethylethylene only.

Calcd. quantities: $<r^2>_0/nl^2$ $\quad d(\ln <r^2>_0)/dT$

A 044

polystyrene, PS

Williams, A. D.; Flory, P. J. *J. Am. Chem. Soc.* **1969**, *91*, 3111.

Bond Length [pm]	Valence Angles [°]	Torsion Angles [°]	ξ (for 343 K)	ξ_0	E_ξ [kJ·mol^{-1}]
C-C : 153	C-C-C : 112	t : 180	$\tau = 0$	1	∞
		g : 60	$\tau^* = 0.80$ or 1.0	1	0.64 or 0
		$\bar{g}$: -60	$\eta = 1.39$	0.50	-2.91
			$\omega = \omega' = \omega'' = 0$		

$$U_p = \begin{bmatrix} \eta\tau^* & 1 & \tau \\ \eta & \omega & \tau \\ \eta & 1 & \tau\omega \end{bmatrix} \quad U_m = \begin{bmatrix} \eta\omega'' & 1 & \eta\omega' \\ \eta & \omega & \tau\omega' \\ \eta\omega' & \omega' & \tau\omega\omega'' \end{bmatrix} \quad U_r = \begin{bmatrix} \eta & \omega' & \tau\omega'' \\ \eta\omega' & 1 & \tau\omega \\ \eta\omega'' & \omega & \tau\omega'^2 \end{bmatrix}$$

$$U_p\,U_m = U_m^{(2)} = \begin{bmatrix} \eta & \eta\tau^* \\ 0 & \eta \end{bmatrix} \quad U_p\,U_r = U_r^{(2)} = \begin{bmatrix} \eta^2\tau^* & 1 \\ \eta^2 & 0 \end{bmatrix}$$

rows and columns: t, g, $\bar{g}$

Comments:
Mixtures of diastereomers of 2,4,6-triphenylheptane are epimerized. The mole fractions of *isotactic*, *heterotactic*, and *syndiotactic* isomers at equilibrium at 343 K are 0.217, 0.499, and 0.284, respectively. There results are interpreted according to the theory of stereochemical equilibrium. The theory of equilibria between isomers and the associated theory of the conformer populations for each isomer provide a mutually consistent interpretation of the two kinds of results, the same arbitrary parameters being used for both. Stereochemical equilibria and conformer population calculated for PS for the same parameters differ considerably from those for the oligomers.

Calcd. quantities: Isomer ratios for oligomers   Conformer ratios in oligomers   Diad and triad composition at stereochemical equilibrium

A 045

polystyrene, PS

Fujiwara, Y.; Flory, P. J. *Macromolecules* **1970**, *3*, 43.

Bond Length [pm]	Valence Angles [°]	Torsion Angles [a)] [°]	ξ (for 343 K)	ξ_0	E_ξ [kJ·mol^{-1}]
C-C : 153	C-C-C : 112	t : 180 – Δφ	$\tau = 0.5$	1	2.0
		g : 60 + Δφ	$\eta \approx 1.5$	1	–1.16
		$\bar{g}$: –60	$\omega = 0.01$ *or* 0.05	1	13.0 or 8.5

$$U_p = \begin{bmatrix} \eta & 1 & \tau \\ \eta & \omega & \tau \\ \eta & 1 & \tau\omega \end{bmatrix} \qquad U_m = \begin{bmatrix} \eta\omega & 1 & \tau\omega \\ \eta & \omega & \tau\omega \\ \eta\omega & \omega & \tau\omega^2 \end{bmatrix} \qquad U_r = \begin{bmatrix} \eta & \omega & \tau\omega \\ \eta\omega & 1 & \tau\omega \\ \eta\omega & \omega & \tau\omega^2 \end{bmatrix}$$

rows and columns: t, g, $\bar{g}$

a) Δφ ≈ 10°.

Comments:
Conformational probabilities are calculated for pentads situated in vinyl chains of varying stereochemical composition according to the RIS model with neighbor dependence. The average conformational probabilities are strongly dependent upon the severity of steric repulsions between groups separated by four bonds, and hence on the average stereochemical composition of the vinyl chain as the whole. The calculations are in qualitative accord with NMR spectra reported for the oligomers, and frequencies observed in the spectrum of atactic PS-d_7 are partially identified. Calculations further indicate that chemical shifts of the pentads should depend appreciably on the stereochemical configurations of diads nearby.

polystyrene, PS

Abe, Y.; Tonelli, A. E.; Flory, P. J. *Macromolecules* **1970**, *3*, 294.

Bond Length [pm]	Valence Angles [°]	Torsion Angles [a) [°]	ξ (for 343 K)	ξ_o	E_ξ [kJ·mol^{-1}]
C-C : 153	C-CH$_2$-C : 112	t : 180 – Δϕ	τ^* = 1.0	1	0
	CH$_2$-C-CH$_2$: 112	g : 60 + Δϕ	τ = 0.5	1	2.1
	C-C*-H : 106.8	$\bar{g}$: –60	η = 1.0	1	0
	H-C-H : 109		ω ≤ 0.05	1	≥ 8.5

$$U_p = \begin{bmatrix} \eta\tau^* & 1 & \tau \\ \eta & \omega & \tau \\ \eta & 1 & \tau\omega \end{bmatrix} \quad U_m = \begin{bmatrix} \eta\omega & 1 & \tau\omega \\ \eta & \omega & \tau\omega \\ \eta\omega & \omega & \tau\omega^2 \end{bmatrix} \quad U_r = \begin{bmatrix} \eta & \omega & \tau\omega \\ \eta\omega & 1 & \tau\omega \\ \eta\omega & \omega & \tau\omega^2 \end{bmatrix}$$

rows and columns: t, g, $\bar{g}$

a) For some of the calculations the rotational states are displaced from their symmetrical locations by Δϕ = 10 or 20°.

Comments:
The theory of strain birefringence is elaborated in terms of the RIS model as applied to vinyl polymer chains. Additivity of the polarizability tensors for constituent groups is assumed. Stress-birefringence coefficients are calculated for PP and for PS. Statistical weight parameters which affect the incidences of various rotational states are varied over ranges consistent with other evidence. The effects of these variations are explored in detail for isotactic and syndiotactic chains.

Calcd. quantities: Strain-birefringence coefficient Γ_2.

polystyrene, PS

Biskup, U.; Cantow, H.-J. *Makromol. Chem.* **1973**, *168*, 315.

Bond Length [pm]	Valence Angles [°]	Torsion Angles [°] a)	ξ b) (for 308 K)	ξ_0	E_ξ [kJ·mol^{-1}]
C-C : 153	C-C-C : 114	t : $180 - \Delta\phi$	η = 1.56	1	−1.13
C-H : 110		g : $60 + \Delta\phi$	τ = 0.2	1	4.15
		$\bar{g}$: -60	τ^* = 1.0	1	0
			ω = 0.038	1	8.4
			ω' = 0.012	1	11.3
			ω'' = 0.012	1	11.3

$$U_p = \begin{bmatrix} \eta\tau^* & 1 & \tau \\ \eta & \omega & \tau \\ \eta & 1 & \tau\omega \end{bmatrix} \quad U_i = \begin{bmatrix} \eta\omega'' & 1 & \tau\omega' \\ \eta & \omega & \tau\omega' \\ \eta\omega' & \omega' & \tau\omega\omega'' \end{bmatrix} \quad U_s = \begin{bmatrix} \eta & \omega' & \tau\omega'' \\ \eta\omega' & 1 & \tau\omega \\ \eta\omega'' & \omega & \tau\omega'^2 \end{bmatrix}$$

rows and columns: t, g, $\bar{g}$

a) $\Delta\phi \approx 7.5°$
b) Other parameter sets are evaluated, too.

Comments:
The unperturbed dimensions and their temperature coefficients are evaluated for poly(*n*-pentene-1), poly(*n*-butene-1), and PS with the RIS model. The calculated values of the unperturbed dimensions for *atactic* and *isotactic* chains are in good agreement with the experimental data. The measured temperature coefficients, however, are described satisfactorily by the model for *atactic* polypentene and polybutene only.

Calcd. quantities: $\langle r^2 \rangle_0 / nl^2$ $d(\ln \langle r^2 \rangle_0)/dT$

polystyrene, PS

Yoon, D. Y.; Sundararajan, P. R.; Flory, P. J. *Macromolecules* **1975**, *8*, 776.

Bond Length [pm]	Valence Angles [°]		Torsion Angles [°]	ξ (for 300 K)	ξ_0	E_ξ [kJ·mol^{-1}]
C-C : 153	C-C$^\alpha$-C : 112		t : 170	η = *1.58*	0.8	−1.7 (±0.4)
	C$^\alpha$-C-C$^\alpha$: 114					
C$^\alpha$-C^{ar} : 151	C-C$^\alpha$-C^{ar} : 112		g : 70	ω = 0.047	1.3	8.3
	C-C$^\alpha$-H : 107					
C-H : 110	C$^\alpha$-C-H : 110			ω' = 0.047	1.3	8.3
	H-C-H : 110					
C^{ar}-C^{ar} : 139	C^{ar}-C^{ar}-H : 120			ω'' = *0.045*	1.8	9.2 (±1.7)
	C^{ar}-C^{ar}-C^{ar} : 120					

$$U_p = \begin{bmatrix} 1 & 1 \\ 1 & 0 \end{bmatrix} \quad U_m = \begin{bmatrix} \omega'' & 1/\eta \\ 1/\eta & \omega/\eta^2 \end{bmatrix} \quad U_r = \begin{bmatrix} 1 & \omega'/\eta \\ \omega'/\eta & 1/\eta^2 \end{bmatrix}$$

rows and columns: t, g

Comments:
Conformational energies of *meso* and *racemic* diads of PS are computed as functions of skeletal bond rotations. Confinement of rotations of the phenyl groups to a small range within which they are nearly perpendicular to the plane defined by the two adjoining skeletal bonds is confirmed. Steric interactions involving the relatively large planar phenyl group virtually preclude $\overline{g}g$ conformations. A simple, two-state RIS scheme is applicable with states at 170° and 70° for both *meso* and *racemic* dyads.

Calcd. quantities: $<r^2>_0 / nl^2$ $d (\ln <r^2>_0) / dT$

A 049

polystyrene, PS

Mays, J. W.; Hadjichristidis, N.; Fetters, L. J. *Macromolecules* **1985**, *18*, 2231.

Bond Length [pm]	Valence Angles [°]		Torsion Angles [°]	ξ (for 300 K)	ξ_0	E_ξ [kJ·mol^{-1}]
C-C : 153	C-C$^\alpha$-C	: 112	t : 170	η = 3.0 to 2.2	0.8	−3.3 to −2.5
	C$^\alpha$-C-C$^\alpha$	: 114				
C$^\alpha$-C^{ar} : 151	C-C$^\alpha$-C^{ar}	: 112	g : 70	ω = 0.047	1.3	8.3
	C-C$^\alpha$-H	: 107				
C-H : 110	C$^\alpha$-C-H	: 110		ω' = 0.047	1.3	8.3
	H-C-H	: 110				
C^{ar}-C^{ar} : 139	C^{ar}-C^{ar}-H	: 120		ω'' = 0.03 to 0.02	1.8	10.0 to 10.8
	C^{ar}-C^{ar}-C^{ar}	: 120				

$$U_p = \begin{bmatrix} 1 & 1 \\ 1 & 0 \end{bmatrix} \qquad U_m = \begin{bmatrix} \omega'' & 1/\eta \\ 1/\eta & \omega/\eta^2 \end{bmatrix} \qquad U_r = \begin{bmatrix} 1 & \omega'/\eta \\ \omega'/\eta & 1/\eta^2 \end{bmatrix}$$

rows and columns: t, g

Comments:
The unperturbed chain dimensions of near-monodisperse *atactic* PS are evaluated from intrinsic viscosity measurements. Negative values for the temperature coefficient of chain dimensions are found. Under conditions where specific solvent effects are eliminated or minimized, measurements yield results in excellent agreement with the theoretical predictions for atactic PS.

Calcd. quantities: $\langle r^2 \rangle_0 / nl^2 \qquad d(\ln \langle r^2 \rangle_0)/dT$

A 050

polystyrene, PS

Rapold, R. F.; Suter, U. W. *Macromol. Theory Simul.* **1994**, *3*, 1.

Bond Length [pm]	Valence Angles [°]		Torsion Angles [°]	ξ (for 300 K)	ξ_0	E_ξ [kJ·mol^{-1}]			
C-C : 153	C-C$^\alpha$-C	: 112	t : 175	η = 2.2	0.5	-3.7			
	C$^\alpha$-C-C$^\alpha$	: 114					$U_p = \begin{bmatrix} 1 & 1 \\ 1 & 0 \end{bmatrix}$	$U_m = \begin{bmatrix} 0 & \eta \\ \eta & \omega \end{bmatrix}$	$U_r = \begin{bmatrix} \eta^2 & \eta\omega' \\ \eta\omega' & 1 \end{bmatrix}$
C$^\alpha$-C^{ar} : 151	C-C$^\alpha$-C^{ar}	: 112	g : 70	ω = 0.078	1.1	6.6			
	C-C$^\alpha$-H	: 107							
C-H : 110	C$^\alpha$-C-H	: 109		ω' = 0.008	1.7	13.3			
	H-C-H	: 106.6					*or:* a) $U_m = \begin{bmatrix} 0 & 1/\eta \\ 1/\eta & \omega/\eta^2 \end{bmatrix}$	$U_r = \begin{bmatrix} 1 & \omega'/\eta \\ \omega'/\eta & 1/\eta^2 \end{bmatrix}$	rows and columns:
C^{ar}-C^{ar} : 140	C^{ar}-C^{ar}-H	: 120		(ω'' = 0)					t, g
	C^{ar}-C^{ar}-C^{ar}	: 120							

a) Assigning a statistical weight of 1 to the *racemic*-tt state.

Comments:
Conformational energies, computed with a force field including coulombic interactions and a simple accounting for the effects of solvents, of *meso* and *racemic* 2,4-diphenylpentane as model substances of PS are computed as functions of the skeletal torsion angles and the phenyl torsion angles.

Calcd. quantities: $<r^2>_0 / nl^2$ $d (\ln <r^2>_0) / dT$ A priori probabilities of diad and triad sequences   NMR coupling constants

*Further calculations on **polystyrene** chains:*

A 051 Tonelli, A. E.; Abe, Y.; Flory, P. J. *Macromolecules* **1970**, *3*, 303.

The depolarization of light scattered at 90° by PP and PS is treated according to RIS theory. Numerical calculations are carried out as functions of the statistical weight parameter ω governing interactions of second order, of the locations of the rotational states, of the chain length expressed by the number x of repeat units, and of the stereochemical composition expressed by the fraction of *racemic* diads.

A 052 Yoon, D. Y.; Flory, P. J. *Macromolecules* **1976**, *9*, 294.

The angular dependence of the intensity of radiation scattering is computed for PE, POE, and PS chains. The scattering functions, $F_x(\mu)$, corresponding to $I\mu^2$ where I is the intensity, are developed for chains of x repeat units in terms of the even moments $<r^{2p}_{ij}>$ of the separation distance between pairs of the units i and j, these moments $<r^{2p}_{ij}>$ up to $p = 4$ being evaluated on the basis of realistic RIS models. The theoretical scattering functions are in agreement with experimental results of small-angle neutron and X-ray scattering by PE in the molten state and by PS in the bulk and in solution.

A 053 Suter, U. W.; Flory, P. J. *J. Chem. Soc.; Farad. II* **1977**, *73*, 1521.

Depolarized Rayleigh scattering of benzene, toluene, cumene, and *tert*.-butylbenzene, 2,4-diphenylpentane (DPP), 2,4,6-triphenylheptane (TPH), and atactic PS with average degrees of polymerization of 21, 38, and 96 are measured. Optical anisotropies of PS and its oligomers DPP and TPH are calculated from the polarizability tensor for cumene. Values of $<\gamma^2>$ computed on this basis by averaging over the conformations of *meso* and *racemic* DPP, of the mixture of isomers comprising TPH, and of *atactic* PS are in good agreement with the results of the DRS measurements.

A 054 Tonelli, A. E. *Macromolecules* **1979**, *12*, 252.

Stereosequence-dependent ^{13}C NMR chemical shifts are calculated for the PS oligomers 2,4-diphenylpentane, 2,4,6-triphenylheptane, and 2,4,6,8-tetraphenylnonane. Calculated chemical shifts are obtained by quantitatively accounting for the number of γ interactions, or *gauche* arrangements, between carbon atoms separated by three bonds. In addition, the effect of magnetic shielding produced by phenyl groups that are first and second neighbors along the chain in either direction from a given carbon atom is considered. Agreement between calculated and observed ^{13}C NMR chemical shifts is good for each of the PS model compounds.

A 055 Tonelli, A. E. *Macromolecules* **1983**, *16*, 604.

^{13}C NMR chemical shifts are calculated for the various stereosequences present in *atactic* PS. Calculated chemical shifts are obtained by quantitatively accounting for the number of γ interactions, or *gauche* arrangements. The effects of magnetic shielding produced by the ring currents from phenyl groups that are first and second neighbors along the chain in either direction from a given carbon atom are also considered.

A 056   Huber, K.; Burchard, W.; Bantle, S. *Polymer* **1987**, *28*, 863.

The structures of linear PS chains as well as of generated RIS backbone chains are discussed in the light of the worm-like chain model. The global dimensions of these chains are represented satisfactorily by the theory of *Kratky* and *Porod*, if the system is under ϑ conditions.

A 057   Bahar, I.; Mattice, W. L. *J. Chem. Phys.* **1989**, *90*, 6775.

The dynamic RIS model, which was proposed before to investigate the dynamics of local conformational transitions in polymers, is elaborated to formulate the increase in the number of excimer-forming sites through rotational sampling. Application of the model to the *meso* and *racemic* diads in PS confirms the fact that conformational mobility of the chain plays a major role in intramolecular excimer formation. Comparison with experiments demonstrates that the decay of the monomer fluorescence in styrene dimers is predominantly governed by the process of conformational transitions.

A 058

poly(p-chlorostyrene), PPCS

Mark, J. E. *J. Chem. Phys.* **1972**, *56*, 458.

Bond Length [pm]	Valence Angles [°]	Torsion Angles [°] [a,b)]	ξ (for 298 K)	ξ_0	E_ξ [kJ · mol^{-1}]
C-C : 153	C-C-C : 112	t : 180 (± $\Delta\phi$)	η = 1.6	1	– 1.2
		g : 60 (+ $\Delta\phi$)	τ = 0.5	1	1.7
		$\bar{g}$: – 60 (– $\Delta\phi$)	ω = 0.0 to 0.05	1	∞ to – 7.4

$$U_p = \begin{bmatrix} \eta & 1 & \tau \\ \eta & \omega & \tau \\ \eta & 1 & \tau\omega \end{bmatrix} \quad U_m = \begin{bmatrix} \eta\omega & 1 & \eta\omega \\ \eta & \omega & \tau\omega \\ \eta\omega & \omega & \tau\omega^2 \end{bmatrix} \quad U_r = \begin{bmatrix} \eta & \omega & \tau\omega \\ \eta\omega & 1 & \tau\omega \\ \eta\omega & \omega & \tau\omega^2 \end{bmatrix}$$

rows and columns: t, g, $\bar{g}$

a) $\Delta\phi$ = 0 or 20°
b) Rotational states for the two skeletal bonds leading into and out of a C$^\alpha$ atom of *d* configuration are located at 180 + $\Delta\phi$, 60, and –60 – $\Delta\phi$°, and 180 – $\Delta\phi$, 60 + $\Delta\phi$, and –60°, respectively. The same two sets of rotational angles pertain to the two skeletal bonds leading, respectively, out of and into a C$^\alpha$ of *l* configuration.

Comments:
Mean-square dimensions and dipole moments of *p*-chlorostyrene chains are calculated as a function of their stereochemical structure, degree of polymerization, and temperature. Theoretical arguments and experimental evidence indicate that *p*-chlorostyrene and styrene chains differ little in conformational energy. Therefore, the present investigation employs conformational energies of styrene chains.

Calcd. quantities: $<r^2>_0 / nl^2$ $\quad d (\ln <r^2>_0) / dT$ $\quad <\mu^2>_0 / xm^2$ $\quad d (\ln <\mu^2>) / dT$

A 059

poly(p-chlorostyrene), PPCS

Saiz, E.; Mark, J. E.; Flory, P. J. *Macromolecules* **1977**, *10*, 967.

Bond Length [pm]	Valence Angles [°]	Torsion Angles [°]	ξ a) (for 300 K)	ξ_o	E_ξ [kJ·mol^{-1}]
C-C : 153	C-CH$_2$-C : 114	t : 180	η = 1.59	0.8	-1.7
	CH$_2$-C-CH$_2$: 112	g : 60	ω = 0.046	1.3	8.3
			ω' = 0.046	1.3	8.3
			ω'' = 0.046	1.8	9.1

$$U_p = \begin{bmatrix} 1 & 1 \\ 1 & 0 \end{bmatrix} \quad U_m = \begin{bmatrix} \omega'' & 1/\eta \\ 1/\eta & \omega/\eta^2 \end{bmatrix} \quad U_r = \begin{bmatrix} 1 & \omega'/\eta \\ \omega'/\eta & 1/\eta^2 \end{bmatrix}$$

rows and columns: t, g

a) The statistical weights are normalized to unity for *racemic*-tt

Comments:
The mean-square dipole moments of PPCS chains are calculated as a function of stereochemical composition using the RIS analysis recently published for PS. The calculations are in good agreement with the average of experimental results for *atactic* PPCS chains estimated to contain ca. 35% *meso* diads. The temperature coefficient is calculated to be negative in agreement with available experiments.

Calcd. quantities: $<\mu^2>_o / xm^2$ $d (\ln <\mu^2>) / dT$

poly(p-chlorostyrene), PPCS

Bahar, I.; Baysal, B. M.; Erman, B. *Macromolecules* **1986**, *19*, 1703.

Bond Length [pm]	Valence Angles [°]	Torsion Angles [°]	ξ (for 300 K)	ξ_o	E_ξ [kJ·mol^{-1}]
C-C : 153	C-CH$_2$-C : 114	t : 170	η = 1.58	0.8	-1.7
	CH$_2$-C-CH$_2$: 112	g : 70	ω = 0.047	1.3	8.3
			ω' = 0.047	1.3	8.3
			ω'' = 0.047	1.8	9.1

$$U_p = \begin{bmatrix} 1 & 1 \\ 1 & 0 \end{bmatrix} \qquad U_m = \begin{bmatrix} \omega'' & 1/\eta \\ 1/\eta & \omega/\eta^2 \end{bmatrix} \qquad U_r = \begin{bmatrix} 1 & \omega'/\eta \\ \omega'/\eta & 1/\eta^2 \end{bmatrix}$$

rows and columns: t, g

Comments:
The effect of solvent on the configurational characteristics of the polymeric chain is studied for the PPCS-benzene system. Among all possible configurations of benzene around the chain unit, two configurations, referred to as the parallel and perpendicular configurations, are found to lead to the lowest energy of interaction. The potential prevailing at the parallel configuration is calculated to be more favorable. Depending on the relative occurrence of these two competing configurations, new energy values are assigned to each rotational isomeric state of the chain structural unit. Configuration-dependent properties such as characteristic ratio, dipole ratio, and their temperature dependence are computed. The reduction in the dipole moment induced by the coupling of benzene with the chlorophenyl group is also included in the calculations. In general, satisfactory agreement with previous experimental measurements is achieved except for the positive temperature coefficient of dipole ratio reported at high temperature.

Calcd. quantities: $\langle r^2 \rangle_o / nl^2 \qquad d(\ln \langle r^2 \rangle_o)/dT \qquad \langle \mu^2 \rangle_o / xm^2 \qquad d(\ln \langle \mu^2 \rangle)/dT$

*Further calculations on **poly(p-halostyrene)** chains:*

A 061   Saiz, E.; Suter, U. W.; Flory, P. J. *J. Chem. Soc. Farad. II* **1977**, *73*, 1538.

The optical anisotropies γ^2 and molar Kerr constants $_mK$ of model compounds, and of poly(p-chlorostyrene) and poly(p-bromostyrene), are determined. Averages $<\gamma^2>$ and $<_mK>$ over all conformations of the polymer chains are calculated as functions of the fraction w_m of the *meso* diads using the RIS model originally developed for PS.

A 062

poly(2-vinylpyridine), P2VP

Tonelli, A. E. *Macromolecules* **1985**, *18*, 2579.

Bond Length [pm]	Valence Angles [°]	Torsion Angles a) [°]	ξ (for 323 K)	ξ_0	E_ξ [kJ·mol^{-1}]
C-C : 153	C-C$^\alpha$-C : 112	t : 170	η = 1.37	1	-0.42
C$^\alpha$-C^{ar} : 151	C$^\alpha$-C-C$^\alpha$: 114	g : 60 or 70	ω = 0.49	1.4	2.8
C-H : 110			ω' = 0.13	1.6	6.8
		χ b) : 180	ω'' = 0.22	0.9	3.8

$$U_p = \begin{bmatrix} 1 & 1 \\ 1 & 0 \end{bmatrix} \quad U_m = \begin{bmatrix} \omega'' & 1/\eta \\ 1/\eta & \omega/\eta^2 \end{bmatrix} \quad U_r = \begin{bmatrix} 1 & \omega'/\eta \\ \omega'/\eta & 1/\eta^2 \end{bmatrix}$$

rows and columns: t, g

a) The critical solvent-interaction distance σ = 4.5 Å; the *racemic (d,l;l,d)*tt state is given a statistical weight of unity.
b) Side-group torsion angle.

Comments:
Conformational energy calculations are coupled with dipole moment measurements to derive a conformational description of P2VP. When an RIS model is used to calculate the dipole moments of P2VP chains with different stereosequences, it is found that the calculated dipole moments are nearly independent of the P2VP stereosequence.

Calcd. quantities: $<r^2>_0 / nl^2$ $d(\ln <r^2>_0)/dT$ $<\mu^2>_0 / xm^2$ $d(\ln <\mu^2>)/dT$

A 063

poly(N-vinyl pyrrolidone), PVP

Tonelli, A. E. *Polymer* **1982**, *23*, 676.

Bond Length [pm]	Valence Angles [°]	Torsion Angles a) [°]	ξ (for 298 K)	ξ_0	E_ξ [kJ·mol^{-1}]
C-C : 154	CH_2-C^α-CH_2: 112	t : 180	ω = 0.092	0.69	5.0
C-H : 110	C^α-CH_2-C^α : 114	g : 70	ω' = 0.085	1.5	7.1
	H-C-H : 110		ω'' = 0.032	1.3	9.2 (±1.3)
	H-C^α-N : 110	χ b) : 0 or 180	η = 0.62	0.73	0.42

$$U_p = \begin{bmatrix} 1 & 1 \\ 1 & 0 \end{bmatrix} \qquad U_m = \begin{bmatrix} \omega'' & 1/\eta \\ 1/\eta & \omega/\eta^2 \end{bmatrix} \qquad U_r = \begin{bmatrix} 1 & \omega'/\eta \\ \omega'/\eta & 1/\eta^2 \end{bmatrix}$$

rows and columns: t, g

a) The critical solvent-interaction distance σ = 5 Å.
b) Side-group torsion angle.

Comments:
Conformational energies are calculated for PVP chains as a function of stereosequence using semiempirical potential energy functions appropriate to peptides and *n*-alkanes. The planar pyrrolidone side groups are permitted to adopt both conformations which result in an eclipsed arrangement of the pyrrolidone N—CH_2 or N—CO and the C^α—H^α bonds. Solvent interactions are considered in the manner used to treat other vinyl polymers bearing planar side groups. Dimensions and dipole moments are calculated using the RIS model developed for PVP from the conformational energies considering both the effects of stereosequence and temperature. The dimensions and dipole moments calculated for atactic PVP chains agree with the dimensions reported in the literature and the dipole moments measured.

Calcd. quantities: $\langle r^2 \rangle_0 / nl^2$ $d(\ln \langle r^2 \rangle_0)/dT$ $\langle \mu^2 \rangle_0 / xm^2$ $d(\ln \langle \mu^2 \rangle)/dT$

*Further calculations on **poly(N-vinyl pyrrolidone)** chains:*

A 064   Tarazona, M. P.; Saiz, E. *Makromol. Chem., Theory Simul.* **1993**, 2, 697.

Conformational energies as function of rotational angles over two consecutive skeletal bonds for both *meso* and *racemic* diads of poly(*N*-vinyl-2-pyrrolidone) are computed. The results of these calculations are used to formulate a statistical model that was then employed to calculate the unperturbed dimensions of this polymer. The conformational energies are sensitive to the Coulombic interactions, which are governed by the dielectric constant of the solvent, and to the size of the solvent molecules. Consequently, the calculated values of the polymeric chain dimensions are strongly dependent on the nature of the solvent, as it was experimentally found before.

A 065

poly(N-vinylcarbazole), PVK

Sundararajan, P. R. *Macromolecules* **1980**, *13*, 512.

Bond Length [pm]	Valence Angles [°]		Torsion Angles a) [°]	ξ (for 298 K)	ξ_o	E_ξ [kJ·mol^{-1}]
C-C : 153	CH_2-C^α-CH_2	:112	for meso:	η = 2.57	1.1	−2.1
C^α-N : 149	C^α-CH_2-C^α	:117 *(tt)*	tt : 160,160			
		:114 *(others)*	gt : 70,180	ω = 0.16	1.0	4.6
C^α-H : 110	C^α-N-C^β	:126	gg : 80, 80			
	C^β-N-$C^{\beta'}$	:108		ω' = 0.34	0.8	2.1
	N-C^β-C^γ	:132	for racemic:			
C^{ar}-N : 139	C^{ar}-C^{ar}-C^{ar}	:120	tt : 170,170	ω'' = 0.09	0.5	4.2
	C^β-C^η-$C^{\eta'}$	:108	gt : 80,165			
C^{ar}-C^{ar} : 139	C^η-$C^{\eta'}$-$C^{\beta'}$	:108	gg : 70, 70			
			χ b) : 0			

$$U_p = \begin{bmatrix} 1 & 1 \\ 1 & 0 \end{bmatrix} \quad U_m = \begin{bmatrix} \omega'' & 1/\eta \\ 1/\eta & \omega/\eta^2 \end{bmatrix} \quad U_r = \begin{bmatrix} 1 & \omega'/\eta \\ \omega'/\eta & 1/\eta^2 \end{bmatrix}$$

rows and columns: t,g

a) The critical solvent-interaction distance σ = 5 Å.
b) Side-group torsion angle.

Comments:
Conformational maps are calculated for the *meso* and *racemic* diads of PVK.

Calcd. quantities: $\langle r^2 \rangle_0 / nl^2$

poly(N-vinylcarbazole), PVK

Abe, A.; Kobayashi, H.; Kawamura, T.; Date, M.; Uryu, T.; Matsuzaki, K. *Macromolecules* **1988**, *21*, 3414.

Bond Length [pm]	Valence Angles [°]	Torsion Angles [°]	ξ (for 300 K)	ξ_0	E_ξ [kJ·mol^{-1}]
C-C : 153	CH$_2$-C$^\alpha$-CH$_2$:112	for meso: tt : 160,160	Set 1: η = 2.5	1	-2.3
	C$^\alpha$-CH$_2$-C$^\alpha$	gt : 70,180	$\omega = \omega' = \omega'' = 0$		
	: 117 (tt)	gg : 80, 80			
	: 114 (others)		Set 2:		
		for racemic: tt : 170,170	η = 2.7	1	-2.5 (±0.4)
			ω = 0.43	1	2.1
		gt : 80,165	ω' = 0.43	1	2.1
		gg : 70, 70	ω'' = 0.005	1	13.4 (±0.8)

$$U_p = \begin{bmatrix} 1 & 1 \\ 1 & 0 \end{bmatrix} \quad U_m = \begin{bmatrix} \eta^2\omega'' & \eta \\ \eta & \omega \end{bmatrix} \quad U_r = \begin{bmatrix} \eta^2 & \eta\omega' \\ \eta\omega' & 1 \end{bmatrix}$$

rows and columns: t, g

Comments:
Model compounds are prepared in high purity and NMR ^{1}H-^{1}H vicinal coupling constants and their temperature dependence are measured. Statistical weight parameters, and thus conformational energies, estimated therefrom are used to elucidate the conformational characteristics of the PVK chain. Experimental values of the characteristic ratios for the unperturbed dimensions and dipole moment are found to be reasonably reproduced by the calculation for moderately *syndiotactic* chains. In the ground state, the so-called "sandwich structure" (*meso*-tt) is a rare occurrence. Configurational statistics of the polymer chain may be quite different in the excited state, where formation of the *meso*-tt form is reported to be very rapid. The conformational energy parameter sets proposed by other research groups are found to overestimate the *meso*-tt state to a considerable degree.

Calcd. quantities: $<r^2>_0 / nl^2$ $\quad <\mu^2>_0 / xm^2$

A 067

poly(methyl vinyl ketone)

Suter, U. W. *J. Am. Chem. Soc.* **1979**, *101*, 6481.

Bond Length [pm]	Valence Angles [°]	Torsion Angles [°]	ξ (for 300 K)	ξ_o	E_ξ [kJ · mol^{-1}]
C-C : 153	C-CH$_2$-C : 114	t : 170	η = 2.7	0.7	- 3.3
C-C* : 151	CH$_2$-C-CH$_2$: 112	g : 70	ω = 0.05	1.0	7.5
C=O : 122	C-CO-C : 116		ω' = 0.02	0.8	9.1
C-H : 110		$\chi^{a)}$: 0 *or* 180	ω'' = 0.008	0.8	11.6

$$U_p = \begin{bmatrix} 1 & 1 \\ 1 & 0 \end{bmatrix} \qquad U_m = \begin{bmatrix} \eta^2\omega'' & \eta \\ \eta & \omega \end{bmatrix} \qquad U_r = \begin{bmatrix} \eta^2 & \eta\omega' \\ \eta\omega' & 1 \end{bmatrix}$$

rows and columns: t, g

a) Side-group torsion angle.

Comments:
A semiempirical force field is constructed for the calculation of conformational potential energies of unstrained, acyclic, aliphatic aldehydes and ketones, taking solvent effects into consideration. Detailed conformational calculations for fitting and testing the necessary parameters are done. The results are incorporated into a RIS scheme for poly(methyl vinyl ketone)s.

Calcd. quantities: $<r^2>_o / nl^2$ $<\mu^2>_o / xm^2$ NMR coupling constants

poly(tert.-butyl vinyl ketone)

Guest, J. A.; Matsuo, K.; Stockmayer, W. H.; Suter, U. W. *Macromolecules* **1980**, *13*, 560.

Bond Length [pm]	Valence Angles [°]	Torsion Angles [°]	ξ [a) (for 300 K)	ξ_0	E_ξ [kJ·mol^{-1}]			
C-C : 153	C-CH$_2$-C : 114	t : 170	η = 1.6	0.6	–2.5 (±0.4)			
	CH$_2$-C-CH$_2$: 112	g : 70	ω^* = 0.23	2.9	6.3 (±0.8)	$U_p = \begin{bmatrix} 1 & 1 \\ 1 & 0 \end{bmatrix}$	$U_m = \begin{bmatrix} 0 & 1 \\ 1 & \omega/\eta \end{bmatrix}$	$U_r = \begin{bmatrix} \eta\omega^* & \omega' \\ \omega' & 1/\eta \end{bmatrix}$
			ω = 0.05	0.9	7.1			
			ω' = 0.02	1.9	11.6			rows and columns: t, g

a) The second-order parameter ω^* expresses the special "around the chain" repulsion between two pivaloyl groups in the *racemic tt* conformation.

Comments:
The characteristic ratio of *atactic* poly(tert.-butyl vinyl ketone) is determined from light scattering and viscosimetry measurements, and at 300 K in benzene the dipole moment ratio and its temperature coefficient are measured. Calculations of C_∞ and D_∞ based on a two-state RIS model, with parameters independently derived from a previously developed semiempirical potential energy surface and from epimerization equilibrium measurements for dimeric and trimeric oligomers, are in excellent agreement with the experimental results. The predicted temperature coefficient is positive but lower in magnitude than that observed.

Calcd. quantities: $<r^2>_0 / nl^2$ $<\mu^2>_0 / xm^2$ $d(\ln <\mu^2>_0)/dT$

A 069

poly(vinyl alcohol), PVA

Wolf, R. M.; Suter, U. W. *Macromolecules* **1984**, *17*, 669.

Bond Length [pm]	Valence Angles [°]	Torsion Angles [°]	ξ (for 300 K)	ξ_0	E_ξ [kJ·mol^{-1}]			
C^α-C : 153	CH_2-C^α-CH_2: 112	t : 175	η = 1.1	1.1	−1.7			
C^α-O : 143	C^α-CH_2-C^α : 114	g : 70	τ = 0.6			$U_p = \begin{bmatrix} 1 & 1 & 1 \\ 1 & 0 & 1 \\ 1 & 1 & 0 \end{bmatrix}$	$U_m = \begin{bmatrix} \eta^2\omega'' & \eta & \eta\tau\omega' \\ \eta & \omega & \tau\omega' \\ \tau\eta\omega' & \tau\omega' & \tau^2\omega\omega'' \end{bmatrix}$	$U_r = \begin{bmatrix} \eta^2 & \eta\omega' & \eta\tau\omega'' \\ \eta\omega' & 1 & \tau\omega \\ \eta\tau\omega'' & \tau\omega & \tau^2\omega/2 \end{bmatrix}$
C^α-H : 110	C^α-O-H : 108.5	$\bar{g}$: −65	ω = 0.08	1.3	7.1			
O-H : 95	C-C^α-O : 111		ω' = 0.27	1.0	3.3			
			ω'' = 0.56	0.9	1.2	rows and columns: t, g, $\bar{g}$		

Comments:
Conformational energies of *meso* and *racemic* 2,4-pentanediol are estimated. For this dimer as well as for PVA in aqueous solution, a simple three-state RIS model is developed.

Calcd. quantities: Vicinal NMR coupling constants (dimer) $<r^2>_0 / nl^2$ (PVA)

poly(vinyl alcohol), PVA

Tonelli, A. E. *Macromolecules* **1985**, *18*, 1086.

Bond Length [pm]	Valence Angles [°]	Torsion Angles [°]	ξ a) (for 323 K)	ξ_o	E_ξ [kJ · mol^{-1}]
C^α-C : 153	CH_2-C^α-CH_2: 112	t : 175	η = 2.14	1.18	–1.6
C^α-O : 143	C^α-CH_2-C^α : 114	g : 70	τ = 0.65	0.65	0
C^α-H : 110	C^α-O-H : 108.5	$\bar{g}$: –65	ω = 0.076	1.34	7.7
O-H : 95	C-C^α-O : 111		ω' = 0.27	1.03	3.6
			ω'' = 0.54	0.87	1.3

$$U_m = \begin{bmatrix} \eta^2\omega'' & \eta & \eta\tau\omega' \\ \eta & \omega & \tau\omega' \\ \eta\tau\omega' & \tau\omega' & \tau^2\omega\omega'' \end{bmatrix} \quad U_r = \begin{bmatrix} \eta^2 & \eta\omega' & \eta\omega'' \\ \tau\omega' & 1 & \tau\omega \\ \eta\tau\omega'' & \tau\omega & \tau^2\omega'^2 \end{bmatrix}$$

$$U_p = \begin{bmatrix} 1 & 1 & 1 \\ 1 & 0 & 1 \\ 1 & 1 & 0 \end{bmatrix}$$

rows and columns: t, g, $\bar{g}$

a) Parameter *set I* for protic solvents such as D_2O; other parameter sets are provided in the paper, too.

Comments:
^{13}C NMR chemical shifts are calculated for the carbon nuclei in PVA to the pentad and hexad levels of stereosequence for the methine and methylene carbons, respectively. The RIS model developed by *Wolf* and *Suter* (**A 069**) is employed to calculate the frequencies. The relative orders of the observed methine pentad and methylene hexad resonances agree with the calculated chemical shifts, in addition to the agreement between the overall chemical shift dispersions measured and predicted for the methylene carbons.

Calcd. quantities: ^{13}C NMR chemical shifts

poly(vinyl acetate), PVAc

Sundararajan, P. R. *Macromolecules* **1978**, *11*, 256.

Bond Length [pm]	Valence Angles [°]	Torsion Angles [°] [a)]	ξ [a)] (for 300 K)	ξ_0	E_ξ [kJ·mol^{-1}]
C^α-C : 153	CH_2-C^α-CH_2 : 109.5	$(tt)_m$: 159 / 159	η = 1.95	0.841	−2.1
		$(gt)_m$: 75 / 170			
C^α-O : 142	C^α-CH_2-C^α : 114	$(\bar{g}t)_m$: −76 / 170	τ = 0.17	0.472	2.5
		$(gg)_m$: 84 / 96			
C^α-H : 110	C^α-O-C : 113	$(g\bar{g})_m$: 89 / −72	ω = 0.15	1.843	6.3
		$(\bar{g}\bar{g})_m$: −78 / −78			
C-O : 136	O-C=O : 118	$(tt)_r$: 172 / 172	ω' = 0.18	1.354	5.0
		$(gt)_r$: 84 / 172			
C=O : 125	O-C-CH_3 : 122	$(\bar{g}t)_r$: −67 / 170	ω'' = 0.14	1.771	6.3
		$(gg)_r$: 76 / 76			
C-CH_3: 152	C-C^α-O : 109.5	$(g\bar{g})_r$: 108 / −72			
		$(\bar{g}\bar{g})_r$: −82 / −82			

$$U_m = \begin{bmatrix} \omega'' & 1/\eta & \tau\omega'/\eta \\ 1/\eta & \omega/\eta^2 & \tau\omega'/\eta^2 \\ \tau\omega'/\eta & \tau\omega'/\eta^2 & \tau^2\omega''/\eta^2 \end{bmatrix} \quad U_r = \begin{bmatrix} 1 & \omega'/\eta & \tau\omega''/\eta \\ \omega'/\eta & 1/\eta^2 & \tau\omega/\eta^2 \\ \tau\omega''/\eta & \tau\omega/\eta^2 & \tau^2\omega'^2/\eta^2 \end{bmatrix}$$

$$U_p = \begin{bmatrix} 1 & 1 & 1 \\ 1 & 0 & 1 \\ 1 & 1 & 0 \end{bmatrix} \quad \text{rows and columns: t, g, } \bar{g}$$

[a)] The critical solvent-interaction distance $\sigma = r_i + r_j + 0.2$ Å ; values for σ = 5Å and $\sigma = \infty$ are given as well in the original paper.

Comments:
Conformational features of *meso* and *racemic* diads of PVAc are examined using energy calculations. In contrast to other vinyl chains bearing planar substituents, the $\bar{g}$ conformation is not prohibited for this polymer. The shifts in the positions of the energy minima from perfect staggering are discussed in terms of the second order interactions. Calculated statistical weight parameters are used to reproduce the experimental data on NMR coupling constants and the characteristic ratios.

Calcd. quantities: $<r^2>_0 / nl^2$ NMR coupling constants

poly(vinyl acetate), PVAc

Riande, E.; Saiz, E.; Mark, J. E. *J. Polym. Sci.; Polym. Phys. Ed.* **1984**, *22*, 863.

Bond Length [pm]	Valence Angles [°]	Torsion Angles [°]	ξ (for 300 K)	ξ_0	E_ξ [kJ·mol^{-1}]			
C-C : 153	C-CH$_2$-C : 114	t : 170	η = 1.66	1	−1.26			
	CH$_2$-C-CH$_2$: 112	g : 70	ω = 0.09	1.3	6.70	$U_p = \begin{bmatrix} 1 & 1 \\ 1 & 0 \end{bmatrix}$	$U_r = \begin{bmatrix} 1 & \omega'/\eta \\ \omega'/\eta & 1/\eta^2 \end{bmatrix}$	$U_m = \begin{bmatrix} \omega'' & 1/\eta \\ 1/\eta & \omega/\eta^2 \end{bmatrix}$
			ω' = 0.10	1.4	6.70			
			ω'' = 0.10	1.2	6.28			rows and columns: t, g

Comments:
Atactic PVAc prepared in a free-radical polymerization is crosslinked by means of benzoyl peroxide. The resulting elastomertic networks are studied in elongation, both unswollen and swollen with triethylbenzene, over the range 273 - 363 K. The most important experimental results obtained are values of the network birefringence, which is negative. Calculations carried out to interpret the birefringence are based on Monte-Carlo simulations of the atactic structure, and on the RIS theory.

Calcd. quantities: NMR coupling constants: $(\Delta J)_m$ = 0.82 (Exptl.: 0.8) $\langle r^2 \rangle_0 / nl^2$ = 9.1 (Exptl.: 9.2)
(for P_r = 0.50) $(\Delta J)_{mm}$ = 1.13 (Exptl.: 1.1) $d(\ln \langle r^2 \rangle_0)/dT$ = 0.52 × 10^{-3} K^{-1}
$(\Delta J)_r$ = 6.40 (Exptl.: 6.5) Optical configuration parameter Δa and its temperature coefficient
$(\Delta J)_{rm}$ = 7.71 (Exptl.: 6.8)

A 073

poly(vinyl acetate), PVAc

Viswanadhan, V. N.; Mattice, W. L. *Makromol. Chem.* **1985**, *186*, 633.

Bond Length [pm]	Valence Angles [°]	Torsion Angles [°]		ξ (for 300 K)	ξ_0	E_ξ [kJ·mol^{-1}]
C^α-C : 153	CH_2-C^α-CH_2:109.5	$(tt)_m$	: 159 / 159	η = 1.95	0.841	−2.1
		$(gt)_m$	: 75 / 170			
C^α-O : 142	C^α-CH_2-C^α : 114	$(\bar{g}t)_m$	: −76 / 170	τ = 0.17	0.472	2.5
		$(gg)_m$	: 84 / 96			
C^α-H : 110	C^α-O-C : 113	$(g\bar{g})_m$	: 89 / −72	ω = 0.15	1.843	6.3
		$(\bar{g}\bar{g})_m$	: −78 / −78			
C-O : 136	O-C=O : 118	$(tt)_r$	: 172 / 172	ω' = 0.18	1.354	5.0
		$(gt)_r$	: 84 / 172			
C=O : 125	O-C-CH_3 : 122	$(\bar{g}t)_r$	: −67 / 170	ω'' = 0.14	1.771	6.3
		$(gg)_r$	: 76 / 76			
C-CH_3 : 152	C-C^α-O : 109.5	$(g\bar{g})_r$	: 108 / −72			
		$(\bar{g}\bar{g})_r$	: −82 / −82			

$$U_m = \begin{bmatrix} \eta\omega'' & 1 & \tau\omega' \\ \eta & \omega & \tau\omega'' \\ \eta\omega' & \omega' & \tau\omega\omega'' \end{bmatrix} \quad U_r = \begin{bmatrix} \eta & \omega' & \tau\omega'' \\ \eta\omega' & 1 & \tau\omega \\ \eta\omega'' & \omega & \tau\omega'^2 \end{bmatrix}$$

$$U_p = \begin{bmatrix} \eta & 1 & \tau \\ \eta & 0 & \tau \\ \eta & 1 & 0 \end{bmatrix} \quad \text{rows and columns: } t, g, \bar{g}$$

a) The critical solvent-interaction distance $\sigma = r_i + r_j + 0.2$ Å.

Comments:
Probabilities of configurations conducive to the intramolecular back-biting abstraction of a hydrogen atom are evaluated for growing unperturbed PVAc chains. A realistic RIS model is used for the chain statistics. Probabilities are found to be smaller than those seen in an earlier treatment of the polyethylene chain. The smaller probabilities of PVAc contribute to the virtual absence of short branches. The present study therefore provides support for the validity of the Roedel mechanism for the formation of short branches in the free radical initiated polymerization of ethylene.

A 074

poly(methyl acrylate), PMA

Flory, P. J. *J. Am. Chem. Soc.* **1967**, *89*, 1798.

Bond Length [pm]	Valence Angles [°]	Torsion Angles [°]	ξ (for 298 K)	ξ_0	E_ξ [kJ · mol^{-1}]
		t : 180	η = 1.85	1	−1.5
		g : 60	τ^* = 0.50	1	1.7
		ω = 0		1	∞

$$U_p = \begin{bmatrix} \eta\tau^* & 1 \\ \eta & \omega \end{bmatrix} \quad U_m = \begin{bmatrix} \eta\omega & 1 \\ \eta & \omega \end{bmatrix} \quad U_r = \begin{bmatrix} \eta & \omega \\ \eta\omega & 1 \end{bmatrix}$$

rows and columns: t, g

Comments:
A general theory of stereochemical equilibria in main chain molecules possessing two or more asymmetric centers is presented.

Calcd. quantities: Equilibrium stereoisomeric composition $\langle r^2 \rangle_0 / nl^2$

A 075

poly(methyl acrylate), PMA

Yoon, D. Y.; Suter, U. W.; Sundararajan, P. R.; Flory, P. J. *Macromolecules* **1975**, *8*, 784.

Bond Length [pm]	Valence Angles [°]		Torsion Angles [°]	ξ (for 300 K)	ξ_o	E_ξ [kJ·mol^{-1}]
C-C : 153	CH$_2$-C$^\alpha$-CH$_2$	: 112	t : 170	η = 1.68	1.0 (±0.1)	−1.3 (±0.4)
C$^\alpha$-C* : 152	C$^\alpha$-CH$_2$-C$^\alpha$	: 114				
C*-O : 136	C-C$^\alpha$-C*	: 112	g : 70	ω = 0.08	1.3 (±0.1)	6.9
C=O : 122	C-C$^\alpha$-H	: 107				
O-CH$_3$: 145	C$^\alpha$-C-H	: 110		ω' = 0.09	1.4 (±0.1)	6.8
C-H : 110	H-C-H	: 110				
	C$^\alpha$-C*=O	: 114		ω'' = 0.10	1.2 (±0.1)	6.3 (±1.3)
	C$^\alpha$-C*-O*	: 121				
	C*-O-CH$_3$	: 113				

$$U_p = \begin{bmatrix} 1 & 1 \\ 1 & 0 \end{bmatrix} \qquad U_m = \begin{bmatrix} \omega'' & 1/\eta \\ 1/\eta & \omega/\eta^2 \end{bmatrix} \qquad U_r = \begin{bmatrix} 1 & \omega'/\eta \\ \omega'/\eta & 1/\eta^2 \end{bmatrix}$$

rows and columns: t, g

Comments:
Conformational energies of *meso* and *racemic* diads of PMA are computed as functions of skeletal bond rotation, the planar ester group being oriented perpendicular to the plane defined by the two adjoining skeletal bonds. Solvent interactions affect the relative energies of various conformations, as found for PS. The conformational energy surfaces are similar to those of PS. The energies E(η) and E(ω''), being affected by solvent interactions, cannot be reliably estimated from the energy calculations. However, agreement with experimental dimensions of the PMA chain and stereochemical equilibria in dimeric and trimeric oligomers is achieved by using the values given above.

Calcd. quantities: $\langle r^2 \rangle_o / nl^2$ NMR coupling parameters

A 076

poly(methyl acrylate), PMA

Ojalvo, E. A.; Saiz, E.; Masegosa, R. M.; Hernández-Fuentes, I. *Macromolecules* **1979**, *12*, 865.

Bond Length [pm]	Valence Angles [°]	Torsion Angles [°]	ξ a,b) (for 300 K)	ξ_0	E_ξ [kJ·mol^{-1}]
C-C : 153	CH_2-C^α-CH_2: 112	t : 170	η = 1.63	0.97	−1.3
C^α-C* : 152	C^α-CH_2-C^α : 114	g : 70	ω = 0.002	0.25	11.7
C*-O : 136	C-C^α-C* : 109.5		ω' = 0.026	1.05	9.2
C=O : 122	C^α-C*-O* : 121		ω_i'' = 0.072	1.06	6.7
O-CH_3 : 145	C*-O-CH_3 : 113		ω_d'' = 0.17	0.94	4.2
C-H : 110		χ c) = 0, 180	ρ = 0.64	1.07	1.3

$$U_p = \begin{bmatrix} 1 & 0 & 1 & 0 \\ 0 & 1 & 0 & 1 \\ 1 & 0 & 0 & 0 \\ 0 & 1 & 0 & 0 \end{bmatrix} \quad U_m = \begin{bmatrix} \omega''_i & \rho\omega''_d & 1/\eta & \rho/\eta \\ \omega''_d & \rho\omega''_i & 1/\eta & \rho/\eta \\ 1/\eta & \rho/\eta & \omega/\eta^2 & \rho\omega/\eta^2 \\ 1/\eta & \rho/\eta & \omega/\eta^2 & \rho\omega/\eta^2 \end{bmatrix} \quad U_r = \begin{bmatrix} 1 & \rho & \omega'/\eta & \rho\omega'/\eta \\ 1 & \rho & \omega'/\eta & \rho\omega'/\eta \\ \omega'/\eta & \rho\omega'/\eta & 1/\eta^2 & \rho/\eta^2 \\ \omega'/\eta & \rho\omega'/\eta & 1/\eta^2 & \rho/\eta^2 \end{bmatrix}$$

rows and columns: t_o, t_π, g_o, g_π

a) A statistical weight parameter, ρ, is introduced to represent the difference between χ = 0 and 180° (π) states.
b) Two different values of the ω'' parameter are introduced in *meso*-tt states: ω_i'' for conformations in which both ester groups have the same orientation and ω_d'' for conformations having different orientations of these groups. This distinction is not needed for the other states because of the large separation between ester groups in these conformations.
c) Side-group rotational angle.

Comments:
The simplified two-rotational-state scheme previously used for PMA is not able to predict satisfactory values for its dipole moment. A more realistic scheme with four rotational states is introduced which allows for the distinguishing between different interactions for different orientations of the ester group lateral to the chain. Values of dimensions, dipole moments, stereochemical equilibria, and NMR coupling constants calculated using this scheme are in agreement with experimental results. However, this scheme fails to reproduce the experimental variation of dipole moment with temperature.

Calcd. quantities: $<r^2>_0 / nl^2$ $<\mu^2> / xm^2$ $d (\ln <\mu^2>) / dT$ Stereochemical equilibrium f_m in oligomers (x = 2,3)   NMR coupling constants

A 077

poly(methyl acrylate), PMA

Tarazona, M. P.; Saiz, E. *Macromolecules* **1983**, *16*, 1128.

Bond Length [pm]	Valence Angles [°]	Torsion Angles [°]	ξ (for 300 K)	ξ_0	E_ξ [kJ·mol^{-1}]
C-C : 153	CH$_2$-C$^\alpha$-CH$_2$: 112	t : 170	η = 1.68	1.0	−1.3 (±0.4)
C$^\alpha$-C* : 152	C$^\alpha$-CH$_2$-C$^\alpha$: 114	g : 70	ω = 0.089	1.3	6.7
C*-O : 136	C-C$^\alpha$-C* : 109.5	χ = 0, 180	ω' = 0.095	1.4	6.7
C=O : 122	C$^\alpha$-C*-O* : 121		ω'' = 0.096	1.2	6.3 (±1.3)
O-CH$_3$: 145	C*-O-CH$_3$: 113				
C-H : 110					

$$U_p = \begin{bmatrix} 1 & 1 \\ 1 & 0 \end{bmatrix} \qquad U_m = \begin{bmatrix} \omega'' & 1/\eta \\ 1/\eta & \omega/\eta^2 \end{bmatrix} \qquad U_r = \begin{bmatrix} 1 & \omega'/\eta \\ \omega'/\eta & 1/\eta^2 \end{bmatrix}$$

rows and columns: t, g

Comments:
Dipole moments of PMA are calculated with the RIS scheme with two states per bond. When the dipole moment of the ester group lateral to the chain is placed in the orientation deduced by semiempirical methods, calculated values of the dipole moment are in excellent agreement with experimental results without any kind of adjustement of the energy parameters obtained by *Flory* and co-workers (**A 075**).

Calcd. quantities: $<r^2>_0/nl^2$ $<\mu^2>/xm^2$ $d(\ln <\mu^2>)/dT$ Stereochemical equilibria   NMR coupling constants

A 078

esters of 2,4-dimethylglutaric acid: dimers of PMA

Saiz, J. S.; San Román, J.; Madruga, E. L.; Riande, E. *Macromolecules* **1989**, *22*, 1330.

Bond Length [pm]	Valence Angles [°]	Torsion Angles [°]	ξ a) (for 303 K)
C-C : 153	CH_3-C^α-CH_2: 112	$(tt)_m$: 164 / 164	ρ = 1.1
C^α-C* : 152	C^α-CH_2-C^α : 114	$(tg)_m$: 177 / 66	β = 1.8
C*-O : 136		$(gt)_m$: 66 / 177	γ_1 = 2.0
C=O : 122		$(tt)_r$: 177 / 177	γ_2 = 1.3
		χ b) = 0, 180	γ = 4.0

$$U_p = \begin{bmatrix} 1 & 0 & 1 & 0 \\ 0 & \rho & 0 & \rho \\ 1 & 0 & 0 & 0 \\ 0 & \rho & 0 & 0 \end{bmatrix} \quad U_r = \begin{bmatrix} 1 & \gamma_1 & 0 & 0 \\ \gamma_1 & \gamma_2 & 0 & 0 \\ 0 & 0 & 0 & 0 \\ 0 & 0 & 0 & 0 \end{bmatrix} \quad U_m = A \times \begin{bmatrix} 1 & \gamma & \beta & \beta \\ \gamma & 1 & \beta & \beta \\ \beta & \beta & 0 & 0 \\ \beta & \beta & 0 & 0 \end{bmatrix}$$

rows and columns: t_o, t_π, g_0, g_π

a) The statistical weight of $\chi = \pi$ relative to $\chi = 0$ orientations is denoted by ρ. The factors γ_1 and γ_2 arise from the differences in Coulombic interactions of $t_o t_\pi$ (or $t_\pi t_o$) and $t_\pi t_\pi$ relative to $t_o t_o$ in the *racemic* diad. In the *meso* diad, γ comes from the difference in Coulombic interactions between $t_o t_\pi$ (or $t_\pi t_o$) and $t_o t_o$ (or $t_\pi t_\pi$); A and β are combinations of statistical weights defined as A = $\omega'' \delta_m/\delta_r$ and $\beta = 1/\eta \delta_m \omega''$ with ω'' representing the weight for the second-order interaction between two ester groups juxtaposed as in the *meso*-tt state, η is the first-order statistical weight for *trans* versus *gauche*, and δ_m and δ_r represent Boltzmann factors of the Coulombic interactions in $t_o t_o$ orientations of *m* and *r* diads, respectively.
b) Side-group rotation angle.

Comments:
The dipole moments of model compounds of the diads of phenyl- and chlorophenyl-substituted acrylate polymers are measured in benzene solution. The results are interpreted in terms of the RIS model. A four-states model used in the analysis of PMA, after adjustment of some parameters, is able to reproduce the experimental values.

Calcd. quantities: $\langle\mu^2\rangle_m$ $\langle\mu^2\rangle_r$ $d(\ln\langle\mu^2\rangle)/dT$

A 079

poly(cyclohexyl acrylate), PCA

Diaz-Calleja, R.; Riande, E.; San Román, J. *Macromolecules* **1992**, *25*, 2875.

Bond Length [pm]	Valence Angles [°]	Torsion Angles [°]	ξ a) (for 303 K)
CH_2-C^α-CH_2: 112.5		$(tt)_m$: 164 / 164	ρ = 1.1
C^α-CH_2-C^α : 112.5		$(tg)_m$: 177 / 66	β = 1.8
		$(gt)_m$: 66, 177	γ_1 = 1.3
		$(tt)_r$: 177 / 177	γ_2 = 2.5 or 1.91
		ψ = ± 140 χ = 0, 180	γ = 4.0

$$U_p = \begin{bmatrix} 1 & 0 & \rho & 0 \\ 0 & \rho & 0 & \rho \\ 1 & 0 & 0 & 0 \\ 0 & \rho & 0 & 0 \end{bmatrix} \quad U_r = \begin{bmatrix} 1 & \gamma_1 & 0 & 0 \\ \gamma_1 & \gamma_2 & 0 & 0 \\ 0 & 0 & 0 & 0 \\ 0 & 0 & 0 & 0 \end{bmatrix} \quad U_m = \begin{bmatrix} 1 & \gamma & \beta & \beta \\ \gamma & 1 & \beta & \beta \\ \beta & \beta & 0 & 0 \\ \beta & \beta & 0 & 0 \end{bmatrix}$$

rows and columns: t_0, t_π, g_0, g_π (see: b))

a) The statistical weight of $\chi = \pi$ relative to $\chi = 0$ orientations is denoted by ρ. The factors γ_1 and γ_2 arise from the differences in Coulombic interactions of $t_0 t_\pi$ (or $t_\pi t_0$) and $t_\pi t_\pi$ relative to $t_0 t_0$ in the *racemic* diad. In the *meso* diad, γ comes from the difference in Coulombic interactions between $t_0 t_\pi$ (or $t_\pi t_0$) and $t_0 t_0$ (or $t_\pi t_\pi$); A and β are combinations of statistical weights defined as A = $\omega'' \delta_m / \delta_r$ and β = $1/\eta \delta_m \omega''$ with ω'' representing the weight for the second-order interaction between two ester groups juxtaposed as in the *meso*-tt state, η is the first-order statistical weight for *trans* versus *gauche*, and δ_m and δ_r represent Boltzmann factors of the Coulombic interactions in $t_0 t_0$ orientations of *m* and *r* diads, respectively.
b) The index corresponds to the value of χ.

Calcd. quantities: Total intramolecular correlation factor g_{intra} $d (\ln g_{intra}) / d T$

A 080

esters of 2,4-dimethylglutaric acid: phenyl ester, PDG

Saiz, J. S.; San Román, J.; Madruga, E. L.; Riande, E. *Macromolecules* **1989**, *22*, 1330.

Bond Length [pm]	Valence Angles [°]	Torsion Angles [°]	ξ a) (for 303 K)			
CH_3-C^α-CH_2: 112		$(tt)_m$: 164 / 164	ρ = 1.1			
C^α-CH_2-C^α : 114		$(tg)_m$: 177 / 66	β = 1.8	$U_p = \begin{bmatrix} 1 & 0 & 1 & 0 \\ 0 & \rho & 0 & \rho \\ 1 & 0 & 0 & 0 \\ 0 & \rho & 0 & 0 \end{bmatrix}$	$U_r = \begin{bmatrix} 1 & \gamma_1 & 0 & 0 \\ \gamma_1 & \gamma_2 & 0 & 0 \\ 0 & 0 & 0 & 0 \\ 0 & 0 & 0 & 0 \end{bmatrix}$	$U_m = A \times \begin{bmatrix} 1 & \gamma & \beta & \beta \\ \gamma & 1 & \beta & \beta \\ \beta & \beta & 0 & 0 \\ \beta & \beta & 0 & 0 \end{bmatrix}$
		$(gt)_m$: 66 / 177	γ_1 = 1.4			
		$(tt)_r$: 177 / 177	γ_2 = 2.5			
		χ = 0, 180	γ = 4.5		rows and columns: t_0, t_π, g_0, g_π	

a) The statistical weight of $\chi = \pi$ relative to $\chi = 0$ orientations is denoted by ρ. The factors γ_1 and γ_2 arise from the differences in Coulombic interactions of $t_0 t_\pi$ (or $t_\pi t_0$) and $t_\pi t_\pi$ relative to $t_0 t_0$ in the *racemic* diad. In the *meso* diad, γ comes from the difference in Coulombic interactions between $t_0 t_\pi$ (or $t_\pi t_0$) and $t_0 t_0$ (or $t_\pi t_\pi$); A and β are combinations of statistical weights defined as $A = \omega'' \delta_m / \delta_r$ and $\beta = 1/\eta \delta_m \omega''$ with ω'' representing the weight for the second-order interaction between two ester groups juxtaposed as in the *meso*-tt state, η is the first-order statistical weight for *trans* versus *gauche*, and δ_m and δ_r represent Boltzmann factors of the Coulombic interactions in $t_0 t_0$ orientations of *m* and *r* diads, respectively.

Comments:
The dipole moments of model compounds of the diads of phenyl- and chlorophenyl-substituted acrylate polymers are measured in benzene solution. The results are interpreted in terms of the RIS model. A four-states model used in the analysis of PMA, after adjustment of some parameters, is able to reproduce the experimental values.

Calcd. quantities: $\langle \mu^2 \rangle_m$ $\langle \mu^2 \rangle_r$ $d(\ln \langle \mu^2 \rangle)/dT$

A 081

poly(phenyl acrylate), PPA

Saiz, J. S.; Riande, E.; San Román, J.; Madruga, E. L. *Macromolecules* **1990**, *23*, 785.

Bond Length [pm]	Valence Angles [°]	Torsion Angles [°]	ξ [a) (for 303 K)
	CH_2-C^α-CH_2: 112	$(tt)_m$: 164 / 164	ρ = 1.1
	C^α-CH_2-C^α : 114	$(tg)_m$: 177 / 66	β = 1.8
		$(gt)_m$: 66 / 177	γ_1 = 1.4
		$(tt)_r$: 177 / 177	γ_2 = 2.5
		χ = 0, 180	γ = 4.5

$$U_p = \begin{bmatrix} 1 & 0 & 1 & 0 \\ 0 & \rho & 0 & \rho \\ 1 & 0 & 0 & 0 \\ 0 & \rho & 0 & 0 \end{bmatrix} \quad U_r = \begin{bmatrix} 1 & \gamma_1 & 0 & 0 \\ \gamma_1 & \gamma_2 & 0 & 0 \\ 0 & 0 & 0 & 0 \\ 0 & 0 & 0 & 0 \end{bmatrix} \quad U_m = A \times \begin{bmatrix} 1 & \gamma & \beta & \beta \\ \gamma & 1 & \beta & \beta \\ \beta & \beta & 0 & 0 \\ \beta & \beta & 0 & 0 \end{bmatrix}$$

rows and columns: t_0, t_π, g_0, g_π

a) The statistical weight of $\chi = \pi$ relative to $\chi = 0$ orientations is denoted by ρ. The factors γ_1 and γ_2 arise from the differences in Coulombic interactions of $t_0 t_\pi$ (or $t_\pi t_0$) and $t_\pi t_\pi$ relative to $t_0 t_0$ in the *racemic* diad. In the *meso* diad, γ comes from the difference in Coulombic interactions between $t_0 t_\pi$ (or $t_\pi t_0$) and $t_0 t_0$ (or $t_\pi t_\pi$); A and β are combinations of statistical weights defined as $A = \omega'' \delta_m/\delta_r$ and $\beta = 1/\eta \delta_m \omega''$ with ω'' representing the weight for the second-order interaction between two ester groups juxtaposed as in the *meso*-tt state, η is the first-order statistical weight for *trans* versus *gauche*, and δ_m and δ_r represent Boltzmann factors of the Coulombic interactions in $t_0 t_0$ orientations of m and r diads, respectively.

Comments:

Dielectric constants ε of benzene solutions of poly(phenyl acrylate) and poly(chlorophenyl acrylates) are measured. The dipole moments of the chains are interpreted by using a four-state scheme obtained by splitting each (t and g) rotational isomer into two rotational states that account for the position *cis* or *trans* of the ester group with respect to the methine hydrogen. The model gives a good account of the experimental results, assuming that the rotational angles about O—Ph bonds are $\psi = \pm 60°$ and $\pm 120°$ for PMCPA and $\psi = \pm 75°$ for POCPA.

Calcd. quantities: $\langle \mu^2 \rangle_m$ $\langle \mu^2 \rangle_r$ $d(\ln \langle \mu^2 \rangle)/dT$

esters of 2,4-dimethylglutaric acid: *ortho*-chlorophenyl ester, OCPDG

Saiz, J. S.; San Román, J.; Madruga, E. L.; Riande, E. *Macromolecules* **1989**, *22*, 1330.

Bond Length [pm]	Valence Angles [°]	Torsion Angles [°]	ξ a) (for 303 K)
	CH_3-C^α-CH_2: 112	$(tt)_m$: 164 / 164	ρ = 1.1
	C^α-CH_2-C^α : 114	$(tg)_m$: 177 / 66	β = 0.3
		$(gt)_m$: 66 / 177	γ_1 = 2.0
		$(tt)_r$: 177 / 177	γ_2 = 0.2
		χ = 0, 180	γ = 2.3

$$U_p = \begin{bmatrix} 1 & 0 & 1 & 0 \\ 0 & \rho & 0 & \rho \\ 1 & 0 & 0 & 0 \\ 0 & \rho & 0 & 0 \end{bmatrix} \quad U_r = \begin{bmatrix} 1 & \gamma_1 & 0 & 0 \\ \gamma_1 & \gamma_2 & 0 & 0 \\ 0 & 0 & 0 & 0 \\ 0 & 0 & 0 & 0 \end{bmatrix} \quad U_m = A \times \begin{bmatrix} 1 & \gamma & \beta & \beta \\ \gamma & 1 & \beta & \beta \\ \beta & \beta & 0 & 0 \\ \beta & \beta & 0 & 0 \end{bmatrix}$$

rows and columns: t_0, t_π, g_0, g_π

a) The statistical weight of $\chi = \pi$ relative to $\chi = 0$ orientations is denoted by ρ. The factors γ_1 and γ_2 arise from the differences in Coulombic interactions of $t_0 t_\pi$ (or $t_\pi t_0$) and $t_\pi t_\pi$ relative to $t_0 t_0$ in the *racemic* diad. In the *meso* diad, γ comes from the difference in Coulombic interactions between $t_0 t_\pi$ (or $t_\pi t_0$) and $t_0 t_0$ (or $t_\pi t_\pi$); A and β are combinations of statistical weights defined as $A = \omega'' \delta_m / \delta_r$ and $\beta = 1/\eta \delta_m \omega''$ with ω'' representing the weight for the second-order interaction between two ester groups juxtaposed as in the *meso*-tt state, η is the first-order statistical weight for *trans* versus *gauche*, and δ_m and δ_r represent Boltzmann factors of the Coulombic interactions in $t_0 t_0$ orientations of m and r diads, respectively.

Comments:
The dipole moments of model compounds of the diads of phenyl- and chlorophenyl-substituted acrylate polymers are measured in benzene solution. The results are interpreted in terms of the RIS model. A four-states model used in the analysis of PMA, after adjustment of some parameters, is able to reproduce the experimental values.

Calcd. quantities: $<\mu^2>_m$ $<\mu^2>_r$ $d (\ln <\mu^2>) / dT$

A 083

poly(ortho-chlorophenyl acrylate), POCPA

Saiz, J. S.; Riande, E.; San Román, J.; Madruga, E. L. *Macromolecules* **1990**, *23*, 785.

Bond Length [pm]	Valence Angles [°]	Torsion Angles [°]	ξ a) (for 303 K)			
CH_2-C^α-CH_2: 112		$(tt)_m$: 164 / 164	ρ = 1.1			
C^α-CH_2-C^α : 114		$(tg)_m$: 177 / 66	β = 0.3	$U_p = \begin{bmatrix} 1 & 0 & 1 & 0 \\ 0 & \rho & 0 & \rho \\ 1 & 0 & 0 & 0 \\ 0 & \rho & 0 & 0 \end{bmatrix}$	$U_r = \begin{bmatrix} 1 & \gamma_1 & 0 & 0 \\ \gamma_1 & \gamma_2 & 0 & 0 \\ 0 & 0 & 0 & 0 \\ 0 & 0 & 0 & 0 \end{bmatrix}$	$U_m = A \times \begin{bmatrix} 1 & \gamma & \beta & \beta \\ \gamma & 1 & \beta & \beta \\ \beta & \beta & 0 & 0 \\ \beta & \beta & 0 & 0 \end{bmatrix}$
		$(gt)_m$: 66 / 177	γ_1 = 2.0			
		$(tt)_r$: 177 / 177	γ_2 = 0.2			
		χ = 0, 180	γ = 2.3			rows and columns: t_0, t_π, g_0, g_π

a) The statistical weight of $\chi = \pi$ relative to $\chi = 0$ orientations is denoted by ρ. The factors γ_1 and γ_2 arise from the differences in Coulombic interactions of $t_0 t_\pi$ (or $t_\pi t_0$) and $t_\pi t_\pi$ relative to $t_0 t_0$ in the *racemic* diad. In the *meso* diad, γ comes from the difference in Coulombic interactions between $t_0 t_\pi$ (or $t_\pi t_0$) and $t_0 t_0$ (or $t_\pi t_\pi$); A and β are combinations of statistical weights defined as $A = \omega'' \delta_m / \delta_r$ and $\beta = 1/\eta \delta_m \omega''$ with ω'' representing the weight for the second-order interaction between two ester groups juxtaposed as in the *meso*-tt state, η is the first-order statistical weight for *trans* versus *gauche*, and δ_m and δ_r represent Boltzmann factors of the Coulombic interactions in $t_0 t_0$ orientations of *m* and *r* diads, respectively.

Comments:
Dielectric constants ε of benzene solutions of poly(phenyl acrylate) and poly(chlorophenyl acrylates) are measured. The dipole moments of the chains are interpreted by using a four-state scheme obtained by splitting each (t and g) rotational isomer into two rotational states that account for the position *cis* or *trans* of the ester group with respect to the methine hydrogen. The model gives a good account of the experimental results, assuming that the rotational angles about O—Ph bonds are $\psi = \pm 60°$ and $\pm 120°$ for PMCPA and $\psi = \pm 75°$ for POCPA.

Calcd. quantities: $<\mu^2>_m$ $<\mu^2>_r$ $d (\ln <\mu^2>) / dT$

esters of 2,4-dimethylglutaric acid: *meta*-chlorophenyl ester, MCPDG

Saiz, J. S.; San Román, J.; Madruga, E. L.; Riande, E. *Macromolecules* **1989**, *22*, 1330.

Bond Length [pm]	Valence Angles [°]	Torsion Angles [°]	ξ a) (for 303 K)			
	CH_3-C^α-CH_2: 112	$(tt)_m$: 164 / 164	ρ = 1.1			
	C^α-CH_2-C^α : 114	$(tg)_m$: 177 / 66	β = 2.4	$U_p = \begin{bmatrix} 1 & 0 & 1 & 0 \\ 0 & \rho & 0 & \rho \\ 1 & 0 & 0 & 0 \\ 0 & \rho & 0 & 0 \end{bmatrix}$	$U_r = \begin{bmatrix} 1 & \gamma_1 & 0 & 0 \\ \gamma_1 & \gamma_2 & 0 & 0 \\ 0 & 0 & 0 & 0 \\ 0 & 0 & 0 & 0 \end{bmatrix}$	$U_m = A \times \begin{bmatrix} 1 & \gamma & \beta & \beta \\ \gamma & 1 & \beta & \beta \\ \beta & \beta & 0 & 0 \\ \beta & \beta & 0 & 0 \end{bmatrix}$
		$(gt)_m$: 66 / 177	γ_1 = 0.2			
		$(tt)_r$: 177 / 177	γ_2 = 3.1			
		χ = 0, 180	γ = 0.2		rows and columns: t_0, t_π, g_0, g_π	

a) The statistical weight of $\chi = \pi$ relative to $\chi = 0$ orientations is denoted by ρ. The factors γ_1 and γ_2 arise from the differences in Coulombic interactions of $t_0 t_\pi$ (or $t_\pi t_0$) and $t_\pi t_\pi$ relative to $t_0 t_0$ in the *racemic* diad. In the *meso* diad, γ comes from the difference in Coulombic interactions between $t_0 t_\pi$ (or $t_\pi t_0$) and $t_0 t_0$ (or $t_\pi t_\pi$); A and β are combinations of statistical weights defined as $A = \omega'' \delta_m/\delta_r$ and $\beta = 1/\eta \delta_m \omega''$ with ω'' representing the weight for the second-order interaction between two ester groups juxtaposed as in the *meso*-tt state, η is the first-order statistical weight for *trans* versus *gauche*, and δ_m and δ_r represent Boltzmann factors of the Coulombic interactions in $t_0 t_0$ orientations of *m* and *r* diads, respectively.

Comments:
The dipole moments of model compounds of the diads of phenyl- and chlorophenyl-substituted acrylate polymers are measured in benzene solution. The results are interpreted in terms of the RIS model. A four-states model used in the analysis of PMA, after adjustment of some parameters, is able to reproduce the experimental values.

Calcd. quantities: $<\mu^2>_m$ $<\mu^2>_r$ $d (\ln <\mu^2>) / dT$

A 085

poly(meta-chlorophenyl acrylate), PMCPA

Saiz, J. S.; Riande, E.; San Román, J.; Madruga, E. L. *Macromolecules* **1990**, *23*, 785.

Bond Length [pm]	Valence Angles [°]	Torsion Angles [°]	ξ [a)] (for 303 K)
	CH_2-C^α-CH_2: 112	$(tt)_m$: 164 / 164	ρ = 1.1
	C^α-CH_2-C^α : 114	$(tg)_m$: 177 / 66	β = 2.4
		$(gt)_m$: 66 / 177	γ_1 = 0.2
		$(tt)_r$: 177 / 177	γ_2 = 3.1
		χ = 0, 180	γ = 0.2

$$U_p = \begin{bmatrix} 1 & 0 & 1 & 0 \\ 0 & \rho & 0 & \rho \\ 1 & 0 & 0 & 0 \\ 0 & \rho & 0 & 0 \end{bmatrix} \quad U_r = \begin{bmatrix} 1 & \gamma_1 & 0 & 0 \\ \gamma_1 & \gamma_2 & 0 & 0 \\ 0 & 0 & 0 & 0 \\ 0 & 0 & 0 & 0 \end{bmatrix} \quad U_m = A \times \begin{bmatrix} 1 & \gamma & \beta & \beta \\ \gamma & 1 & \beta & \beta \\ \beta & \beta & 0 & 0 \\ \beta & \beta & 0 & 0 \end{bmatrix}$$

rows and columns: t_0, t_π, g_0, g_π

a) The statistical weight of $\chi = \pi$ relative to $\chi = 0$ orientations is denoted by ρ. The factors γ_1 and γ_2 arise from the differences in Coulombic interactions of $t_0 t_\pi$ (or $t_\pi t_0$) and $t_\pi t_\pi$ relative to $t_0 t_0$ in the *racemic* diad. In the *meso* diad, γ comes from the difference in Coulombic interactions between $t_0 t_\pi$ (or $t_\pi t_0$) and $t_0 t_0$ (or $t_\pi t_\pi$); A and β are combinations of statistical weights defined as $A = \omega'' \delta_m / \delta_r$ and $\beta = 1/\eta \delta_m \omega''$ with ω'' representing the weight for the second-order interaction between two ester groups juxtaposed as in the *meso*-tt state, η is the first-order statistical weight for *trans* versus *gauche*, and δ_m and δ_r represent Boltzmann factors of the Coulombic interactions in $t_0 t_0$ orientations of *m* and *r* diads, respectively.

Comments:
Dielectric constants ε of benzene solutions of poly(phenyl acrylate) and poly(chlorophenyl acrylates) are measured. The dipole moments of the chains are interpreted by using a four-state scheme obtained by splitting each (t and g) rotational isomer into two rotational states that account for the position *cis* or *trans* of the ester group with respect to the methine hydrogen. The model gives a good account of the experimental results, assuming that the rotational angles about O–Ph bonds are $\psi = \pm 60°$ and $\pm 120°$ for PMCPA and $\psi = \pm 75°$ for POCPA.

Calcd. quantities: $<\mu^2>_m$ $<\mu^2>_r$ $d (\ln <\mu^2>) / dT$

esters of 2,4-dimethylglutaric acid: *para*-chlorophenyl ester, PCPDG

Saiz, J. S.; San Román, J.; Madruga, E. L.; Riande, E. *Macromolecules* **1989**, *22*, 1330.

Bond Length [pm]	Valence Angles [°]	Torsion Angles [°]	ξ [a)] (for 303 K)
CH_3-C^α-CH_2: 112		$(tt)_m$: 164 / 164	ρ = 1.1
C^α-CH_2-C^α : 114		$(tg)_m$: 177 / 66	β = 5.0
		$(gt)_m$: 66 / 177	γ_1 = 1.4
		$(tt)_r$: 177 / 177	γ_2 = 2.3
		χ = 0, 180	γ = 0.2

$$U_p = \begin{bmatrix} 1 & 0 & 1 & 0 \\ 0 & \rho & 0 & \rho \\ 1 & 0 & 0 & 0 \\ 0 & \rho & 0 & 0 \end{bmatrix} \quad U_r = \begin{bmatrix} 1 & \gamma_1 & 0 & 0 \\ \gamma_1 & \gamma_2 & 0 & 0 \\ 0 & 0 & 0 & 0 \\ 0 & 0 & 0 & 0 \end{bmatrix} \quad U_m = A \times \begin{bmatrix} 1 & \gamma & \beta & \beta \\ \gamma & 1 & \beta & \beta \\ \beta & \beta & 0 & 0 \\ \beta & \beta & 0 & 0 \end{bmatrix}$$

rows and columns: t_0, t_π, g_0, g_π

a) The statistical weight of $\chi = \pi$ relative to $\chi = 0$ orientations is denoted by ρ. The factors γ_1 and γ_2 arise from the differences in Coulombic interactions of $t_0 t_\pi$ (or $t_\pi t_0$) and $t_\pi t_\pi$ relative to $t_0 t_0$ in the *racemic* diad. In the *meso* diad, γ comes from the difference in Coulombic interactions between $t_0 t_\pi$ (or $t_\pi t_0$) and $t_0 t_0$ (or $t_\pi t_\pi$); A and β are combinations of statistical weights defined as $A = \omega''\delta_m/\delta_r$ and $\beta = 1/\eta\delta_m\omega''$ with ω'' representing the weight for the second-order interaction between two ester groups juxtaposed as in the *meso*-tt state, η is the first-order statistical weight for *trans* versus *gauche*, and δ_m and δ_r represent Boltzmann factors of the Coulombic interactions in $t_0 t_0$ orientations of *m* and *r* diads, respectively.

Comments:
The dipole moments of model compounds of the diads of phenyl- and chlorophenyl-substituted acrylate polymers are measured in benzene solution. The results are interpreted in terms of the RIS model. A four-states model used in the analysis of PMA, after adjustment of some parameters, is able to reproduce the experimental values.

Calcd. quantities: $\langle\mu^2\rangle_m$ $\quad$ $\langle\mu^2\rangle_r$ $\quad$ $d(\ln\langle\mu^2\rangle)/dT$

A 087

poly(para-chlorophenyl acrylate), PPCPA

Saiz, J. S.; Riande, E.; San Román, J.; Madruga, E. L. *Macromolecules* **1990**, *23*, 785.

Bond Length [pm]	Valence Angles [°]	Torsion Angles [°]	ξ [a)] (for 303 K)
	CH_2-C^α-CH_2: 112	$(tt)_m$: 164 / 164	ρ = 1.1
	C^α-CH_2-C^α : 114	$(tg)_m$: 177 / 66	β = 5.0
		$(gt)_m$: 66 / 177	γ_1 = 1.4
		$(tt)_r$: 177 / 177	γ_2 = 2.3
		χ = 0, 180	γ = 0.2

$$U_p = \begin{bmatrix} 1 & 0 & 1 & 0 \\ 0 & \rho & 0 & \rho \\ 1 & 0 & 0 & 0 \\ 0 & \rho & 0 & 0 \end{bmatrix} \quad U_r = \begin{bmatrix} 1 & \gamma_1 & 0 & 0 \\ \gamma_1 & \gamma_2 & 0 & 0 \\ 0 & 0 & 0 & 0 \\ 0 & 0 & 0 & 0 \end{bmatrix} \quad U_m = A \times \begin{bmatrix} 1 & \gamma & \beta & \beta \\ \gamma & 1 & \beta & \beta \\ \beta & \beta & 0 & 0 \\ \beta & \beta & 0 & 0 \end{bmatrix}$$

rows and columns: t_o, t_π, g_o, g_π

a) The statistical weight of $\chi = \pi$ relative to $\chi = 0$ orientations is denoted by ρ. The factors γ_1 and γ_2 arise from the differences in Coulombic interactions of $t_o t_\pi$ (or $t_\pi t_o$) and $t_\pi t_\pi$ relative to $t_o t_o$ in the *racemic* diad. In the *meso* diad, γ comes from the difference in Coulombic interactions between $t_o t_\pi$ (or $t_\pi t_o$) and $t_o t_o$ (or $t_\pi t_\pi$); A and β are combinations of statistical weights defined as $A = \omega'' \delta_m / \delta_r$ and $\beta = 1/\eta \delta_m \omega''$ with ω'' representing the weight for the second-order interaction between two ester groups juxtaposed as in the *meso*-tt state, η is the first-order statistical weight for *trans* versus *gauche*, and δ_m and δ_r represent Boltzmann factors of the Coulombic interactions in $t_o t_o$ orientations of *m* and *r* diads, respectively.

Comments:
Dielectric constants ϵ of benzene solutions of poly(phenyl acrylate) and poly(chlorophenyl acrylates) are measured. The dipole moments of the chains are interpreted by using a four-state scheme obtained by splitting each (t and g) rotational isomer into two rotational states that account for the position *cis* or *trans* of the ester group with respect to the methine hydrogen. The model gives a good account of the experimental results, assuming that the rotational angles about O–Ph bonds are $\psi = \pm 60°$ and $\pm 120°$ for PMCPA and $\psi = \pm 75°$ for POCPA.

Calcd. quantities: $<\mu^2>_m$ $<\mu^2>_r$ $d (\ln <\mu^2>) / dT$

A 088

4-biphenyl 2,4-dimethylglutarate, PPG

Saiz, J. S.; Riande, E.; San Román, J.; Madruga, E. L. *Macromolecules* **1990**, *23*, 3491.

Bond Length [pm]	Valence Angles [°]	Torsion Angles [°]	ξ [a] (for 303 K)
	CH_3-C^α-CH_2: 112	$(tt)_m$: 164 / 164	ρ = 1.1
	C^α-CH_2-C^α : 112	$(tg)_m$: 177 / 66	β = 1.8
		$(gt)_m$: 66 / 177	γ_1 = 1.4
		$(tt)_r$: 177 / 177	γ_2 = 2.5
		χ = 0, 180	γ = 4.5
			$\alpha \approx 0$

$$U_p = \begin{bmatrix} 1 & 0 & 1 & 0 \\ 0 & \rho & 0 & \rho \\ 1 & 0 & 0 & 0 \\ 0 & \rho & 0 & 0 \end{bmatrix} \quad U_r = \begin{bmatrix} 1 & \gamma_1 & 0 & 0 \\ \gamma_1 & \gamma_2 & 0 & 0 \\ 0 & 0 & \alpha & \alpha \\ 0 & 0 & \alpha & \alpha \end{bmatrix} \quad U_m = A \times \begin{bmatrix} 1 & \gamma & \beta & \beta \\ \gamma & 1 & \beta & \beta \\ \beta & \beta & 0 & 0 \\ \beta & \beta & 0 & 0 \end{bmatrix}$$

rows and columns: t_0, t_π, g_0, g_π

[a] The statistical weight of $\chi = \pi$ relative to $\chi = 0$ orientations is denoted by ρ. The factors γ_1 and γ_2 arise from the differences in Coulombic interactions of $t_0 t_\pi$ (or $t_\pi t_0$) and $t_\pi t_\pi$ relative to $t_0 t_0$ in the *racemic* diad. In the *meso* diad, γ comes from the difference in Coulombic interactions between $t_0 t_\pi$ (or $t_\pi t_0$) and $t_0 t_0$ (or $t_\pi t_\pi$); A and β are combinations of statistical weights defined as $A = \omega'' \delta_m / \delta_r$ and $\beta = 1/\eta \delta_m \omega''$ with ω'' representing the weight for the second-order interaction between two ester groups juxtaposed as in the *meso*-tt state, η is the first-order statistical weight for *trans* versus *gauche*, and δ_m and δ_r represent Boltzmann factors of the Coulombic interactions in $t_0 t_0$ orientations of *m* and *r* diads, respectively. α governs the stability of gg versus tt in the *racemic* diad: $\alpha = [(1 + 2\gamma_1 \rho + \gamma_2 \rho^2) f_{gg}] / [(1 + 2\rho + \rho^2)(1-f_{gg})]$, $f_{gg} \approx 0.18$.

Comments:
The dipole moment of model compounds of the repeating unit of polyacrylates with rigid side groups are measured in benzene. The experimental results of the diesters are interpreted using a four-state RIS scheme. Good agreement between experimental and theoretical values is found by using the statistical weight parameters utilized in the theoretical calculations of the diads of phenyl and chlorophenyl esters of acrylic acid.

Calcd. quantities: $<\mu^2>_m$ $<\mu^2>_r$ $d (\ln <\mu^2>) / dT$

A 089

4-phenoxyphenyl 2,4-dimethylglutarate and
4-benzoylphenyl 2,4-dimethylglutarate

Saiz, J. S.; Riande, E.; San Román, J.; Madruga, E. L. *Macromolecules* **1990**, *23*, 3491.

Bond Length [pm]	Valence Angles [°]	Torsion Angles [°]	ξ a) (for 303 K)			
	CH_3-C^α-CH_2: 112	$(tt)_m$: 164 / 164	$\rho = 1.1$			
	C^α-CH_2-C^α : 112	$(tg)_m$: 177 / 66	$\beta = 5.0$	$U_p = \begin{bmatrix} 1 & 0 & 1 & 0 \\ 0 & \rho & 0 & \rho \\ 1 & 0 & 0 & 0 \\ 0 & \rho & 0 & 0 \end{bmatrix}$	$U_r = \begin{bmatrix} 1 & \gamma_1 & 0 & 0 \\ \gamma_1 & \gamma_2 & 0 & 0 \\ 0 & 0 & \alpha & \alpha \\ 0 & 0 & \alpha & \alpha \end{bmatrix}$	$U_m = A \times \begin{bmatrix} 1 & \gamma & \beta & \beta \\ \gamma & 1 & \beta & \beta \\ \beta & \beta & 0 & 0 \\ \beta & \beta & 0 & 0 \end{bmatrix}$
		$(gt)_m$: 66 / 177	$\gamma_1 = 1.4$			
		$(tt)_r$: 177 / 177	$\gamma_2 = 2.3$			
			$\gamma = 0.2$		rows and columns: t_0, t_π, g_0, g_π	
		$\chi = 0, 180$	$\alpha \approx 0$			

a) The statistical weight of $\chi = \pi$ relative to $\chi = 0$ orientations is denoted by ρ. The factors γ_1 and γ_2 arise from the differences in Coulombic interactions of $t_0 t_\pi$ (or $t_\pi t_0$) and $t_\pi t_\pi$ relative to $t_0 t_0$ in the *racemic* diad. In the *meso* diad, γ comes from the difference in Coulombic interactions between $t_0 t_\pi$ (or $t_\pi t_0$) and $t_0 t_0$ (or $t_\pi t_\pi$); A and β are combinations of statistical weights defined as $A = \omega'' \delta_m/\delta_r$ and $\beta = 1/\eta \delta_m \omega''$ with ω'' representing the weight for the second-order interaction between two ester groups juxtaposed as in the *meso*-tt state, η is the first-order statistical weight for *trans* versus *gauche*, and δ_m and δ_r represent Boltzmann factors of the Coulombic interactions in $t_0 t_0$ orientations of *m* and *r* diads, respectively. α governs the stability of gg versus tt in the *racemic* diad: $\alpha = [(1 + 2\gamma_1 \rho + \gamma_2 \rho^2) f_{gg}] / [(1 + 2\rho + \rho^2)(1-f_{gg})]$, $f_{gg} \approx 0.18$.

Comments:

The dipole moment of model compounds of the repeating unit of polyacrylates with rigid side groups are measured in benzene. The experimental results of the diesters are interpreted using a four-state RIS scheme. Good agreement between experimental and theoretical values is found by using the statistical weight parameters utilized in the theoretical calculations of the diads of phenyl and chlorophenyl esters of acrylic acid.

Calcd. quantities: $<\mu^2>_m$ $<\mu^2>_r$ $d (\ln <\mu^2>) / dT$

poly(4-phenoxyphenyl acrylate), PPOA and 4-phenoxyphenyl-2,4-dimethylglutarate, PPOGD

Díaz-Calleja, R.; Riande, E.; San Román, J. *Polymer* **1993**, *34*, 3456.

Bond Length [pm]	Valence Angles [°]	Torsion Angles [°]	ξ (for 303 K)
	CH_2-C^α-CH_2: 111.5	$(tt)_m$: 164 / 164	$\rho = 1.1$
	C^α-CH_2-C^α : 111.5	$(tg)_m$: 177 / 66	$\beta = 3.68$
		$(tt)_r$: 177 / 177	$\gamma_1 = 1.4$
			$\gamma_2 = 1.90$
		$\chi = 0, 180$	
			$\gamma = 0.2$

$$U_p = \begin{bmatrix} 1 & 0 & 1 & 0 \\ 0 & \rho & 0 & \rho \\ 1 & 0 & 0 & 0 \\ 0 & \rho & 0 & 0 \end{bmatrix} \quad U_r = \begin{bmatrix} 1 & \gamma_1 & 0 & 0 \\ \gamma_1 & \gamma_2 & 0 & 0 \\ 0 & 0 & 0 & 0 \\ 0 & 0 & 0 & 0 \end{bmatrix} \quad U_m = \begin{bmatrix} 1 & \gamma & \beta & \beta \\ \gamma & 1 & \beta & \beta \\ \beta & \beta & 0 & 0 \\ \beta & \beta & 0 & 0 \end{bmatrix}$$

rows and columns: t_0, t_π, g_0, g_π

Comments:
Dielectric measurements are carried out on PPOA and PPODG. The dielectric spectrum of PPOA in the bulk presents a prominent glass-rubber relaxation followed by a subglass absorption. The low-molecular-weight compound only exhibits a prominent glass-liquid absorption followed by a diffuse and weak subglass relaxation. This behaviour cannot be explained in terms of only intramolecular interactions, and therefore intermolecular interactions must play an important role in this process.

Calcd. quantities: Total intramolecular correlation factor $g_{intra} = (<\mu^2> / x) / (<\mu^2>_0)$ $d (\ln g_{intra}) / dT$

A 091

poly(α-methylstyrene), PMS

Sundararajan, P. R. *Macromolecules* **1977**, *10*, 623.

Bond Length [pm]	Valence Angles [°]	Torsion Angles [°] [b)]	ξ [a)] (for 300 K)	ξ_o	E_ξ [kJ·mol^{-1}]
C-C : 153	CH$_2$-C$^\alpha$-CH$_2$: 109.5	for meso: tt : 173,173	α = 0.26	0.6	2.1
C-C^{ar} : 151	C$^\alpha$-CH$_2$-C$^\alpha$: 122	gt : 75,166 gg : 60, 60	β = 4.85	0.9	−4.2
C-H : 110	CH$_2$-C$^\alpha$-C^{ar} : 109.5				
C^{ar}-C^{ar}: 134	C^{ar}-C^{ar}-C^{ar} : 120	for racemic: tt : 166,166			
C$^\alpha$-CH$_3$: 153	C^{ar}-C^{ar}-H : 120				

$$U_p = \begin{bmatrix} 1 & 1 \\ 1 & 0 \end{bmatrix} \quad U_m = \begin{bmatrix} \eta^2\omega'' & \eta\omega\omega' \\ \eta\omega\omega' & \omega^2 \end{bmatrix} \quad U_r = \begin{bmatrix} \eta^2\omega'/2 & \eta\omega\omega' \\ \eta\omega\omega' & \omega^2 \end{bmatrix}$$

After normalizing with respect to the *meso*-tt state:

$$U_m = \begin{bmatrix} 1 & \alpha \\ \alpha & \alpha^2/\beta \end{bmatrix} \quad U_r = \begin{bmatrix} \beta & \alpha \\ \alpha & \alpha^2/\beta \end{bmatrix}$$

rows and columns: t, g

a) $\alpha = \omega'/\eta\omega''$; $\beta = \omega'^2/\omega\omega''$
b) The critical solvent-interaction distance $\sigma = 5$ Å

Comments:
Energy differences between the various conformational states of the *meso* and *racemic* diads of PMS are estimated using semiempirical methods. A crude method of including the conformation-dependent solvent interactions is applied in estimating the energies. The tt state of the *racemic* diad is about 4 kJ · mol^{-1} lower in energy than that of the *meso* diad. The gt state of *meso* and *racemic* diads are almost of equal energy to the *meso*-tt state. The characteristic ratios are calculated for *isotactic*, *syndiotactic*, and *stereoirregular* chains. The calculated values show fair agreement with the experimental results. The temperature dependence of the characteristic ratio for *atactic* PMS is evaluated and it parallels experimental observations.

Calcd. quantities: $<r^2>_o / nl^2$ $\quad d(\ln <r^2>)/dT$

poly(methyl methacrylate), PMMA

Sundararajan, P. R.; Flory, P. J. *J. Am. Chem. Soc.* **1974**, *96*, 5025.

Bond Length [pm]	Valence Angles [°]	Torsion Angles [°]	ξ (for 300 K)	ξ_0	E_ξ [kJ·mol^{-1}]
C^α-CH_2 : 153	CH_2-C^α-CH_2: 110	t : 180	$\alpha = 0.25$	1.6	4.6
C^α-CH_3 : 153	C^α-CH_2-C^α : 122	g : 60	$\beta = 3.81$	1.4	−2.5
C^α-C^* : 152	CH_2-C^α-C^* : 109.5				
C^*-O : 136	C^α-C^*-O : 114		$\alpha = \omega'/\eta\omega''$		
C^*=O^* : 122	C^α-C^*=O^* : 121		$\beta = \omega'^2/\omega\omega''$		
O-CH_3 : 145	C^*-O-CH_3 : 113				

$$U_p = \begin{bmatrix} 1 & 1 \\ 1 & 0 \end{bmatrix} \qquad U_m = \begin{bmatrix} 1 & \alpha \\ \alpha & \alpha^2/\beta \end{bmatrix} \qquad U_r = \begin{bmatrix} \beta & \alpha \\ \alpha & \alpha^2/\beta \end{bmatrix}$$

rows and columns: t, g

Comments:
Conformational energy calculations indicate the $\bar{g}$ state to be at least 30 kJ·mol^{-1} higher than the t and the g states. With the exclusion of the former conformation, all interactions of long range are eliminated, and the statistical weight matrices for the respective bond pairs reduce to 2 × 2 order.

Calcd. quantities: $\langle r^2 \rangle_0 / nl^2$ $\qquad d(\ln \langle r^2 \rangle_0)/dT$

A 093

poly(methyl methacrylate), PMMA

Vacatello, M.; Flory, P. J. *Macromolecules* **1986**, *19*, 405.

Bond Length [pm]	Valence Angles [°]	Torsion Angles [°]	ξ a) (for 300 K)	ξ_0	E_ξ [kJ · mol^{-1}]
C-C : 153	CH_2-C^α-CH_2 :	t_- : −160	α = 2.33	1.09	−1.9
C^α-C^* : 152	106 (for tt)				
C^*-O : 136	111 (for tg)	t_+ : 170	α^- = 3.46	0.96	−3.2
C=O : 122	116 (for gg)				
O-CH_3 : 145		g_- : 80	β = 358.3	1.16	−14.3
C-H : 110	C^α-CH_2-C^α : 114				
		g_+ : 55	ρ^* = 0.59	0.53	−0.25
	C-C-H : 109.5				
	H-C-H : 107.5	$\bar{g}_-$: −55	ρ^- = 23.6	0.64	−9.0
	C-C=O : 122				
	C-C-O : 114	$\bar{g}_+$: −80			
	O-C=O : 124				
	C-O-C : 110				

$$U_m = \begin{bmatrix} 0 & 1 & 0 & \alpha\beta & 0 & \alpha\alpha^- \\ 0 & \alpha\beta & 0 & \alpha\alpha^- & 0 & \\ 0 & \beta^2 & 0 & \alpha^-\beta & & \\ 0 & \alpha^-\beta & 0 & & & \\ \alpha^{-2} & 0 & & & & \\ 0 & & & & & \end{bmatrix} \qquad U_r = \begin{bmatrix} \alpha^2 & 0 & \alpha\beta & 0 & \rho^- & 0 \\ \alpha^2 & 0 & \alpha\beta & 0 & \rho^- & \\ \beta^2 & 0 & \alpha^2\beta & 0 & & \\ \beta^2 & 0 & \alpha^-\beta & & & \\ \alpha^{-2} & 0 & & & & \\ \alpha^{-2} & & & & & \end{bmatrix}$$

rows and columns: $t_-, t_+, g_-, g_+, \bar{g}_-, \bar{g}_+$

For U_p, see: b)

a) The statistical weights may be normalized to unity for the *meso* $t_- t_+$ diad. Then, the substitution $\alpha = v/(\rho\omega)^{1/2}$, $\alpha^- = v'/(\rho\omega)^{1/2}$, $\beta = (\omega/\rho)^{1/2}$, $\rho^- = \rho'/\rho$, $\rho^* = \rho''/\rho$ is performed.
b) Values for the interdiad statistical weight matrix, U_p, in which all first-order interactions are included, are given in the original paper, Table VI.

Comments:
Conformational energy calculations are carried out for monomeric and trimeric oligomers of PMMA and for four-bond segments embedded in stereoregular PMMA chains. All incident interactions are taken into account.

Calcd. quantities: *A priori* probabilities $\langle r^2 \rangle_0 / nl^2$

poly(methyl methacrylate), PMMA

Sundararajan, P. R. *Macromolecules* **1986**, *19*, 415.

Bond Length [pm]	Valence Angles [°]	Torsion Angles a) [°]		ξ b) (for 300 K)	ξ_0	E_ξ [kJ·mol^{-1}]
C-C : 153	CH$_2$-C$^\alpha$-CH$_2$: 112	(tt)$_m$	: −177.2 / −177.2	α = 0.16	0.85	4.15
	C$^\alpha$-CH$_2$-C$^\alpha$: 124	(tg)$_m$	: 178.2 / 67.2			
C$^\alpha$-C* : 152	C-C$^\alpha$-C* : 112	(t$\bar{g}$)$_m$	: −175.9 / −64.2	β = 1.4	0.77	−1.50
	C-C$^\alpha$-H : 107	(gg)$_m$	: 60.7 / 60.7			
C*-O : 136	C$^\alpha$-C-H : 110	(g$\bar{g}$)$_m$	: 67 / −70.6	ρ = 0.16	0.87	4.25
	H-C-H : 110	($\bar{g}\,\bar{g}$)$_m$	: −61.7 / −61.7			
C=O : 122	C$^\alpha$-C*=O : 114	(tt)$_r$	: 178.3 / 178.3	ψ = 0.015	1	10.5
	C$^\alpha$-C*-O* : 121	(tg)$_r$	: −175.8 / 60.8			
O-CH$_3$: 145	C*-O-CH$_3$: 113	(t$\bar{g}$)$_r$	: 176.5 / −70.1			
		(gg)$_r$	: 68.1 / 68.1			
		(g$\bar{g}$)$_r$	: 56 / −68.9			
		($\bar{g}\,\bar{g}$)$_r$	: −68.1 / −68.1			

$$U_p = \begin{bmatrix} 1 & 1 & 1 \\ 1 & 0 & \psi \\ 1 & \psi & 0 \end{bmatrix} \quad \text{rows and columns: t, g, } \bar{g}$$

$$U_m = \begin{bmatrix} \eta^2 \omega\omega'' & \eta\omega\omega' & \eta\tau\omega'^2 \\ \eta\omega\omega' & \omega^2 & \tau\omega\omega' \\ \eta\tau\omega'^2 & \tau\omega\omega' & \tau^2\omega\omega'' \end{bmatrix} = \begin{bmatrix} 1 & \alpha & \beta\rho \\ \alpha & \alpha^2/\beta & \alpha\rho \\ \beta\rho & \alpha\rho & \rho^2 \end{bmatrix}$$

$$U_r = \begin{bmatrix} \eta^2\omega'^2 & \eta\omega\omega' & \eta\tau\omega\omega'' \\ \eta\omega\omega' & \omega^2 & \tau\omega\omega' \\ \eta\tau\omega\omega'' & \tau\omega\omega' & \tau^2\omega'^2 \end{bmatrix} = \begin{bmatrix} \beta & \alpha & \rho \\ \alpha & \alpha^2/\beta & \alpha\rho \\ \rho & \alpha\rho & \beta\rho^2 \end{bmatrix}$$

a) Other combinations are tested as well in the original paper.
b) $\alpha = \omega' / \eta\omega''$; $\beta = \omega'^2 / \omega\omega''$; $\rho = \tau / \eta$

Comments:
Conformational energies associated with the various states of the *racemic* and *meso* diads of PMMA are estimated. The skeletal bond angle at the methylene carbon atom and the side-group torsion angles are varied in order to minimize the energy at each of the skeletal conformations. In contrast to the previous calculations with fixed side-group torsion angles, the present results show that the $\bar{g}$ conformation is accessible to the chain. Agreement between the experimental and theoretical values of the characteristic ratio, for the isotactic chain, is achieved if the tt state is treated with 170,−165°, leading to helical segments with a large pitch. The energy differences between the different states estimated from the calculations are compared with the values derived from FT-IR experiments.

Calcd. quantities: $\quad <r^2>_0 / nl^2 \quad\quad d (\ln <r^2>) / dT$

A 095

poly(ethyl methacrylate), PEMA

Kuntman, A.; Bahar, I.; Baysal, B. M. *Macromolecules* **1990**, *23*, 4959.

Bond Length [pm]	Valence Angles [°]	Torsion Angles [°]		ξ a) (for 300 K)	ξ_0	E_ξ [kJ·mol^{-1}]			
C-C : 153	CH$_2$-C$^\alpha$-CH$_2$: 110	$(tt)_m$	: $-177.2 / -177.2$	$\alpha = 0.22$	0.85	3.35			
		$(tg)_m$	: $178.2 / 67.2$						
C$^\alpha$-C* : 152	C$^\alpha$-CH$_2$-C$^\alpha$: 124	$(t\bar{g})_m$	: $-175.9 / -64.2$	$\beta = 1.79$	0.77	-2.1	$U_p = \begin{bmatrix} 1 & 1 & 1 \\ 1 & 0 & \psi \\ 1 & \psi & 0 \end{bmatrix}$	rows and columns: t, g, $\bar{g}$	
		$(gg)_m$	: $60.7 / 60.7$						
C*-O : 136		$(g\bar{g})_m$	: $67 / -70.6$	$\rho = 0.37$	0.87	2.1			
		$(\bar{g}g)_m$	: $-61.7 / -61.7$						
C=O : 122		$(tt)_r$	: $178.3 / 178.3$	$\psi = 0.015$	1	10.5			
		$(tg)_r$	: $-175.8 / 60.8$						
		$(t\bar{g})_r$	: $176.5 / -70.1$				$U_m = \begin{bmatrix} 1 & \alpha & \beta\rho \\ \alpha & \alpha^2/\beta & \alpha\rho \\ \beta\rho & \alpha\rho & \rho^2 \end{bmatrix}$	$U_r = \begin{bmatrix} \beta & \alpha & \rho \\ \alpha & \alpha^2/\beta & \alpha\rho \\ \rho & \alpha\rho & \beta\rho^2 \end{bmatrix}$	
		$(gg)_r$	: $68.1 / 68.1$						
		$(g\bar{g})_r$	: $56 / -68.9$						
		$(\bar{g}g)_r$	: $-68.1 / -68.1$						

a) $\alpha = \omega' / \eta\omega''$; $\beta = \omega'^2 / \omega\omega''$; $\rho = \tau / \eta$

Comments:
Dipole measurements of PEMA are performed in different solvents, and experimental measurements are theoretically analyzed on the basis of the RIS formalism. Energetic and geometric parameters proposed for PMMA are found to give a reasonable account of the experimentally measured values, in general. However, the change in average dipoles with temperature is found to exhibit a substantial dependence on the type of solvent, which indicates the importance of specific solvent-polymer interaction on the conformational characteristics of the chain.

Calcd. quantities: $<r^2>_0 / nl^2$ $<\mu^2> / xm^2$ $d (\ln <\mu^2>) / dT$

*Further calculations on **poly(methyl methacrylate)** chains:*

A 096   Jenkins, R.; Porter, R. S. *Polymer* **1982**, *23*, 105.

The results of measurements of unperturbed dimensions for a series of stereoregular PMMA are reported. Comparison of the results obtained with those predicted by previous RIS models show good agreement.

A 097   Vacatello, M.; Yoon, D. Y.; Flory, P. J. *Macromolecules* **1990**, *23*, 1993.

The molecular scattering functions of *isotactic*, *syndiotactic*, and *atactic* PMMA chains are calculated according to the three different RIS models in the literature [two-state model (**A 092**), three-state model (**A 094**), and six-state model (**A 093**)] and compared with experiments. Comparison with the neutron scattering experiments on glassy (*atactic*) PMMA shows that only the most rigorous six-state model predicts all the features exhibited by the experimental results over the entire scattering vector. In particular, the two-state and the three-state model predict the occurrence of a second maximum in the absolute-scale Kratky plot at values of q that are considerably smaller than the experimental results. In contrast, the scattering function calculated in the framework of the six-state model shows the two maxima and the intervening minimum in the same locations of q as those in the experimental curve. The six-state RIS model also predicts the molecular scattering functions in good agreement with the available neutron scattering results on the *isotactic* and *syndiotactic* PMMA in the bulk, as well as the X-ray scattering experiments on *syndiotactic* PMMA in solution.

A 098   Mora, M. A.; Rubio-Arroyo, M. F.; Salcedo, R. *Polymer* **1994**, *35*, 1078.

MO calculations are carried out for monomeric, dimeric and trimeric oligomers of PMMA. The minimum-energy geometry is achieved in all cases, and the geometrical and electronic parameters obtained are compared with experimental data. A tttg$^+$ conformation along the main-chain skeletal sequence is found.

A 099

alternating poly(ethylene-propylene) copolymers, PEP

Zirkel, A.; Richter, D. Pyckhout-Hintzen, W.; Fetters, L. J. *Macromolecules* **1992**, *25*, 954.

Bond Length [pm]	Valence Angles [°]	Torsion Angles [°]	ξ (for 298 K)	ξ_0	E_ξ [kJ·mol^{-1}]
C-C : 154	C-C-C : 112	t : 180	$\eta = 1$	1	0
	($\Delta\theta = 3°$)	g$^+$: 60	$\omega = 0.017$	1	10.0
		g$^-$: −60	$\sigma = 0.36$	1	2.5
			$\tau = 0.43$	1	2.1

$$U_e = \begin{bmatrix} 1 & \sigma & \sigma \\ 1 & \sigma & \sigma\omega \\ 1 & \sigma\omega & \sigma \end{bmatrix} \quad U_d = \begin{bmatrix} \eta & 1 & \tau \\ \eta & 1 & \tau\omega \\ \eta & \omega & \tau \end{bmatrix}$$

$$U_{de} = \begin{bmatrix} 1 & \sigma\omega & \sigma \\ 1 & \sigma & \sigma\omega \\ 1 & \sigma\omega & \sigma\omega \end{bmatrix} \quad U_{ed} = \begin{bmatrix} \eta & \tau & 1 \\ \eta & \tau\omega & \omega \\ \eta\omega & \tau\omega & 1 \end{bmatrix}$$

rows and columns: t, g$^+$, g$^-$

Comments:
The unperturbed dimensions of alternating atactic PEP are investigated by small-angle neutron scattering (SANS) over a wide temperature range. The results are compared with a RIS calculation. The PEP samples are prepared by the hydrogenation of essentially 1,4-polyisoprene. Experimental and theoretical results show good agreement as far as low and intermediate temperatures are concerned. Moreover, the temperature coefficient for chain dimensions of $-1.16 (\pm 0.03) \times 10^{-3}$ K^{-1} extracted from the data in the range of 298 - 453 K is in good agreement with the value of $-1 (\pm 0.2) \times 10^{-3}$ K^{-1} obtained from the measurements in ϑ solution over a smaller temperature range. Above 453 K, discrepancies emerge between theory and experiment.

Calcd. quantities: $<r^2>_0 / nl^2$ $<s^2>_0 / nl^2$ Temperature coefficients

branch formation in the free radical initiated polymerization of vinyl monomers

Mattice, W. L.; Viswanadhan, V. N. *Macromolecules* **1986**, *19*, 568.

Bond Length [pm]	Valence Angles [°]	Torsion Angles [°]
	C-C-C : 112	t : 180
		g : 60, 55, *or* 50
		$\bar{g}$: −60, −55, *or* −50

$$U_1 = [\eta^* \ 1 \ \eta^*]$$

$$U_2 = \begin{bmatrix} \eta\omega'' & 1 & \tau\omega' \\ \eta & 1 & \tau \\ \eta & \omega' & \tau\omega'' \end{bmatrix} \qquad U_p = \begin{bmatrix} \eta & 1 & \tau \\ \eta & \omega & \tau \\ \eta & 1 & \tau\omega \end{bmatrix}$$

$$U_m = \begin{bmatrix} \eta\omega'' & 1 & \tau\omega' \\ \eta & \omega & \tau\omega' \\ \eta\omega' & \omega' & \tau\omega\omega'' \end{bmatrix} \qquad U_r = \begin{bmatrix} \eta & \omega' & \tau\omega'' \\ \eta\omega' & 1 & \tau\omega \\ \eta\omega' & \omega & \tau\omega'^2 \end{bmatrix}$$

rows and columns: t, g, $\bar{g}$

Parameter sets:

Polymer	η	τ	ω	ω'	ω''
Poly(vinyl bromide)	1.3	0.43	0.015	0.035	0.0024
Poly(styrene)	1.56	0	0.046	0.046	0.046
Poly(alkyl vinyl ether)	2.1	0.43	0.035	0.12	0.16
Poly(N-vinyl carbazole)	2.5	0	0.16	0.35	0.09
Poly(methyl vinyl ketone)	2.7	0	0.05	0.02	0.008
Poly(vinyl chloride)	4.2	0.45	0.032	0.071	0.032
average	2.4	0.22	0.056	0.11	0.056

$\eta^* = 0.43\eta$ in each case.

Comments:
Isolated butyl branches in low-density polyethylene are formed by an intrachain radical rearrangement that is followed by repeated addition of ethylene without further rearrangement. Here, stereochemical selectivity during the formation of $CH_2R-CH_2-CHR-CH_2$-branches in the free radical initiated polymerization of monosubstituted vinyl monomers is investigated. The configuration partition functions are denoted by Z_m and Z_r, respectively. They can be written as $Z_m = U_1 \ U_2 \ U_p \ U_m \ U_p \ [v_1 \ v_2 \ v_3]^T$ and $Z_r = U_1 \ U_2 \ U_p \ U_r \ U_p \ [v_1 \ v_2 \ v_3]^T$. Numerical values for v depend on the degree of polymerization and stereochemical composition of the remainder of the chain.

A 101

polyoxypropylene, poly(propylene oxide),
polymethyloxirane, PPO, POP

poly((R)-oxypropylene)

Abe, A.; Hirano, T.; Tsuruta, T. *Macromolecules* **1979**, *12*, 1092.

Bond Length [pm]	Valence Angles [°]	Torsion Angles [a,b,c] [°]	ξ [a,b,c] (for 303 K)	ξ_0	E_ξ [kJ·mol^{-1}]
C-C : 153	C-O-C : 111.5	For C-C bonds:	α = 1.65	1	−1.26 both
		t : 180			
C-O : 143	C-C-O : 111.5	g$^+$: 60 or 70	β = 0.56 or 0.44	1	1.46 or 2.09
		g$^-$: −60 or −70			
C-H : 110	C-C-C : 110	For C-O bonds:	σ = 0.12 or 0.22	1	5.4 or 3.8
		t : −160 or −170			
	C-C-H : 110	g$^+$: 60			
		g$^-$: −80 or −70	ω = 0.51	1	1.7 both
		For O-C bonds:			
		t : 180			
		g$^+$: 80 or 70			
		g$^-$: −80 or −70			

$$U_a^R = \begin{bmatrix} 1 & \alpha & \beta \\ 0 & \alpha & \beta\omega \\ 1 & \alpha\omega & 0 \end{bmatrix} \quad U_b^R = \begin{bmatrix} 1 & 0 & 1 \\ 1 & 0 & \omega \\ 1 & 0 & 1 \end{bmatrix} \quad U_c^R = \begin{bmatrix} 1 & \sigma & 0 \\ 1 & 0 & 0 \\ 1 & 0 & \sigma \end{bmatrix}$$

$$U_i^S = Q \, U_i^R \, Q, \text{ where } Q = \begin{bmatrix} 1 & 0 & 0 \\ 0 & 0 & 1 \\ 0 & 1 & 0 \end{bmatrix} \quad (\text{see: } ^{d)})$$

rows and columns: t, g$^+$, g$^-$

a) Two different parameter sets are used.
b) The two different values given throughout correspond to parameter *sets I and II*, respectively, of the original paper.
c) The g$_\alpha$ conformation is identified with g$^+$, and g$_\beta$ with g$^-$.
d) Superscripts R and S, respectively, represent the stereochemical configuration at the asymmetric C atoms.

Comments:
Conformational energies of the POP chains are calculated by using semiempirical potential energy functions. Experimental values of the same energies are also obtained from the RIS analysis of the unperturbed dimension, dipole moment, and bond conformations observed for the isotactic samples. The magnitude of stabilization energies associated with the *gauche* O−C−C−O arrangements is estimated from the difference between theoretical and experimental energies thus established. Conformational energy parameters thus estimated are used to calculate the characteristic ratio $<r^2>_0/nl^2$, the dipole moment ratio $<\mu^2>/xm^2$, the bond conformations for the skeletal C−C bond, and their temperature coefficients for *isotactic, syndiotactic*, and *atactic* chains. The effects of atypical head-to-head and tail-to-tail placements are also examined as well as *stereoirregular* chains.

Calcd. quantities:

$<r^2>_0/nl^2$	= 6.1 or 6.0	(for 323 K)	(*Exptl.*: 6.0)
$d(\ln <r^2>_0)/dT$	= −1.59 × 10^{-3} K^{-1}	(for 323 K, set I)	
$<\mu^2>/xm^2$	= 0.50	(for 303 K)	(*Exptl.*: 0.49, 0.54)
	0.43 × 10^{-3} K^{-1}	(for 303 K, set I)	

poly(3-methyl tetrahydrofuran), PMTHF

Saiz, E.; Tarazona, M. P.; Riande, E.; Guzmán, J. *J. Polym. Sci.; Polym. Phys. Ed.* **1984**, 22, 2165.

Bond Length [pm]	Valence Angles [°]	Torsion Angles [°]	ξ (for 303 K)	ξ_0	E_ξ [kJ·mol^{-1}]
C-C : 153	C-C-O: 110 or 112	t : 180	η = 0.79	1	0.59
C-O : 143	C-C-C: 110 or 114	For b,d,g,i: g$^+$: 60 g$^-$: -60	ω = 0.03 σ = 0.03	1 1	8.8 8.8
	C-O-C: 110	For a,e,f,j,k: g$^+$: 70 g$^-$: -70	σ' = 1.52 σ'' = 0.14	1 1	-1.05 5.0
		For c,h: g$^+$: 67.5 g$^-$: -67.5			

$$U_a = \begin{bmatrix} 1 & \sigma'' & \sigma'' \\ 0 & 0 & 0 \\ 1 & 0 & 0 \end{bmatrix} \quad U_b = \begin{bmatrix} 1 & \sigma' & \sigma'\omega \\ 1 & \sigma' & 0 \\ 1 & 0 & \sigma'\omega \end{bmatrix} \quad U_c = \begin{bmatrix} 1 & \sigma & 1 \\ 1 & \sigma\omega & \omega \\ 1 & \sigma\omega & 1 \end{bmatrix} \quad U_d = \begin{bmatrix} 1 & 1 & \eta \\ 1 & 1 & \eta\omega \\ 1 & \omega & \eta \end{bmatrix}$$

$$U_e = \begin{bmatrix} 1 & 0 & \sigma'' \\ 1 & \sigma'' & 0 \\ 1 & 0 & 0 \end{bmatrix} \quad U_f = \begin{bmatrix} 1 & \sigma'' & 0 \\ 0 & 0 & 0 \\ 1 & 0 & 0 \end{bmatrix} \quad U_g = \begin{bmatrix} 1 & \eta & 1 \\ 1 & 0 & 0 \\ 0 & 0 & 1 \end{bmatrix} \quad U_h = \begin{bmatrix} 1 & 1 & \sigma \\ 1 & 1 & \sigma\omega \\ 1 & \omega & \sigma \end{bmatrix}$$

$$U_i = \begin{bmatrix} 1 & \sigma'\omega & \sigma' \\ 1 & \sigma' & \sigma'\omega \\ 1 & \sigma'\omega & \sigma'\omega \end{bmatrix} \quad U_j = \begin{bmatrix} 1 & \sigma'' & \sigma'' \\ 1 & \sigma''\omega & 0 \\ 1 & 0 & \sigma'' \end{bmatrix} \quad U_k = \begin{bmatrix} 1 & \sigma'' & \sigma'' \\ 1 & \sigma'' & 0 \\ 1 & 0 & \sigma'' \end{bmatrix}$$

rows and columns: t, g$^+$, g$^-$

Comments:
The stress-optical behaviour of an unswollen elastomeric network of PMTHF is measured for different elongation ratios at several temperatures. Values of $\Delta\alpha$ range from 2.4 to 2.8 in units of 10^{-24} cm^3, in the temperature range studied. Theoretical calculations carried out with the RIS model give values of $\Delta\alpha$ noticeably smaller than the experimental results; however, a small increase in the backbone valence angles improves the theoretical results. Theoretical and experimental values of the temperature coefficient of $\Delta\alpha$ are in clear disagreement; a qualitative explanation for this discrepancy is discussed.

Calcd. quantities: (for x = 100)

$\Delta\alpha$ = 2.0 × 10^{-24} cm^3 (Exptl.: ~ 2.5 × 10^{-24} cm^3)

$d(\ln \Delta\alpha)/dT$ = -0.3 × 10^{-3} K^{-1} (Exptl.: 3.1 × 10^{-3} K^{-1})

A 103

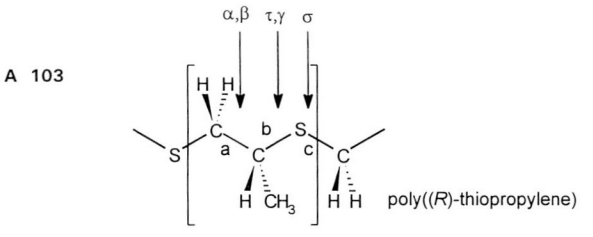

poly((R)-thiopropylene)

polythiopropylene, poly(propylene sulfide), PTP

Abe, A. *Macromolecules* **1980**, *13*, 541.

Bond Length [pm]	Valence Angles [°]	Torsion Angles [°]	ξ (for 298 K)	ξ_0	E_ξ [kJ·mol^{-1}]
C-C : 153	C*-S-C : 100	*For C-C bonds:*	$\alpha = 0.57$	1	1.38
C-S : 181.5	S-C-C* : 114	t : 171			
		g$^+$: 69	$\beta = 0.11$	1	5.4
		g$^-$: −60			
C-H : 110	C-C*-S : 110		$\gamma = 0.85$	1	0.4
		For C-S bonds:*			
	C-C*-C : 110	t : −168	$\tau = 0.13$	1	5.0
		g$^+$: 60			
	C*-C-H : 110	g$^-$: −72	$\omega = 0.079$	1	6.3
		For S-C bonds:	$\omega' = 0.16$	1	4.6
		t : 180			
		g$^+$: 67	$\omega'' = 0.50$	1	1.7
		g$^-$: −67			
			$\sigma = 1.09$	1	−0.21

$$U_a^R = \begin{bmatrix} 1 & \alpha & \beta \\ \omega & \alpha & \beta\omega' \\ 1 & \alpha\omega' & \beta\omega \end{bmatrix} \quad U_b^R = \begin{bmatrix} 1 & \tau & \gamma \\ 1 & \tau & \gamma\omega' \\ 1 & \tau\omega' & \gamma \end{bmatrix}$$

$$U_c^R = \begin{bmatrix} 1 & \sigma & \sigma\omega'' \\ 1 & \sigma\omega'' & \sigma\omega'' \\ 1 & \sigma\omega'' & \sigma \end{bmatrix} \quad \text{rows and columns: t, g}^+\text{, g}^-$$

Comments:
Conformational energies associated with PTP chains are calculated by using semiempirical potential energy functions. Reliability of these functions is tested against the known values of conformational energies of various simple alkyl sulfides. The magnitude of the *gauche* sulfur effect is estimated from the RIS analysis of the experimental values of the unperturbed dimension, dipole moment, and their temperature coefficients observed for atactic samples of PTP.

Calcd. quantities: (for x = 200)	isotactic	syndiotactic	atactic	isotactic [atactic]
$<r^2>_0 / nl^2$	= 4.3	3.7	4.0	(Exptl.: / [4.0])
$d(\ln <r^2>_0)/dT$	= −1.1 × 10^{-3} K^{-1}	−0.24 × 10^{-3} K^{-1}	−0.63 × 10^{-3} K^{-1}	(Exptl.: −2.8 [−2.0] × 10^{-3} K^{-1})
$<\mu^2>/xm^2$	= 0.41	0.38	0.39	(Exptl.: 0.33 or 0.39 [0.37 or 0.44])
$d(\ln <\mu^2>)/dT$	= 2.2 × 10^{-3} K^{-1}	2.3 × 10^{-3} K^{-1}	2.2 × 10^{-3} K^{-1}	(Exptl.: 2.1 (or 2.0) [4.0 (or1.5)] × 10^{-3} K^{-1})

*Further calculations on **poly(propylene sulfide)** chains:*

A 104 Riande, E.; Boileau, S.; Hemery, P.; Mark, J. E. *J. Chem. Phys.* **1979**, *71*, 4206.

Dielectric constant measurements in carbon tetrachloride and in benzene are used to obtain dipole moments of atactic poly(propylene sulfide) chains of sufficiently low molecular weight that excluded volume effects are absent. The results are in very good agreement with RIS calculations. Comparison with previously reported experimental results on the high molecular weight polymer confirm the existence of a significant specific solvent effect on the dipole moments.

A 105 Riande, E.; Boileau, S.; Hemery, P.; Mark, J. E. *Macromolecules* **1979**, *12*, 702.

Poly(propylene sulfide) is synthesized in both the iosotactic and atactic stereochemical forms. They are investigated by means of dielectric constant measurements in benzene and CCl_4. The experimental results are in good agreement with values predicted from RIS calculations.

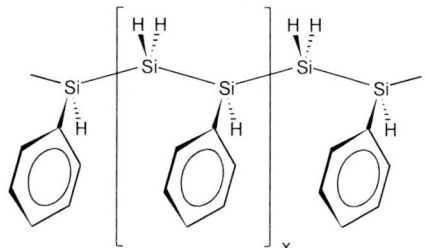

polysilastyrene

Welsh, W. J.; Damewood Jr., J. R.; West, R. C. *Macromolecules* **1989**, *22*, 2947.

Bond Length [pm]	Valence Angles [a] [°]	Torsion Angles [b] [°]	ξ [b] (for 298 K)	ξ_0	E_ξ [kJ·mol^{-1}]	
Si-Si : 234	Si-Si-Si : 109.2	t : 180	σ = 0.17 / 10.10	1	4.4 / −5.7	$U = \begin{bmatrix} 1 & \sigma & \lambda \\ 1 & \sigma\psi & \lambda\omega' \\ 1 & \sigma\omega & \lambda\psi' \end{bmatrix}$
Si-H : 149		g : 63.6 / 60	λ = 0.02 / 3.10	1	9.7 / −2.8	
Si-C$_{ar}$: 188		$\bar{g}$: −63.6 / −60	ψ = 0.48 / 4.65	1	1.8 / −3.8	
			ψ' = 0.05 / 12.58	1	7.4 / −6.3	rows and columns: t, g, $\bar{g}$
			$\omega = \omega'$ = 0 / 0	1	∞	

[a] Calculated bond angles in general vary with conformation more so than do bond length due to the "softer" nature of their deformation energy functions.
[b] Values correspond to *iso / syn* placements, respectively.

Comments:
Molecular mechanics techniques are employed to calculate the molecular structure and conformational energies of model compounds for polyphenylmethylsilylene and polysilastyrene, in both *isotactic* and *syndiotactic* stereochemical forms. The structural and conformational energy data provided are used to calculate, by application of the RIS theory, the unperturbed chain dimensions, given as the characteristic ratio, and its temperature coefficient.

Calcd. quantities: $<r^2>_0 / nl^2$ $d(\ln <r^2>_0)/dT$

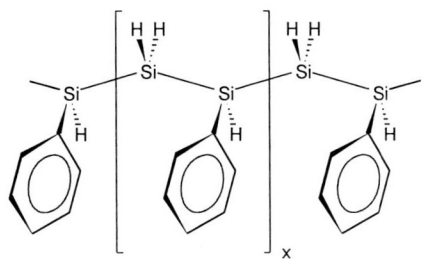

polysilastyrene

Sundararajan, P. R. *Macromolecules* **1991**, *24*, 1420.

Bond Length [pm]	Valence Angles [°]	Torsion Angles [°]		ξ (for 300 K)	ξ_0	E_ξ [kJ·mol^{-1}]
Si-Si : 234	Si-Si-Si : 109.5	(tt)$_m$	: 180 / 160	$\eta = 0.53$	1.52	2.64
		(tg)$_m$	: 160 / 100			
Si-H : 149		(t$\bar{g}$)$_m$	: 140 / −50	$\tau = 0.02$	2.22	11.72
		(gg)$_m$	: 50 / 70			
Si-C^{ar} : 187		(g$\bar{g}$)$_m$	: 70 / −70	$\omega'' = 5.17$	1.00	−4.10
		($\bar{g}\bar{g}$)$_m$	: −50 / −70			
C^{ar}-C^{ar} : 139		(tt)$_r$	: − / −	$\omega' = 3.31$	1.15	−2.64
		(tg)$_r$	: 180 / 70			
C^{ar}-H : 110		(t$\bar{g}$)$_r$	: 180 / −50	$\omega = 1.34$	1.04	−0.63
		(gg)$_r$	: 70 / 70			
		(g$\bar{g}$)$_r$	: 80 / −60			
		($\bar{g}\bar{g}$)$_r$	: −50 / −40			

$$U_p = \begin{bmatrix} 1 & 1 & 1 \\ 1 & 0 & 1 \\ 1 & 1 & 0 \end{bmatrix} \quad U_m = \begin{bmatrix} \omega'' & 1/\eta & \tau\omega'/\eta \\ 1/\eta & \omega/\eta^2 & \tau\omega'/\eta^2 \\ \tau\omega'/\eta & \tau\omega'/\eta^2 & \tau^2\omega''\omega/\eta^2 \end{bmatrix}$$

$$U_r = \begin{bmatrix} 1 & \omega'/\eta & \tau\omega''/\eta \\ \omega'/\eta & 1/\eta^2 & \tau\omega/\eta^2 \\ \tau\omega''/\eta & \tau\omega/\eta^2 & \tau^2\omega'^2/\eta^2 \end{bmatrix}$$

rows and columns: t, g, $\bar{g}$

Calcd. quantities: $\langle r^2 \rangle_0 / nl^2$ $d(\ln \langle r^2 \rangle_0)/dT$

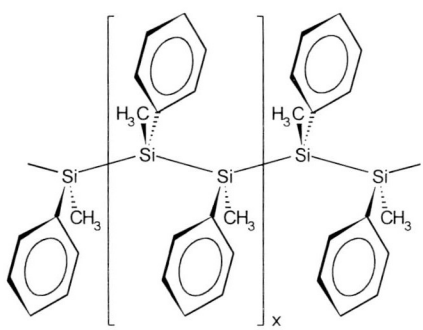

polymethylphenylsilylene

Sundararajan, P. R. *Macromolecules* **1988**, *21*, 1256.

Bond Length [pm]	Valence Angles [°]	Torsion Angles [a,b] [°]			
Si-Si : 234	Si-Si-Si : 116	For up,up,up: tt : $-171.4, -171.3$	For up,down,up: tt : $-146.4, -159.1$		
Si-C : 187	Si-Si-C : 109	t$\bar{g}$: $-168.7, -66.8$	t$\bar{g}$: 178.6, -73	$U_{(up,up,up)} = \begin{bmatrix} 1 & 0.017 \\ 0.017 & 0 \end{bmatrix}$	$U_{(up,down,up)} = \begin{bmatrix} 0.144 & 0.044 \\ 0.044 & 0.076 \end{bmatrix}$
Si-C^{ar} : 187	Si-Si-C^{ar} : 109		$\bar{g}$t : $-73.7, -178.5$		
C^{ar}-C^{ar} : 139	C-Si-C : 104.2	For up,down,up: tt : $-171.9, -171.7$	$\bar{g}\bar{g}$: $-58.3, -60.2$	$U_{(up,up,down)} = \begin{bmatrix} 0.073 & 0.447 \\ 0.044 & 0.005 \end{bmatrix}$	
C^{ar}-H : 110	C^{ar}-C^{ar}-C^{ar} : 120	t$\bar{g}$: $-165.6, -78.9$			
		$\bar{g}\bar{g}$: $-56, -56$			rows and columns: t, $\bar{g}$

a) With respect to the plane of the skeletal bonds in the planar *all-trans* conformation, the phenyl groups on Si$_{i-1}$, Si$_i$, and Si$_{i+1}$ can be all up, all down, or alternate. A total of eight configurations are possible.

b) Description of the statistical weights in terms of symbols is not attempted here, due to their large number.

Comments:
The conformational energies and helix parameters are calculated for the various states of polymethylphenylsilylene chain. The calculations are performed for the three different relative dispositions of the phenyls attached to three successive silicon atoms. The minima are invariably shifted from perfectly staggered positions.

Calcd. quantities: First and second order *a priori* bond probabilities $\quad <r^2>_0 / nl^2$

A 109

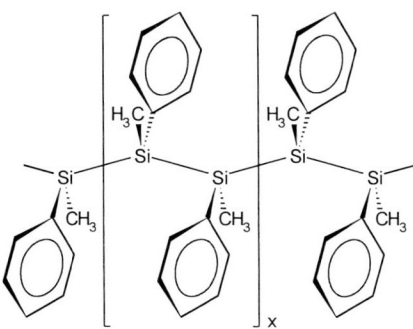

polymethylphenylsilylene

Welsh, W. J.; Damewood Jr., J. R.; West, R. C. *Macromolecules* **1989**, *22*, 2947.

Bond Length [pm]	Valence Angles [a) [°]	Torsion Angles [°]	ξ [b) (for 298 K)	ξ_o	E_ξ [kJ·mol^{-1}]	
Si-Si : 235	Si-Si-Si : 113.9	t : 180	σ = 0 / 0	1	∞	
Si-C : 187		g : 64	λ = 0 / 0	1	∞	$U = \begin{bmatrix} 1 & \sigma & \lambda \\ 1 & \sigma\psi & \lambda\omega' \\ 1 & \sigma\omega & \lambda\psi' \end{bmatrix}$
Si-C$_{ar}$: 185		$\bar{g}$: -64	ψ = 0.31 / 0.05	1	2.9 / 7.4	
			ψ' = 0.03 / 0	1	8.7 / ∞	
			$\omega = \omega'$ = 0 / 0	1	∞	rows and columns: t, g, $\bar{g}$

a) Calculated bond angles in general vary with conformation more so than do bond length due to the "softer" nature of their deformation energy functions.
b) Values correspond to *iso / syn* placements, respectively.

Comments:
Molecular mechanics techniques are employed to calculate the molecular structure and conformational energies of model compounds for polymethylphenylsilylene and polysilastyrene, in both *isotactic* and *syndiotactic* stereochemical forms. The structural and conformational energy data provided are used to calculate, by application of the RIS theory, the unperturbed chain dimensions, given as the characteristic ratio, and its temperature coefficient.

Calcd. quantities: $<r^2>_o / nl^2$ $d (\ln <r^2>_o) / dT$

polysilapropylene, PSP

Sundararajan, P. R. *Macromolecules* **1990**, *23*, 3179.

Bond Length [pm]	Valence Angles [°]	Torsion Angles [°]		ξ (for 300 K)	ξ_o	E_ξ [kJ·mol^{-1}]
Si-C : 187	C-Si-C : 109.5	$(tt)_m$	: 175 / 175	η = 1.00	0.983	− 0.017
		$(tg)_m$	: − 179.5 / 62.9			
C-H : 110	Si-C-Si : 118	$(t\bar{g})_m$	: 171.2 / − 68.9	τ = 1.52	0.879	− 1.36
		$(gg)_m$	: 72.1 / 72.1			
		$(g\bar{g})_m$	: 69.1 / − 55.5	ω'' = 1.00	0.887	− 0.31
		$(\bar{g}\bar{g})_m$	: − 65 / − 65			
		$(tt)_r$	: − 179.1 / − 179.1	ω' = 0.79	0.782	− 0.13
		$(tg)_r$	: 174.1 / 72.2			
		$(t\bar{g})_r$	: 168.5 / − 54.8	ω = 0.69	0.786	0.31
		$(gg)_r$	: 61 / 61			
		$(g\bar{g})_r$	: 75.5 / − 70.4			
		$(\bar{g}\bar{g})_r$	: − 58.1 / − 58.1			

$$U_p = \begin{bmatrix} 1 & 1 & 1 \\ 1 & 0 & 1 \\ 1 & 1 & 0 \end{bmatrix}$$

$$U_m = \begin{bmatrix} \omega'' & 1/\eta & \tau\omega'/\eta \\ 1/\eta & \omega/\eta^2 & \tau\omega'/\eta^2 \\ \tau\omega'/\eta & \tau\omega'/\eta^2 & \tau^2\omega''\omega/\eta^2 \end{bmatrix}$$

$$U_r = \begin{bmatrix} 1 & \omega'/\eta & \tau\omega''/\eta \\ \omega'/\eta & 1/\eta^2 & \tau\omega/\eta^2 \\ \tau\omega''/\eta & \tau\omega/\eta^2 & \tau^2\omega'^2/\eta^2 \end{bmatrix}$$

rows and columns: t, g, $\bar{g}$

Comments:

In an effort to correlate the conformational features of polysilane derivatives with their properties, calculations are performed to determine the relative stabilities of the conformational states of the *meso* and *racemic* diads of polysilapropylene. Energy maps are constructed in terms of internal rotation angles to calculate the average properties of the chain. The calculations show that the difference in energy between the various states of the *meso* and *racemic* diad is small. Hence, PSP can be considered to be more flexible than the analogous carbon polymer, PP. The characteristic ratios of the unperturbed end-to-end distances for the *iso-* and *syndiotactic* PSP are less than those for the PP of corresponding tacticity.

Calcd. quantities: $<r^2>_0 / nl^2$ $d (\ln <r^2>_0) / dT$

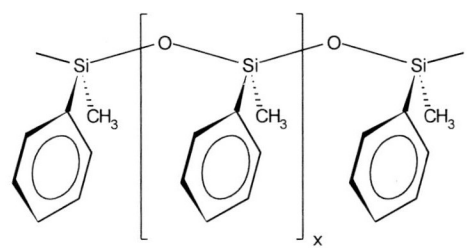

polymethylphenylsiloxane, PMPS

Mark, J. E.; Ko, J. H. *J. Polym. Sci.; Polym. Phys. Ed.* **1975**, *13*, 2221.

Bond Length [pm]	Valence Angles [°]	Torsion Angles [°]	ξ (for 338 K)	ξ₀	E_ξ [kJ·mol⁻¹]
Si-O : 164	Si-O-Si : 143	t : 180	σ = 0.58	1	1.5
Si-C : 190	O-Si-O : 109.5	g⁺ : 60	ω' = 44.4	1	−10.5
C-H : 109		g⁻ : −60	ω'' = 3.9	1	−3.8
C^{ar}-C^{ar} : 139			ω''' = 2.1	1	−2.1
			ω = 0.09	1	6.7
			δ = 3.3 to 6.1	1	−3.3 to −5.0

$$U_p = \begin{bmatrix} 1 & \sigma & \sigma \\ 1 & \sigma & 0 \\ 1 & 0 & \sigma \end{bmatrix}$$

$$U_{dd} = \begin{bmatrix} \omega'\omega''' & \sigma\omega''\delta & \sigma\omega'' \\ \omega'' & \sigma\delta & \sigma\omega\omega''' \\ \omega''\delta & \sigma\omega\omega' & \sigma\delta \end{bmatrix}$$

$$U_{dl} = \begin{bmatrix} \omega''^2 & \sigma\omega'''\delta & \sigma\omega' \\ \omega'''\delta & \sigma & \sigma\omega\omega''' \\ \omega' & \sigma\omega\omega'' & \sigma\delta^2 \end{bmatrix}$$

rows and columns: t, g⁺, g⁻

Comments:
RIS theory is used to study the unperturbed dimensions of PMPS chains as a function of their stereochemical structure. The required conformational energies are obtained from semi-empirical, interatomic potential energy functions and from known results on PDMS.

Calcd. quantities: $\langle r^2 \rangle_0 / nl^2$ $d(\ln \langle r^2 \rangle_0)/dT$

cis-1,4-polybutadiene, *cis*-1,4-PBD

Mark, J. E. *J. Am. Chem. Soc.* **1966**, *88*, 4354.

Bond Length [pm]	Valence Angles [°]	Torsion Angles [°]	ξ (for 343 K)	ξ_0	E_ξ [kJ·mol^{-1}]
C-C : 153	CH$_2$-CH=CH : 125	For CH$_2$-CH$_2$:	γ = 10	1	−6.7
		t : 180			
C=C : 134	CH-CH$_2$CH$_2$: 112	g$^+$: 60	σ = 1.4	1	−0.8
		g$^-$: −60			
C-H : 110	CH$_2$-C-H : 117.5	For CH-CH$_2$:			
$l_a = l_b$: 270	CH$_2$CH-H : 110	c : 0			
		s$^+$: 120			
	θ_a : 70	s$^-$: −120			

$$U_a = \begin{bmatrix} 1 & 1 & 0 & 0 & 0 & 1 \\ 0 & 0 & 0 & 0 & 0 & 0 \\ 1 & 1 & 0 & 0 & 0 & 1 \\ 0 & 0 & 0 & 0 & 0 & 0 \\ 1 & 1 & 0 & 0 & 0 & 1 \\ 0 & 0 & 0 & 0 & 0 & 0 \end{bmatrix} \quad U_b = \begin{bmatrix} 0 & 1 & 0 & 0 & 0 & 1 \\ 1 & \gamma & 0 & 0 & 0 & \gamma \\ 0 & 0 & 0 & 0 & 0 & 0 \\ 0 & 0 & 0 & 0 & 0 & 0 \\ 0 & 0 & 0 & 0 & 0 & 0 \\ 1 & \gamma & 0 & 0 & 0 & \gamma \end{bmatrix} \quad U_c = \begin{bmatrix} 1 & 0 & \sigma & 0 & \sigma & 0 \\ 1 & 0 & \sigma & 0 & \sigma & 0 \\ 0 & 0 & 0 & 0 & 0 & 0 \\ 0 & 0 & 0 & 0 & 0 & 0 \\ 0 & 0 & 0 & 0 & 0 & 0 \\ 1 & 0 & \sigma & 0 & \sigma & 0 \end{bmatrix}$$

rows and columns: t, s$^+$, g$^+$, c, g$^-$, s$^-$

Comments:
The RIS model with neighbor dependence is used to calculate random-coil dimensions for the *cis*-forms of PBD and PIP in the limit of large x. Comparison of calculated and experimental values of the characteristic ratio and its temperature coefficient is used to determine intramolecular energies of various conformational sequences of the chain backbone.

Calcd. quantities:

$\langle r^2 \rangle_0 / nl^2$ = 4.9 (*Exptl.*: 4.9 ± 0.2)

$d (\ln \langle r^2 \rangle_0) / dT$ = 0.40 × 10^{-3} K^{-1} (*Exptl.*: 0.40 × 10^{-3} K^{-1})

cis-1,4-polyisoprene, *cis*-1,4-PIP

Mark, J. E. *J. Am. Chem. Soc.* **1966**, *88*, 4354.

Bond Length [pm]	Valence Angles [°]	Torsion Angles [°]	ξ (for 343 K)	ξ_0	E_ξ [kJ·mol^{-1}]
C-C : 153	CH$_2$-CR=CH : 125	For CH$_2$-CH$_2$: t : 180	γ = 10	1	−6.7
C=C : 134	CH-CH$_2$CH$_2$: 112	g$^+$: 60 g$^-$: −60	σ = 2.46	1	−2.5
C-H : 110	CH$_2$-C-R : 117.5	For CH-CH$_2$: c : 0			
l$_a$ = l$_b$: 270	CH$_2$CH-H : 110	s$^+$: 120 s$^-$: −120			
	θ_a : 70				

$$U_a = \begin{bmatrix} 1 & 1 & 0 & 0 & 0 & 1 \\ 0 & 0 & 0 & 0 & 0 & 0 \\ 1 & 1 & 0 & 0 & 0 & 1 \\ 0 & 0 & 0 & 0 & 0 & 0 \\ 1 & 1 & 0 & 0 & 0 & 1 \\ 0 & 0 & 0 & 0 & 0 & 0 \end{bmatrix} \quad U_b = \begin{bmatrix} 0 & 1 & 0 & 0 & 0 & 1 \\ 1 & \gamma & 0 & 0 & 0 & \gamma \\ 0 & 0 & 0 & 0 & 0 & 0 \\ 0 & 0 & 0 & 0 & 0 & 0 \\ 0 & 0 & 0 & 0 & 0 & 0 \\ 1 & \gamma & 0 & 0 & 0 & \gamma \end{bmatrix} \quad U_c = \begin{bmatrix} 1 & 0 & \sigma & 0 & \sigma & 0 \\ 1 & 0 & \sigma & 0 & \sigma & 0 \\ 0 & 0 & 0 & 0 & 0 & 0 \\ 0 & 0 & 0 & 0 & 0 & 0 \\ 0 & 0 & 0 & 0 & 0 & 0 \\ 1 & 0 & \sigma & 0 & \sigma & 0 \end{bmatrix}$$

rows and columns: t, s$^+$, g$^+$, c, g$^-$, s$^-$

Comments:
The RIS model with neighbor dependence is used to calculate random-coil dimensions for the *cis*-forms of PBD and PIP in the limit of large *x*. Comparison of calculated and experimental values of the characteristic ratio and its temperature coefficient is used to determine intramolecular energies of various conformational sequences of the chain backbone.

Calcd. quantities:

$\langle r^2 \rangle_0 / nl^2$ = 4.7 (*Exptl.:* 4.7)

$d(\ln \langle r^2 \rangle_0)/dT$ = 0.56 × 10^{-3} K^{-1} (*Exptl.:* 0.41 (*to* 0.56) × 10^{-3} K^{-1})

trans-1,4-polybutadiene, trans-1,4-PBD

Mark, J. E. *J. Am. Chem. Soc.* **1967**, *89*, 6829.

Bond Length [pm]	Valence Angles [°]	Torsion Angles [°]	ξ (for 373 K)	ξ_0	E_ξ [kJ·mol^{-1}]
C-C : 153	CH$_2$-CH=CH: 125	For CH$_2$-CH$_2$:	σ = 1	1	0
		t : 180	α = 0.96	1	0.13
C=C : 134	CH-CH$_2$CH$_2$: 112	g$^+$: 60	β = 0.24	1	4.6
		g$^-$: −60			
C-H : 110	CH$_2$-C-H : 117.5				
		For CH$_2$-CH :			
	CH$_2$CH-H : 110	c : 0			
		s$^+$: 120			
		s$^-$: −120			
		For C=C :			
		t : 180			

rows and columns: t, s$^+$, g$^+$, c, g$^-$, s$^-$

$$U_a = \begin{bmatrix} 0 & 1 & 0 & \alpha & 0 & 1 \\ 0 & 0 & 0 & 0 & 0 & 0 \\ 0 & 1 & 0 & \beta\alpha & 0 & 1 \\ 0 & 0 & 0 & 0 & 0 & 0 \\ 0 & 1 & 0 & \beta\alpha & 0 & 1 \\ 0 & 0 & 0 & 0 & 0 & 0 \end{bmatrix} \quad U_b = \begin{bmatrix} 0 & 0 & 0 & 0 & 0 & 0 \\ 1 & 0 & 0 & 0 & 0 & 0 \\ 0 & 0 & 0 & 0 & 0 & 0 \\ 1 & 0 & 0 & 0 & 0 & 0 \\ 0 & 0 & 0 & 0 & 0 & 0 \\ 1 & 0 & 0 & 0 & 0 & 0 \end{bmatrix} \quad U_c = \begin{bmatrix} 0 & 1 & 0 & \alpha & 0 & 1 \\ 0 & 0 & 0 & 0 & 0 & 0 \\ 0 & 0 & 0 & 0 & 0 & 0 \\ 0 & 0 & 0 & 0 & 0 & 0 \\ 0 & 0 & 0 & 0 & 0 & 0 \\ 0 & 0 & 0 & 0 & 0 & 0 \end{bmatrix} \quad U_d = \begin{bmatrix} 0 & 0 & 0 & 0 & 0 & 0 \\ 1 & 0 & \sigma & 0 & \sigma & 0 \\ 0 & 0 & 0 & 0 & 0 & 0 \\ 1 & 0 & \beta\sigma & 0 & \beta\sigma & 0 \\ 0 & 0 & 0 & 0 & 0 & 0 \\ 1 & 0 & \sigma & 0 & \sigma & 0 \end{bmatrix}$$

Comments:
Results of calculations based on a RIS model are used to interpret experimental values of the chain dimensions and their temperature coefficient for *trans*-1,4-PBD and *trans*-1,4-PIP. Similarities and differences in conformational and configurational properties between these polymers and the corresponding *cis*-forms are elucidated.

Calcd. quantities:

$\langle r^2 \rangle_0 / nl^2$ = 5.8 (*Exptl.*: 5.8 ± 0.2)

$d (\ln \langle r^2 \rangle_0) / dT$ = −0.65 × 10^{-3} K^{-1} (*Exptl.*: −0.65 × 10^{-3} K^{-1})

trans-1,4-polyisoprene, trans-1,4-PIP

Mark, J. E. *J. Am. Chem. Soc.* **1967**, *89*, 6829.

Bond Length [pm]	Valence Angles [°]	Torsion Angles [°]	ξ (for 338 K)	ξ_0	E_ξ [kJ·mol^{-1}]
C-C : 153	CH$_2$-CR=CH : 125	For CH$_2$-CH$_2$: t : 180	σ' = 0.543	1	1.67
C=C: 134	CH-CH$_2$CH$_2$: 112	g$^+$: 60	α = 0.96	1	0.13
C-H : 110	CH$_2$-C-R : 117.5	g$^-$: −60	β = 0.217	1	4.6
	CH$_2$CH-H : 110	For CH$_2$-CH : c : 0 s$^+$: 120 s$^-$: −120			
	R = CH$_3$	For C=C : t : 180			

rows and columns: t, s$^+$, g$^+$, c, g$^-$, s$^-$

$$U_a = \begin{bmatrix} 0 & 1 & 0 & \alpha & 0 & 1 \\ 0 & 0 & 0 & 0 & 0 & 0 \\ 0 & 1 & 0 & \beta\alpha & 0 & 1 \\ 0 & 0 & 0 & 0 & 0 & 0 \\ 0 & 1 & 0 & \beta\alpha & 0 & 1 \\ 0 & 0 & 0 & 0 & 0 & 0 \end{bmatrix} \quad U_b = \begin{bmatrix} 0 & 0 & 0 & 0 & 0 & 0 \\ 1 & 0 & 0 & 0 & 0 & 0 \\ 0 & 0 & 0 & 0 & 0 & 0 \\ 1 & 0 & 0 & 0 & 0 & 0 \\ 0 & 0 & 0 & 0 & 0 & 0 \\ 1 & 0 & 0 & 0 & 0 & 0 \end{bmatrix} \quad U_c = \begin{bmatrix} 0 & 1 & 0 & \alpha & 0 & 1 \\ 0 & 0 & 0 & 0 & 0 & 0 \\ 0 & 0 & 0 & 0 & 0 & 0 \\ 0 & 0 & 0 & 0 & 0 & 0 \\ 0 & 0 & 0 & 0 & 0 & 0 \\ 0 & 0 & 0 & 0 & 0 & 0 \end{bmatrix} \quad U_d = \begin{bmatrix} 0 & 0 & 0 & 0 & 0 & 0 \\ 1 & 0 & \sigma' & 0 & \sigma' & 0 \\ 0 & 0 & 0 & 0 & 0 & 0 \\ 1 & 0 & 0 & 0 & 0 & 0 \\ 0 & 0 & 0 & 0 & 0 & 0 \\ 1 & 0 & \sigma' & 0 & \sigma' & 0 \end{bmatrix}$$

Comments:
Results of calculations based on a RIS model are used to interpret experimental values of the chain dimensions and their temperature coefficient for *trans*-1,4-PBD and *trans*-1,4-PIP. Similarities and differences in conformational and configurational properties between these polymers and the corresponding *cis*-forms are elucidated.

Calcd. quantities:

$\langle r^2 \rangle_0 / nl^2$ = 7.35 (Exptl.: 7.35)

$d(\ln \langle r^2 \rangle_0)/dT$ = −1.4 × 10^{-3} K^{-1} (Exptl.: −0.27 × 10^{-3} K^{-1})

cis-1,4-polybutadiene, *cis*-1,4-PBD

Ishikawa, T.; Nagai, K. *J. Polym. Sci., Part A-2* **1969**, *7*, 1123.

Bond Length [pm]	Valence Angles [°]	Torsion Angles [°]	ξ a) (for 293 K)	ξ_0	E_ξ [kJ · mol^{-1}]	Transition probability matrices, **p**, are given; $p_c = E_3$		
CH$_2$-CH$_2$: 153	CH$_2$-CH=CH: 125	For CH$_2$-CH$_2$: t : 180 g$^+$: 60 g$^-$: −60	$\gamma = \infty$ / 30 / 30 $\delta = 0$ / 0.322 / 0 $\sigma = 0.81$ / 1 / 1	1 1 1	$-\infty$ / −8.3 / −8.3 ∞ / 2.76 / ∞ 0.50 / 0 / 0	$p_a = \begin{bmatrix} 1 & 1 & 1 \\ \sigma^{1/2} & \sigma^{1/2} & \sigma^{1/2} \\ \sigma^{1/2} & \sigma^{1/2} & \sigma^{1/2} \end{bmatrix}$	$p_b = \begin{bmatrix} \delta & 1 & 1 \\ 1 & \gamma & \gamma \\ 1 & \gamma & \gamma \end{bmatrix}$	$p_d = \begin{bmatrix} 1 & \sigma^{1/2} & \sigma^{1/2} \\ 1 & \sigma^{1/2} & \sigma^{1/2} \\ 1 & \sigma^{1/2} & \sigma^{1/2} \end{bmatrix}$
CH$_2$-CH : 151	CH-CH$_2$CH$_2$: 112							
C=C : 134	CH$_2$-C-H : 117.5	For CH-CH$_2$: t : 180 s$^+$: 120 s$^-$: −120			rows: columns:	t, g$^+$, g$^-$ t, s$^+$, s$^-$	t, s$^+$, s$^-$ t, s$^+$, s$^-$	t, s$^+$, s$^-$ t, g$^+$, g$^-$
	H-C-H : 109.5	For CH=CH : c : 0						

a) Three alternative parameter sets are given.

Calcd. quantities:

$\langle r^2 \rangle_0 / nl^2$ = 4.902 / 4.909 / 4.908 (*Exptl.*: 4.9)

$d (\ln \langle r^2 \rangle_0) / dT$ = −0.060 / 0.081 / 0.081 (× 10^{-3} K^{-1}) (*Exptl.*: 0.4 or −0.4 or 0.65 or −1.0 (× 10^{-3} K^{-1}))

Optical anisotropy $\Delta\Gamma$ = 5.580 / 5.534 / 5.531 (Å^3) (*Exptl.*: 5.8 or 7.5 (Å^3))

$d (\ln \Delta\Gamma) / dT$ = −0.117 / 0.109 / 0.109 (× 10^{-3} K^{-1}) (*Exptl.*: −3.5 × 10^{-3} K^{-1})

cis-1,4-polyisoprene, *cis*-1,4-PIP

Ishikawa, T.; Nagai, K. *J. Polym. Sci., Part A-2* **1969**, *7*, 1123.

Bond Length [pm]	Valence Angles [°]	Torsion Angles [°]	ξ a) (for 293 K)	ξ_0	E_ξ [kJ·mol^{-1}]	Transition probability matrices, **p**, are given; $\mathbf{p}_c = \mathbf{E}_3$
CH$_2$-CH$_2$: 153	CH$_2$-CR=CH : 125	For CH$_2$-CH$_2$: t : 180	γ = 10 / 30 / 30	1	−5.6 / −8.3 / −8.3	$\mathbf{p}_a = \begin{bmatrix} 1 & 1 & 1 \\ \sigma^{1/2} & \sigma^{1/2} & \sigma^{1/2} \\ \sigma^{1/2} & \sigma^{1/2} & \sigma^{1/2} \end{bmatrix}$ $\mathbf{p}_b = \begin{bmatrix} \delta & 1 & 1 \\ 1 & \gamma & \gamma \\ 1 & \gamma & \gamma \end{bmatrix}$ $\mathbf{p}_d = \begin{bmatrix} 1 & \sigma^{1/2} & \sigma^{1/2} \\ 1 & \sigma^{1/2} & \sigma^{1/2} \\ 1 & \sigma^{1/2} & \sigma^{1/2} \end{bmatrix}$
CH$_2$-CH/R : 151	CH/R-CH$_2$CH$_2$: 112	g$^+$: 60 g$^-$: −60	δ = 0.0322	1	8.4	
C=C : 134	CH$_2$-C-H : 117.5	For CH-CH$_2$: t : 180	σ = 3.0 / 1.8 / 1.4	1	−2.68 / −1.47 / −0.75	
	H-C-H : 109.5	s$^+$: 120 s$^-$: −120				rows: t, g$^+$, g$^-$ t, s$^+$, s$^-$ t, s$^+$, s$^-$ columns: t, s$^+$, s$^-$ t, s$^+$, s$^-$ t, g$^+$, g$^-$
	R = CH$_3$	For CH=CH : c : 0				

a) Three alternative parameter sets are given.

Calcd. quantities:

$\langle r^2 \rangle_0 / nl^2$ = 4.695 / 4.695 / 4.709 (Exptl.: 4.7)

$d(\ln \langle r^2 \rangle_0)/dT$ = 0.450 / 0.218 / 0.076 (× 10^{-3} K^{-1}) (Exptl.: 0.41 or 0.56 or 5.3–5.7 or 4.7 or 5.3 (× 10^{-3} K^{-1}))

Optical anisotropy $\Delta\Gamma$ = 0.764 / 0.428 / 0.152 (Å^3) (Exptl.: 0.72 Å^3)

trans-1,4-polybutadiene, *trans*-1,4-PBD

Ishikawa, T.; Nagai, K. *Polymer J.* **1970**, *1*, 116.

Bond Length [pm]	Valence Angles [°]	Torsion Angles [°]	ξ a) (for 373 K)	ξ_0	E_ξ [kJ·mol^{-1}]	Transition probability matrices, **p**, are given; (see: b,c))
CH_2-CH_2 : 153	$CH_2-CH=CH$: 125	For CH_2-CH_2: t : 180 g$^+$: 60 g$^-$: −60	$\alpha = 1/1/1$ $\beta = 0.25/0.45/0.08$ $\sigma = 1/0.763/1.31$	1 1 1	0/0/0 4.3/2.5/7.8 0/0.84/−0.84	
CH_2-CH : 151	$CH-CH_2CH_2$: 112					
$C=C$: 134	CH_2-C-H : 117.5 H-C-H : 109.5	For $CH-CH_2$: c : 0 s$^+$: 120 s$^-$: −120	$\lambda \, \mathbf{p_a} = \begin{bmatrix} 1 & 1 & \alpha^{1/2} \\ \sigma^{1/2} & \sigma^{1/2} & (\alpha\sigma)^{1/2}\beta \\ \sigma^{1/2} & \sigma^{1/2} & (\alpha\sigma)^{1/2}\beta \end{bmatrix}$		$\mathbf{p_b} = \begin{bmatrix} 1 & 0 & 0 \\ 1 & 0 & 0 \\ \alpha^{1/2} & 0 & 0 \end{bmatrix}$	$\mathbf{p_c} = \begin{bmatrix} 1 & 1 & \alpha^{1/2} \\ 0 & 0 & 0 \\ 0 & 0 & 0 \end{bmatrix}$ $\mathbf{p_d} = \begin{bmatrix} 1 & \sigma^{1/2} & \sigma^{1/2} \\ 1 & \sigma^{1/2} & \sigma^{1/2} \\ \alpha^{1/2} & (\alpha\sigma)^{1/2}\beta & (\alpha\sigma)^{1/2}\beta \end{bmatrix}$
		For CH=CH: t : 180	rows: columns:	t, g$^+$, g$^-$ s$^+$, s$^-$, c	s$^+$, s$^-$, c t, a$^+$, a$^-$	t, a$^+$, a$^-$ s$^+$, s$^-$, c	s$^+$, s$^-$, c t, g$^+$, g$^-$

a) Three alternative parameters are given.
b) λ is the largest eigenvalue of $\lambda \mathbf{P_1 P_2 P_3 P_4}$ and is introduced to normalize $\mathbf{P_1 P_2 P_3 P_4}$ so that its largest eigenvalue becomes unity.
c) a$^+$ and a$^-$ are arbitrary states which are introduced to make both $\mathbf{p_b}$ and $\mathbf{p_c}$ square matrices of order three.

Calcd. quantities:

$<r^2>_0 / nl^2$	=	5.804 / 5.850 / 5.847	(*Exptl.*: 5.8)
$d(\ln <r^2>_0)/dT$	=	−0.667 / −1.012 / −0.149 (× 10^{-3} K^{-1})	(*Exptl.*: −0.65 × 10^{-3} K^{-1})
Optical anisotropy $\Delta\Gamma$	=	6.136 / 6.284 / 6.066 (Å^3)	(*Exptl.*: 6.1 Å^3)
$d(\ln \Delta\Gamma)/dT$	=	−0.753 / −2.786 / −0.101 (× 10^{-3} K^{-1})	(*Exptl.*: 4.93 − 10^{-3} K^{-1})

trans-1,4-polyisoprene, *trans*-1,4-PIP

Ishikawa, T.; Nagai, K. *Polymer J.* **1970**, *1*, 116.

Bond Length [pm]	Valence Angles [°]	Torsion Angles [°]	ξ a) (for 333 K)	ξ_o	E_ξ [kJ·mol^{-1}]	
CH_2-CH_2 : 153	CH_2-CR=CH : 125	For CH_2-CH_2:	α = 1 / 1 / 1.46	1	0 / 0 / −1.05	
		t : 180	β = 0.25 / 0.25 / 0	1	3.85 / 3.85 / ∞	
CH_2-CH/R : 151	CH/RCH_2CH_2 : 112	g^+ : 60	σ = 0.67 / 0.536 / 0.942	1	0.9 / 1.72 / 0.17	
		g^- : −60	$\zeta \approx$ 0.75	1	$\approx$ 0.8	
C=C : 134	CH_2-C-H : 117.5					
	H-C-H : 109.5	For CH/R-CH_2:				
		c : 0				
		s^+ : 120				
		s^- : −120				
		For CH=CH :				
	R = CH_3	t : 180				

Transition probability matrices, **p**, are given; (see: b,c,d))

$$\lambda\,\mathbf{p_a} = \begin{bmatrix} 1 & 1 & \alpha^{1/2} \\ \sigma^{1/2} & \sigma^{1/2} & (\alpha\sigma)^{1/2}\beta \\ \sigma^{1/2} & \sigma^{1/2} & (\alpha\sigma)^{1/2}\beta \end{bmatrix}$$

rows: t, g^+, g^-
columns: s^+, s^-, c

$$\mathbf{p_b} = \begin{bmatrix} 1 & 0 & 0 \\ 1 & 0 & 0 \\ \alpha^{1/2} & 0 & 0 \end{bmatrix}$$

s^+, s^-, c
t, a^+, a^-

$$\mathbf{p_c} = \begin{bmatrix} 1 & 1 & 0 \\ 0 & 0 & 0 \\ 0 & 0 & 0 \end{bmatrix}$$

t, a^+, a^-
s^+, s^-, c

$$\mathbf{p_d} = \begin{bmatrix} 1 & \sigma^{1/2} & \sigma^{1/2} \\ 1 & \sigma^{1/2} & \sigma^{1/2} \\ 0 & 0 & 0 \end{bmatrix}$$

s^+, s^-, c
t, g^+, g^-

a) Three alternative parameters are given.
b) λ is the largest eigenvalue of $\lambda P_1 P_2 P_3 P_4$ and is introduced to normalize $P_1 P_2 P_3 P_4$ so that its largest eigenvalue becomes unity.
c) a^+ and a^- are arbitrary states which are introduced to make both $\mathbf{p_b}$ and $\mathbf{p_c}$ square matrices of order three.
d) For *trans*-PIP, $\mathbf{p_a}$ may also be written as: (for details, see original paper)

$$\mathbf{p_a} = \lambda^{-1} \begin{bmatrix} 1 & 1 & \alpha^{1/2} \\ \sigma^{1/2}\zeta & \sigma^{1/2} & (\alpha\sigma)^{1/2}\beta \\ \sigma^{1/2} & \sigma^{1/2}\zeta & (\alpha\sigma)^{1/2}\beta \end{bmatrix}$$

Calcd. quantities:

$<r^2>_o / nl^2$	= 6.544 / 7.353 / 6.537	(Exptl.: 7.35)
$d (\ln <r^2>_o) / dT$	= −1.01 / −1.45 / −0.07 (× 10^{-3} K^{-1})	(Exptl.: −0.27 × 10^{-3} K^{-1})
Optical anisotropy $\Delta\Gamma$	= 6.206 / 7.194 / 5.959 (Å^3)	(Exptl.: 5.8 Å^3)
$d (\ln \Delta\Gamma) / dT$	= −1.18 / −1.71 / −0.09 (× 10^{-3} K^{-1})	(Exptl.: 4.34 × 10^{-3} K^{-1})

1,4-polybutadiene, PBD; the *cis,cis* unit

Abe, Y.; Flory, P. J. *Macromolecules* **1971**, *4*, 219.

Bond Length [pm]	Valence Angles [°]	Torsion Angles [°]	ξ a) (for 323 K)	ξ_o	E_ξ [kJ · mol^{-1}]
-C=C- : 134	C=C-CH$_2$: 125	For C=C: c : 0	$\sigma = 1$	1	0
=C-C- : 151	C=C-H : 119	For CH$_2$-CH$_2$:	$\zeta = 0.101$	0.3	2.93
-C-C- : 153	C-C-C : 112	t : 180 g$^+$: 60			
C-H : 110	H-C-H : 109	g$^-$: -60			
		For CH-CH$_2$: t : 180 s$^+$: 120 s$^-$: -120			

$$U_a = \begin{bmatrix} \zeta & 1 & 1 \end{bmatrix} \quad U_b = \begin{bmatrix} 1 & \sigma & \sigma \\ 1 & \sigma & \sigma \\ 1 & \sigma & \sigma \end{bmatrix} \quad U_c = \begin{bmatrix} \zeta & 1 & 1 \\ \zeta & 1 & 1 \\ \zeta & 1 & 1 \end{bmatrix} \quad U_d = \begin{bmatrix} 1 \\ 1 \\ 1 \end{bmatrix}$$

rows:	c	t, s$^+$, s$^-$	t, g$^+$, g$^-$	t, s$^+$, s$^-$
columns:	t, s$^+$, s$^-$	t, g$^+$, g$^-$	t, s$^+$, s$^-$	c

a) Calculations are also carried out for E(σ) = + 0.42 and - 0.42 kJ · mol^{-1}, giving σ = 0.86 and 1.17 at 323 K.

Comments:
1,4-Polybutadiene chains are treated in terms of structural units CH—CH$_2$—CH$_2$—CH= consisting of three single bonds bounded by double bonds. Although conformational interactions differ markedly depending on the steric configuration, *cis* or *trans*, of the adjoining double bonds, those within a given unit are essentially independent of the conformations of neighboring units in all cases. The units thus defined may be treated, therefore, as statistically independent. Statistical weight parameters are chosen in the light of conformational energy calculations and results of spectroscopic investigations on low molecular analogs.

Calcd. quantities:

$<r^2>_o / nl^2$	= 5.08	(*Exptl.*: 4.9 ($\pm$ 0.2))
$d (\ln <r^2>_o) / dT$	= 0.16 × 10^{-3} K^{-1}	(*Exptl.*: 0.4 × 10^{-3} K^{-1})
Stress optical coefficient Γ_2	= 3.53 × 10^{-24} cm^3	(*Exptl.*: 3.0 (or 4.5) × 10^{-24} cm^3)
A priori probabilities for bond conformations, p_η		

V 010

1,4-polybutadiene, PBD; the *trans,trans* unit

Abe, Y.; Flory, P. J. *Macromolecules* **1971**, *4*, 219.

Bond Length [pm]	Valence Angles [°]	Torsion Angles [°]	ξ a) (for 323 K)	ξ_0	E_ξ [kJ·mol^{-1}]
-C=C- : 134	C=C-CH$_2$: 125	For C=C: t : 180	σ = 1	1	0
=C-C- : 151	C=C-H : 119	For CH$_2$-CH$_2$:	ρ = 0.627	1	1.26
-C-C- : 153	C-C-C : 112	t : 180, g$^+$: 60	β = 0.0446	1	8.37
C-H : 110	H-C-H : 109	g$^-$: -60			
		For CH-CH$_2$: c : 0, s$^+$: 120, s$^-$: -120			

$$U_a = \begin{bmatrix} \rho & 1 & 1 \end{bmatrix}$$

$$U_b = \begin{bmatrix} 1 & 0 & 0 & \sigma & 0 & 0 & \sigma & 0 & 0 \\ 0 & 1 & 0 & 0 & \sigma & 0 & 0 & \sigma & 0 \\ 0 & 0 & 1 & 0 & 0 & \sigma & 0 & 0 & \sigma \end{bmatrix}$$

$$U_c = \begin{bmatrix} \rho\beta & \beta & \beta \\ \rho\beta & 1 & 1 \\ \rho\beta & 1 & 1 \\ \rho\beta & 1 & 1 \\ \rho\beta & \beta & \beta \\ \rho\beta & 1 & 1 \\ \rho\beta & 1 & 1 \\ \rho\beta & 1 & 1 \\ \rho\beta & 1 & 1 \end{bmatrix}$$

$$U_d = \begin{bmatrix} 1 \\ 1 \\ 1 \end{bmatrix}$$

	U$_a$	U$_b$	U$_c$	U$_d$
rows:	t	c, s$^+$, s$^-$	ct, s$^+$t, s$^-$t, cg$^+$, s$^+$g$^+$, s$^-$g$^+$, cg$^-$, s$^+$g$^-$, s$^-$,g$^-$	c, s$^+$, s$^-$
columns:	c, s$^+$, s$^-$	ct, s$^+$t, s$^-$t, cg$^+$, s$^+$g$^+$, s$^-$g$^+$, cg$^-$, s$^+$g$^-$, s$^-$g$^-$	c, s$^+$, s$^-$	t

a) Calculations are also carried out for $E(\sigma)$ = + 0.42 and – 0.42 kJ·mol^{-1}, giving σ = 0.86 and 1.17 at 323 K.

Comments:
1,4-Polybutadiene chains are treated in terms of structural units CH–CH$_2$–CH$_2$–CH= consisting of three single bonds bounded by double bonds. Although conformational interactions differ markedly depending on the steric configuration, *cis* or *trans*, of the adjoining double bonds, those within a given unit are essentially independent of the conformations of neighboring units in all cases. The units thus defined may be treated, therefore, as statistically independent. Statistical weight parameters are chosen in the light of conformational energy calculations and results of spectroscopic investigations on low molecular analogs.

Calcd. quantities:

$\langle r^2 \rangle_0 / nl^2$ = 6.20   (*Exptl.*: 5.8 ± 0.2)

$d(\ln \langle r^2 \rangle_0)/dT$ = – 0.11 × 10^{-3} K^{-1} (*Exptl.*: – 0.65 × 10^{-3} K^{-1})

Stress optical coefficient Γ_2 = 4.00 × 10^{-24} cm^3 (*Exptl.*: 3.5 (*or* 5.1) × 10^{-24} cm^3)

A priori probabilities for bond conformations, p_η

V 011

1,4-polybutadiene, PBD; the *cis,trans* and *trans,cis* units [a]

Abe, Y.; Flory, P. J. *Macromolecules* **1971**, *4*, 219.

the cis-trans unit:

$$U_a = \begin{bmatrix} \zeta & 1 & 1 \end{bmatrix} \qquad U_b = \begin{bmatrix} 1 & 0 & 0 & \sigma & 0 & 0 & \sigma & 0 & 0 \\ 0 & 1 & 0 & 0 & \sigma & 0 & 0 & \sigma & 0 \\ 0 & 0 & 1 & 0 & 0 & \sigma & 0 & 0 & \sigma \end{bmatrix} \qquad U_c = \begin{bmatrix} \rho & 1 & 1 \\ \rho & 1 & 1 \\ \rho & 1 & 1 \\ \rho\beta^{1/2} & \beta & \beta \\ \rho\beta & 1 & 1 \\ \rho\beta & 1 & 1 \\ \rho\beta^{1/2} & \beta & \beta \\ \rho\beta & 1 & 1 \\ \rho\beta & 1 & 1 \end{bmatrix} \qquad U_d = \begin{bmatrix} 1 \\ 1 \\ 1 \end{bmatrix}$$

the trans,cis unit:

$$U_a = \begin{bmatrix} \rho & 1 & 1 \end{bmatrix} \qquad U_b = \begin{bmatrix} 1 & 0 & 0 & \sigma & 0 & 0 & \sigma & 0 & 0 \\ 0 & 1 & 0 & 0 & \sigma & 0 & 0 & \sigma & 0 \\ 0 & 0 & 1 & 0 & 0 & \sigma & 0 & 0 & \sigma \end{bmatrix} \qquad U_c = \begin{bmatrix} \zeta & 1 & 1 \\ \zeta & 1 & 1 \\ \zeta & 1 & 1 \\ \zeta\beta^{1/2} & \beta & \beta \\ \zeta & 1 & 1 \\ \zeta & 1 & 1 \\ \zeta\beta^{1/2} & \beta & \beta \\ \zeta & 1 & 1 \\ \zeta & 1 & 1 \end{bmatrix} \qquad U_d = \begin{bmatrix} 1 \\ 1 \\ 1 \end{bmatrix}$$

	U_a:	U_b:	U_c:	U_d:		U_a:	U_b:	U_c:	U_d:
rows:	c	t, s^+, s^-	tt, s^+t, s^-t, tg^+, s^+g^+, s^-g^+, tg^-, s^+g^-, s^-g^-	c, s^+, s^-	rows:	t	c, s^+, s^-	ct, s^+t, s^-t, cg^+, s^+g^+, s^-g^+, cg^-, s^+g^-, s^-g^-	t, s^+, s^-
columns:	c, s^+, s^-	tt, s^+t, s^-t, tg^+, s^+g^+, s^-g^+, tg^-, s^+g^-, s^-g^-	c, s^+, s^-	t	columns:	c, s^+, s^-	ct, s^+t, s^-t, cg^+, s^+g^+, s^-g^+, cg^-, s^+g^-, s^-g^-	t, s^+, s^-	c

[a] Parameters used are those given for the *cis,cis* and *trans,trans* units (**V 009** and **V010**).

Comments:
1,4-Polybutadiene chains are treated in terms of structural units $CH-CH_2-CH_2-CH=$ consisting of three single bonds bounded by double bonds. Although conformational interactions differ markedly depending on the steric configuration, *cis* or *trans*, of the adjoining double bonds, those within a given unit are essentially independent of the conformations of neighboring units in all cases. The units thus defined may be treated, therefore, as statistically independent. Statistical weight parameters are chosen in the light of conformational energy calculations and results of spectroscopic investigations on low molecular analogs.

V 012

1,4-polyisoprene, PIP; the *cis* unit

Abe, Y.; Flory, P. J. *Macromolecules* **1971**, *4*, 230.

Bond Length [pm]	Valence Angles [°]	Torsion Angles [°]	ξ (for 323 K)	ξ_0	E_ξ [kJ·mol^{-1}]
-C=C- : 134	C=C-CH$_2$: 125	For bond a: t : 180 s$^+$: 60 s$^-$: -60	$\sigma = 1$	1	0
=C-C- : 151	C=C-H : 119		$\zeta = 0.101$	0.3	2.93
-C-C- : 153	C-C-C : 112	$\omega = 0.154$	1	5.02	
C-H : 110	H-C-H : 109	For bond b: t : 180 g$^+$: 60 g$^-$: -60			
	C=C-CH$_3$: 124	For bond c: s$^+$: 120 or 110 s$^-$: -120 or -110			

$U_a = [\zeta \; 1 \; 1]$	$U_b = \begin{bmatrix} 1 & \sigma & \sigma \\ 1 & \sigma & \sigma \\ 1 & \sigma & \sigma \end{bmatrix}$	$U_c = \begin{bmatrix} 1 & \sigma & \sigma \\ \omega & 1 & \\ 1 & & \omega \end{bmatrix}$	$U_d = \begin{bmatrix} 1 \\ 1 \end{bmatrix}$	
rows: c	t, s$^+$, s$^-$	t, g$^+$, g$^-$	s$^+$, s$^-$	
columns: t, s$^+$, s$^-$	t, g$^+$, g$^-$	s$^+$, s$^-$	c	

Comments:
The configurational characteristics of 1,4-polyisoprene chains are treated in terms of the conformationally independent units. Properties of PIP chains are calculated for the *cis* and *trans* stereoregular polymers and for those of intermediate composition measured by the fraction of *cis* residues. In the case of the structural unit whose double bond is *cis*, the c state for the third single bond is precluded by a severe steric overlap. Energy calculations indicate $\phi(s^\pm) \approx \pm 110°$ for the third single bond, for both *cis* and *trans* units, instead of $\pm 120°$. Properties of *cis*-PIP, unlike those of *trans*-PIP, are insensitive to the statistical weights; they depend strongly on certain of the geometrical parameters, however. Calculations on *cis*-PIP and *trans*-PIP are compared with experimentally observed characteristic ratios, temperature coefficients, stress-birefringence coefficients, and, in the case of *cis*-PIP, with strain dichroism observed for infrared bands.

Calcd. quantities: [a]

$<r^2>_0 / nl^2$	= 4.55 / 3.84	(Exptl.: 4.7)
$d (\ln <r^2>_0) / dT$	= 0.18 / 0.16 ($\times 10^{-3}$ K^{-1})	(Exptl.: 0.41 or 0.56 ($\times 10^{-3}$ K^{-1}))
Stress optical coefficient Γ_2	= 2.84 / 2.25 ($\times 10^{-24}$ cm^3)	(Exptl.: 2.4 or 2.9 ($\times 10^{-24}$ cm^3))
A priori probabilities for bond conformations, p_η		
Strain dichroism observed for infrared bands		

[a] The first value is for $\phi(s^\pm) = \pm 120°$ and the second one for $\phi(s^\pm) = \pm 110°$ for bond c.

V 013

1,4-polyisoprene, PIP: the *trans* unit

Abe, Y.; Flory, P. J. *Macromolecules* **1971**, *4*, 230.

Bond Length [pm]	Valence Angles [°]	Torsion Angles [°]	ξ (for 323 K)	ξ_0	E_ξ [kJ·mol^{-1}]
-C=C- : 134	C=C-CH$_2$: 125	*For bond a:*	$\sigma = 1$	1	0
		t : 180	$\zeta = 0.101$	1	2.93
=C-C- : 151	C=C-H : 119	s$^+$: 120	$\rho = 1.60$	1	–1.26
		s$^-$: –120	$\beta = 0.0446$	1	8.37
-C-C- : 153	C-C-C : 112		$\omega = 0.154$	1	5.02
C-H : 110	H-C-H : 109	*For bond b:*			
		t : 180			
	C=C-CH$_3$: 124	g$^+$: 60			
		g$^-$: –60			
		For bond c:			
		c : 0			
		s$^+$: 120 *or* 110			
		s$^-$: –120 *or* –110			

$U_a = [\zeta\ 1\ 1]$

$U_b = \begin{bmatrix} 1 & 0 & 0 & \sigma & 0 & 0 & \sigma & 0 & 0 \\ 0 & 1 & 0 & 0 & \sigma & 0 & 0 & \sigma & 0 \\ 0 & 0 & 1 & 0 & 0 & \sigma & 0 & 0 & \sigma \end{bmatrix}$

$U_c = \begin{bmatrix} \rho & 1 & 1 \\ \rho & 1 & 1 \\ \rho & 1 & 1 \\ \rho\beta^{1/2} & \omega & 1 \\ \rho\beta & \omega & 1 \\ \rho\beta & \omega & 1 \\ \rho\beta^{1/2} & 1 & \omega \\ \rho\beta & 1 & \omega \\ \rho\beta & 1 & \omega \end{bmatrix}$

$U_d = \begin{bmatrix} 1 \\ 1 \\ 1 \end{bmatrix}$

	U_a:	U_b:	U_c:	U_d:
rows:	t	t, s$^+$, s$^-$	ct, s$^+$t, s$^-$t; cg$^+$, s$^+$g$^+$, s$^-$g$^+$; cg$^-$, s$^+$g$^-$, s$^-$g$^-$	c, s$^+$, s$^-$
columns:	t, s$^+$, s$^-$	ct, s$^+$t, s$^-$t; cg$^+$, s$^+$g$^+$, s$^-$g$^+$; cg$^-$, s$^+$g$^-$, s$^-$g$^-$	c, s$^+$, s$^-$	t

---

Comments:
The configurational characteristics of 1,4-polyisoprene chains are treated in terms of the conformationally independent units. Properties of PIP chains are calculated for the *cis* and *trans* stereoregular polymers and for those of intermediate composition measured by the fraction of *cis* residues. Conformational energy calculations confirm the choise of t and s$^\pm$ rotational states for the first single bond of the structural unit, and c and s$^\pm$ states for the third, the second single bond being assigned the usual t and g$^\pm$ states. In the case of the structural unit whose double bond is *cis*, the c state for the third single bond is precluded by a severe steric overlap. Energy calculations indicate $\phi(s^\pm) \approx \pm 110°$ for the third single bond, for both *cis* and *trans* units, instead of $\pm 120°$. Properties of *cis*-PIP, unlike those of *trans*-PIP, are insensitive to the statistical weights; they depend strongly on certain of the geometrical parameters, however. Calculations on *cis*-PIP and *trans*-PIP are compared with experimentally observed characteristic ratios, temperature coefficients, stress-birefringence coefficients, and, in the case of *cis*-PIP, with strain dichroism observed for infrared bands.

Calcd. quantities: [a]

$\langle r^2 \rangle_0 / nl^2$	= 6.95 / 6.60	(Exptl.: 6.6 or 7.4)
$d(\ln \langle r^2 \rangle_0)/dT$	= –0.96 / –1.04 (× 10^{-3} K^{-1})	
Stress optical coefficient Γ_2	= 4.59 / 4.60 (× 10^{-24} cm^3)	(Exptl.: 3.5 or 5.5 (× 10^{-24} cm^3))
A priori probabilities for bond conformations, p_η		

[a] The first value is for $\phi(s^\pm) = \pm 120°$ and the second one for $\phi(s^\pm) = \pm 110°$ for bond c.

V 014

cis- and trans-1,4-polybutadiene, PBD

Tanaka, S.; Nakajima, A. *Polymer J.* **1972**, *3*, 500.

Bond Length [pm]	Valence Angles [°]	Torsion Angles [°]	ξ (for 323 K)	ξ_0	E_ξ [kJ·mol^{-1}]				
-C=C- : 134	C=C-CH$_2$: 125	For bonds a,c: c : 0 or t : 180 s$^+$: 120 s$^-$: -120	$\sigma = 1$	1	0		$U_a^{cis} = \begin{bmatrix} 1 & 1 & 1 \\ 1 & 1 & 1 \\ 1 & 1 & 1 \end{bmatrix}$	$U_c^{cis} = \begin{bmatrix} \delta & 1 & 1 \\ 1 & \gamma & \gamma \\ 1 & \gamma & \gamma \end{bmatrix}$	$U_d^{cis} = \begin{bmatrix} 1 & \sigma & \sigma \\ 1 & \sigma & \sigma \\ 1 & \sigma & \sigma \end{bmatrix}$
=C-C- : 151	C=C-H : 119		$\delta = 6$	1	-4.8				
-C-C- : 153	C-C-C : 112		$\gamma = 30$	1	-9.1				
C-H : 110	H-C-H : 109	For bond b: c : 0 or t : 180	$\alpha = 1$	1	0	rows: columns:	t, g$^+$, g$^-$ t, s$^+$, s$^-$	t, s$^+$, s$^-$ t, s$^+$, s$^-$	t, s$^+$, s$^-$ t, g$^+$, g$^-$
			$\beta = 0.25$	1	3.7		$U_a^{trans} = \begin{bmatrix} 1 & \alpha & 1 \\ 1 & \beta\alpha & 1 \\ 1 & \beta\alpha & 1 \end{bmatrix}$	$U_c^{trans} = \begin{bmatrix} 1 & \alpha & 1 \\ 1 & \alpha & 1 \\ 1 & \alpha & 1 \end{bmatrix}$	$U_d^{trans} = \begin{bmatrix} 1 & \sigma & \sigma \\ 1 & \beta\sigma & \sigma \\ 1 & \beta\sigma & \sigma \end{bmatrix}$
		For bond d: t : 180 g$^+$: 60 g$^-$: -60				rows: columns:	t, g$^+$, g$^-$ s$^+$, c, s$^-$	s$^+$, c, s$^-$ s$^+$, c, s$^-$	s$^+$, c, s$^-$ t, g$^+$, g$^-$

Comments:

The characteristic ratios of stereoirregular 1,4-polybutadiene and 1,4-polyisoprene chains are theoretically investigated by the Monte Carlo procedure in accordance with the model proposed by *Mark* (**V 001** and **V 003**). It is pointed out that the presence of discrete *cis* units in *trans*-rich chains significantly reduces the characteristic ratio while that of discrete *trans* units in *cis*-rich chains has little effect on the characteristic ratio. The characteristic ratio and its dependence on both the *trans* and *cis* contents and their sequence distribution is calculated for stereoirregular polymers in accordance with the interdependent RIS model proposed by *Mark* (**V 001** and **V 003**), and *Ishikawa* and *Nagai* (**V 005** and **V 007**).

Calcd. quantities: $<r^2>_0 / nl^2$ = 4.85 (*cis*) / 5.77 (*trans*) (Exptl.: 4.9 (*cis*) / 5.8 (*trans*))

V 015

cis- and trans-1,4-polyisoprene, PIP

Tanaka, S.; Nakajima, A. *Polymer J.* **1972**, *3*, 500.

Bond Length [pm]	Valence Angles [°]	Torsion Angles [°]	ξ (for 323 K)	ξ_0	E_ξ [kJ · mol^{-1}]			
-C=C- : 134	C=C-CH$_2$: 125	For bonds a,c: c : 0 or	σ' = 0.536	1	1.67			
=C-C- : 151	C=C-H : 119	t : 180 s$^+$: 120	δ = 0	1	∞		$U_a^{cis} = \begin{bmatrix} 1 & 1 & 1 \\ 1 & 1 & 1 \\ 1 & 1 & 1 \end{bmatrix}$	$U_c^{cis} = \begin{bmatrix} \delta & 1 & 1 \\ 1 & \gamma & \gamma \\ 1 & \gamma & \gamma \end{bmatrix}$	$U_d^{cis} = \begin{bmatrix} 1 & \sigma' & \sigma' \\ 1 & \sigma' & \sigma' \\ 1 & \sigma' & \sigma' \end{bmatrix}$
-C-C- : 153	C-C-C : 112	s$^-$: -120	γ = 30	1	-9.1			
						rows:	t, g$^+$, g$^-$	t, s$^+$, s$^-$	t, s$^+$, s$^-$
C-H : 110	H-C-H : 109	For bond b: c : 0 or	α = 1	1	0	columns:	t, s$^+$, s$^-$	t, s$^+$, s$^-$	t, g$^+$, g$^-$
	C=C-CH$_3$: 124	t : 180	β = 0.25	1	3.7		$U_a^{trans} = \begin{bmatrix} 1 & \alpha & 1 \\ 1 & \beta\alpha & 1 \\ 1 & \beta\alpha & 1 \end{bmatrix}$	$U_c^{trans} = \begin{bmatrix} 1 & 0 & 1 \\ 1 & 0 & 1 \\ 1 & 0 & 1 \end{bmatrix}$	$U_d^{trans} = \begin{bmatrix} 1 & \sigma' & \sigma' \\ 0 & 0 & 0 \\ 1 & \sigma' & \sigma' \end{bmatrix}$
		For bond d: t : 180 g$^+$: 60 g$^-$: -60				rows: columns:	t, g$^+$, g$^-$ s$^+$, c, s$^-$	s$^+$, c, s$^-$ s$^+$, c, s$^-$	s$^+$, c, s$^-$ t, g$^+$, g$^-$

Comments:
The characteristic ratios of stereoirregular 1,4-polybutadiene and 1,4-polyisoprene chains are theoretically investigated by the Monte Carlo procedure in accordance with the model proposed by *Mark* (**V 002** and **V 004**). It is pointed out that the presence of discrete *cis* units in *trans*-rich chains significantly reduces the characteristic ratio while that of discrete *trans* units in *cis*-rich chains has little effect on the characteristic ratio. The characteristic ratio and its dependence on both the *trans* and *cis* contents and their sequence distribution is calculated for stereoirregular polymers in accordance with the interdependent RIS model proposed by *Mark* (**V 002** and **V 004**), and *Ishikawa* and *Nagai* (**V 006** and **V 008**).

Calcd. quantities: $<r^2>_0 / nl^2$ 4.92 (*cis*) / 7.30 (*trans*)   (Exptl.: 4.7 (*cis*) / 7.35 (*trans*))

V 016

cis-1,4-polybutadiene, cis-PBD, and cis-1,4-polyisoprene, cis-PIP

Kajiwara, K.; Burchard, W. *Macromolecules* **1984**, *17*, 2669.

R = a) H
b) CH_3

Bond Length [pm]	Valence Angles [°]	Torsion Angles [°]	ξ a) (for 323 K)	ξ_0	E_ξ [kJ·mol^{-1}]
-C=C- : 134	C=C-CH_2 : 125	For bonds 1,3: t : 180	σ = 1.4 / 0.543	1	– 0.9 / 1.64
=C-C- : 151	C=C-H : 119	s$^+$: 120 s$^-$: –120	γ = 10	1	6.2
-C-C- : 153	C-C-C : 112	For bond 2:			
C-H : 110	H-C-H : 109	c : 0			
	C=C-CH_3 : 124	For bond 4: t : 180 g$^+$: 60 g$^-$: –60			

$$U_a = \begin{bmatrix} 1 & 1 & 0 & 0 & 0 & 1 \\ 0 & 0 & 0 & 0 & 0 & 0 \\ 1 & 1 & 0 & 0 & 0 & 1 \\ 0 & 0 & 0 & 0 & 0 & 0 \\ 1 & 1 & 0 & 0 & 0 & 1 \\ 0 & 0 & 0 & 0 & 0 & 0 \end{bmatrix} \quad U_b = \begin{bmatrix} 0 & 1 & 0 & 0 & 0 & 1 \\ 1 & \gamma & 0 & 0 & 0 & \gamma \\ 0 & 0 & 0 & 0 & 0 & 0 \\ 0 & 0 & 0 & 0 & 0 & 0 \\ 0 & 0 & 0 & 0 & 0 & 0 \\ 1 & \gamma & 0 & 0 & 0 & \gamma \end{bmatrix} \quad U_c = \begin{bmatrix} 1 & 0 & \sigma & 0 & \sigma & 0 \\ 1 & 0 & \sigma & 0 & \sigma & 0 \\ 0 & 0 & 0 & 0 & 0 & 0 \\ 0 & 0 & 0 & 0 & 0 & 0 \\ 0 & 0 & 0 & 0 & 0 & 0 \\ 1 & 0 & \sigma & 0 & \sigma & 0 \end{bmatrix}$$

rows and columns: t, s$^+$, g$^+$, c, g$^-$, s$^-$

a) Values correspond to PBD and PIP, respectively.

Comments:
Monte Carlo simulations are applied to estimate the characteristic ratios and ρ parameters from the RIS models for PE, PBD, and PIP. Here the ρ parameter is defined as the ratio of the radius of gyration to the hydrodynamic radius. The ρ parameters of these real chains in the unperturbed state show only a slight dependence on the microconformation in the limit of large molecular weights and are found close to 1.504, which is the value of an idealized Gaussian chain. The estimated ρ parameters of the real chains appear to be correlated to the chain stiffness and increase with the characteristic ratios.

Calcd. quantities: Bond conformation probabilities $\rho = \langle S^2 \rangle^{1/2} / R_H$ ($\langle S^2 \rangle^{1/2}$: radius of gyration, R_H : hydrodynamic radius)

The characteristic ratios are estimated as 3.76 and 3.95, respectively, for cis-PBD and cis-PIP, which are approximately 30% lower than the corresponding experimentally observed values. The ρ_∞ parameters 1.45 and 1.44 are close to that of a Gaussian chain.

trans-1,4-polybutadiene, trans-PBD, and trans-1,4-polyisoprene, trans-PIP

R = a) H
b) CH_3

Kajiwara, K.; Burchard, W. *Macromolecules* **1984**, *17*, 2669.

Bond Length [pm]	Valence Angles [°]	Torsion Angles [°]	ξ [a] (for 323 K)	ξ_0	E_ξ [kJ·mol^{-1}]
-C=C- : 134	C=C-CH_2 : 125	For bonds a,c: c : 0	$\sigma = 1 / 0.543$	1	0 / 1.64
=C-C- : 151	C=C-H : 119	s$^+$: 120 s$^-$: -120	$\alpha = 0.96$	1	0.11
-C-C- : 153	C-C-C : 112	For bond b:	$\beta = 0.24 / 0.124$	1	3.8 / 5.6
C-H : 110	H-C-H : 109	t : 180			
	C=C-CH_3 : 124	For bond d: t : 180 g$^+$: 60 g$^-$: -60			

$$U_a = \begin{bmatrix} 0 & 1 & 0 & \alpha & 0 & 1 \\ 0 & 0 & 0 & 0 & 0 & 0 \\ 0 & 1 & 0 & \beta\alpha & 0 & 1 \\ 0 & 0 & 0 & 0 & 0 & 0 \\ 0 & 1 & 0 & \beta\alpha & 0 & 1 \\ 0 & 0 & 0 & 0 & 0 & 0 \end{bmatrix} \quad U_b = \begin{bmatrix} 0 & 0 & 0 & 0 & 0 & 0 \\ 1 & 0 & 0 & 0 & 0 & 0 \\ 0 & 0 & 0 & 0 & 0 & 0 \\ 1 & 0 & 0 & 0 & 0 & 0 \\ 0 & 0 & 0 & 0 & 0 & 0 \\ 1 & 0 & 0 & 0 & 0 & 0 \end{bmatrix} \quad U_c = \begin{bmatrix} 0 & 1 & 0 & \alpha & 0 & 1 \\ 0 & 0 & 0 & 0 & 0 & 0 \\ 0 & 0 & 0 & 0 & 0 & 0 \\ 0 & 0 & 0 & 0 & 0 & 0 \\ 0 & 0 & 0 & 0 & 0 & 0 \\ 0 & 0 & 0 & 0 & 0 & 0 \end{bmatrix}$$

$$U_d^{PBD} = \begin{bmatrix} 0 & 0 & 0 & 0 & 0 & 0 \\ 1 & 0 & \sigma & 0 & \sigma & 0 \\ 0 & 0 & 0 & 0 & 0 & 0 \\ 1 & 0 & \beta\sigma & 0 & \beta\sigma & 0 \\ 0 & 0 & 0 & 0 & 0 & 0 \\ 1 & 0 & \sigma & 0 & \sigma & 0 \end{bmatrix} \quad U_d^{PIP} = \begin{bmatrix} 0 & 0 & 0 & 0 & 0 & 0 \\ 1 & 0 & \sigma & 0 & \sigma & 0 \\ 0 & 0 & 0 & 0 & 0 & 0 \\ 1 & 0 & 0 & 0 & 0 & 0 \\ 0 & 0 & 0 & 0 & 0 & 0 \\ 1 & 0 & \sigma & 0 & \sigma & 0 \end{bmatrix}$$

rows and columns: t, s$^+$, g$^+$, c, g$^-$, s$^-$

[a] Values correspond to PBD and PIP, respectively.

Comments:
Monte Carlo simulations are applied to estimate the characteristic ratios and ρ parameters from the RIS models for PE, PBD, and PIP. Here the ρ parameter is defined as the ratio of the radius of gyration to the hydrodynamic radius. The ρ parameters of these real chains in the unperturbed state show only a slight dependence on the microconformation in the limit of large molecular weights and are found close to 1.504, which is the value of an idealized Gaussian chain. The estimated ρ parameters of the real chains appear to be correlated to the chain stiffness and increase with the characteristic ratios.

Calcd. quantities: Bond conformation probabilities $\rho = <S^2>^{1/2} / R_H$ ($<S^2>^{1/2}$: radius of gyration, R_H : hydrodynamic radius)

The characteristic ratios are estimated as 5.80 and 7.26, respectively, for trans-PBD and trans-PIP and are consistent with those found experimentally. The ρ_∞ parameters are estimated as 1.46 and 1.49, respectively.

polycyclobutene, *cis*- and *trans*-1,4-polybutadiene, PBD

Suter, U. W.; Höcker, H. *Makromol. Chem.* **1988**, *189*, 1603.

Bond Length [pm]	Valence Angles [°]	Torsion Angles [°]	ξ (for 323 K)	ξ_0	E_ξ [kJ · mol^{-1}]					
-C=C- : 134	C=C-CH$_2$: 125	For bond a: t : 180 or	β = 0.045	1	8.314	$U_a^{cis} = \begin{bmatrix} \zeta & 1 & 1 \end{bmatrix}$	columns: t, s$^+$, s$^-$		$U_a^{trans} = \begin{bmatrix} \rho & 1 & 1 \end{bmatrix}$	columns: c, s$^+$, s$^-$
=C-C- : 151	C-C-C : 112	c : 0 s$^+$: 120	ζ = 0.101	0.3	2.910	$U_b^{cis} = \begin{bmatrix} 1 & 1 & 1 \\ 1 & 1 & 1 \\ 1 & 1 & 1 \end{bmatrix}$	rows: t, s$^+$, s$^-$ columns: t, g$^+$, g$^-$		$U_b^{trans} = \begin{bmatrix} 1 & \beta & \beta \\ 1 & 1 & 1 \\ 1 & 1 & 1 \end{bmatrix}$	rows: c, s$^+$, s$^-$ columns: t, g$^+$, g$^-$
-C-C- : 153		s$^-$: -120 For bond b:	ρ = 0.627	1	1.247					
		t : 180 g$^+$: 60	σ = 0.459	1	2.079					
		g$^-$: -60 For bond c:	ω = 0.044	1	8.314	$U_c^{cis} = \begin{bmatrix} \zeta & 1 & 1 \\ \zeta & 1 & 1 \\ \zeta & 1 & 1 \end{bmatrix}$	rows: t, g$^+$, g$^-$ columns: t, s$^+$, s$^-$		$U_c^{trans} = \begin{bmatrix} \rho & 1 & 1 \\ \rho\beta & 1 & 1 \\ \rho\beta & 1 & 1 \end{bmatrix}$	rows: t, g$^+$, g$^-$ columns: c, s$^+$, s$^-$
		t : 180 or c : 0 s$^+$: 120 s$^-$: -120 For bond d: c : 0 or t : 180				$U_d^{cis} = \begin{bmatrix} 1 \\ 1 \\ 1 \end{bmatrix}$	rows: t, s$^+$, s$^-$		$U_d^{trans} = \begin{bmatrix} 1 \\ 1 \\ 1 \end{bmatrix}$	rows: c, s$^+$, s$^-$

Comments:
Macrocyclization equilibrium data from metathesis reactions of cycloolefins are compared with predictions of a novel, simple RIS scheme for polyalkenes with at least three single bonds between adjacent double bonds. The predictions agree with experiment within the combined experimental errors, supporting the conformational models and confirming that macrocyclization equilibrium has indeed been established in the metathesis reactions.

Calcd. quantites: Macrocyclization equilibrium constants

Polycyclooctene

Suter, U. W.; Höcker, H. *Makromol. Chem.* **1988**, *189*, 1603.

Bond Length [pm]	Valence Angles [°]	Torsion Angles [°]	ξ (for 323 K)	ξ_0	E_ξ [kJ·mol^{-1}]
-C=C- : 134	C=C-CH$_2$: 125	For bond a: t : 180 or c : 0 s$^+$: 120 s$^-$: -120	β = 0.045	1	8.314
=C-C- : 151	C-C-C : 112		ζ = 0.101	0.3	2.910
-C-C- : 153			ρ = 0.627	1	1.247
		For bond b: t : 180 g$^+$: 60 g$^-$: -60	σ = 0.459	1	2.079
		For bond c: t : 180 or c : 0 s$^+$: 120 s$^-$: -120	ω = 0.044	1	8.314
		For bond d: c : 0 or t : 180			
		For bonds 1-4: t : 180 g$^+$: 60 g$^-$: -60			

$$U_a^{cis} = \begin{bmatrix} \zeta & 1 & 1 \end{bmatrix} \quad \text{columns: t, s}^+\text{, s}^-$$

$$U_b^{cis} = \begin{bmatrix} 1 & 1 & 1 \\ 1 & 1 & 1 \\ 1 & 1 & 1 \end{bmatrix} \quad \begin{array}{l}\text{rows: t, s}^+\text{, s}^- \\ \text{columns: t, g}^+\text{, g}^-\end{array}$$

$$U_c^{cis} = \begin{bmatrix} \zeta & 1 & 1 \\ \zeta & 1 & 1 \\ \zeta & 1 & 1 \end{bmatrix} \quad \begin{array}{l}\text{rows: t, g}^+\text{, g}^- \\ \text{columns: t, s}^+\text{, s}^-\end{array}$$

$$U_d^{cis} = \begin{bmatrix} 1 \\ 1 \\ 1 \end{bmatrix} \quad \text{rows: t, s}^+\text{, s}^-$$

$$U_{1-3}^{cis/trans} = \begin{bmatrix} 1 & \sigma & \sigma \\ 1 & \sigma & \sigma\omega \\ 1 & \sigma\omega & \sigma \end{bmatrix} \quad \begin{array}{l}\text{rows: t, g}^+\text{, g}^- \\ \text{columns: t, g}^+\text{, g}^-\end{array}$$

$$U_a^{trans} = \begin{bmatrix} \rho & 1 & 1 \end{bmatrix} \quad \text{columns: c, s}^+\text{, s}^-$$

$$U_b^{trans} = \begin{bmatrix} 1 & \beta & \beta \\ 1 & 1 & 1 \\ 1 & 1 & 1 \end{bmatrix} \quad \begin{array}{l}\text{rows: c, s}^+\text{, s}^- \\ \text{columns: t, g}^+\text{, g}^-\end{array}$$

$$U_c^{trans} = \begin{bmatrix} \rho & 1 & 1 \\ \rho\beta & 1 & 1 \\ \rho\beta & 1 & 1 \end{bmatrix} \quad \begin{array}{l}\text{rows: t, g}^+\text{, g}^- \\ \text{columns: c, s}^+\text{, s}^-\end{array}$$

$$U_d^{trans} = \begin{bmatrix} 1 \\ 1 \\ 1 \end{bmatrix} \quad \text{rows: t, s}^+\text{, s}^-$$

$$U_4^{cis/trans} = \begin{bmatrix} 1 & 1 & 1 \\ 1 & 1 & \omega \\ 1 & \omega & 1 \end{bmatrix} \quad \begin{array}{l}\text{rows: t, g}^+\text{, g}^- \\ \text{columns: t, g}^+\text{, g}^-\end{array}$$

Comments:
Macrocyclization equilibrium data from metathesis reactions of cycloolefins are compared with predictions of a novel, simple RIS scheme for polyalkenes with at least three single bonds between adjacent double bonds. The predictions agree with experiment within the combined experimental errors, supporting the conformational models and confirming that macrocyclization equilibrium has indeed been established in the metathesis reactions.

Calcd. quantites: Macrocyclization equilibrium constants

V 020

Polycyclododecene

Suter, U. W.; Höcker, H. *Makromol. Chem.* **1988**, *189*, 1603.

Bond Length [pm]	Valence Angles [°]	Torsion Angles [°]	ξ (for 323 K)	ξ_o	E_ξ [kJ·mol^{-1}]
-C=C- : 134	C=C-CH$_2$: 125	For bond a: t : 180 or	β = 0.045	1	8.314
=C-C- : 151	C-C-C : 112	c : 0 s$^+$: 120	ζ = 0.101	0.3	2.910
-C-C- : 153		s$^-$: -120 For bond b:	ρ = 0.627	1	1.247
		t : 180 g$^+$: 60	σ = 0.459	1	2.079
		g$^-$: -60 For bond c: t : 180 or c : 0	ω = 0.044	1	8.314
		s$^+$: 120 s$^-$: -120 For bond d: c : 0 or t : 180			
		For bonds 1-8: t : 180 g$^+$: 60 g$^-$: -60			

$$U_a^{cis} = \begin{bmatrix} \zeta & 1 & 1 \end{bmatrix} \quad \text{columns: } t, s^+, s^- \qquad U_a^{trans} = \begin{bmatrix} \rho & 1 & 1 \end{bmatrix} \quad \text{columns: } c, s^+, s^-$$

$$U_b^{cis} = \begin{bmatrix} 1 & 1 & 1 \\ 1 & 1 & 1 \\ 1 & 1 & 1 \end{bmatrix} \quad \begin{array}{l}\text{rows: } t, s^+, s^- \\ \text{columns: } t, g^+, g^-\end{array} \qquad U_b^{trans} = \begin{bmatrix} 1 & \beta & \beta \\ 1 & 1 & 1 \\ 1 & 1 & 1 \end{bmatrix} \quad \begin{array}{l}\text{rows: } c, s^+, s^- \\ \text{columns: } t, g^+, g^-\end{array}$$

$$U_c^{cis} = \begin{bmatrix} \zeta & 1 & 1 \\ \zeta & 1 & 1 \\ \zeta & 1 & 1 \end{bmatrix} \quad \begin{array}{l}\text{rows: } t, g^+, g^- \\ \text{columns: } t, s^+, s^-\end{array} \qquad U_c^{trans} = \begin{bmatrix} \rho & 1 & 1 \\ \rho\beta & 1 & 1 \\ \rho\beta & 1 & 1 \end{bmatrix} \quad \begin{array}{l}\text{rows: } t, g^+, g^- \\ \text{columns: } c, s^+, s^-\end{array}$$

$$U_d^{cis} = \begin{bmatrix} 1 \\ 1 \\ 1 \end{bmatrix} \quad \text{rows: } t, s^+, s^- \qquad U_d^{trans} = \begin{bmatrix} 1 \\ 1 \\ 1 \end{bmatrix} \quad \text{rows: } t, s^+, s^-$$

$$U_{1-7}^{cis/trans} = \begin{bmatrix} 1 & \sigma & \sigma \\ 1 & \sigma & \sigma\omega \\ 1 & \sigma\omega & \sigma \end{bmatrix} \quad \begin{array}{l}\text{rows: } t, g^+, g^- \\ \text{columns: } t, g^+, g^-\end{array} \qquad U_8^{cis/trans} = \begin{bmatrix} 1 & 1 & 1 \\ 1 & 1 & \omega \\ 1 & \omega & 1 \end{bmatrix} \quad \begin{array}{l}\text{rows: } t, g^+, g^- \\ \text{columns: } t, g^+, g^-\end{array}$$

Comments:
Macrocyclization equilibrium data from metathesis reactions of cycloolefins are compared with predictions of a novel, simple RIS scheme for polyalkenes with at least three single bonds between adjacent double bonds. The predictions agree with experiment within the combined experimental errors, supporting the conformational models and confirming that macrocyclization equilibrium has indeed been established in the metathesis reactions.

Calcd. quantites: Macrocyclization equilibrium constants

Further calculations on polymeric olefins:

V 021   Eliezer, I.; Hayman, H. J. G. *J. Polym. Sci.* **1957**, *23*, 387.

The mean-square length and mean-square radius of gyration of *cis*-1,4-polybutadiene and of *trans*-1,4-polybutadiene are calculated, the results being expressed in terms of the degree of polymerization and of other parameters of physical significance connected with the rotations about the various single bonds in these molecules. The calculations are based on the simplifying assumptions that no correlation exists between rotations about bonds separated by at least one double bond. The excluded volume effect is neglected throughout.

V 022   Allegra, G.; Flisi, U.; Crespi, G. *Makromol. Chem.* **1964**, *75*, 189.

Matrix calculations are carried out for theoretical calculation of the unperturbed dimensions of *cis*-1,4-polybutadiene which takes into account the particular nature of the energy barriers to the rotation around single C—C bonds adjacent to the double bonds.

V 023   Allegra, G. *Makromol. Chem.* **1967**, *110*, 58.

Contrary to the usual procedure which considers only the local conformations of minimum energy, the mean-square end-to-end distance of *cis*-1,4-polybutadiene is recalculated taking into account the whole continuum of conformational states; elastic strain on the C—C=C bond angles is also allowed, but only the values which minimize the conformational energy are retained, for every set of rotation angles.

V 024   Tosi, C.; Ciampelli, F. *Europ. Polym. J.* **1969**, *5*, 759.

The energy difference ΔE between *gauche* and *trans* conformers, resulting from rotation about single bonds of the chain backbone in *trans*-1,4-polybutadiene, *cis*-1,4-polybutadiene and 1,5-hexadiene, is evaluated from IR measurements on the bands characteristic of bending vibration of CH_2 groups. Consideration of the experimental results, on the basis of the RIS model, leads to the conclusion that a value of 0.4 to 0.8 kJ · mol^{-1} is the best estimate of ΔE.

V 025   Allegra, G.; Brückner, S.; Schmidt, M.; Wegner, G. *Macromolecules* **1986**, *19*, 399.

Light-scattering results from dissolved polydiacetylene chains may be quite satisfactorily interpreted on the basis of a single rotational isomer undergoing twisting and bending fluctuations, their amplitude being directely derived from IR force constants. The usual rotation-matrix alogithm is adoped. Randomly introducing a 1/6 fraction of ±90° rotations around the triple bonds of the structure -(CR=CR—C≡C-)$_x$ to account for the specifically observed conjugation length, the mean-square chain size shrinks by less than 7 % (the persistence length decreases from 190 to 177 Å); inaccuracy in the force constant adoped for twisting around the C=C bond and perhaps excluded-volume effects may account for the difference. It is shown that both models are very well represented by an ideal wormlike chain with the proper persistence length. No significant amount of either the *cis* conformation of the double bonds or the cumulene structure =(CR—CR=C=C)$_x$ is compatible with observation, in that they would cause a drastic reduction of the chain size.

V 026   Zhou, Z.; Yan, D.; Alfonso, G. C. *Makromol. Chem., Theory Simul.* **1993**, *2*, 721.

A method based on matrix algebra and on the RIS scheme to study the configurational-conformational entropy of 1,4-polydienes with geometrical isomerism is developed. Bernoullian and first- and second-order Markovian statistics for the sequences of *cis* and *trans* units along the chains is derived. Calculations performed by using available experimental data for the configurational parameters and the conformational partition function of 1,4-polybutadiene and 1,4-polyisoprene show that entropy is a monotonic function of the geometrical isomer composition. While the entropy of polybutadiene increases with the content of *cis* units, the reverse is true for polyisoprene. The conformational parameters suggested by *Abe* and *Flory* (V 009 - V 013) and modified by *Hadjichristidis et al.* [*J. Polym. Sci., Polym. Phys. Ed.* **1982**, *20*, 743] are used to calculate the conformational partition function. The indices of the energies given here (as PBD / PIP; given in kJ · mol^{-1}) have the same meaning as in the papers of the above authors: E(σ) = – 0.42 / 0.42, E(ζ) = 2.93 / 2.93, E(ρ) = 1.26 / – 1.26, E(β) = 8.4 / 8.4, E(ω) = -- / 5.02; statistical weights, ξ, are calculated as $\xi = \exp(-E(\xi)/(RT))$.

V 027

poly(ϵ-aminocaproamide), polyamide 6

Flory, P. J.; Williams, A. D. *J. Polym. Sci.: Part A-2* **1967**, *5*, 399.

Bond Length [pm]	Valence Angles [°]	Torsion Angles [°]	ξ (for 298 K)	ξ_0	E_ξ [kJ·mol^{-1}]
C-C : 153	θ_1 : 123	t : 180	σ = 0.43	1	2.1
N-C : 146	θ_k : 112 (k = 2-6)	g^+ : 60	σ_α = 1	1	0
			σ_δ = 1	1	0
C*-C : 151		g^- : −60	σ_β = 1	1	0
	θ_7 : 114		σ_γ = 1	1	0
C*-N : 133			ω = 0.034	1	8.4
	C-C-H : 109		$\omega_{\alpha\beta}$ = 0.1	1	5.7
C-H : 109			ω_δ = 0.1	1	5.7
	O*-C*-C : 121		ω_β = 0.1	1	5.7
C*=O* : 122			$\omega_{\gamma\delta}$ = 0.1	1	5.7
N-H : 104					

$$U_1 = \begin{bmatrix} 1 & 0 & 0 \\ 1 & 0 & 0 \\ 1 & 0 & 0 \end{bmatrix} \quad U_2 = \begin{bmatrix} 1 & \sigma_\alpha & \sigma_\alpha \\ 0 & 0 & 0 \\ 0 & 0 & 0 \end{bmatrix} \quad U_3 = \begin{bmatrix} 1 & \sigma_\beta & \sigma_\beta \\ 1 & \sigma_\beta & \sigma_\beta\omega_{\alpha\beta} \\ 1 & \sigma_\beta\omega_{\alpha\beta} & \sigma_\beta \end{bmatrix}$$

$$U_4 = \begin{bmatrix} 1 & \sigma & \sigma \\ 1 & \sigma & \sigma\omega_\beta \\ 1 & \sigma\omega_\beta & \sigma \end{bmatrix} \quad U_5 = \begin{bmatrix} 1 & \sigma & \sigma \\ 1 & \sigma & \sigma\omega \\ 1 & \sigma\omega & \sigma \end{bmatrix} \quad U_6 = \begin{bmatrix} 1 & \sigma_\delta & \sigma_\delta \\ 1 & \sigma_\delta & \sigma_\delta\omega_\delta \\ 1 & \sigma_\delta\omega_\delta & \sigma_\delta \end{bmatrix}$$

$$U_7 = \begin{bmatrix} 1 & \sigma_\gamma & \sigma_\gamma \\ 1 & \sigma_\gamma & \sigma_\gamma\omega_{\gamma\delta} \\ 1 & \sigma_\gamma\omega_{\gamma\delta} & \sigma_\gamma \end{bmatrix}$$

rows and columns: t, g^+, g^-

Comments:

The partial double-bond character of the CO—NH bond renders the amide group planar, the *trans* form being preferred. This preference is enhanced by steric repulsions encountered in the *cis* form when the amide group occurs within a chain molecule.

Calcd. quantities: $(\langle r^2 \rangle_0 / nl^2)_\infty$ = 6.08

V 028

poly(6-aminocaproamide), polyamide 6, PACA

Mutter, M.; Suter, U. W.; Flory, P. J. *J. Am. Chem. Soc.* **1976**, *98*, 5745.

Bond Length [pm]	Valence Angles [°]	Torsion Angles [°]	(for 525 K)
C-C : 153	θ_1 : 123	For bonds 2-7:	
		t : 180	
N-C : 146	θ_k : 112	g$^+$: 60	
	(k = 2-6)	g$^-$: −60	
C-C* : 151			
	θ_7 : 114	For bond 1:	
C*-N : 133		t : 180	

$$U_1 = \begin{bmatrix} 1 \\ 1 \\ 1 \end{bmatrix} \qquad U_2 = \begin{bmatrix} 1 & 1 & 1 \end{bmatrix} \qquad U_3 = U_6 = U_7 = \begin{bmatrix} 1 & 1 & 1 \\ 1 & 1 & 0.27 \\ 1 & 0.27 & 1 \end{bmatrix} \qquad U_4 = U_5 = \begin{bmatrix} 1 & 0.62 & 0.62 \\ 1 & 0.62 & 0.09 \\ 1 & 0.09 & 0.62 \end{bmatrix}$$

rows and columns: t, g$^+$, g$^-$

Comments:
The probability density function $W(0)$ at $r = 0$ and the directional correlation factors for homologous PACA sequences with x = 2 to 7 units are evaluated. The influences of these factors on the cyclization equilibria constants K_x are determined. Agreement of theory with experimental results for x = 3 - 6 is within limits set by uncertainties in the calculations combined with experimental errors, i.e., within about 15 %.

Calcd. quantities: Cyclization equilibria constants, K_x, for x = 3 - 7.

V

The following model requires two pages.
Therefore, this side has been left blank

poly(hexamethylene adipamide), polyamide 66

Flory, P. J.; Williams, A. D. *J. Polym. Sci.; Part A-2* **1967**, *5*, 399.

Bond Length [pm]	Valence Angles [°]	Torsion Angles [°]	ξ (for 298 K)	ξ_0	E_ξ [kJ·mol^{-1}]
C-C : 153	$\theta_1 = \theta_8$: 123	t : 180	σ = 0.43	1	2.1
			σ_α = 1	1	0
N-C : 146	θ_k : 112	g^+ : 60	σ_δ = 1	1	0
	(k = 2-7, 10-13)		σ_β = 1	1	0
C*-C : 151		g^- : -60	σ_γ = 1	1	0
	$\theta_9 = \theta_{14}$: 114				
C*-N : 133			ω = 0.034	1	8.4
	C-C-H : 109		$\omega_{\alpha\beta}$ = 0.10	1	5.7
C-H : 109			ω_δ = 0.10	1	5.7
	O*-C*-C : 121		ω_β = 0.10	1	5.7
C*=O* : 122			$\omega_{\gamma\delta}$ = 0.10	1	5.7
N-H : 104					

$$U_1 = U_9 = \begin{bmatrix} 1 & 0 & 0 \\ 1 & 0 & 0 \\ 1 & 0 & 0 \end{bmatrix} \quad U_2 = \begin{bmatrix} 1 & \sigma_\alpha & \sigma_\alpha \\ 0 & 0 & 0 \\ 0 & 0 & 0 \end{bmatrix} \quad U_3 = \begin{bmatrix} 1 & \sigma_\beta & \sigma_\beta \\ 1 & \sigma_\beta & \sigma_\beta\omega_{\alpha\beta} \\ 1 & \sigma_\beta\omega_{\alpha\beta} & \sigma_\beta \end{bmatrix}$$

$$U_4 = \begin{bmatrix} 1 & \sigma & \sigma \\ 1 & \sigma & \sigma\omega_\beta \\ 1 & \sigma\omega_\beta & \sigma \end{bmatrix} \quad U_5 = U_6 = \begin{bmatrix} 1 & \sigma & \sigma \\ 1 & \sigma & \sigma\omega \\ 1 & \sigma\omega & \sigma \end{bmatrix} \quad U_7 = \begin{bmatrix} 1 & \sigma_\beta & \sigma_\beta \\ 1 & \sigma_\beta & \sigma_\beta\omega_\beta \\ 1 & \sigma_\beta\omega_\beta & \sigma_\beta \end{bmatrix}$$

$$U_8 = \begin{bmatrix} 1 & \sigma_\alpha & \sigma_\alpha \\ 1 & \sigma_\alpha & \sigma_\alpha\omega_{\alpha\beta} \\ 1 & \sigma_\alpha\omega_{\alpha\beta} & \sigma_\alpha \end{bmatrix} \quad U_{10} = \begin{bmatrix} 1 & \sigma_\gamma & \sigma_\gamma \\ 0 & 0 & 0 \\ 0 & 0 & 0 \end{bmatrix} \quad U_{11} = \begin{bmatrix} 1 & \sigma_\delta & \sigma_\delta \\ 1 & \sigma_\delta & \sigma_\delta\omega_{\gamma\delta} \\ 1 & \sigma_\delta\omega_{\gamma\delta} & \sigma_\delta \end{bmatrix}$$

rows and columns: t, g$^+$, g$^-$

(see: a))

$$U_{12} = \begin{bmatrix} 1 & \sigma & \sigma \\ 1 & \sigma & \sigma\omega_\delta \\ 1 & \sigma\omega_\delta & \sigma \end{bmatrix} \quad U_{13} = \begin{bmatrix} 1 & \sigma_\delta & \sigma_\delta \\ 1 & \sigma_\delta & \sigma_\delta\omega_\delta \\ 1 & \sigma_\delta\omega_\delta & \sigma_\delta \end{bmatrix} \quad U_{14} = \begin{bmatrix} 1 & \sigma_\gamma & \sigma_\gamma \\ 1 & \sigma_\gamma & \sigma_\gamma\omega_\gamma_\delta \\ 1 & \sigma_\gamma\omega_\gamma_\delta & \sigma_\gamma \end{bmatrix}$$

---

a) There is a misprint in the original paper. Here, U_{10} is replaced by U_9, and U_{11} is replaced by U_{10}.

Comments:
The partial double-bond character of the CO—NH bond renders the amide group planar, the *trans* form being preferred. This preference is enhanced by steric repulsions encountered in the *cis* form when the amide group occurs within a chain molecule.

Calcd. quantities: $(\langle r^2 \rangle_0 / nl^2)_\infty$ = 6.10   (*Exptl.*: 5.95)

Further calculations on **aliphatic polyester** and **polyamide** chains:

V 030   Tonelli, A. E. *J. Chem. Phys.* **1971**, *54*, 4637.

The inter- and intramolecular contributions to the entropy and energy of fusion are calculated for several linear aliphatic polyesters and polyamides assuming the fusion process consists of two independent contributions: the volume expansion (intermolecular contribution) and the increase in the conformational freedom of each polymer chain on melting (intramolecular contribution). The intramolecular entropy and energy contributions are obtained from the configurational partition function and its temperature coefficient calculated for an isolated, unperturbed polymer chain using the RIS approximation.

V 031   Tonelli, A. E. *Europ. Polym. J.* **1978**, *14*, 305.

The Flory-Williams RIS models describing the conformational characteristics of polyesters and polyamides are combined to calculate the dimensions of the alternating polyesteramide PEA. Excellent agreement is found with experimental values of the unperturbed dimensions. New determinations of molecular dipole moments are reported for dimethyl-*trans*-1,4-cyclohexanedicarboxylate and for *trans*-1,4-cyclohexanediol diacetate.

V 032

poly(L-lactic acid)

Brant, D. A.; Tonelli, A. E.; Flory, P. J. *Macromolecules* **1969**, 2, 225, 228.

Bond Length [pm]	Valence Angles [°]	Torsion Angle Pairs [a] ϕ_i, ψ_i [°]			stat. weight
$O\text{-}C^\alpha$: 144	$O\text{-}C^\alpha\text{-}C$: 110	I : g^+g^+	: 73, 46	→	0.368
$C^\alpha\text{-}C$: 152	$C^\alpha\text{-}C\text{-}O$: 114	III : g^+t	: 73, −158	→	0.550
$C\text{-}O$: 134	$C\text{-}O\text{-}C^\alpha$: 113	I' : tg^+	:160, 50	→	0.040
$C=O$: 122	$C^\alpha\text{-}C=O$: 121	III' : tt	:160, −161	→	0.042
$C^\alpha\text{-}C^\beta$: 154	$O\text{-}C^\alpha\text{-}C^\beta$: 110				
$C^\alpha\text{-}H^\alpha$: 107	$O\text{-}C^\alpha\text{-}H^\alpha$: 110				
l_{vb} : 370	η : 18.9				
	ξ : 19.9				

(for 358 K)

Average transformation matrix:

$$\langle T_i \rangle = w_I \langle T_I \rangle + w_{III} \langle T_{III} \rangle + w_{I'} \langle T_{I'} \rangle + w_{III'} \langle T_{III'} \rangle = \begin{bmatrix} 0.332 & 0.167 & 0.017 \\ -0.273 & 0.085 & 0.257 \\ 0.778 & -0.355 & 0.082 \end{bmatrix}$$

Because of the major contribution to z from minima I and III, calcs. are also carried out using averaged matrix:

$$\langle T_i \rangle = (w_I \langle T_I \rangle + w_{III} \langle T_{III} \rangle) / (w_I + w_{III}) = \begin{bmatrix} 0.305 & 0.186 & 0.032 \\ -0.356 & 0.117 & 0.264 \\ 0.827 & -0.359 & 0.102 \end{bmatrix}$$

[a] Moderate steric conflict of $(CO)_{i-1}$ with H_i, in conjunction with a maximum in $V_\phi(\phi)$ produces a ridge at $\phi_i = 60°$ which divides the allowed regions into the subsidiary minima at I, I' and III, III'. A similar ridge in the L-Ala map is much less pronounced because of the larger bond angle at the nitrogen atom.

Comments:
The conformational energy of the L-lactyl residue of poly(L-lactic acid) is calculated as a function of rotation angles ϕ and ψ about the $O-C^\alpha$ and $C^\alpha-C$ bonds, respectively, the ester bond being planar *trans*. Methods of calculation correspond to those applied previously to various polypeptides. Dipolar interactions, though much smaller than in polypeptides, are significantly important. The conformational energy contour map over ϕ and ψ is dominated by four well-defined minima. The two of lowest energy, situated approximately at the g^+t and g^+g^+ conformations, suffice for interpretation of the configurational characteristics of the chain.

Calcd. quantities: $\langle r^2 \rangle_0 / nl^2$ = 2.02 or 1.92   (Exptl.: 2.13)

$d (\ln \langle r^2 \rangle_0) / dT$

V 033

poly(L-lactic acid)

Tonelli, A. E. *Macromolecules* **1992**, *25*, 3581.

Bond Length [pm]	Valence Angles [°]	Torsion Angle Pairs [a,b] ϕ_i, ψ_i [°]			stat. weight (for 358 K)
$O-C^\alpha$: 144	$O-C^\alpha-C$: 110	I : g^+g^+	: 73, 48	→	0.368
$C^\alpha-C$: 152	$C^\alpha-C-O$: 114	III : g^+t	: 73, −160	→	0.550
$C-O$: 134	$C-O-C^\alpha$: 113	I' : tg^+	: 160, 48	→	0.040
$C=O$: 122	$C^\alpha-C=O$: 121	III' : tt	: 160, −160	→	0.042
$C^\alpha-C^\beta$: 154	$O-C^\alpha-C^\beta$: 110				
$C^\alpha-H^\alpha$: 107	$O-C^\alpha-H^\alpha$: 110				
l_{vb} : 370	η : 18.9				
	ξ : 19.9				

$$U_a = \begin{bmatrix} 1 & 0 \\ 1 & 0 \end{bmatrix} \quad U_b = \begin{bmatrix} 1 & 1 \\ 0 & 0 \end{bmatrix} \quad U_c = \begin{bmatrix} 0.042 & 0.040 \\ 0.550 & 0.368 \end{bmatrix}$$

rows and columns: t, g^+

a) Moderate steric conflict of $(CO)_{i-1}$ with H_i, in conjunction with a maximum in $V_\phi(\phi)$ produces a ridge at $\phi_i = 60°$ which divides the allowed regions into the subsidiary minima at I, I' and III, III'. A similar ridge in the L-Ala map is much less pronounced because of the larger bond angle at the nitrogen atom.

b) By symmetry, a D-lactyl residue energy map also displays four energy minima located at $(\phi,\psi)_D = (-\phi, -\psi)_L$.

Comments:
The conformations of stereoregular poly(L-lactide) and regularly alternating poly(L,D-lactide) confined to occupy cylinders of varying radii are determined in an effort to learn if either polylactide could be incorporated in the narrow channels of its inclusion compound with urea.

The following model requires three pages.
Therefore, this side has been left blank

dimethyl malonate

Abe, A. *J. Am. Chem. Soc.* **1984**, *106*, 14.

Bond Length [pm]	Valence Angles [°]	Torsion Angles for C^*-C^1; [°] b)	ξ (for 298 K)	ξ_0	E_ξ [kJ·mol^{-1}]	Matrices of unnormalized *a priori* statistical weights for *I* (U_{M-I}) and *II* (U_{M-II}) c)	
C-C* : 153	C-C-C* : 112.0	Model I:					
		g$^+$: 57.3	α = 1 (to 2)	1	0 (to –1.7)	$U_{M-I} = \begin{bmatrix} 1.00 & 0.22 & 0.50 & 0.54 & 0.79 & 0.01 \\ 0.22 & 1.00 & 0.50 & 0.79 & 0.54 & 0.01 \\ 0.50 & 0.50 & 0.04 & 0.34 & 0.34 & 0.00 \\ 0.54 & 0.79 & 0.34 & 0.69 & 0.36 & 0.03 \\ 0.79 & 0.54 & 0.34 & 0.36 & 0.69 & 0.03 \\ 0.01 & 0.01 & 0.00 & 0.03 & 0.03 & 0.00 \end{bmatrix}$	rows and columns: g^+, g^-, t, g^{+*}, g^{-*}, c (see: a))
C-H : 110	C-C-H : 110.0	g$^-$: –57.3					
		t : 180					
C*-O : 135	C-C*-O : 111.4	g^{+*} : 122.7					
		g^{-*} : –122.7					
C*=O* : 120	C-C*=O* : 126.3	c : 0					
O-CH$_3$: 143	C*-O-CH$_3$: 116.7	Model II:					
		g$^+$: 60	α = 0	1	∞	$U_{M-II} = \begin{bmatrix} 0.26 & 0.06 & 1.00 \\ 0.06 & 0.26 & 1.00 \\ 1.00 & 1.00 & 0.53 \end{bmatrix}$	rows and columns: g^+, g^-, t
		g$^-$: –60					
		t : 180	β = 7.52	1	–5.0		

a) Matrices are composed of *a priori* statistical weights. The conformational energies of the molecule are obtained as a function of bond rotations around C*–C and C–C*.
b) The ester group was assumed to be in the planar, *trans* configuration, thus O–C eclipsing C*–O*.
c) For the rotation around the C*–C bond, two models are examined. In model I, a six-state scheme is adopted, in which the ester groups are allowed to take the reversed orientation to a certain extent, besides the regular C*–O*/C–C and C*=O*/C–H eclipsings. In the *trans* C*–C conformation (180°), the carbonyl group C*=O* is opposed by the C_1-C_2 bond. The reverse conformations, in which the ester C*–O bond eclipses one of the vicinal bonds (thus C*=O* being *trans* to these bonds), may be defined at three rotational angles: i.e. ±120 and 0°. To account for the possible existence of the intrinsic energy difference, E(α) associated with these reversal of conformations, a statistical weight of α is assigned to all of the reversed states. In model II, the regular three-state scheme, C*=O* eclipses C_1-C_2 and two C_1–H bonds. The conformation in which C*=O* is *cis* to C_1-C_2 is energetically more favorable than the other two states. Thus, a statistical weight parameter β is assigned to the C*=O*/C_1-C_2 eclipsed form. Here E(β) (≥0) represents the *stabilization* energy inherent to this conformation. Energy calculations are performed for all rotational states defined as above around the C*-C^1 bond. An identical model is assumed for the C_n–C* bond.

Models analogous to that of dimethyl malonate are also developed for dimethyl succinate and dimethyl glutarate, respectively. For the C—C bonds joining two methylene carbons, here rotational energy minima are determined by calculating the conformational energies as a function of the rotational angles. A detailled discussion is given in the original paper.

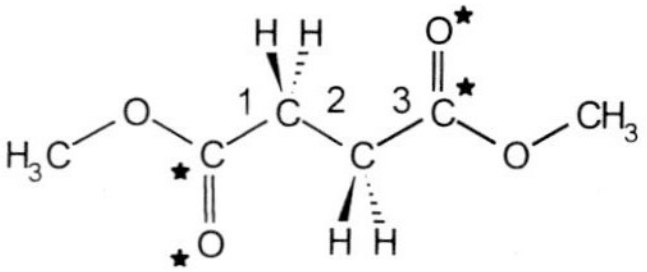

dimethyl succinate

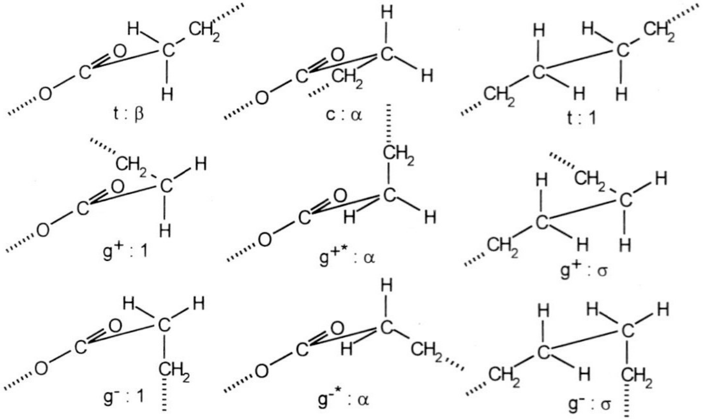

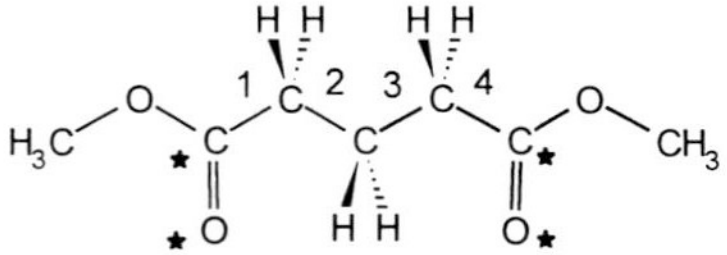

dimethyl glutarate

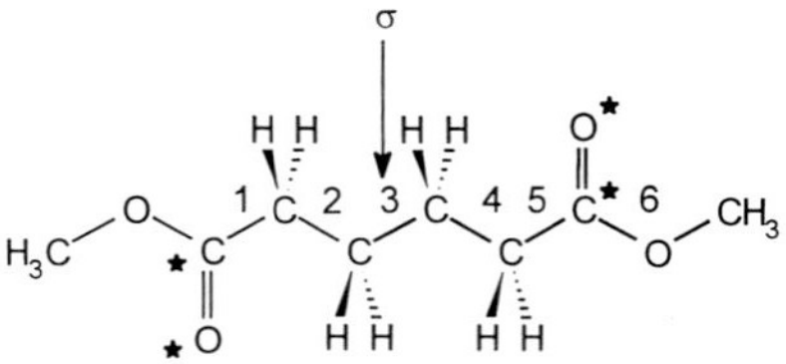

dimethyl adipate

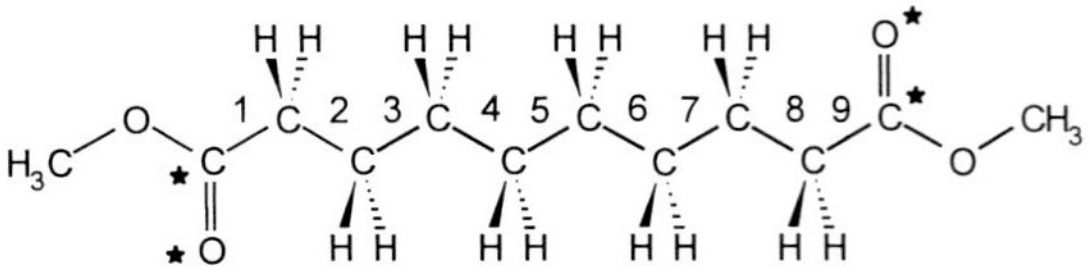

dimethyl sebacate

t : β

c : α

t : 1

g⁺ : 1

g⁺* : α

g⁺ : σ

g⁻ : 1

g⁻* : α

g⁻ : σ

Bond Length [pm]	Valence Angles [°]	Torsion Angles for C^*-C^1; [°] [a]	ξ (for 298 K)	ξ_0	E_ξ [kJ·mol^{-1}]	The results of the six-state scheme for C^*–C^1 with $\alpha = 1.0$ are given. [b]
C-C : 153	C-C-C : 112.0	g^+ : 57.3	$\alpha = 1.0$	1	0	
		g^- : – 57.3	$\beta = 1.0$	1	0	
C-C* : 153	C-C-C* : 112.0	t : 180	$\eta_1 = 1.0$	1	0	
		g^{+*} : 122.7	$\eta_2 = 0.74$	1	0.75	
C-H : 110	C-C-H : 110.0	g^{-*} : – 122.7	$\eta_3 = 0.86$	1	0.38	
C*-O : 135	C-C*-O : 111.4	$\xi_1 = 1.27$		1	– 0.59	
		$\xi_3 = 0.52$		1	1.63	
C*=O* : 120	C-C*=O* : 126.3	$\xi_4 = 0.30$		1	2.98	
		$\xi_5 = 1.27$		1	– 0.59	
O-CH$_3$: 143	C*-O-CH$_3$: 116.7					
		$\sigma = 0.43$		1	2.1	

$$U_1 = [1 \ 1 \ \beta\eta_2 \ \alpha\eta_3 \ \alpha\eta_3] \quad U_2 = \begin{bmatrix} 1 & \xi_1 & 0 \\ 1 & 0 & \xi_1 \\ 1 & \xi_3 & \xi_3 \\ 1 & \xi_4 & \xi_5 \\ 1 & \xi_5 & \xi_4 \end{bmatrix} \quad U_3 = \begin{bmatrix} 1 & \sigma & \sigma \\ 1 & \sigma & 0 \\ 1 & 0 & \sigma \end{bmatrix} \quad U_4 = \begin{bmatrix} 1 & 1 & 1 \\ 1 & 1 & 0 \\ 1 & 0 & 1 \end{bmatrix}$$

$$U_5 = U_2^T$$

rows and / or columns : g^+, g^-, t, g^{+*}, g^{-*} and / or: t, g^+, g^- ; for U_1 : row: t

[a] The ester group is assumed to be in the planar *trans* configuration, thus O–C eclipsing C^*–O^*.

[b] In these molecules, the two terminal ester groups are separated by more than $n = 5$ C–C bonds. Short-range interactions between these two groups inevitably precipitate severe second-order steric conflicts along the chain, except when n is very large. For the dialkyl esters with $n \geq 4$, therefore, the spatial arrangements of the ester groups located on both terminals may be treated independently from each other. Conformational energies depending on rotations about the first two bonds, i.e., C^*–C^1 and C^1–C^2, are estimated from calculations for methyl pentanoate, a monoester model compound.

Comments:
Conformational energies of the dimethyl esters of a series of dicarboxylic acids are calculated by using semiempirical potential energy functions. Dipole moments of these compounds are evaluated and compared with the experimental results. Various models which have been proposed for the conformations around the CC–(CO)O bond in the literature, are taken into consideration. The observed values of the dipole moment are found to be reasonably reproduced by two models: One expressed in the six-state scheme and the other in the three-state scheme. The latter model provides a better explanation for the dielectric behaviour of diethyl succinate, which is known to be abnormal. Both models, however, fail to give an agreement with experiment in the *trans-gauche* energy difference for diethyl succinate. An analogous procedure is also given for dimethyl sebacate. Here, the internal C–C bonds are treated in the usual manner.

Calcd. quantities: $\langle \mu^2 \rangle^{1/2}$ $d(\ln \langle \mu^2 \rangle^{1/2})/dT$

V 035

dimethyl succinate

Abe, A.; Miura, I.; Furuya, H. *J. Phys. Chem.* **1987**, *91*, 6496.

Bond Length [pm]	Valence Angles [°]	Torsion Angles [°]	ξ (for 298 K)	ξ_0	E_ξ [kJ · mol^{-1}]
C-C : 153	C-C-C : 112.0	For O-C* bonds: t : 180	σ = 2.74	1	– 2.5
C-C* : 153	C-C-C* : 112.0		σ' = 5.45 - 0.18	1	– 4.2 to + 4.2
C-H : 110	C-C-H : 110.0	For C*-C$^\alpha$ bonds: t : 180 g$^+$: 57.3 g$^-$: – 57.3			
C*-O : 135	C-C*-O : 111.4				
C*=O* : 120	C-C*=O* : 126.3	For C-C bonds: t : 180 g$^+$: 67.5 g$^-$: – 67.5			
O-CH$_3$: 143	C*-O-CH$_3$: 116.7				

Comments:

The conformational characteristics of dimethyl esters of dicarboxylic acids are studies by the NMR and dipole moment method. Conformational energies of the internal CH$_2$–CH$_2$ bonds are determined from the observed ^{1}H-^{1}H vicinal coupling constants. Preferred conformations around the C–C* bond are elucidated from the RIS analysis of dipole moments. With the RIS parameters thus established, the orientational correlation between the terminal ester groups is examined. The analysis provides the reason why the odd-even effect in the dipole moment is moderate, and attenuates rapidly with n in the α,ω-diester series.

Calcd. quantities: ^{1}H-^{1}H NMR vicinal coupling constants $<\mu^2>^{1/2}$ $d (\ln <\mu^2>^{1/2}) / dT$

dimethyl glutarate

Abe, A.; Miura, I.; Furuya, H. *J. Phys. Chem.* **1987**, *91*, 6496.

Bond Length [pm]	Valence Angles [°]	Torsion Angles [°]		ξ (for 298 K)	ξ_0	E_ξ [kJ·mol^{-1}]
C-C : 153	C-C-C : 112.0	For O-C* bonds: t : 180		σ = 0.60	1	1.26 ± 0.4
C-C* : 153	C-C-C* : 112.0			σ' = 5.45 - 0.18	1	- 4.2 to + 4.2
		For C*-C$^\alpha$ bonds:				
C-H : 110	C-C-H : 110.0	t : 180		ω = 0.034	1	8.4
		g^+ : 57.3				
C*-O : 135	C-C*-O: 111.4	g^- : - 57.3		ω' = 0.034	1	8.4
C*=O* : 120	C-C*=O* : 126.3	For C-C bonds:				
		t : 180				
O-CH$_3$: 143	C*-O-CH$_3$: 116.7	g^+ : 67.5				
		g^- : - 67.5				

Comments:
The conformational characteristics of dimethyl esters of dicarboxylic acids are studies by the NMR and dipole moment method. Conformational energies of the internal CH$_2$–CH$_2$ bonds are determined from the observed ^{1}H-^{1}H vicinal coupling constants. Preferred conformations around the C–C* bond are elucidated from the RIS analysis of dipole moments. With the RIS parameters thus established, the orientational correlation between the terminal ester groups is examined. The analysis provides the reason why the odd-even effect in the dipole moment is moderate, and attenuates rapidly with n in the α,ω-diester series.

Calcd. quantities: ^{1}H-^{1}H NMR vicinal coupling constants $<\mu^2>^{1/2}$ $d(\ln <\mu^2>^{1/2})/dT$

dimethyl adipate

Abe, A.; Miura, I.; Furuya, H. *J. Phys. Chem.* **1987**, *91*, 6496.

Bond Length [pm]	Valence Angles [°]	Torsion Angles [°]	ξ (for 298 K)	ξ_0	E_ξ [kJ·mol^{-1}]
C-C : 153	C-C-C : 112.0	For O-C* bonds: t : 180	$\sigma = 0.26$	1	3.34 ± 0.4
C-C* : 153	C-C-C* : 112.0		$\sigma' = 0.034 - 30$	1	−8.4 to + 8.4
C-H : 110	C-C-H : 110.0	For C*-C$^\alpha$ bonds: t : 180 g$^+$: 57.3	$\tau = 0.43$	1	2.1 ± 0.4
C*-O : 135	C-C*-O : 111.4	g$^-$: − 57.3	$\omega = 0.43 - 0.034$	1	2.1 to 8.4
C*=O* : 120	C-C*=O* : 126.3	For C-C bonds: t : 180	$\omega' = 0.43 - 0.034$	1	2.1 to 8.4
O-CH$_3$: 143	C*-O-CH$_3$: 116.7	g$^+$: 67.5 g$^-$: − 67.5			

Comments:
The conformational characteristics of dimethyl esters of dicarboxylic acids are studies by the NMR and dipole moment method. Conformational energies of the internal CH$_2$−CH$_2$ bonds are determined from the observed ^{1}H-^{1}H vicinal coupling constants. Preferred conformations around the C−C* bond are elucidated from the RIS analysis of dipole moments. With the RIS parameters thus established, the orientational correlation between the terminal ester groups is examined. The analysis provides the reason why the odd-even effect in the dipole moment is moderate, and attenuates rapidly with n in the α,ω-diester series.

Calcd. quantities: ^{1}H-^{1}H NMR vicinal coupling constants $\quad <\mu^2>^{1/2} \quad d(\ln <\mu^2>^{1/2})/dT$

poly(neopentyl glycol succinate), PNS

Riande, E.; Guzmán, J. *J. Chem. Soc. Perkin Trans. II* **1988**, 299.

Bond Length [pm]	Valence Angles [°]	Torsion Angles [°]	ξ (for 300 K)	ξ_o	E_ξ [kJ·mol^{-1}]
C-C : 153	O-C*-C : 114	Bonds 3,4,7-9:	$\sigma_\alpha = 1$	1	0
C-C* : 151	C*-O-C : 113	t : 180 g$^+$: 60 g$^-$: −60	$\sigma_\beta = 0.37$	1	2.5
C-O : 143	O-C-C : 110		$\sigma' = 2.72$	1	−2.5
O-C* : 134	C-C-C : 110	For bonds 2,5: t : 180 c$^+$: 105 c$^-$: −105	$\sigma'' = 0.13$	1	5.0
C*=O* : 122	O-C*=O* : 125		$\omega = 0.37$	1	2.5
C-H : 109		For bonds 1,6: t : 180			

$$U_2 = \begin{bmatrix} 1 & \sigma'' & \sigma'' \end{bmatrix} \quad U_3 = \begin{bmatrix} 1 & \sigma' & \sigma' \\ 1 & \sigma' & 0 \\ 1 & 0 & \sigma' \end{bmatrix} \quad U_4 = \begin{bmatrix} 1 & \sigma' & \sigma' \\ 1 & \sigma' & \sigma'\omega \\ 1 & \sigma'\omega & \sigma' \end{bmatrix} \quad U_5 = \begin{bmatrix} 1 & \sigma'' & \sigma'' \\ 1 & \sigma'' & 0 \\ 1 & 0 & \sigma'' \end{bmatrix}$$

$$U_7 = \begin{bmatrix} 1 & \sigma_\alpha & \sigma_\alpha \end{bmatrix} \quad U_8 = \begin{bmatrix} 1 & \sigma_\beta & \sigma_\beta \\ 1 & \sigma_\beta & 0 \\ 1 & 0 & \sigma_\beta \end{bmatrix} \quad U_9 = \begin{bmatrix} 1 & \sigma_\alpha & \sigma_\alpha \\ 1 & \sigma_\alpha & 0 \\ 1 & 0 & \sigma_\alpha \end{bmatrix}$$

For 3 × 3 matrices: rows and columns: t, g$^+$, g$^-$
For U_2, U_7: rows: t, columns: t, g$^+$, g$^-$

Comments:
Values of the dipole moment ratio of PNS are obtained from dielectric measurements. From thermoelastic experiments, performed on polymer networks, the temperature coefficient of the unperturbed dimensions is determined. Analysis of these results using the RIS model is performed leading to the parameters given above.

Calcd. quantities: $\langle r^2 \rangle_o / nl^2$ $d(\ln \langle r^2 \rangle_o)/dT$ $\langle \mu^2 \rangle / xm^2$ $d(\ln \langle \mu^2 \rangle)/dT$

V 039

polyoxyneopentyleneoxyadipoyl, PNA

Riande, E.; Guzmán, J.; Adabbo, H. *Macromolecules* **1986**, *19*, 2567.

Bond Length [pm]		Valence Angles [°]		Torsion Angles [°]		ξ (for 300 K)	ξ_0	E_ξ [kJ·mol^{-1}]
C-C	: 153	O-C*-C	: 114	Bonds 3,4,7-11:		$\sigma_\alpha = 1$	1	0
				t : 180				
C-C*	: 151	C*-O-C	: 113	g$^+$: 60		$\sigma_\beta = 1$	1	0
				g$^-$: -60				
C-O	: 143	O-C-C	: 110			$\sigma = 0.43$	1	2.1
				For bonds 2,5:				
O-C*	: 134	C-C-C	: 110	t : 180		$\sigma' = 2.72$	1	-2.5
				c$^+$: 105				
C*=O*	: 122	O-C*=O*	: 125	c$^-$: -105		$\sigma'' = 0.19$	1	4.2
C-H	: 109			For bonds 1,6:		$\omega = 0.37$	1	2.5
				t : 180				

$$U_2 = \begin{bmatrix} 1 & \sigma'' & \sigma'' \end{bmatrix} \quad U_3 = \begin{bmatrix} 1 & \sigma' & \sigma' \\ 1 & \sigma' & 0 \\ 1 & 0 & \sigma' \end{bmatrix} \quad U_4 = \begin{bmatrix} 1 & \sigma' & \sigma' \\ 1 & \sigma' & \sigma'\omega \\ 1 & \sigma'\omega & \sigma' \end{bmatrix} \quad U_5 = \begin{bmatrix} 1 & \sigma'' & \sigma'' \\ 1 & \sigma'' & 0 \\ 1 & 0 & \sigma'' \end{bmatrix}$$

$$U_7 = \begin{bmatrix} 1 & \sigma_\alpha & \sigma_\alpha \end{bmatrix} \quad U_8 = U_{10} = \begin{bmatrix} 1 & \sigma_\beta & \sigma_\beta \\ 1 & \sigma_\beta & 0 \\ 1 & 0 & \sigma_\beta \end{bmatrix} \quad U_9 = \begin{bmatrix} 1 & \sigma & \sigma \\ 1 & \sigma & 0 \\ 1 & 0 & \sigma \end{bmatrix} \quad U_{11} = \begin{bmatrix} 1 & \sigma_\alpha & \sigma_\alpha \\ 1 & \sigma_\alpha & 0 \\ 1 & 0 & \sigma_\alpha \end{bmatrix}$$

For 3 × 3 matrices: rows and columns: t, g$^+$, g$^-$
For U_2, U_7: rows: t, columns: t, g$^+$, g$^-$

Comments:
Values of the mean-square dipole moment $<\mu^2>$ of PNA are determined from measurements of dielectric constants and refractive indices of the polymer in benzene. The dipole moment ratio and the temperature coefficient of both the dipole moment and the unperturbed dimensions are critically interpreted using the RIS model. Good agreement between theory and experiment is obtained by assuming that the *gauche* states about C(CH$_3$)$_2$–CH$_2$ bonds have an energy 2.5 kJ·mol^{-1} lower than the alternative *trans* states.

Calcd. quantities: $<r^2>_0 / nl^2$ $\quad d (\ln <r^2>_0) / dT$ $\quad <\mu^2> / xm^2$ $\quad d (\ln <\mu^2>) / dT$

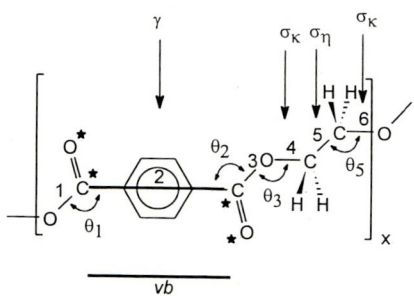

poly(ethylene terephthalate), PET

Williams, A. D.; Flory, P. J. *J. Polym. Sci.: Part A-2* **1967**, *5*, 417.

Bond Length [pm]	Valence Angles [°]	Torsion Angles [°]	ξ a) (for 303 K)	ξ_0	E_ξ [kJ · mol^{-1}]
C-C : 153	$\theta_1 = \theta_2$: 114	For bonds 1,3-6: t : 180	σ_κ = 0.50	1	1.7
C-O : 144	$\theta_3 = \theta_6$: 113	g^+ : 60 g^- : -60	σ_η = 1.5	1	-1.0
C*-O : 134	$\theta_4 = \theta_5$: 110	For bond 2:	$\omega_{\eta\kappa}$ = 0.10	1	5.8
C-H : 109	C-C-H : 109	t : 180 c : 0	γ = 1	1	0
C*=O* : 122	O-C-O : 125		For 3 × 3 matrices: rows and columns: t, g^+, g^-		
l_{vb} : 574			Otherwise: rows : t, c ; columns: t, g^+, g^- (or vice versa)		

$$U_1 = \begin{bmatrix} 1 & 0 & 0 \\ 1 & 0 & 0 \\ 1 & 0 & 0 \end{bmatrix} \quad U_2 = \begin{bmatrix} 1 & \gamma \\ 0 & 0 \\ 0 & 0 \end{bmatrix} \quad U_3 = \begin{bmatrix} 1 & 0 & 0 \\ 1 & 0 & 0 \end{bmatrix} \quad U_4 = \begin{bmatrix} 1 & \sigma_x & \sigma_x \\ 0 & 0 & 0 \\ 0 & 0 & 0 \end{bmatrix}$$

$$U_5 = \begin{bmatrix} 1 & \sigma_\eta & \sigma_\eta \\ 1 & \sigma_\eta & \sigma_\eta\omega_{\eta\kappa} \\ 1 & \sigma_\eta\omega_{\eta\kappa} & \sigma_\eta \end{bmatrix} \quad U_6 = \begin{bmatrix} 1 & \sigma_\kappa & \sigma_\kappa \\ 1 & \sigma_\kappa & \sigma_\kappa\omega_{\eta\kappa} \\ 1 & \sigma_\kappa\omega_{\eta\kappa} & \sigma_\kappa \end{bmatrix}$$

a) γ weights the *cis* relative to the *trans* conformation of the terephthaloyl residue.

Comments:
All ester groups in the PET chain are assigned confidently to be planar *trans*. The restriction of bonds 1 and 3 to the *trans* states and bond 2 to a choice between *cis* and *trans* leads to the simple statistical weight matrices. The length of the span of the terephthaloyl residue in PET guarantees independence of the conformations of successive repeating units of the chain.

Calcd. quantities: Mean-square dimension ratio, $(\langle r^2 \rangle_0 / M)_\infty$ = 0.93 Å^2 · mol · g^{-1} (*Exptl.*: 1.05 Å^2 · mol · g^{-1})

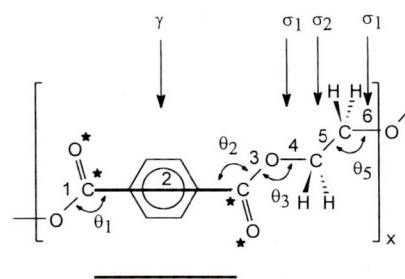

poly(ethylene terephthalate), PET

Walker, G. R.; Semlyen, J. A. *Polymer* **1970**, *11*, 472.

Bond Length [pm]	Valence Angles [°]	Torsion Angles [°]	ξ a) (for 303 K)	ξ_0	E_ξ [kJ·mol^{-1}]
C-C : 153	$\theta_1 = \theta_2$: 114	For bond 5:	$\sigma_1 = 0.73$	1	0.778
		t : 180	$\sigma_2 = 1.5$	1	−1.0
C-O : 144	$\theta_3 = \theta_6$: 113	g$^+$: 60			
		g$^-$: −60	$\omega_1 = 1.82$	1	1.51
C*-O : 134	$\theta_4 = \theta_5$: 110		$\omega_2 = 0.92$	1	0.21
		For bonds 4,6:	$\omega_3 = 2.10$	1	−1.9
C-H : 109	C-C-H : 109	t : 180	$\omega_4 = 0.64$	1	1.1
		g$^+$: 85			
C*=O* : 122	O-C-O : 125	g$^-$: −85	$\delta_1 = 1.17$	1	−0.40
			$\delta_2 = 0.60$	1	1.29
l_{vb} : 574		For bond 2:	$\delta_3 = 0.57$	1	1.42
		t : 180	$\delta_4 = 1.62$	1	−1.22
		c : 0	$\delta_5 = 1.12$	1	−0.29
			$\delta_6 = 1.35$	1	−0.76
			$\delta_7 = 0.63$	1	1.16
			$\delta_8 = 2.72$	1	−2.52
			$\delta_9 = 0.04$	1	8.11
			$\gamma = 1$	1	0

$$U_1 = \begin{bmatrix} 1 & 0 & 0 \\ 0 & 1 & 0 \\ 0 & 0 & 1 \\ 1 & 0 & 0 \\ 0 & 1 & 0 \\ 0 & 0 & 1 \\ 1 & 0 & 0 \\ 0 & 1 & 0 \\ 0 & 0 & 1 \end{bmatrix} \quad U_2 = \begin{bmatrix} 1 & \gamma \\ 1 & \gamma \\ 1 & \gamma \end{bmatrix} \quad U_3 = \begin{bmatrix} 1 & 0 \\ 0 & 1 \end{bmatrix} \quad U_4 = \begin{bmatrix} 1 & \sigma_1 & \sigma_1 \\ 1 & \sigma_1 & \sigma_1 \end{bmatrix}$$

$$U_5 = \begin{bmatrix} 1 & \sigma_2\omega_1 & \sigma_2\omega_1 & 0 & 0 & 0 & 0 & 0 & 0 \\ 0 & 0 & 0 & \omega_2 & \sigma_2\omega_3 & \sigma_2\omega_4 & 0 & 0 & 0 \\ 0 & 0 & 0 & 0 & 0 & 0 & \omega_2 & \sigma_2\omega_4 & \sigma_2\omega_3 \end{bmatrix}$$

$$U_6 = \begin{bmatrix} 1 & \sigma_1\omega_2\delta_1 & \sigma_1\omega_2\delta_1 & 0 & 0 & 0 & 0 & 0 & 0 \\ \delta_1 & \sigma_1\omega_2\delta_5 & \sigma_1\omega_2\delta_6 & 0 & 0 & 0 & 0 & 0 & 0 \\ \delta_1 & \sigma_1\omega_2\delta_6 & \sigma_1\omega_2\delta_5 & 0 & 0 & 0 & 0 & 0 & 0 \\ 0 & 0 & 0 & \omega_1\delta_2 & \sigma_1\omega_3\delta_3 & \sigma_1\omega_4\delta_4 & 0 & 0 & 0 \\ 0 & 0 & 0 & \omega_1\delta_3 & \sigma_1\omega_3\delta_7 & \sigma_1\omega_3\delta_8 & 0 & 0 & 0 \\ 0 & 0 & 0 & \omega_1\delta_4 & \sigma_1\omega_3\delta_8 & \sigma_1\omega_4\delta_9 & 0 & 0 & 0 \\ 0 & 0 & 0 & 0 & 0 & 0 & \omega_1\delta_2 & \sigma_1\omega_4\delta_4 & \sigma_1\omega_3\delta_3 \\ 0 & 0 & 0 & 0 & 0 & 0 & \omega_1\delta_3 & \sigma_1\omega_4\delta_9 & \sigma_1\omega_3\delta_8 \\ 0 & 0 & 0 & 0 & 0 & 0 & \omega_1\delta_3 & \sigma_1\omega_4\delta_8 & \sigma_1\omega_3\delta_7 \end{bmatrix}$$

Matrix:	U_1	U_2	U_3	U_4	U_5
rows:	tt, tg$^+$, tg$^-$ g$^+$t, g$^+$g$^+$, g$^+$g$^-$, g$^-$t, g$^-$g$^+$, g$^-$g$^-$	tt, g$^+$t, g$^-$t	tt, tc	tt, ct	tt, tg$^+$, tg$^-$
columns:	tt, g$^+$t, g$^-$t	tt, tc	tt, ct	tt, tg$^+$, tg$^-$	tt, tg$^+$, tg$^-$, g$^+$t, g$^+$g$^+$, g$^+$g$^-$, g$^-$t, g$^-$g$^+$, g$^-$g$^-$

U_6 : rows and columns: tt, tg$^+$, tg$^-$, g$^+$t, g$^+$g$^+$, g$^+$g$^-$, g$^-$t, g$^-$g$^+$, g$^-$g$^-$

a) γ weights the *cis* relative to the *trans* conformation of the terephthaloyl residue. ω-Parameters correspond to interactions arising from rotations about **pairs** of skeletal bonds, and δ parameters correspond to interactions arising from rotations about **triads** of skeletal bonds. The statistical weight parameters ω and δ are assigned values based on estimates of the steric and electrostatic attractions and repulsions between the nonbonded carbon and oxygen atoms of the ester groups.

Calcd. quantities: Cyclic trimer concentration (*given in:* per cent by weight, 570 K) [b]: w_3 = 0.45 or 0.12   (*Exptl.:* 1.4)

Mean-square dimension ratio (*given in:* Å$^2 \cdot$ mol $\cdot$ g^{-1} at 303 K) [b]: $(<r^2>_0 / M)_\infty$ = 0.55 or 0.92   (*Exptl.:* 1.0)

b) Two different methods of calculation are applied; for more details, see original paper.

poly(ethylene terephthalate), PET

Mendicuti, F.; Viswanadhan, V. N.; Mattice, W. L. *Polymer* **1988**, *29*, 875.

Bond Length [pm]	Valence Angles [°]	Torsion Angles [°]	ξ (for 298 K)	ξ_0	E_ξ [kJ·mol^{-1}]				
C-C : 153	$\theta_1 = \theta_2 : 114$	For bonds 4-6: t : 180	$\sigma_\kappa = 0.489$	1	1.78	$U_1 = \begin{bmatrix} 1 \\ 1 \\ 1 \end{bmatrix}$	$U_2 = \begin{bmatrix} 1 & \gamma \end{bmatrix}$	$U_3 = \begin{bmatrix} 1 \\ 1 \end{bmatrix}$	$U_4 = \begin{bmatrix} 1 & \sigma_\kappa & \sigma_\kappa \end{bmatrix}$
C-O : 146	$\theta_3 = \theta_6 : 113$	g^+ : 60 (and ±20) g^- : −60 (and ±20)	$\sigma_\eta = 1.96$	1	−1.67				
O-C* : 134	$\theta_4 = \theta_5 : 110$		$\omega_{\eta\kappa} = 0.092$	1	5.90				For matrices U_5 - U_6: rows and columns: t, g^+, g^-
l_{vb} : 574		For bonds 1,3: t : 180	$\gamma = 1$	1	0	$U_5 = \begin{bmatrix} 1 & \sigma_\eta & \sigma_\eta \\ 1 & \sigma_\eta & \sigma_\eta\omega_{\eta\kappa} \\ 1 & \sigma_\eta\omega_{\eta\kappa} & \sigma_\eta \end{bmatrix}$		$U_6 = \begin{bmatrix} 1 & \sigma_\kappa & \sigma_\kappa \\ 1 & \sigma_\kappa & \sigma_\kappa\omega_{\eta\kappa} \\ 1 & \sigma_\kappa\omega_{\eta\kappa} & \sigma_\kappa \end{bmatrix}$	
C*=O* : 122		For bond 2: t : 180 c : 0							

Matrix indices for rows/columns are: U_1 : t,g^+,g^-/t ; U_2 : t/t,c ; U_3 : t,c/t ; U_4 : t/t,g^+,g^-

Comments:
Steady-state emission and excitation spectra are measured for the dimethyl terephthalate and three polyesters in four solvents. The polymers have the repeating unit AB_m, where A is $-CO-C_6H_4-COO-$, B is $-(CH_2-CH_2-O)_m-$ and $m = 1,2,3$. An RIS treatment of the unperturbed polymers identifies the conformations that should be conducive to excimer formation by nearest-neighbor aromatic rings. The population of such conformations is maximal in the polyesters in which $m = 2$.

V 043

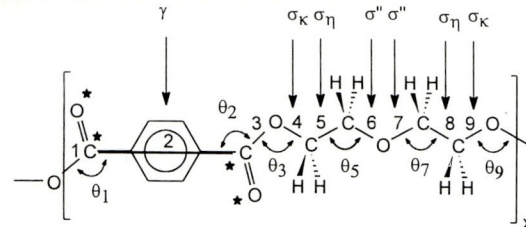

poly(diethyleneglycol terephthalate), PDET

Riande, E. *J. Polym. Sci.; Polym. Phys. Ed.* **1977**, *15*, 1397.

Bond Length [pm]	Valence Angles [°]	Torsion Angles [°]	ξ (for 308 K)	ξ_0	E_ξ [kJ·mol^{-1}]
C-C : 153	$\theta_1 = \theta_2$: 114	For bonds 4-9: t : 180	σ_κ = 0.50	1	1.78
C-O : 144	$\theta_3 = \theta_9$: 113	g$^+$: 60 g$^-$: -60	σ_η = 1.5	1	-1.04
O-C* : 134	θ_k : 110 (k = 4,5,7,8)	For bonds 1,3:	$\omega_{\eta\kappa}$ = 0.10	1	5.90
l_{vb} : 574		t : 180	σ'' = 0.22	1	3.88
	θ_6 : 111	For bond 2: t : 180 c : 0	ω = 0.56 γ = 1	1 1	1.48 0

$$U_1 = \begin{bmatrix} 1 \\ 1 \\ 1 \end{bmatrix} \quad U_2 = \begin{bmatrix} 1 & \gamma \end{bmatrix} \quad U_3 = \begin{bmatrix} 1 \\ 1 \end{bmatrix} \quad U_4 = \begin{bmatrix} 1 & \sigma_\kappa & \sigma_\kappa \end{bmatrix}$$

$$U_5 = \begin{bmatrix} 1 & \sigma_\eta & \sigma_\eta \\ 1 & \sigma_\eta & \sigma_\eta\omega_{\eta\kappa} \\ 1 & \sigma_\eta\omega_{\eta\kappa} & \sigma_\eta \end{bmatrix} \quad U_6 = \begin{bmatrix} 1 & \sigma'' & \sigma'' \\ 1 & \sigma'' & \sigma''\omega \\ 1 & \sigma''\omega & \sigma'' \end{bmatrix} \quad U_7 = \begin{bmatrix} 1 & \sigma'' & \sigma'' \\ 1 & \sigma'' & 0 \\ 1 & 0 & \sigma'' \end{bmatrix}$$

$$U_8 = \begin{bmatrix} 1 & \sigma_\eta & \sigma_\eta \\ 1 & \sigma_\eta & \sigma_\eta\omega \\ 1 & \sigma_\eta\omega & \sigma_\eta \end{bmatrix} \quad U_9 = \begin{bmatrix} 1 & \sigma_\kappa & \sigma_\kappa \\ 1 & \sigma_\kappa & \sigma_\kappa\omega_{\eta\kappa} \\ 1 & \sigma_\kappa\omega_{\eta\kappa} & \sigma_\kappa \end{bmatrix}$$

For matrices $U_5 - U_9$: rows and columns: t, g$^+$, g$^-$

Matrix indices for rows/columns are: U_1 : t,g$^+$,g$^-$/t ; U_2 : t/t,c ; U_3 : t,c/t ; U_4 : t/t,g$^+$,g$^-$

Comments:
Dielectric constants are determined for a fraction of PDET in benzene. The data indicate that the dipole moment ratio is somewhat higher than that of PEO, and its temperature coefficient is in the vicinity of zero.

Calcd. quantities:

Mean-square dimension ratio	$(<r^2>_0 / M)_\infty$	= 0.80 Å^2·mol·g^{-1}	
Dipole moment ratio	$<\mu^2> / Nm^2$	= 0.697	(Exptl.: 0.688)
	$d(\ln <\mu^2>)/dT$	= 3.3 × 10^{-5} K^{-1}	(Exptl.: ≈ 0)

V 044

diethylene glycol dibenzoate, ODB

San Román, J.; Guzmán, J.; Riande, E.; Santoro, J.; Rico, M. *Macromolecules* **1982**, *15*, 609.

Bond Length [pm]	Valence Angles [°]	Torsion Angles [°]	ξ (for 303 K)	ξ_0	E_ξ [kJ·mol^{-1}]
C-C : 153	$\theta_1 = \theta_7$: 113	For bond 2-7: t : 180	σ_κ = 0.51	1	1.7
O-C : 144	θ_k : 110 (k = 2,3,5,6)	g^+ : 60 g^- : −60	σ' = 3.7 to 7.3	1	−3.3 to −5.0
O-C* : 134	θ_4 : 111	For bonds 1,8: t : 180	σ'' = 0.22	1	3.8
			$\omega_{\eta\kappa}$ = 0.12	1	5.4
			ω = 0.51	1	1.7
			ω' = 0	1	∞

$$U_1 = \begin{bmatrix} 1 \\ 1 \end{bmatrix} \qquad U_2 = \begin{bmatrix} 1 & \sigma_\kappa & \sigma_\kappa \end{bmatrix}$$

$$U_3 = \begin{bmatrix} 1 & \sigma' & \sigma' \\ 1 & \sigma' & \sigma'\omega_{\eta\kappa} \\ 1 & \sigma'\omega_{\eta\kappa} & \sigma' \end{bmatrix} \quad U_4 = \begin{bmatrix} 1 & \sigma'' & \sigma'' \\ 1 & \sigma'' & \sigma''\omega \\ 1 & \sigma''\omega & \sigma'' \end{bmatrix} \quad U_5 = \begin{bmatrix} 1 & \sigma'' & \sigma'' \\ 1 & \sigma'' & \sigma''\omega' \\ 1 & \sigma''\omega' & \sigma'' \end{bmatrix}$$

$$U_6 = \begin{bmatrix} 1 & \sigma' & \sigma' \\ 1 & \sigma' & \sigma'\omega \\ 1 & \sigma'\omega & \sigma' \end{bmatrix} \quad U_7 = \begin{bmatrix} 1 & \sigma_\kappa & \sigma_\kappa \\ 1 & \sigma_\kappa & \sigma_\kappa\omega_{\eta\kappa} \\ 1 & \sigma_\kappa\omega_{\eta\kappa} & \sigma_\kappa \end{bmatrix}$$

For matrices U_3 - U_7: rows and columns: t, g^+, g^-

Matrix indices for rows/columns are: U_1 : t,c/t ; U_2 : t/t,g^+,g^-

Comments:

The rotameric probability about CH_2-CH_2 bonds in poly(diethylene glycol terephthalate) (PDET) and poly(thiodiethylene glycol terephthalate) (PSET) is obtained from ^{1}H NMR studies of the respective low molecular weight analogues diethylene glycol dibenzoate and thiodiethylene glycol dibenzoate. The results, interpreted in terms of the RIS model, suggest that *gauche* states about CH_2-CH_2 bonds in both PDET and PSET have an energy ca. 2.1 kJ·mol^{-1} lower than similar states about these bonds in POE and PTDG, respectively.

Calcd. quantities: Molar fractions of the *gauche* states, X_g ; $\langle\mu^2\rangle^{1/2}$ = 2.66 D (*Exptl.*: 2.72 D)

V 045

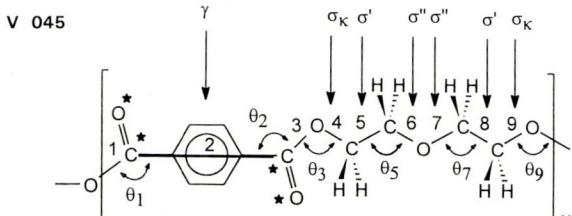

poly(diethylene glycol terephthalate), PDET

Riande, E.; Guzmán, J.; Llorente, M. A. *Macromolecules* **1982**, *15*, 298.

Bond Length [pm]	Valence Angles [°]	Torsion Angles [°]	ξ (for 343 K)	ξ_0	E_ξ [kJ·mol^{-1}]
C-C : 153	$\theta_1 = \theta_2$: 114	For bond 4-9: t : 180	σ_K = 0.55	1	1.7
O-C : 144	$\theta_3 = \theta_9$: 113	g^+ : 60 g^- : -60	σ' = 2.7 to 4.4	1	-2.8 to -4.2
O-C* : 134	θ_k : 110 (k = 4,5,7,8)	For bond 2:	σ'' = 0.27	1	3.7
l_{vb} : 574		t : 180 c : 0	ω_K = 0.14	1	5.6
	θ_6 : 111		ω = 0.60	1	1.46
		For bonds 1,3: t : 180	ω' = 0	1	∞

$$U_1 = \begin{bmatrix} 1 \\ 1 \\ 1 \end{bmatrix} \quad U_2 = \begin{bmatrix} 1 & 1 \end{bmatrix} \quad U_3 = \begin{bmatrix} 1 \\ 1 \end{bmatrix} \quad U_4 = \begin{bmatrix} 1 & \sigma_K & \sigma_K \end{bmatrix}$$

$$U_5 = \begin{bmatrix} 1 & \sigma' & \sigma' \\ 1 & \sigma' & \sigma'\omega_K \\ 1 & \sigma'\omega_K & \sigma' \end{bmatrix} \quad U_6 = \begin{bmatrix} 1 & \sigma'' & \sigma'' \\ 1 & \sigma'' & \sigma''\omega \\ 1 & \sigma''\omega & \sigma'' \end{bmatrix} \quad U_7 = \begin{bmatrix} 1 & \sigma'' & \sigma'' \\ 1 & \sigma'' & \sigma''\omega' \\ 1 & \sigma''\omega' & \sigma'' \end{bmatrix}$$

$$U_8 = \begin{bmatrix} 1 & \sigma' & \sigma' \\ 1 & \sigma' & \sigma'\omega \\ 1 & \sigma'\omega & \sigma' \end{bmatrix} \quad U_9 = \begin{bmatrix} 1 & \sigma_K & \sigma_K \\ 1 & \sigma_K & \sigma_K\omega_K \\ 1 & \sigma_K\omega_K & \sigma_K \end{bmatrix}$$

For matrices $U_5 - U_9$: rows and columns: t, g^+, g^-

Matrix indices for rows/columns are: U_1 : t,g^+,g^-/t ; U_2 : t/t,c ; U_3 : t,c/t ; U_4 : t/t,g^+,g^-

Comments:

Hydroxyl-terminated PDET chains are end-linked into noncrystallizable trifunctional networks using an aromatic triisocyanate. The networks thus obtained are studied with regard to their stress-strain isotherms. The analysis of the temperature coefficient of PDET in terms of the RIS model confirms the results obtained from ^{1}H NMR studies, according to which the *gauche* states about CH_2-CH_2 bonds in the polymer chain have an energy significantly lower than these states about similar bonds in POE.

Calcd. quantities: $\langle r^2 \rangle_0 / nl^2$; $d(\ln \langle r^2 \rangle_0)/dT$; $\langle \mu^2 \rangle / xm^2$; $d(\ln \langle \mu^2 \rangle)/dT$

V 046

poly(diethylene glycol terephthalate), PDET

Riande, E.; Guzmán, J.; Tarazona, M. P.; Saiz, E. *J. Polym. Sci.; Polym. Phys. Ed.* **1984**, *22*, 917.

Bond Length [pm]	Valence Angles [°]	Torsion Angles [°]	ξ a) (for 308 K)	ξ_0	E_ξ [kJ·mol^{-1}]
C-C : 153	$\theta_1 = \theta_2$: 114	For bonds 4-9: t : 180	σ_κ = 0.50	1	1.78
C-O : 144	$\theta_3 = \theta_9$: 113	g$^+$: 60 g$^-$: −60	σ_η = 3.8	1	−3.35
O-C* : 134	θ_k : 110 (k = 4,5,7,8)	For bond 2: t : 180 c : 0	$\omega_{\eta\kappa}$ = 0.10	1	5.90
C*=O* : 122	θ_6 : 111		σ'' = 0.21	1	3.88
l_{vb} : 574	C*-C*=O* : 125	For bonds 1,3: t : 180	ω = 0.55	1	1.49
			γ = 0.92	1	0.21

$$U_1 = \begin{bmatrix} 1 \\ 1 \\ 1 \end{bmatrix} \quad U_2 = \begin{bmatrix} 1 & \gamma \end{bmatrix} \quad U_3 = \begin{bmatrix} 1 \\ 1 \end{bmatrix} \quad U_4 = \begin{bmatrix} 1 & \sigma_\kappa & \sigma_\kappa \end{bmatrix}$$

$$U_5 = \begin{bmatrix} 1 & \sigma_\eta & \sigma_\eta \\ 1 & \sigma_\eta & \sigma_\eta\omega_{\eta\kappa} \\ 1 & \sigma_\eta\omega_{\eta\kappa} & \sigma_\eta \end{bmatrix} \quad U_6 = \begin{bmatrix} 1 & \sigma'' & \sigma'' \\ 1 & \sigma'' & \sigma''\omega \\ 1 & \sigma''\omega & \sigma'' \end{bmatrix} \quad U_7 = \begin{bmatrix} 1 & \sigma'' & \sigma'' \\ 1 & \sigma'' & 0 \\ 1 & 0 & \sigma'' \end{bmatrix}$$

$$U_8 = \begin{bmatrix} 1 & \sigma_\eta & \sigma_\eta \\ 1 & \sigma_\eta & \sigma_\eta\omega \\ 1 & \sigma_\eta\omega & \sigma_\eta \end{bmatrix} \quad U_9 = \begin{bmatrix} 1 & \sigma_\kappa & \sigma_\kappa \\ 1 & \sigma_\kappa & \sigma_\kappa\omega_{\eta\kappa} \\ 1 & \sigma_\kappa\omega_{\eta\kappa} & \sigma_\kappa \end{bmatrix}$$

For matrices $U_5 - U_9$: rows and columns: t, g$^+$, g$^-$

Matrix indices for rows/columns are: U_1 : t,g$^+$,g$^-$/t ; U_2 : t/t,c ; U_3 : t,c/t ; U_4 : t/t,g$^+$,g$^-$

a) E(γ), governing the occurrence of *cis* versus *trans* states for the virtual C*–C* bond, was taken to be 0.210 kJ·mol^{-1}; i.e., the *trans* orientation of the two ester groups is slightly preferred owing to the dipole-dipole interactions.

Comments:
The stress-optical behaviour of an elastomeric network of PDET is measured over a wide range of elongation ratios and temperatures. Theoretical calculations are carried out with the RIS model. For $\Delta\alpha$, no reasonable modification of the conformational energies or contributions to the anisotropic part of the polarizability tensor would achieve agreement between theory and experiments. The discrepancy between theoretical and experimental results may be qualitatively explained by intermolecular interactions. Agreement between theory and experiment is only obtained assuming the unlikely value of about + 4.2 kJ·mol^{-1} for E(σ_η).

Calcd. quantities: $<r^2_0>/M$ $<\mu^2>/Nm^2$ $d(\ln <\mu^2>)/dT$ optical configuration parameter $\Delta\alpha$

poly(triethylene glycol terephthalate)

Riande, E.; Guzmán, J. *J. Polym. Sci., Polym. Phys. Ed.* **1985**, *23*, 1235.

Bond Length [pm]	Valence Angles [°]	Torsion Angles [°]	ξ (for 303 K)	ξ_0	E_ξ [kJ·mol^{-1}]
C-C : 153	O-C-C : 114	t : 180	$\sigma'' = 0.22$	1	3.77
C-O : 143	C-O-CH$_2$: 113	g$^+$: 60	$\sigma_\eta = 3.78$	1	-3.35
C*-O : 133	O-CH$_2$-CH$_2$: 110	g$^-$: -60	$\sigma' = 2.29$	1	-2.09
l$_{vb}$: 574	CH$_2$-O-CH$_2$: 111	ϕ : 0 or 180	$\omega = 0.37$	1	2.51
			$\sigma_k = 0.52$	1	1.67
			$\nu = 0.92$	1	0.21
			$\omega_{\eta k} = 0.04$	1	8.37

Comments:
The quantities Δa and $d(\ln \Delta a)/dT$ are calculated by means of the RIS model. Better agreement between the theoretical and experimental values of these parameters is found for poly(triethylene glycol terephthalate) than for poly(diethylene glycol terephthalate). Since the polarities of these two chains are similar, intermolecular interactions involving terephthaloyl residues may be responsible for the discrepancies observed between theory and experiment for Δa in aromatic polyesters.

Calcd. quantities: Δa $d(\ln \Delta a)/dT$

V 048

poly(diethylene glycol terephthalate), PDET

Mendicuti, F.; Viswanadhan, V. N.; Mattice, W. L. *Polymer* **1988**, *29*, 875.

Bond Length [pm]	Valence Angles [°]	Torsion Angles [°]	ξ (for 298 K)	ξ_o	E_ξ [kJ·mol^{-1}]
C–C : 153	$\theta_1 = \theta_2 : 114$	For bonds 4-9:	$\sigma_\kappa = 0.49$	1	1.78
C–O : 146	$\theta_3 = \theta_9 : 113$	t : 180	$\sigma_\eta = 1.96$	1	–1.67
		g$^+$: 60 (and ±20)			
		g$^-$: –60 (and ±20)			
O–C* : 134	θ_k : 110 (k = 4,5,7,8)	For bonds 1,3:	$\omega_{\eta\kappa} = 0.10$	1	5.90
l_{vb} : 574		t : 180	$\sigma'' = 0.21$	1	3.88
	θ_6 : 111				
C*=O* : 122		For bond 2:	$\omega = 0.55$	1	1.49
		t : 180	$\gamma = 1$	1	0
		c : 0			

$$U_1 = \begin{bmatrix} 1 \\ 1 \\ 1 \end{bmatrix} \quad U_2 = \begin{bmatrix} 1 & \gamma \end{bmatrix} \quad U_3 = \begin{bmatrix} 1 \\ 1 \end{bmatrix} \quad U_4 = \begin{bmatrix} 1 & \sigma_\kappa & \sigma_\kappa \end{bmatrix}$$

$$U_5 = \begin{bmatrix} 1 & \sigma_\eta & \sigma_\eta \\ 1 & \sigma_\eta & \sigma_\eta\omega_{\eta\kappa} \\ 1 & \sigma_\eta\omega_{\eta\kappa} & \sigma_\eta \end{bmatrix} \quad U_6 = \begin{bmatrix} 1 & \sigma'' & \sigma'' \\ 1 & \sigma'' & \sigma''\omega \\ 1 & \sigma''\omega & \sigma'' \end{bmatrix} \quad U_7 = \begin{bmatrix} 1 & \sigma'' & \sigma'' \\ 1 & \sigma'' & 0 \\ 1 & 0 & \sigma'' \end{bmatrix}$$

$$U_8 = \begin{bmatrix} 1 & \sigma_\eta & \sigma_\eta \\ 1 & \sigma_\eta & \sigma_\eta\omega \\ 1 & \sigma_\eta\omega & \sigma_\eta \end{bmatrix} \quad U_9 = \begin{bmatrix} 1 & \sigma_\kappa & \sigma_\kappa \\ 1 & \sigma_\kappa & \sigma_\kappa\omega_{\eta\kappa} \\ 1 & \sigma_\kappa\omega_{\eta\kappa} & \sigma_\kappa \end{bmatrix}$$

For matrices U_5 – U_9: rows and columns: t, g$^+$, g$^-$

Matrix indices for rows/columns are: U_1 : t,g$^+$,g$^-$/t ; U_2 : t/t,c ; U_3 : t,c/t ; U_4 : t/t,g$^+$,g$^-$

Comments:

Steady-state emission and excitation spectra are measured for the dimethyl terephthalate and three polyesters in four solvents. The polymers have the repeating unit AB$_m$, where A is –CO–C$_6$H$_4$–COO–, B is –(CH$_2$–CH$_2$–O)$_m$– and m = 1,2,3. An RIS treatment of the unperturbed polymers identifies the conformations that should be conducive to excimer formation by nearest-neighbor aromatic rings. The population of such conformations is maximal in the polyesters in which m = 2.

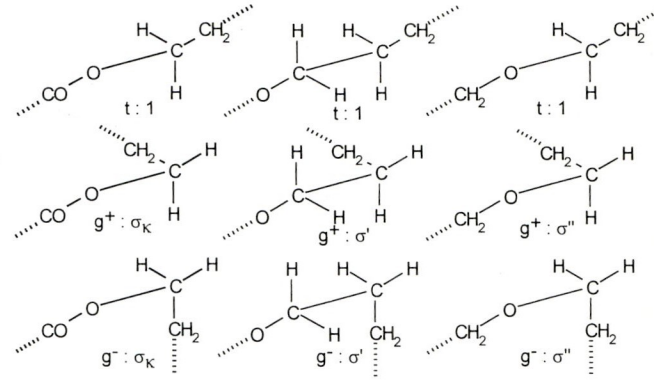

poly(ditrimethylene glycol terephthalate), PDTT.

Gonzáles, C. C.; Riande, E.; Bello, A.; Pereña, J. M. *Macromolecules* **1988**, *21*, 3230.

Bond Length [pm]	Valence Angles [°]	Torsion Angles [°]	ξ (for 303 K)	ξ_0	E_ξ [kJ·mol^{-1}]				
C*–O : 133	θ_3, θ_{11} : 113	For bonds 4–11: t : 180	$\sigma_\kappa = 0.61$	1	1.30	$U_1 = \begin{bmatrix}1\\1\\1\end{bmatrix}$	$U_2 = \begin{bmatrix}1 & \gamma\end{bmatrix}$	$U_3 = \begin{bmatrix}1\\1\end{bmatrix}$	$U_4 = \begin{bmatrix}1 & \sigma_\kappa & \sigma_\kappa\end{bmatrix}$
C–O : 143	θ_1, θ_2 : 114	g^+ : 60 g^- : –60	$\sigma' = 1.28$	1	–0.63				
C–C : 153	θ_k : 110 (k = 4,6,8,10)	For bond 2:	$\sigma'' = 0.22$	1	3.77	$U_5 = \begin{bmatrix}1 & \sigma' & \sigma'\\1 & \sigma' & \sigma'\omega_{\eta\kappa}\\1 & \sigma'\omega_{\eta\kappa} & \sigma'\end{bmatrix}$	$U_6 = U_{10} = \begin{bmatrix}1 & \sigma' & \sigma'\\1 & \sigma' & \sigma'\omega\\1 & \sigma'\omega & \sigma'\end{bmatrix}$	$U_7 = U_8 = \begin{bmatrix}1 & \sigma'' & \sigma''\\1 & \sigma'' & 0\\1 & 0 & \sigma''\end{bmatrix}$	
l_{vb} : 574	θ_7 : 111	t : 180 c : 0	$\omega = 0.40$	1	2.30				
	θ_5, θ_9 : 112	For bonds 1,3: t : 180	$\omega_{\eta\kappa} = 0.098$	1	5.86				
			$\gamma = 1$	1	0	$U_9 = \begin{bmatrix}1 & \sigma' & \sigma'\\1 & \sigma' & 0\\1 & 0 & \sigma'\end{bmatrix}$		$U_{11} = \begin{bmatrix}1 & \sigma_\kappa & \sigma_\kappa\\1 & \sigma_\kappa & \sigma_\kappa\omega_{\eta\kappa}\\1 & \sigma_\kappa\omega_{\eta\kappa} & \sigma_\kappa\end{bmatrix}$	For matrices $U_5 - U_{11}$: rows and columns: t, g^+, g^-

Matrix indices for rows/columns are: U_1 : t,g^+,g^-/t ; U_2 : t/t,c ; U_3 : t,c/t ; U_4 : t/t,g^+,g^-

Comments:
PDTT, prepared by the melt-phase procedure from dimethyl terephthalate and ditrimethylene glycol, is studied with regard to its unperturbed dimensions and polarity. Both the unperturbed dimensions and the dipole moments are interpreted by using the RIS model. The principal conclusion of this analysis is that these conformational properties are extremely dependent on the *gauche* population about CH_2-CH_2 bonds of the glycol residue and the dimensions are also very sensitive to the second-order interactions arising from *gauche* rotations of different sign about two consecutive CH_2-CH_2 skeletal bonds. Stabilizing *gauche* effects about CH_2-CH_2 bonds reported in poly(trimethylene oxide) are not detected in the present system.

Calcd. quantities: $<r^2>_0 / M$ $d(\ln <r^2>_0) / dT$ $<\mu^2> / xm^2$ $d(\ln <\mu^2>) / dT$

V 050

poly(thiodiethylene glycol terephthalate), PSDET

Riande, E.; Guzmán, J.; San Román, J. *J. Chem. Phys.* **1980**, *72*, 5263.

Bond Length [pm]	Valence Angles [°]	Torsion Angles [°]	ξ (for 313 K)	ξ_0	E_ξ [kJ·mol^{-1}]
C-C : 153	$\theta_1 = \theta_2$: 114	For bond 4-9: t : 180	$\sigma_K = 0.5$	1	1.8
O-C : 144	$\theta_3 = \theta_9$: 113	g^+ : 60 g^- : -60	$\sigma' = 0.6$	1	1.3
O-C* : 134	$\theta_5 = \theta_7$: 114	For bond 2:	$\sigma'' = 1.20$	1	-0.5
C-S : 181.5	θ_6 : 100	t : 180 c : 0	$\omega'_K = 0.05$	1	7.8
	$\theta_4 = \theta_8$: 110		$\omega = 0.56$	1	1.5
		For bonds 1,3: t : 180	$\omega'' = 0$	1	∞

$$U_1 = \begin{bmatrix} 1 \\ 1 \\ 1 \end{bmatrix} \quad U_2 = \begin{bmatrix} 1 & 1 \end{bmatrix} \quad U_3 = \begin{bmatrix} 1 \\ 1 \end{bmatrix} \quad U_4 = \begin{bmatrix} 1 & \sigma_K & \sigma_K \end{bmatrix}$$

$$U_5 = \begin{bmatrix} 1 & \sigma' & \sigma' \\ 1 & \sigma' & \sigma'\omega'_K \\ 1 & \sigma'\omega'_K & \sigma' \end{bmatrix} \quad U_6 = \begin{bmatrix} 1 & \sigma'' & \sigma'' \\ 1 & \sigma'' & \sigma''\omega \\ 1 & \sigma''\omega & \sigma'' \end{bmatrix} \quad U_7 = \begin{bmatrix} 1 & \sigma'' & \sigma'' \\ 1 & \sigma'' & \sigma''\omega'' \\ 1 & \sigma''\omega'' & \sigma'' \end{bmatrix}$$

$$U_8 = \begin{bmatrix} 1 & \sigma' & \sigma' \\ 1 & \sigma' & \sigma'\omega \\ 1 & \sigma'\omega & \sigma' \end{bmatrix} \quad U_9 = \begin{bmatrix} 1 & \sigma_K & \sigma_K \\ 1 & \sigma_K & \sigma_K\omega'_K \\ 1 & \sigma_K\omega'_K & \sigma_K \end{bmatrix}$$

For matrices $U_5 - U_9$: rows and columns: t, g^+, g^-

Matrix indices for rows/columns are: U_1 : t,g^+,g^-/t ; U_2 : t/t,c ; U_3 : t,c/t ; U_4 : t/t,g^+,g^-

---

Comments:
The dipole moments of PSDET chains are determined as a function of temperature by means of dielectric constant measurements in dioxane. The experimental results are found to be in fair agreement with theoretical results based on an RIS model in which the required conformational energies are obtained from previous configurational analyses on poly(ethylene terephthalate), poly(diethylene terephthalate), and poly(thiodiethylene glycol).

Calcd. quantities:	Mean-square dimension ratio	$(\langle r^2 \rangle_0 / M)_\infty$	$= 1.00$ Å^2·mol·g^{-1}	
	Dipole moment ratio	$\langle \mu^2 \rangle / Nm^2$	$= 0.61$	(Exptl.: 0.54)
		$d (\ln \langle \mu^2 \rangle) / dT$	$\approx 2 \times 10^{-4}$ K^{-1}	(Exptl.: $\approx 6 \times 10^{-4}$ K^{-1})

thiodiethylene glycol dibenzoate, SDB

San Román, J.; Guzmán, J.; Riande, E.; Santoro, J.; Rico, M. *Macromolecules* **1982**, *15*, 609.

Bond Length [pm]	Valence Angles [°]	Torsion Angles [°]	ξ (for 303 K)	ξ_0	E_ξ [kJ·mol^{-1}]
C-C : 153	$\theta_1 = \theta_7$: 113	For bond 2-7: t : 180	$\sigma_\kappa = 0.51$	1	1.7
O-C : 144	θ_k : 110 (k = 2,3,5,6)	g^+ : 60 g^- : -60	$\sigma' = 1$	1	0
O-C* : 134			$\sigma'' = 1.18$	1	-0.42
C-S : 181.5	θ_4 : 111	For bonds 1,8: t : 180	$\omega_{\eta\kappa} = 0.12$	1	5.4
			$\omega = 0.51$	1	1.7
			$\omega' = 0$	1	∞

Matrix indices for rows/columns are: U_1 : t,c/t ; U_2 : t/t,g$^+$,g$^-$

$$U_1 = \begin{bmatrix} 1 \\ 1 \end{bmatrix} \quad U_2 = \begin{bmatrix} 1 & \sigma_\kappa & \sigma_\kappa \end{bmatrix}$$

$$U_3 = \begin{bmatrix} 1 & \sigma' & \sigma' \\ 1 & \sigma' & \sigma'\omega_{\eta\kappa} \\ 1 & \sigma'\omega_{\eta\kappa} & \sigma' \end{bmatrix} \quad U_4 = \begin{bmatrix} 1 & \sigma'' & \sigma'' \\ 1 & \sigma'' & \sigma''\omega \\ 1 & \sigma''\omega & \sigma'' \end{bmatrix} \quad U_5 = \begin{bmatrix} 1 & \sigma'' & \sigma'' \\ 1 & \sigma'' & \sigma''\omega' \\ 1 & \sigma''\omega' & \sigma'' \end{bmatrix}$$

$$U_6 = \begin{bmatrix} 1 & \sigma' & \sigma' \\ 1 & \sigma' & \sigma'\omega \\ 1 & \sigma'\omega & \sigma' \end{bmatrix} \quad U_7 = \begin{bmatrix} 1 & \sigma_\kappa & \sigma_\kappa \\ 1 & \sigma_\kappa & \sigma_\kappa\omega_{\eta\kappa} \\ 1 & \sigma_\kappa\omega_{\eta\kappa} & \sigma_\kappa \end{bmatrix}$$

For matrices U_3 - U_7: rows and columns: t, g$^+$, g$^-$

Comments:
The rotameric probability about CH$_2$—CH$_2$ bonds in poly(diethylene glycol terephthalate) (PDET) and poly(thiodiethylene glycol terephthalate) (PSET) is obtained from ^{1}H NMR studies of the respective low molecular weight analogues diethylene glycol dibenzoate and thiodiethylene glycol dibenzoate. The results, interpreted in terms of the RIS model, suggest that *gauche* states about CH$_2$—CH$_2$ bonds in both PDET and PSET have an energy ca. 2.1 kJ·mol^{-1} lower than similar states about these bonds in POE and PTDG, respectively.

Calcd. quantities: Molar fractions of the *gauche* states, X_g $<\mu^2>^{1/2}$ = 2.73 D   (*Exptl.*: 2.88 D)

poly(triethylene glycol terephthalate), PTET

Mendicuti, F.; Viswanadhan, V. N.; Mattice, W. L. *Polymer* **1988**, *29*, 875.

Bond Length [pm]	Valence Angles [°]	Torsion Angles [°]	ξ (for 298 K)	ξ_0	E_ξ [kJ·mol^{-1}]
C-C : 153	$\theta_1 = \theta_2$: 114	For bonds 4-12: t : 180 g$^+$: 60 (and ±20) g$^-$: -60 (and ±20)	$\sigma_\kappa = 0.49$	1	1.78
C-O : 146	$\theta_3 = \theta_{12}$: 113		$\sigma_\eta = 1.96$	1	-1.67
O-C* : 134	θ_k : 110 (k = 4,5,7, 8,10,11)	For bonds 1,3: t : 180	$\omega_{\eta\kappa} = 0.10$	1	5.90
l_{vb} : 574			$\sigma'' = 0.21$	1	3.88
C*=O* : 122	$\theta_6 = \theta_9$: 111	For bond 2: t : 180 c : 0	$\omega = 0.55$	1	1.49
			$\gamma = 1$	1	0

$U_1 = \begin{bmatrix} 1 \\ 1 \\ 1 \end{bmatrix}$ $U_2 = [1\ \gamma]$ $U_3 = \begin{bmatrix} 1 \\ 1 \end{bmatrix}$ $U_4 = [1\ \sigma_\kappa\ \sigma_\kappa]$

$U_5 = \begin{bmatrix} 1 & \sigma_\eta & \sigma_\eta \\ 1 & \sigma_\eta & \sigma_\eta\omega_{\eta\kappa} \\ 1 & \sigma_\eta\omega_{\eta\kappa} & \sigma_\eta \end{bmatrix}$ $U_6 = U_9 = \begin{bmatrix} 1 & \sigma'' & \sigma'' \\ 1 & \sigma'' & \sigma''\omega \\ 1 & \sigma''\omega & \sigma'' \end{bmatrix}$ $U_7 = U_{10} = \begin{bmatrix} 1 & \sigma'' & \sigma'' \\ 1 & \sigma'' & 0 \\ 1 & 0 & \sigma'' \end{bmatrix}$

$U_8 = U_{11} = \begin{bmatrix} 1 & \sigma_\eta & \sigma_\eta \\ 1 & \sigma_\eta & \sigma_\eta\omega \\ 1 & \sigma_\eta\omega & \sigma_\eta \end{bmatrix}$ $U_{12} = \begin{bmatrix} 1 & \sigma_\kappa & \sigma_\kappa \\ 1 & \sigma_\kappa & \sigma_\kappa\omega_{\eta\kappa} \\ 1 & \sigma_\kappa\omega_{\eta\kappa} & \sigma_\kappa \end{bmatrix}$

For matrices U_5 - U_{12}: rows and columns: t, g$^+$, g$^-$

Matrix indices for rows/columns are: U_1 : t, g$^+$, g$^-$/t ; U_2 : t/t,c ; U_3 : t,c/t ; U_4 : t/t,g$^+$,g$^-$

Comments:

Steady-state emission and excitation spectra are measured for the dimethyl terephthalate and three polyesters in four solvents. The polymers have the repeating unit AB$_m$, where A is –CO–C$_6$H$_4$–COO–, B is –(CH$_2$–CH$_2$–O)$_m$– and m = 1,2,3. An RIS treatment of the unperturbed polymers identifies the conformations that should be conducive to excimer formation by nearest-neighbor aromatic rings. The population of such conformations is maximal in the polyesters in which m = 2.

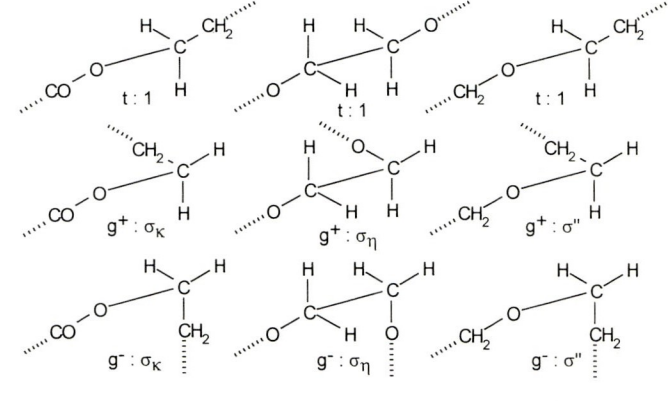

poly(diethylene glycol isophthalate), PDEI

Riande, E.; Guzmán, J.; de Abajo, J. *Makromol. Chem.* **1984**, *185*, 1943.

Bond Length [pm]	Valence Angles [°]	Torsion Angles [°]	ξ (for 303 K)	ξ_0	E_ξ [kJ · mol^{-1}]
C-C : 153	$\theta_1 = \theta_5$: 114	For bonds 7 - 12: t : 180	σ_κ = 0.40	1	2.34
C-O : 146	$\theta_6 = \theta_{12}$: 113	g$^+$: 60 g$^-$: -60	σ_η = 1.9 to 3.8	1	-1.67 to -3.35
O-C* : 151	θ_k : 110 (k = 7,8,10,11)	For bonds 2,5:	$\omega_{\eta\kappa}$ = 0.10	1	5.86
C^{ar}-C^{ar} : 139.5	θ_9 : 111	t : 180 c : 0	σ'' = 0.14	1	5.02
C^{ar}-C* : 148		For bonds 1,3,4,6: t : 180	ω = 0.37	1	2.51

$$U_1 = \begin{bmatrix} 1 \\ 1 \\ 1 \end{bmatrix} \quad U_2 = U_5 = [1\ 1] \quad U_4 = [1] \quad U_3 = U_6 = \begin{bmatrix} 1 \\ 1 \end{bmatrix} \quad U_7 = [1\ \sigma_\kappa\ \sigma_\kappa]$$

$$U_8 = \begin{bmatrix} 1 & \sigma_\eta & \sigma_\eta \\ 1 & \sigma_\eta & \sigma_\eta\omega_{\eta\kappa} \\ 1 & \sigma_\eta\omega_{\eta\kappa} & \sigma_\eta \end{bmatrix} \quad U_9 = \begin{bmatrix} 1 & \sigma'' & \sigma'' \\ 1 & \sigma'' & \sigma''\omega \\ 1 & \sigma''\omega & \sigma'' \end{bmatrix} \quad U_{10} = \begin{bmatrix} 1 & \sigma'' & \sigma'' \\ 1 & \sigma'' & 0 \\ 1 & 0 & \sigma'' \end{bmatrix}$$

$$U_{11} = \begin{bmatrix} 1 & \sigma_\eta & \sigma_\eta \\ 1 & \sigma_\eta & \sigma_\eta\omega \\ 1 & \sigma_\eta\omega & \sigma_\eta \end{bmatrix} \quad U_{12} = \begin{bmatrix} 1 & \sigma_\kappa & \sigma_\kappa \\ 1 & \sigma_\kappa & \sigma_\kappa\omega_{\eta\kappa} \\ 1 & \sigma_\kappa\omega_{\eta\kappa} & \sigma_\kappa \end{bmatrix}$$

For matrices U_8 - U_{12}: rows and columns are t, g$^+$, g$^-$

Matrix indices for rows/columns are: U_1 : t, g$^+$,g$^-$/t ; U_2, U_5 : t/t,c ; U_6 : t,c/t ; U_7 : t/t,g$^+$,g$^-$

Comments:
Values of the mean-square dipole moment, $<\mu^2>$, of PDEI are determined as a function of temperature. The value of the dipole moment ratio is 0.697 at 303 K. Trifunctional model networks are prepared. From thermoelastic experiments performed on the networks over a temperature range 293 - 353 K, it is found that the value of the temperature coefficient of the unperturbed dimensions amounts to 1.05 ± 0.17 K^{-1}. The dipole moments and the temperature coefficients of both the dipole moments and the unperturbed dimensions are critically interpreted in terms of the RIS model, and are found to be in a reasonable agreement.

Calcd. quantities: $<\mu^2> / xm^2$ $d(\ln <\mu^2>) / dT$ $d(\ln <r^2>) / dT$

V 054

poly(diethylene glycol phthalate), PDEP

Riande, E.; de la Campa, J. G.; Schlereth, D. D.; de Abajo, J.; Guzmán, J. *Macromolecules* **1987**, *20*, 1641.

Bond Length [pm]		Valence Angles [°]		Torsion Angles [°] [a)]		ξ [b)]	ξ_o	E_ξ [kJ·mol^{-1}]	(see: [c)])
C-C	: 153	C^{ar}-C^{ar}-C^{ar}	: 120	*For bonds 6-11:* t	: 180	σ_κ = 0.51	1	1.7	
C-O	: 143	C^*-C^{ar}-C^{ar}	: 112	g^+ : 60 g^- : -60		σ' = 3.77	1	-3.4	
C^*-C^{ar}	: 148	C-C-H	: 110			σ'' = 0.23	1	3.8	
C-H	: 109	C^{ar}-C^*-O^*	: 125	*For bonds 2,4:* g^+ : 90 g^- : -90		ω = 0.38	1	2.5	
C^*-O	: 133	C^*-O-C	: 113			$\omega_{\eta\kappa}$ = 0.044	1	8.0	
C^*=O^*	: 122	C^{ar}-C^*-O	: 114	*For bonds 1,5:* t	: 180	γ = 0.46	1	2.0	
C^{ar}-C^{ar}	: 139	O-C-C	: 118.5	*For bond 3:* c	: 0				

$$U_1 = \begin{bmatrix} 1 \\ 1 \\ 1 \end{bmatrix} \quad U_2 = [1\ \gamma] \quad U_2' = [\gamma\ 1] \quad U_3 = U_5 = \begin{bmatrix} 1 \\ 1 \end{bmatrix} \quad U_4 = [1\ 0] \quad U_4' = [0\ 1]$$

$$U_6 = [1\ \sigma_\kappa\ \sigma_\kappa] \quad U_7 = \begin{bmatrix} 1 & \sigma' & \sigma' \\ 1 & \sigma' & \sigma'\omega_{\eta\kappa} \\ 1 & \sigma'\omega_{\eta\kappa} & \sigma' \end{bmatrix} \quad U_8 = \begin{bmatrix} 1 & \sigma'' & \sigma'' \\ 1 & \sigma'' & \sigma''\omega \\ 1 & \sigma''\omega & \sigma'' \end{bmatrix} \quad U_9 = \begin{bmatrix} 1 & \sigma'' & \sigma'' \\ 1 & \sigma'' & 0 \\ 1 & 0 & \sigma'' \end{bmatrix}$$

$$U_{10} = \begin{bmatrix} 1 & \sigma' & \sigma' \\ 1 & \sigma' & \sigma'\omega \\ 1 & \sigma'\omega & \sigma' \end{bmatrix} \quad U_{11} = \begin{bmatrix} 1 & \sigma_\kappa & \sigma_\kappa \\ 1 & \sigma_\kappa & \sigma_\kappa\omega_{\eta\kappa} \\ 1 & \sigma_\kappa\omega_{\eta\kappa} & \sigma_\kappa \end{bmatrix}$$

For matrices U_7 - U_{11}: rows and columns: t, g^+, g^-

Matrix indices for rows/columns are: U_1 : t, g^+, g^-/t ; U_2 : t/g^+,g^- ; U_3 : g^+,g^-/c ; U_4 : c/g^+,g^- ; U_5 : g^+,g^-/t ; U_6 : t/t,g^+,g^-

a) Agreement between theory and experiment was also obtained by applying torsion angles about bonds 2 and 4 of ± 110°, and E(γ) = 5.02 kJ·mol^{-1}.
b) The energy E(γ) accounts for the relative fraction of $g^\pm$ c $g^\mp$ / $g^\pm$ c $g^\pm$ conformations corresponding to bonds 2 through 4.
c) Z = [U_1($U_2 U_3 U_4$ + $U'_2 U'_3 U'_4$) U_5 U J]^x; the matrix U embodies the product of the 1 × 3 statistical weight matrix corresponding to bonds 6 and the 3 × 3 matrices corresponding to bonds 7 through 11.

Comments:
Dielectric measurements are performed on solutions of PDEP in benzene at different temperatures. Conformational energy calculations indicate that when the two ester groups of phthaloyl residue are coplanar to the phenyl groups, the interaction energy is strongly repulsive. However, the energy becomes attractive when the rotational angles about C^{ar}—C^* bonds place the ester groups on a plane perpendicular to the phenyl group. By use of the information derived from these calculations, a three-rotational-state model gives values of the dipole moment and its temperature coefficient in good agreement with the experimental results.

Calcd. quantities: $<\mu^2> / xm^2$ = 0.624   (*Exptl.*: 0.638)
$d (\ln <\mu^2>) / dT$ = 1.55 × 10^{-3} K^{-1} (*Exptl.*: 2.1 × 10^{-3} K^{-1})

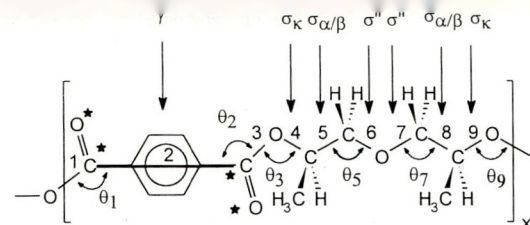

poly(dipropylene glycol terephthalate), PDPT

Diaz-Calleja, R.; Riande, E.; Guzmán, J. *Macromolecules* **1989**, *22*, 3654.

Bond Length [pm]	Valence Angles [°]	Torsion Angles [°]	ξ (for 303 K)	ξ_0	E_ξ [kJ·mol^{-1}]
C*-O : 134	C-O-C* : 113	For bond 4/9:	σ_α = 3.78	1	−3.35
		t : −/+ 160			
C-O : 146	C^{ar}-C*-O: 114	g : −/+ 80	σ_β = 1.18	1	−0.42
		For bond 5,8:			
C-C : 153	O-C-C : 110	t : 175	σ_κ = 1.18	1	−0.42
		g$^+$: 60			
l$_{vb}$: 574	C-O-C : 111	g$^-$: −85	σ'' = 0.14	1	5.02
		For bond 6,7:			
	C-C-C : 112	t : 180	ω = 0.47	1	1.88
		g$^+$: 60			
		g$^-$: −85	ω' = 0.19	1	4.19
		For bond 1,3:			
		t : 180			
		For bond 2:			
		t : 180			
		c : 0			

$$U_1 = \begin{bmatrix} 1 \\ 1 \\ 1 \end{bmatrix} \quad U_2 = \begin{bmatrix} 1 & 1 \end{bmatrix} \quad U_3 = \begin{bmatrix} 1 \\ 1 \end{bmatrix} \quad U_4 = \begin{bmatrix} 1 & 0 & \sigma_\kappa \end{bmatrix} \quad \text{(see: a)}$$

$$U_5 = \begin{bmatrix} 1 & \sigma_\alpha & \sigma_\beta \\ 0 & 0 & 0 \\ 1 & \sigma_\alpha \omega' & \sigma_\beta \end{bmatrix} \quad U_6 = \begin{bmatrix} 1 & 0 & \sigma'' \\ 1 & \sigma'' & \sigma\omega \\ 1 & \sigma''\omega & 0 \end{bmatrix} \quad U_7 = \begin{bmatrix} 1 & \sigma'' & 0 \\ 0 & 0 & 0 \\ 1 & 0 & 0 \end{bmatrix}$$

$$U_8 = \begin{bmatrix} 1 & \sigma_\beta & \sigma_\alpha \\ 1 & 0 & \sigma_\alpha\omega \\ 0 & \sigma_\beta\omega & \sigma_\alpha \end{bmatrix} \quad U_9 = \begin{bmatrix} 1 & \sigma_\kappa & 0 \\ 1 & \sigma_\kappa & 0 \\ 1 & \sigma_\kappa\omega' & 0 \end{bmatrix}$$

For matrices U_5 - U_9: rows and columns: t, g$^+$, g$^-$

Matrix indices for rows/columns are: U_1 : t,g$^+$,g$^-$/t ; U_2 : t/t,c ; U_3 : t,c/t ; U_4 : t/t,g$^+$,g$^-$

a) The composition of several **U** depends on the stereochemistry of the attachment of the CH$_3$ to the chain.

Comments:
Poly(propylene glycol terephthalate) (PPT) and PDPT are obtained by condensation of propylene glycol and dipropylene glycol with dimethyl terephthalate, respectively. The dielectric constants of benzene solutions of PDPT chains are measured in the temperature range 303 - 333 K. The theoretical analysis of the dipole moment ratio suggests that *gauche* states about −CH(CH$_3$)−CH$_2$− bonds, which place an oxygen atom between a methyl group and another oxygen atom, have higher energy than the alternative gauche states. The critical interpretation of the dielectric and mechanical results seems to suggest that the β subglass absorptions are caused by conformational transitions about −CH(CH$_3$)−CH$_2$− bonds of the glycol residue.

Calcd. quantities: $<\mu^2>_0 / xm^2$ $d(\ln <\mu^2>)/dT$ Conformational entropy S

V 056   Riande, E.; Guzmán, J.; de la Campa, J. G.; de Abajo, J. *J. Polym. Sci.; Part B: Polym. Phys.* **1987**, *25*, 2403.

Authors give a short description of a RIS model developed for poly(propylene glycol terephthalate), PPT. This model is based on the known RIS models for PET. Statistical weight parameters and torsion angles of the chain are discussed. Simple scrutiny of structural characteristics of PPT chains indicate that certain conformations that are accessible in PET are suppressed in PPT. In this work, the dipole moment of these chains and its temperature coefficient are measured and the results compared with those calculated using statistical mechanics. A comparative study of the polarity of PPT and PET is also made.

V 057   Mendicuti, F.; Patel, B.; Waldeck, D. H.; Mattice, W. L. *Polymer* **1989**, *30*, 1680.

Steady-state fluorescence spectra are obtained for dilute solutions of 16 aromatic polyesters with the repeating unit AB_m. The chromophore, A, is isophthaloyl or phthaloyl, and the flexible spacer, B, is methylene or oxymethylene. The number of methylene units is 2 - 6, and the number of oxyethylene 1 - 4. Results obtained are interpreted within the frame of the RIS method.

V 058   Bahar, I.; Mattice, W. L. *J. Chem. Phys.* **1989**, *90*, 6783.

The dynamic RIS formalism is used to calculate the rate of first passage from non-excimer-forming conformations to excimer-forming conformations in seven aromatic polyesters with different flexible spacers between the aromatic rings. The equilibrium chain statistics provides a good description of the relative excimer population for these polyesters, even at times where the dynamic contribution is significant.

POLA [a]

Tan, J. S. *J. Polym. Sci.; Polym. Phys. Ed.* **1974**, *12*, 175.

Bond Length [pm]	Valence Angles [°]	Torsion Angles [°]	ξ [b] (for 298 K)	ξ_0	E_ξ [kJ · mol^{-1}]
l_1 : 567	θ_1 : 112	t : 180	σ = 0.01	1	11.4
l_2 : 567	θ_2 : 113	c : 0	α = 1	1	0
l_3 : 136	θ_3 : 114		β = 1	1	0
l_4 : 579	θ_4 : 92				
l_5 : 510	θ_5 : 143				
l_6 : 136	θ_6 : 113				

$$U_1 = \begin{bmatrix} 1 \\ 1 \end{bmatrix} \quad U_2 = [1] \quad U_3 = [1 \ \sigma]$$

$$U_4 = \begin{bmatrix} 1 & \alpha \\ 1 & \alpha \end{bmatrix} \quad U_5 = \begin{bmatrix} 1 & \beta \\ 1 & \beta \end{bmatrix} \quad U_6 = \begin{bmatrix} 1 & \sigma \\ 1 & \sigma \end{bmatrix}$$

Matrix indices for rows/columns are:
U_1 : t,c/t ; U_2 : t/t ; U_3 : t/t,c ; U_4 : t,c/t,c ; U_5 : t,c/t,c ; U_6 : t,c/t,c .

[a] Polyester prepared from 4′,5-(1,1,3-trimethyl-3-phenylindan)dicarboxylic acid and 2,2-bis(4′-hydroxyphenyl)propane.
[b] For U_3, U_6, σ weights the *cis* relative to the *trans* conformation; for U_4 and U_5, a statistical weight α and β, respectively, is assigned for the *cis* isomer relative to the *trans*.

Comments:
RIS theory is applied to investigate chain configuration of POLA. Independent conformations for each repeat monomer unit of the chain are assumed in the calculations of the unperturbed dimensions. Rotations about the oxygen-phenylene-carbon bonds are considered to be free with twofold symmetric potentials. The *trans* and *cis* conformations of the carbonyl-phenylene-carbon and the indan-carbonyl residues are assumed to have equal probability. The bond vectors for this model lie in a plane because every torsion angle is 0° or 180°.

Calcd. quantities: Mean-square dimension ratio $(<r^2>_0 / M)_\infty$ = 0.73 Å^2 · mol · g^{-1} (*Exptl.*: 0.72 (± 0.02) Å^2 · mol · g^{-1})

poly(ethylene terephthalate), polyester liquid crystals, $m = 2$ **(ethylene), PET**

Mendicuti, F.; Patel, B.; Viswanadhan, V. N.; Mattice, W. L. *Polymer* **1988**, *29*, 1669.

Bond Length [pm]	Valence Angles [°]	Torsion Angles [°]	ξ (for 300 K)	ξ_0	E_ξ [kJ · mol^{-1}]
C^{ar}-C^{ar} : 140	C^{ar}-C^{ar}-C^{ar}: 120	For bond 0: t : 180	σ_1 = 0.51	1	1.68
C^{ar}-C^* : 147	C^{ar}-C^{ar}-C^* : 120		σ_2 = 3.9	1	−3.39
C^*-O : 134	C^{ar}-C^*-O : 125	For bonds 1-3: t : 180 g$^+$: 60	ω_1 = 0.034	1	8.4
O-C : 144	C^*-O-C : 113	g$^-$: −60			
C-C : 153	O-C-C : 110				
	C-C-C : 112				

$$U_1 = [1 \; 2\sigma_1] \qquad U_4^T = [1 \; 1] \qquad U_i = \begin{bmatrix} 1 & 2\sigma_i \\ 1 & \sigma_i(1+\omega_i) \end{bmatrix} \quad i = 2,3$$

rows / columns:

t / t, g$^+$ + g$^-$ $\qquad$ t / t, g$^+$ + g$^-$ $\qquad$ t, g$^+$ + g$^-$ / t, g$^+$ + g$^-$

Comments:
Excimer formation is studied in polyesters in which the repeating unit is represented by the formula AB$_m$, where A is p-OOC−Ph−COO, B is CH$_2$ and m = 2,...,6. In dichloroethane, dioxane and ethyl acetate, the ratio of excimer to monomer emission, I_D / I_M, exhibits an odd-even effect with the largest value seen in the polyester with m = 3. The conformations conducive to excimer formation by nearest neighbor aromatic rings are identified with the aid of an RIS analysis of Ph−COO-B$_m$−OOC−Ph. A single species completely dominates the excimer formation when m = 2. As m increases more species can adopt the excimer-forming geometry. When m = 6, no conformation accounts for more than ca. 1/7 of the excimers present. The equilibrium RIS analysis can rationalize the occurrence of the largest value of I_D / I_M at m = 3 and the existence of an odd-even effect at m = 2,...,5. However, it overestimates the importance of the excimers when m = 6.

1,2-ethanediol dibenzoate

Mendicuti, F.; Rodrigo, M. M.; Tarazona, M. P.; Saiz, E. *Macromolecules* **1990**, *23*, 1139.

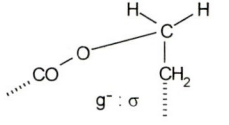

Bond Length [pm]	Valence Angles [°]		Torsion Angles [°]		ξ a) (for 298 K)	ξ_o	E_ξ [kJ·mol^{-1}]
C-C : 153	C^{ar}-C^{ar}-C^{ar}: 120		For C*-O bonds:		$\sigma = 0.31$	1	2.93
			t : 180				
C*-C^{ar} : 139	C*-C^{ar}-C^{ar} : 112				$\sigma_1 = 2.74$	1	-2.5
			For O-C bonds:				
C-H : 109	C-C-H : 110		t : 180		$\omega = 0.034$	1	8.4
			g$^+$: 80				
C*-O : 133	C^{ar}-C*-O* : 125		g$^-$: -80				
C*=O* : 122	C*-O-C : 113		For C-C bonds:				
			t : 180				
C-O : 143	C^{ar}-C*-O : 114		g$^+$: 60				
			g$^-$: -60				
	O-C-C : 118.5						

$$U_1 = \begin{bmatrix} 1 & \sigma & \sigma \end{bmatrix} \quad U_2 = \begin{bmatrix} 1 & \sigma_1 & \sigma_1 \\ 1 & \sigma_1 & \sigma_1\omega \\ 1 & \sigma_1\omega & \sigma_1 \end{bmatrix} \quad U_3 = \begin{bmatrix} 1 & \sigma & \sigma \\ 1 & \sigma & \sigma\omega \\ 1 & \sigma\omega & \sigma \end{bmatrix}$$

rows / columns: t / t, g$^+$, g$^-$

For 3 × 3 matrices: rows and columns: t, g$^+$, g$^-$

a) The model given here is based on an RIS model originally developed by *Mendicuti et al.* [*Polymer* **1988**, *29*, 875]; a second set of conformational energies based on a model of *Bahar* and *Mattice* [*J. Chem. Phys.* **1989**, *90*, 6783] is discussed as well.

Comments:
Dipole moments and Kerr constants of dibenzoates of several diols are measured at 298 K. Theoretical analysis is performed with standard methods of the RIS model. Comparison of theory with experiment indicates that $<\mu^2>$ is almost insensitive to the conformational energies, particularly for $m > 3$. Kerr constants are much more sensitive to the conformational energies. Good agreement between theoretical and experimental values of both dipole moments and Kerr constants of all these compounds is achieved by adjustement of the optical parameters and the position of the rotational isomers.

Calcd. quantities:
Dipole moments $<\mu^2>$ = 8.11 D^2 (*Exptl.*: 7.58 D^2)
Molar Kerr constants $_mK$ = 27.9 × 10^{-25} m^5 V^{-2} mol^{-1} (*Exptl.*: 29.1 (± 1.6) × 10^{-25} m^5 V^{-2} mol^{-1})

1,3-propanediol dibenzoate

Mendicuti, F.; Rodrigo, M. M.; Tarazona, M. P.; Saiz, E. *Macromolecules* **1990**, *23*, 1139.

Bond Length [pm]		Valence Angles [°]		Torsion Angles [°]		ξ a) (for 298 K)	ξ_o	E_ξ [kJ · mol^{-1}]
C-C	: 153	C^{ar}-C^{ar}-C^{ar}	: 120	For C*-O bonds:		$\sigma = 0.31$	1	2.93
C^*-C^{ar}	: 139	C^*-C^{ar}-C^{ar}	: 112	t : 180		$\sigma_1 = 1.40$	1	−0.84
				For O-C bonds:				
C-H	: 109	C-C-H	: 110	t : 180		$\omega_1 = 0$	1	∞
				g^+ : 40				
C^*-O	: 133	C^{ar}-C^*-O^*	: 125	g^- : −40		$\omega_2 = 0.51$	1	1.67
C^*=O^*	: 122	C^*-O-C	: 113	For C-C bonds:				
				t : 180				
C-O	: 143	C^{ar}-C^*-O	: 114	g^+ : 80				
				g^- : −80				
		O-C-C	: 118.5					

$$U_1 = \begin{bmatrix} 1 & \sigma & \sigma \end{bmatrix} \quad U_2 = \begin{bmatrix} 1 & \sigma_1 & \sigma_1 \\ 1 & \sigma_1 & 0 \\ 1 & 0 & \sigma_1 \end{bmatrix} \quad U_3 = \begin{bmatrix} 1 & \sigma_1 & \sigma_1 \\ 1 & \sigma_1 & \sigma_1\omega_2 \\ 1 & \sigma_1\omega_2 & \sigma_1 \end{bmatrix} \quad U_4 = \begin{bmatrix} 1 & \sigma & \sigma \\ 1 & \sigma & 0 \\ 1 & 0 & \sigma \end{bmatrix}$$

rows / columns: t / t, g^+, g^-

For 3 × 3 matrices: rows and columns: t, g^+, g^-

a) The model given here is based on an RIS model originally developed by *Mendicuti et al.* [*Polymer* **1988**, *29*, 875]; a second set of conformational energies based on a model of *Bahar* and *Mattice* [*J. Chem. Phys.* **1989**, *90*, 6783] is discussed as well.

Comments:
Dipole moments and Kerr constants of dibenzoates of several diols are measured at 298 K. Theoretical analysis is performed with standard methods of the RIS model. Comparison of theory with experiment indicates that $<\mu^2>$ is almost insensitive to the conformational energies, particularly for $m > 3$. Kerr constants are much more sensitive to the conformational energies. Good agreement between theoretical and experimental values of both dipole moments and Kerr constants of all these compounds is achieved by adjustement of the optical parameters and the position of the rotational isomers.

Calcd. quantities:

Dipole moments $<\mu^2>$ = 6.78 D^2 (*Exptl.*: 6.71 D^2)

Molar Kerr constants $_mK$ = 9.6 × 10^{-25} m^5 V^{-2} mol^{-1} (*Exptl.*: 6.9 (± 3.9) × 10^{-25} m^5 V^{-2} mol^{-1})

poly(propylene terephthalate), polyester liquid crystals, $m = 3$ (trimethylene)

Mendicuti, F.; Patel, B.; Viswanadhan, V. N.; Mattice, W. L. *Polymer* **1988**, *29*, 1669.

Bond Length [pm]	Valence Angles [°]	Torsion Angles [°]	ξ (for 300 K)	ξ_o	E_ξ [kJ·mol^{-1}]
$C^{ar}-C^{ar}$: 140	$C^{ar}-C^{ar}-C^{ar}$: 120	For bond 0: t : 180	σ_1 = 0.51	1	1.68
$C^{ar}-C^*$: 147	$C^{ar}-C^{ar}-C^*$: 120		σ_2 = 2.0	1	-1.73
C^*-O : 134	$C^{ar}-C^*-O$: 125	For bonds 1-4: t : 180	ω_1 = 0	1	∞
O-C : 144	C^*-O-C : 113	g^+ : 60 g^- : -60	ω_2 = 0.51	1	1.68
C-C : 153	O-C-C : 110				
	C-C-C : 112				

$U_1 = [1 \ 2\sigma_1]$ $U_5^T = [1 \ 1]$ $U_i = \begin{bmatrix} 1 & 2\sigma_i \\ 1 & \sigma_i(1+\omega_i) \end{bmatrix}$ $i = 2-4$

rows / columns:
t / t, $g^+ + g^-$ t / t, $g^+ + g^-$ t, $g^+ + g^- $ / t, $g^+ + g^-$

Comments:
Excimer formation is studied in polyesters in which the repeating unit is represented by the formula AB_m, where A is p-OOC—Ph—COO, B is CH_2 and $m = 2,...,6$. In dichloroethane, dioxane and ethyl acetate, the ratio of excimer to monomer emission, I_D / I_M, exhibits an odd-even effect with the largest value seen in the polyester with $m = 3$. The conformations conducive to excimer formation by nearest neighbor aromatic rings are identified with the aid of an RIS analysis of Ph—COO-B_m—OOC—Ph. A single species completely dominates the excimer formation when $m = 2$. As m increases more species can adopt the excimer-forming geometry. When $m = 6$, no conformation accounts for more than ca. 1/7 of the excimers present. The equilibrium RIS analysis can rationalize the occurrence of the largest value of I_D / I_M at $m = 3$ and the existence of an odd-even effect at $m = 2,...,5$. However, it overestimates the importance of the excimers when $m = 6$.

V 064

1,4-butanediol dibenzoate

Mendicuti, F.; Rodrigo, M. M.; Tarazona, M. P.; Saiz, E. *Macromolecules* **1990**, *23*, 1139.

Bond Length [pm]	Valence Angles [°]	Torsion Angles [°]	ξ a) (for 298 K)	ξ_0	E_ξ [kJ·mol^{-1}]
C-C : 153	$C^{ar}-C^{ar}-C^{ar}$: 120	For C*-O bonds:	$\sigma = 0.31$	1	2.93
C*-C^{ar} : 139	C*-C^{ar}-C^{ar} : 112	t : 180	$\sigma_1 = 1.40$	1	−0.84
C-H : 109	C-C-H : 110	For O-C bonds:	$\sigma_2 = 0.43$	1	2.1
		t : 180			
C*-O : 133	C^{ar}-C*-O* : 125	g$^+$: 40	$\omega_1 = 0$	1	∞
		g$^-$: −40			
C*=O* : 122	C*-O-C : 113	For OC-CC bonds:	$\omega_3 = 0.36$	1	2.51
		t : 180			
C-O : 143	C^{ar}-C*-O : 114	g$^+$: 80			
		g$^-$: −80			
	O-C-C : 118.5				
		For CC-CC bonds:			
		t : 180			
		g$^+$: 60			
		g$^-$: −60			

$U_1 = [1 \ \sigma \ \sigma]$

$$U_2 = \begin{bmatrix} 1 & \sigma_1 & \sigma_1 \\ 1 & \sigma_1 & \sigma_1\omega_1 \\ 1 & \sigma_1\omega_1 & \sigma_1 \end{bmatrix} \quad U_3 = \begin{bmatrix} 1 & \sigma_2 & \sigma_2 \\ 1 & \sigma_2 & \sigma_2\omega_3 \\ 1 & \sigma_2\omega_3 & \sigma_2 \end{bmatrix}$$

rows / columns: t / t, g$^+$, g$^-$

$$U_4 = \begin{bmatrix} 1 & \sigma_1 & \sigma_1 \\ 1 & \sigma_1 & \sigma_1\omega_3 \\ 1 & \sigma_1\omega_3 & \sigma_1 \end{bmatrix} \quad U_5 = \begin{bmatrix} 1 & \sigma & \sigma \\ 1 & \sigma & \sigma\omega_1 \\ 1 & \sigma\omega_1 & \sigma \end{bmatrix}$$

For 3 × 3 matrices: rows and columns t, g$^+$, g$^-$

a) The model given here is based on an RIS model originally developed by *Mendicuti et al.* [*Polymer* **1988**, *29*, 875]; a second set of conformational energies based on a model of *Bahar* and *Mattice* [*J. Chem. Phys.* **1989**, *90*, 6783] is discussed as well.

Comments:
Dipole moments and Kerr constants of dibenzoates of several diols are measured at 298 K. Theoretical analysis is performed with standard methods of the RIS model. Comparison of theory with experiment indicates that $\langle \mu^2 \rangle$ is almost insensitive to the conformational energies, particularly for $m > 3$. Kerr constants are much more sensitive to the conformational energies. Good agreement between theoretical and experimental values of both dipole moments and Kerr constants of all these compounds is achieved by adjustement of the optical parameters and the position of the rotational isomers.

Calcd. quantities:

Dipole moments $\langle \mu^2 \rangle$ = 6.93 D^2 (*Exptl.*: 7.20 D^2)

Molar Kerr constants $_mK$ = 14.3 × 10^{-25} m^5 V^{-2} mol^{-1} (*Exptl.*: 11.6 (± 1.5) × 10^{-25} m^5 V^{-2} mol^{-1})

poly(butylene terephthalate), polyester liquid crystals, $m = 4$ (tetramethylene)

Mendicuti, F.; Patel, B.; Viswanadhan, V. N.; Mattice, W. L. *Polymer* **1988**, *29*, 1669.

Bond Length [pm]	Valence Angles [°]	Torsion Angles [°]	ξ (for 300 K)	ξ_0	E_ξ [kJ·mol^{-1}]
C^{ar}-C^{ar} : 140	C^{ar}-C^{ar}-C^{ar} : 120	For bond 0: t : 180	σ_1 = 0.51	1	1.68
C^{ar}-C^* : 147	C^{ar}-C^{ar}-C^* : 120		σ_2 = 1.4	1	−0.84
C^*-O : 134	C^{ar}-C^*-O : 125	For bonds 1-5: t : 180	σ_3 = 0.43	1	2.1
O-C : 144	C^*-O-C : 113	g^+ : 60 g^- : −60	ω_1 = 0	1	∞
C-C : 153	O-C-C : 110		ω_2 = 0.39	1	2.35
	C-C-C : 112				

$U_1 = [1 \; 2\sigma_1]$ $\quad U_6^T = [1 \; 1]$ $\quad U_i = \begin{bmatrix} 1 & 2\sigma_i \\ 1 & \sigma_i(1+\omega_i) \end{bmatrix}$ $\quad i = 2\text{-}5$

rows / columns:
$t / t, g^+ + g^-$ $\quad\quad t / t, g^+ + g^-$ $\quad\quad t, g^+ + g^- / t, g^+ + g^-$

Comments:
Excimer formation is studied in polyesters in which the repeating unit is represented by the formula AB_m, where A is p-OOC−Ph−COO, B is CH_2 and $m = 2,...,6$. In dichloroethane, dioxane and ethyl acetate, the ratio of excimer to monomer emission, I_D / I_M, exhibits an odd-even effect with the largest value seen in the polyester with $m = 3$. The conformations conducive to excimer formation by nearest neighbor aromatic rings are identified with the aid of an RIS analysis of Ph−COO-B_m−OOC−Ph. A single species completely dominates the excimer formation when $m = 2$. As m increases more species can adopt the excimer-forming geometry. When $m = 6$, no conformation accounts for more than ca. 1/7 of the excimers present. The equilibrium RIS analysis can rationalize the occurrence of the largest value of I_D / I_M at $m = 3$ and the existence of an odd-even effect at $m = 2,...,5$. However, it overestimates the importance of the excimers when $m = 6$.

V 066

1,5-pentanediole dibenzoate

Mendicuti, F.; Rodrigo, M. M.; Tarazona, M. P.; Saiz, E. *Macromolecules* **1990**, *23*, 1139.

Bond Length [pm]	Valence Angles [°]	Torsion Angles [°]	ξ [a] (for 298 K)	ξ_0	E_ξ [kJ·mol^{-1}]
C-C : *153*	C^{ar}-C^{ar}-C^{ar} : *120*	For C*-O bonds: t : 180	σ = 0.31	1	2.93
C*-C^{ar} : *139*	C*-C^{ar}-C^{ar} : *112*		σ_1 = 1.40	1	−0.84
C-H : *109*	C-C-H : *110*	For O-C bonds: t : 180 g$^+$: 40	σ_2 = 0.43	1	2.1
C*-O : *133*	C^{ar}-C*-O* : *125*	g$^-$: −40	ω_1 = 0	1	∞
C*=O* : *122*	C*-O-C : *113*	For OC-CC bonds: t : 180 g$^+$: 80	ω_3 = 0.36	1	2.51
C-O : *143*	C^{ar}-C*-O : *114*	g$^-$: −80	ω_4 = 0.034	1	8.37
	O-C-C : *118.5*	For CC-CC bonds: t : 180 g$^+$: 60 g$^-$: −60			

$$U_1 = [1\ \sigma\ \sigma]$$

rows / columns: t / t, g$^+$, g$^-$

$$U_2 = \begin{bmatrix} 1 & \sigma_1 & \sigma_1 \\ 1 & \sigma_1 & \sigma_1\omega_1 \\ 1 & \sigma_1\omega_1 & \sigma_1 \end{bmatrix} \quad U_3 = \begin{bmatrix} 1 & \sigma_2 & \sigma_2 \\ 1 & \sigma_2 & \sigma_2\omega_3 \\ 1 & \sigma_2\omega_3 & \sigma_2 \end{bmatrix}$$

$$U_4 = \begin{bmatrix} 1 & \sigma_2 & \sigma_2 \\ 1 & \sigma_2 & \sigma_2\omega_4 \\ 1 & \sigma_2\omega_4 & \sigma_2 \end{bmatrix} \quad U_5 = \begin{bmatrix} 1 & \sigma_1 & \sigma_1 \\ 1 & \sigma_1 & \sigma_1\omega_3 \\ 1 & \sigma_1\omega_3 & \sigma_1 \end{bmatrix}$$

$$U_6 = \begin{bmatrix} 1 & \sigma & \sigma \\ 1 & \sigma & \sigma\omega_1 \\ 1 & \sigma\omega_1 & \sigma \end{bmatrix}$$

For 3 × 3 matrices: rows and columns: t, g$^+$, g$^-$

[a] The model given here is based on an RIS model originally developed by *Mendicuti et al.* [*Polymer* **1988**, *29*, 875]; a second set of conformational energies based on a model of *Bahar* and *Mattice* [*J. Chem. Phys.* **1989**, *90*, 6783] is discussed as well.

Comments:
Dipole moments and Kerr constants of dibenzoates of several diols are measured at 298 K. Theoretical analysis is performed with standard methods of the RIS model. Comparison of theory with experiment indicates that $\langle \mu^2 \rangle$ is almost insensitive to the conformational energies, particularly for $m > 3$. Kerr constants are much more sensitive to the conformational energies. Good agreement between theoretical and experimental values of both dipole moments and Kerr constants of all these compounds is achieved by adjustement of the optical parameters and the position of the rotational isomers.

Calcd. quantities:

Dipole moments $\langle \mu^2 \rangle$ = 7.11 D^2 (Exptl.: 7.43 D^2)

Molar Kerr constants $_mK$ = 12.3 × 10^{-25} m^5 V^{-2} mol^{-1} (Exptl.: 9.4 (± 2.4) · 10^{-25} m^5 V^{-2} mol^{-1})

poly(pentamethylene terephthalate), polyester liquid crystals, $m = 5$ (pentamethylene).

Mendicuti, F.; Patel, B.; Viswanadhan, V. N.; Mattice, W. L. *Polymer* **1988**, *29*, 1669.

Bond Length [pm]	Valence Angles [°]	Torsion Angles [°]	ξ (for 300 K)	ξ_0	E_ξ [kJ·mol^{-1}]
C^{ar}-C^{ar} : 140	C^{ar}-C^{ar}-C^{ar} : 120	For bond 0: t : 180	$\sigma_1 = 0.51$	1	1.68
C^{ar}-C^* : 147	C^{ar}-C^{ar}-C^* : 120	For bonds 1-6:	$\sigma_2 = 1.4$	1	-0.84
C^*-O : 134	C^{ar}-C^*-O : 125	t : 180 g$^+$: 60	$\sigma_3 = 0.43$	1	2.1
O-C : 144	C^*-O-C : 113	g$^-$: -60	$\omega_1 = 0$	1	∞
C-C : 153	O-C-C : 110		$\omega_2 = 0.39$	1	2.35
	C-C-C : 112		$\omega_3 = 0.034$	1	8.4

$$U_1 = \begin{bmatrix} 1 & 2\sigma_1 \end{bmatrix} \quad U_7^T = \begin{bmatrix} 1 & 1 \end{bmatrix} \quad U_i = \begin{bmatrix} 1 & 2\sigma_i \\ 1 & \sigma_i(1+\omega_i) \end{bmatrix} \quad i = 2\text{-}6$$

rows / columns:

t / t, g$^+$ +g$^-$ t / t, g$^+$ +g$^-$ t, g$^+$ +g$^-$ / t, g$^+$ +g$^-$

Comments:
Excimer formation is studied in polyesters in which the repeating unit is represented by the formula AB$_m$, where A is p-OOC—Ph—COO, B is CH$_2$ and $m = 2,...,6$. In dichloroethane, dioxane and ethyl acetate, the ratio of excimer to monomer emission, I_D / I_M, exhibits an odd-even effect with the largest value seen in the polyester with $m = 3$. The conformations conducive to excimer formation by nearest neighbor aromatic rings are identified with the aid of an RIS analysis of Ph—COO-B$_m$—OOC—Ph. A single species completely dominates the excimer formation when $m = 2$. As m increases more species can adopt the excimer-forming geometry. When $m = 6$, no conformation accounts for more than ca. 1/7 of the excimers present. The equilibrium RIS analysis can rationalize the occurrence of the largest value of I_D / I_M at $m = 3$ and the existence of an odd-even effect at $m = 2,...,5$. However, it overestimates the importance of the excimers when $m = 6$.

V 068

1,6-hexanediol dibenzoate

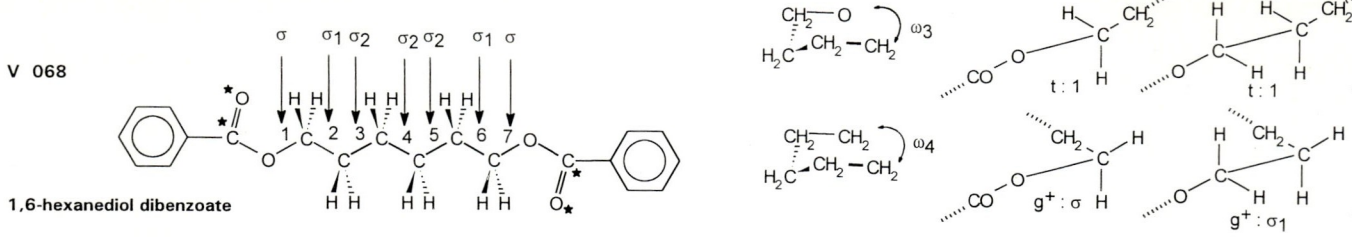

Mendicuti, F.; Rodrigo, M. M.; Tarazona, M. P.; Saiz, E. *Macromolecules* **1990**, *23*, 1139.

Bond Length [pm]	Valence Angles [°]	Torsion Angles [°]	ξ a) (for 298 K)	ξ_0	E_ξ [kJ·mol^{-1}]
C-C : 153	C^{ar}-C^{ar}-C^{ar} : 120	For C*-O bonds: t : 180	σ = 0.31	1	2.93
C*-C^{ar} : 139	C*-C^{ar}-C^{ar} : 112		σ_1 = 1.40	1	-0.84
C-H : 109	C-C-H : 110	For O-C bonds: t : 180 g$^+$: 40 g$^-$: -40	σ_2 = 0.43 ω_1 = 0	1 1	2.1 ∞
C*-O : 133	C^{ar}-C*-O* : 125				
C*=O* : 122	C*-O-C : 113	For OC-CC bonds: t : 180 g$^+$: 80 g$^-$: -80	ω_3 = 0.36 ω_4 = 0.034	1 1	2.51 8.37
C-O : 143	C^{ar}-C*-O : 114				
	O-C-C : 118.5	For CC-CC bonds: t : 180 g$^+$: 60 g$^-$: -60			

$U_1 = \begin{bmatrix} 1 & \sigma & \sigma \end{bmatrix}$

rows / columns: t / t, g$^+$, g$^-$

$U_2 = \begin{bmatrix} 1 & \sigma_1 & \sigma_1 \\ 1 & \sigma_1 & \sigma_1\omega_1 \\ 1 & \sigma_1\omega_1 & \sigma_1 \end{bmatrix}$

$U_3 = \begin{bmatrix} 1 & \sigma_2 & \sigma_2 \\ 1 & \sigma_2 & \sigma_2\omega_3 \\ 1 & \sigma_2\omega_3 & \sigma_2 \end{bmatrix}$

$U_4 = \begin{bmatrix} 1 & \sigma_2 & \sigma_2 \\ 1 & \sigma_2 & \sigma_2\omega_4 \\ 1 & \sigma_2\omega_4 & \sigma_2 \end{bmatrix}$

$U_5 = \begin{bmatrix} 1 & \sigma_2 & \sigma_2 \\ 1 & \sigma_2 & \sigma_2\omega_4 \\ 1 & \sigma_2\omega_4 & \sigma_2 \end{bmatrix}$

For 3 × 3 matrices: rows and columns: t, g$^+$, g$^-$

$U_6 = \begin{bmatrix} 1 & \sigma_1 & \sigma_1 \\ 1 & \sigma_1 & \sigma_1\omega_3 \\ 1 & \sigma_1\omega_3 & \sigma_1 \end{bmatrix}$

$U_7 = \begin{bmatrix} 1 & \sigma & \sigma \\ 1 & \sigma & \sigma\omega_1 \\ 1 & \sigma\omega_1 & \sigma \end{bmatrix}$

a) The model given here is based on an RIS model originally developed by *Mendicuti et al.* [*Polymer* **1988**, *29*, 875]; a second set of conformational energies based on a model of *Bahar* and *Mattice* [*J. Chem. Phys.* **1989**, *90*, 6783] is discussed as well.

Comments:

Dipole moments and Kerr constants of dibenzoates of several diols are measured at 298 K. Theoretical analysis is performed with standard methods of the RIS model. Comparison of theory with experiment indicates that $<\mu^2>$ is almost insensitive to the conformational energies, particularly for $m > 3$. Kerr constants are much more sensitive to the conformational energies. Good agreement between theoretical and experimental values of both dipole moments and Kerr constants of all these compounds is achieved by adjustement of the optical parameters and the position of the rotational isomers.

Calcd. quantities:	Dipole moments $<\mu^2>$	= 7.11 D^2	(*Exptl.*: 7.74 D^2)
	Molar Kerr constants $_mK$	= 13.6 × 10^{-25} m^5 V^{-2} mol^{-1}	(*Exptl.*: 12.3 (± 2.7) × 10^{-25} m^5 V^{-2} mol^{-1})

poly(hexamethylene terephthalate), polyester liquid crystals; $m = 6$ (hexamethylene)

Mendicuti, F.; Patel, B.; Viswanadhan, V. N.; Mattice, W. L. *Polymer* **1988**, *29*, 1669.

Bond Length [pm]	Valence Angles [°]	Torsion Angles [°]	ξ (for 300 K)	ξ_0	E_ξ [kJ·mol^{-1}]
C^{ar}-C^{ar} : 140	C^{ar}-C^{ar}-C^{ar}: 120	For bond 0: t : 180	σ_1 = 0.51	1	1.68
C^{ar}-C^* : 147	C^{ar}-C^{ar}-C^* : 120		σ_2 = 1.4	1	−0.84
		For bonds 1-7:			
C^*-O : 134	C^{ar}-C^*-O : 125	t : 180 g^+ : 60	σ_3 = 0.43	1	2.1
O-C : 144	C^*-O-C : 113	g^- : −60	σ_4 = 0.43	1	2.1
C-C : 153	O-C-C : 110		ω_1 = 0	1	∞
	C-C-C : 112		ω_2 = 0.39	1	2.35
			ω_3 = 0.034	1	8.4

$$U_1 = \begin{bmatrix} 1 & 2\sigma_1 \end{bmatrix} \quad U_8^T = \begin{bmatrix} 1 & 1 \end{bmatrix} \quad U_i = \begin{bmatrix} 1 & 2\sigma_i \\ 1 & \sigma_i(1+\omega_i) \end{bmatrix} \quad i = 2\text{-}7$$

rows / columns:
t / t, $g^+ + g^-$ t / t, $g^+ + g^-$ t, $g^+ + g^-$ / t, $g^+ + g^-$

Comments:
Excimer formation is studied in polyesters in which the repeating unit is represented by the formula AB_m, where A is *p*-OOC−Ph−COO, B is CH_2 and $m = 2,...,6$. In dichloroethane, dioxane and ethyl acetate, the ratio of excimer to monomer emission, I_D / I_M, exhibits an odd-even effect with the largest value seen in the polyester with $m = 3$. The conformations conducive to excimer formation by nearest neighbor aromatic rings are identified with the aid of an RIS analysis of Ph−COO-B_m−OOC−Ph. A single species completely dominates the excimer formation when $m = 2$. As m increases more species can adopt the excimer-forming geometry. When $m = 6$, no conformation accounts for more than ca. 1/7 of the excimers present. The equilibrium RIS analysis can rationalize the occurrence of the largest value of I_D / I_M at $m = 3$ and the existence of an odd-even effect at $m = 2,...,5$. However, it overestimates the importance of the excimers when $m = 6$.

di(2-naphthyl)succinate

Mendicuti, F.; Saiz, E.; Zúñiga, I.; Patel, B.; Mattice, W. L. *Polymer* **1992**, *33*, 2031.

Bond Length [pm]		Valence Angles [°]		Torsion Angles [a)] [°]		ξ (for 298 K)	ξ_o	E_ξ [kJ·mol^{-1}]
C^{ar}-C^{ar}	: 140	C^{ar}-C^{ar}-C^{ar}	: 120	For C^{ar}-O bonds:		σ = 5.4 to 0.034	1	-4.2 to 8.4
				t^+ : 120				
C^{ar}-O	: 130	C^{ar}-C^{ar}-O	: 120	t^- : -120		σ_1 = 0.59	1	1.3
				c^+ : 60				
C^*-O	: 132	C^{ar}-O-C^*	: 124	c^- : -60		ω_1 = 0.034	1	8.4
C^*-CH_2	: 150	O-C^*-CH_2	: 118	For C-C, C^*-C:				
				t : 180 ($\pm\Delta\phi$)				
CH_2-CH_2	: 153	C^*-CH_2-CH_2	:109	g^+ : 60 ($\pm\Delta\phi$)				
				g^- : -60 ($\pm\Delta\phi$)				
		CH_2-CH_2-CH_2	: 112					

a) Three states denoted by t, g^+, g^- are used for each C^*−C and C−C bond. Each of these states is assigned three dihedral angles. The central of these three dihedral angles is 180° and ±60°. The other two dihedral angles for each state are located ±Δϕ from the central angle. Differences in conformational energy produced by rotations of ±Δϕ are ignored. The purpose of Δϕ is to mimic the effect of the rapid torsional oscillation that can occur within each rotational isomeric potential well. The value is uniformly assigned as Δϕ = 20°.

Comments:
The prospensity for the formation of bends in the flexible spacer in polyesters containing naphthyl units is examined by the study of the fluorescence of a series of diesters. The dependence of the degree of intramolecular excimer formation on the length of the aliphatic spacer, under circumstances where the dynamics of rotational isomerism in the flexible spacer is suppressed, is evaluated by extrapolation of the measurements to infinite viscosity η. The extrapolated results exhibit an odd-even effect, with the more intense excimer emission being observed when the number of methylene groups is odd. The odd-even effect is rationalized by an RIS analysis of the diesters.

di(2-naphthyl)glutarate

Mendicuti, F.; Saiz, E.; Zúñiga, I.; Patel, B.; Mattice, W. L. *Polymer* **1992**, *33*, 2031.

Bond Length [pm]	Valence Angles [°]	Torsion Angles a) [°]	ξ (for 298 K)	ξ_o	E_ξ [kJ · mol^{-1}]
C^{ar}-C^{ar} : 140	C^{ar}-C^{ar}-C^{ar} : 120	For C^{ar}-O bonds: t^+ : 120	σ = 5.4 − 0.034	1	− 4.2 to 8.4
C^{ar}-O : 130	C^{ar}-C^{ar}-O : 120	t^- : − 120 c^+ : 60	σ_1 = 0.59	1	1.3
C^*-O : 132	C^{ar}-O-C^* : 122.9	c^- : − 60	ω_1 = 0.36	1	2.5
C^*-CH_2 : 150	O-C^*-CH_2 : 118.3	For C-C, C^*-C: t : 180 ($\pm\Delta\phi$) g^+ : 60 ($\pm\Delta\phi$) g^- : − 60 ($\pm\Delta\phi$)	ω_2 = 0.26	1	3.3
CH_2-CH_2 : 153	C^*-CH_2-CH_2 : 109.9				
	CH_2-CH_2-CH_2 : 112				

a) Three states denoted by t, g^+, g^- are used for each C^*−C and C−C bond. Each of these states is assigned three dihedral angles. The central of these three dihedral angles is 180° and ±60°. The other two dihedral angles for each state are located ±Δϕ from the central angle. Differences in conformational energy produced by rotations of ±Δϕ are ignored. The purpose of Δϕ is to mimic the effect of the rapid torsional oscillation that can occur within each rotational isomeric potential well. The value is uniformly assigned as Δϕ = 20°.

Comments:
The prospensity for the formation of bends in the flexible spacer in polyesters containing naphthyl units is examined by the study of the fluorescence of a series of diesters. The dependence of the degree of intramolecular excimer formation on the length of the aliphatic spacer, under circumstances where the dynamics of rotational isomerism in the flexible spacer is suppressed, is evaluated by extrapolation of the measurements to infinite viscosity η. The extrapolated results exhibit an odd-even effect, with the more intense excimer emission being observed when the number of methylene groups is odd. The odd-even effect is rationalized by an RIS analysis of the diesters.

di(2-naphthyl)adipate

Mendicuti, F.; Saiz, E.; Zúñiga, I.; Patel, B.; Mattice, W. L. *Polymer* **1992**, *33*, 2031.

Bond Length [pm]	Valence Angles [°]	Torsion Angles a) [°]	ξ (for 298 K)	ξ_0	E_ξ [kJ · mol^{-1}]
C^{ar}-C^{ar} : 140	C^{ar}-C^{ar}-C^{ar} : 120	For C^{ar}-O bonds: t^+ : 120	σ = 5.4 - 0.034	1	– 4.2 to 8.4
C^{ar}-O : 130	C^{ar}-C^{ar}-O : 120	t^- : – 120 c^+ : 60	σ_1 = 0.59	1	1.3
C^*-O : 132	C^{ar}-O-C^* : 122.9	c^- : – 60	σ_2 = 0.43	1	2.1
C^*-CH_2 : 150	O-C^*-CH_2 : 118.3	For C-C, C^*-C: t : 180 ($\pm \Delta\phi$)	ω_1 = 0.36	1	2.5
CH_2-CH_2 : 153	C^*-CH_2-CH_2 : 109.9	g^+ : 60 ($\pm \Delta\phi$) g^- : – 60 ($\pm \Delta\phi$)	ω_2 = 0	1	∞
	CH_2-CH_2-CH_2 : 112				

a) Three states denoted by t, g^+, g^- are used for each C^*—C and C—C bond. Each of these states is assigned three dihedral angles. The central of these three dihedral angles is 180° and ±60°. The other two dihedral angles for each state are located $\pm \Delta\phi$ from the central angle. Differences in conformational energy produced by rotations of $\pm \Delta\phi$ are ignored. The purpose of $\Delta\phi$ is to mimic the effect of the rapid torsional oscillation that can occur within each rotational isomeric potential well. The value is uniformly assigned as $\Delta\phi$ = 20°.

Comments:

The prospensity for the formation of bends in the flexible spacer in polyesters containing naphthyl units is examined by the study of the fluorescence of a series of diesters. The dependence of the degree of intramolecular excimer formation on the length of the aliphatic spacer, under circumstances where the dynamics of rotational isomerism in the flexible spacer is suppressed, is evaluated by extrapolation of the measurements to infinite viscosity η. The extrapolated results exhibit an odd-even effect, with the more intense excimer emission being observed when the number of methylene groups is odd. The odd-even effect is rationalized by an RIS analysis of the diesters.

di(2-naphthyl)pimelate

Mendicuti, F.; Saiz, E.; Zúñiga, I.; Patel, B.; Mattice, W. L. *Polymer* **1992**, *33*, 2031.

Bond Length [pm]	Valence Angles [°]	Torsion Angles a) [°]	ξ (for 298 K)	ξ_o	E_ξ [kJ·mol^{-1}]
C^{ar}-C^{ar} : 140	C^{ar}-C^{ar}-C^{ar} : 120	For C^{ar}-O bonds: t^+ : 120	σ = 5.4 - 0.034	1	-4.2 to 8.4
C^{ar}-O : 130	C^{ar}-C^{ar}-O : 120	t^- : -120 c^+ : 60	σ_1 = 0.59	1	1.3
C^*-O : 132	C^{ar}-O-C^* : 122.9	c^- : -60	σ_2 = 0.43	1	2.1
C^*-CH_2 : 150	O-C^*-CH_2 : 118.3	For C-C, C^*-C: t : 180 ($\pm\Delta\phi$)	ω = 0.034	1	8.4
CH_2-CH_2 : 153	C^*-CH_2-CH_2 : 109.9	g^+ : 60 ($\pm\Delta\phi$) g^- : -60 ($\pm\Delta\phi$)	ω_1 = 0.36	1	2.5
	CH_2-CH_2-CH_2 : 112		ω_2 = 0	1	∞

a) Three states denoted by t, g^+, g^- are used for each C^*-C and C-C bond. Each of these states is assigned three dihedral angles. The central of these three dihedral angles is 180° and ±60°. The other two dihedral angles for each state are located $\pm\Delta\phi$ from the central angle. Differences in conformational energy produced by rotations of $\pm\Delta\phi$ are ignored. The purpose of $\Delta\phi$ is to mimic the effect of the rapid torsional oscillation that can occur within each rotational isomeric potential well. The value is uniformly assigned as $\Delta\phi$ = 20°.

Comments:

The prospensity for the formation of bends in the flexible spacer in polyesters containing naphthyl units is examined by the study of the fluorescence of a series of diesters. The dependence of the degree of intramolecular excimer formation on the length of the aliphatic spacer, under circumstances where the dynamics of rotational isomerism in the flexible spacer is suppressed, is evaluated by extrapolation of the measurements to infinite viscosity η. The extrapolated results exhibit an odd-even effect, with the more intense excimer emission being observed when the number of methylene groups is odd. The odd-even effect is rationalized by an RIS analysis of the diesters.

di(2-naphthyl)suberate

Mendicuti, F.; Saiz, E.; Zúñiga, I.; Patel, B.; Mattice, W. L. *Polymer* **1992**, *33*, 2031.

Bond Length [pm]	Valence Angles [°]	Torsion Angles [a)] [°]	ξ (for 298 K)	ξ_0	E_ξ [kJ·mol^{-1}]
C^{ar}-C^{ar} : 140	C^{ar}-C^{ar}-C^{ar} : 120	For C^{ar}-O bonds:	σ = 5.4 to 0.034	1	–4.2 to 8.4
		t^+ : 120			
C^{ar}-O : 130	C^{ar}-C^{ar}-O : 120	t^- : –120	σ_1 = 0.59	1	1.3
		c^+ : 60			
C^*-O : 132	C^{ar}-O-C^* : 122.9	c^- : –60	σ_2 = 0.43	1	2.1
C^*-CH$_2$: 150	O-C^*-CH$_2$: 118.3	For C-C, C^*-C:	ω = 0.034	1	8.4
		t : 180 ($\pm\Delta\phi$)			
CH$_2$-CH$_2$: 153	C^*-CH$_2$-CH$_2$: 109.9	g^+ : 60 ($\pm\Delta\phi$)	ω_1 = 0.36	1	2.5
		g^- : –60 ($\pm\Delta\phi$)			
	CH$_2$-CH$_2$-CH$_2$: 112		ω_2 = 0	1	∞

a) Three states denoted by t, g^+, g^- are used for each C^*—C and C—C bond. Each of these states is assigned three dihedral angles. The central of these three dihedral angles is 180° and ±60°. The other two dihedral angles for each state are located $\pm\Delta\phi$ from the central angle. Differences in conformational energy produced by rotations of $\pm\Delta\phi$ are ignored. The purpose of $\Delta\phi$ is to mimic the effect of the rapid torsional oscillation that can occur within each rotational isomeric potential well. The value is uniformly assigned as $\Delta\phi$ = 20°.

Comments:

The prospensity for the formation of bends in the flexible spacer in polyesters containing naphthyl units is examined by the study of the fluorescence of a series of diesters. The dependence of the degree of intramolecular excimer formation on the length of the aliphatic spacer, under circumstances where the dynamics of rotational isomerism in the flexible spacer is suppressed, is evaluated by extrapolation of the measurements to infinite viscosity η. The extrapolated results exhibit an odd-even effect, with the more intense excimer emission being observed when the number of methylene groups is odd. The odd-even effect is rationalized by an RIS analysis of the diesters.

polyesters from 2,6-naphthalene dicarboxylic acid, y = 1-5

Mendicuti, F.; Saiz, E.; Mattice, W. L. *Polymer* **1992**, *33*, 4908.

Bond Length [pm]	Valence Angles [°]	Torsion Angles [°] [a]		
C^{ar}-C^{ar} : 140	C^{ar}-C^{ar}-C^{ar} : 120	For C^{ar}-C^* bonds:		
		t :	180	
C^{ar}-C^* : 144	C^{ar}-C^{ar}-C^* : 120	c :	0	
C^*-O : 134	C^{ar}-C^*-O : 119.1	For C-C, O-C:		
		t :	180 ($\pm\Delta\phi$)	
CH_2-CH_2 : 153	C^*-O-CH_2 : 114.4	g^+ :	60 ($\pm\Delta\phi$)	
		g^- :	$-$60 ($\pm\Delta\phi$)	
C^*-CH_2 : 151	O-CH_2-CH_2 : 110.0			
	CH_2-CH_2-CH_2 : 112	Statistical weight parameters were those developed for the corresponding diesters (**V 070 - V 074**)		

[a] Three states denoted by t, g^+, g^- are used for each C^*—C and C—C bond. Each of these states is assigned three dihedral angles. The central of these three dihedral angles is 180° and $\pm 60°$. The other two dihedral angles for each state are located $\pm\Delta\phi$ from the central angle. Differences in conformational energy produced by rotations of $\pm\Delta\phi$ are ignored. The purpose of $\Delta\phi$ is to mimic the effect of the rapid torsional oscillation that can occur within each rotational isomeric potential well. The value is uniformly assigned as $\Delta\phi = 20°$.

Comments:
The fluorescence is measured in dilute solution and in glassy PMMA for polyesters in which 2,6-naphthalene dicarboxylate is the rigid unit, and $(CH_2)_{y+1}$ is the flexible spacer. The anisotropy in the rigid medium demonstrates the existence of intramolecular energy migration, which becomes more important as y decreases from 5 to 1. The Förster radius is about 12 Å in the bichromophoric compounds and 14 Å in the polyesters.

V 076

poly{2,2-bis[4-(2-hydroxyethoxy)phenyl]-propane adipate}, PDA

Guzmán, J.; Riande, E.; Salvador, R.; de Abajo, J. *Macromolecules* **1991**, *24*, 5357.

Bond Length [pm]	Valence Angles [°]	Torsion Angles [°]	ξ (for 303 K)	ξ_0	E_ξ [kJ·mol^{-1}]
C-C : 153	O-C*-C : 144	For bonds 6-9:	$\sigma = 0.22$	1	3.77
C-C* : 151	C*-O-C : 111	g$^+$: 45	$\sigma_1 = 3.78$	1	−3.35
		t : 135			
		t$^-$: −135			
O-C* : 130	others: 111.5	g$^-$: −45	$\sigma_2 = 0.61$	1	1.26
C-O : 143		For bonds 2,13:	$\sigma_\alpha = 1 - 0.31$	1	0 − 5.02
		t : 180			
l$_{vb}$: 300			$\sigma_3 = 0.61 - 0.31$	1	1.26 − 2.93
		For bond 10:			
		t : 180	$\sigma_4 = 0.43$	1	2.1
		g$^+$: 60			
		g$^-$: −60			
		For other bonds:			
		t : 180			
		g$^+$: 70			
		g$^-$: −70			

$$U_6 = \begin{bmatrix} 1 & 1 & 1 \\ 0 & 1 & 0 \\ 1 & 0 & 1 \end{bmatrix} \quad U_7 = \begin{bmatrix} 1 & 1 & 1 \\ 1 & 1 & 1 \\ 1 & 1 & 1 \end{bmatrix} \quad U_8 = \begin{bmatrix} 1 & 0 & 1 & 0 \\ 0 & 1 & 0 & 1 \\ 1 & 0 & 1 & 0 \\ 0 & 1 & 0 & 1 \end{bmatrix}$$

$$U_9 = \begin{bmatrix} 1 & 1 & 1 \\ 1 & 1 & 1 \\ 1 & 1 & 1 \\ 1 & 1 & 1 \end{bmatrix} \quad U_{10} = \begin{bmatrix} 1 & 0 & \sigma \\ 1 & \sigma & 0 \\ 1 & 0 & \sigma \\ 1 & \sigma & 0 \end{bmatrix}$$

rows and/or columns:
g$^+$, t$^+$, t$^-$, g$^-$
and/or:
t, g$^+$, g$^-$

Comments:
The dipole moment ratio and the temperature coefficient of both the dipole moment and the unperturbed dimensions of the polyesters PDA and PDS are measured. The experimental value of $d(\ln \langle r^2 \rangle_0)/dT$ shows an anomalous dependence on the elongation ratio of the networks at which the thermoelastic measurements are performed. Although the rotational states scheme gives a fairly good account of the polarity of the chains, it fails in reproducing the experimental values of $d(\ln \langle r^2 \rangle_0)/dT$; the causes of this disagreement are discussed.

Calcd. quantities: $\langle r^2 \rangle_0 / nl^2$ $\qquad d(\ln \langle r^2 \rangle)/dT \qquad \langle \mu^2 \rangle / xm^2 \qquad d(\ln \langle \mu^2 \rangle)/dT$

V 077

poly{2,2-bis[4-(2-hydroxyethoxy)phenyl]-propane sebacate}, PDS

Guzmán, J.; Riande, E.; Salvador, R.; de Abajo, J. *Macromolecules* **1991**, *24*, 5357.

Bond Length [pm]	Valence Angles [°]	Torsion Angles [°]	ξ (for 303 K)	ξ_0	E_ξ [kJ · mol^{-1}]
C-C : 153	O-C*-C : 144	For bonds 6-9: g^+ : 45	$\sigma = 0.22$	1	3.77
C-C* : 151	C*-O-C : 111	t^+ : 135 t^- : -135	$\sigma_1 = 3.78$	1	-3.35
O-C* : 130	others: 111.5	g^- : -45	$\sigma_2 = 0.61$	1	1.26
C-O : 143		For bonds 2,13: t : 180	$\sigma_\alpha = 1 - 0.31$	1	0 - 5.02
l_{vb} : 300			$\sigma_3 = 0.61 - 0.31$	1	1.26 - 2.93
		For bond 10: t : 180 g^+ : 60 g^- : -60	$\sigma_4 = 0.43$	1	2.1
		For other bonds: t : 180 g^+ : 70 g^- : -70			

$$U_6 = \begin{bmatrix} 1 & 1 & 1 & 1 \\ 0 & 1 & 0 & 1 \\ 1 & 0 & 1 & 0 \end{bmatrix} \quad U_7 = \begin{bmatrix} 1 & 1 & 1 & 1 \\ 1 & 1 & 1 & 1 \\ 1 & 1 & 1 & 1 \end{bmatrix} \quad U_8 = \begin{bmatrix} 1 & 0 & 1 & 0 \\ 0 & 1 & 0 & 1 \\ 1 & 0 & 1 & 0 \\ 0 & 1 & 0 & 1 \end{bmatrix}$$

$$U_9 = \begin{bmatrix} 1 & 1 & 1 & 1 \\ 1 & 1 & 1 & 1 \\ 1 & 1 & 1 & 1 \\ 1 & 1 & 1 & 1 \end{bmatrix} \quad U_{10} = \begin{bmatrix} 1 & 0 & \sigma \\ 1 & \sigma & 0 \\ 1 & 0 & \sigma \\ 1 & \sigma & 0 \end{bmatrix}$$

rows and/or columns: g^+, t^+, t^-, g^-
and/or: t, g^+, g^-

Comments:
The dipole moment ratio and the temperature coefficient of both the dipole moment and the unperturbed dimensions of the polyesters PDA and PDS are measured. The experimental value of $d\,(\ln <r^2>_o)\,/\,dT$ shows an anomalous dependence on the elongation ratio of the networks at which the thermoelastic measurements are performed. Although the rotational states scheme gives a fairly good account of the polarity of the chains, it fails in reproducing the experimental values of $d\,(\ln <r^2>_o)\,/\,dT$; the causes of this disagreement are discussed.

Calcd. quantities: $<r^2>_o\,/\,nl^2$ $d\,(\ln <r^2>)\,/\,dT$ $<\mu^2>\,/\,xm^2$ $d\,(\ln <\mu^2>)\,/\,dT$

liquid crystalline polyesters, type I

Abe, A. *Macromolecules* **1984**, *17*, 2280.

Bond Length [pm]	Valence Angles [°]	Torsion Angles for C^*-C^1; [°]	
O-C* : 135	C^{ar}-O-C* : 116.7	t	: 180
		g^{+*}	: 122.7
C*-C : 153	C-C-C : 112	g^+	: 57.3
		c	: 0
C-C : 153	O-C*-C : 111.4	g^{-*}	: −122.7
		g^-	: −57.3
	C*-C-C : 112		

Parameters required for the description of polyesters I are taken from a recent paper (Abe, A. *J. Am. Chem. Soc.* **1984**, *106*, 14) which dealt with the dipole moments of dialkyl esters of dicarboxylic acids. Since the ester groups are all assumed to be in the *trans* configuration, short-range interactions between consecutive rigid cores are unimportant. As for the rotation around the O-C*—C-C bond, the six-state scheme (termed model I in the above paper) is employed. The statistical weight parameter α representing the relative importance of the *reversed* ester conformations with respect to the normal ones is set equal to unity. The three-state scheme (termed model II) proposed alternatively in the above reference is examined for chains with $n = 5$ and 6 for comparison. In this model the C*O*/CC eclipsed form is assumed to be intrinsically more stable than the C*O*/CH form: a stabilzation energy E(β) of $5.0 \text{ kJ} \cdot \text{mol}^{-1}$ is adopted.

When the number y of intervening methylene groups is less than three in polyester I, the two adjoining ester groups may interact with each other in close proximity. The interaction energies are very much dependent on the rotation around the neighboring C*—C and C—C* bonds as well. For $y \geq 3$, therefore, statistical weights are deduced for the individual conformations of the residue from the calculated total energies. In these cases, contributions from each of the constituent bonds are not estimated separately. For polyester I with $n \geq 4$, the conventional matrix multiplication method is employed.

V

The following model requires two pages.
Therefore, this side has been left blank

V 079   (see also V 078)

liquid crystalline polyesters, type II

y = 1 to 8 (see also the following page)

Abe, A. *Macromolecules* **1984**, *17*, 2280.

Bond Length [pm]	Valence Angles [°]	Torsion Angles [°]	ξ (for 500 K)	ξ_0	E_ξ [kJ·mol^{-1}]
O-C* : 137	C^{ar}-C*-O: 110.9	Bond type 1:	$\sigma_1 = 0.71$	1	1.4
	t : 180	$\sigma_2 = 1.7$	1	-2.2
C-O : 144	C-C-C : 112	g$^+$: 76	$\sigma_2' = 1.0$	1	0
		g$^-$: -76	$\sigma_3 = 0.6$	1	2.1
C-C : 153	C*-O-C : 118.3	Bond type 2:	$\omega_1 = 0.1$	1	9.6
		t : 180	$\omega_1' = 0.0$	1	∞
	O-C-C : 110	g$^+$: 60	$\omega_2 = 0.4$	1	3.8
		g$^-$: -60	$\omega_2' = 0.6$	1	2.1
		Bond type 3:	$\omega_3 = 0.13$	1	8.5
		t : 180			
		g$^+$: 68.5			
		g$^-$: -68.5			

Geometrical parameters employed for polyesters II are those used in the analysis of aromatic polyesters by *Erman, Flory,* and *Hummel* (*Macromolecules* **1980**, *13*, 484). Statistical weight matrices may be formulated for any given residue by the usual procedure.

Comments:
Conformational analysis is performed on semiflexible polyesters having repeat units such as $-(X-O-CO-(CH_2)_y-CO-O)-$ (type I) and $-(X-CO-O-(CH_2)_y-O-CO)-$ (type II). These polymers are known to exhibit thermotropic mesophases when (aromatic) rigid cores X are sufficiently anisotropic. Spatial orientations of a given core are elucidated in a Cartesian coordinate system fixed to the preceding core. The angle θ defined by unit vectors attached to two successive rigid cores is evaluated for each conformation of the intervening flexible segment. When the number of methylene units y in the flexible segment is even, the angle θ is found to be distributed in the range 0 - 30° (30-40%) and 85 - 130° (60-70%). For polymers with y = odd, the major portion of the calculated angle θ is located in the region 50-90°, and to a varying degree (0-20%), orientations are also permitted in the range θ > 160°. Only the y = even polymers conform to the concept of an ordinary nematic ordering. Based on these results, an explanation is offered for the observed odd-even oscillation in the entropy change ΔS_{ni} and the isotropization temperature.

V 080   (see also **V 081** and **V 082**)

thermotropic polyesters: polyester I ; $y = 4, 5$

Yoon, D. Y.; Bruckner, S. *Macromolecules* **1985**, *18*, 651.

Bond Length [pm]	Valence Angles [°]	Torsion Angles [°]	ξ (for 500 K)	ξ_0	E_ξ [kJ · mol^{-1}]
$C^{ar}-C^{ar}$: 139	$C^{ar}-C^{ar}-C^{ar}$: 120	ϕ_1 : 180 ϕ^+_2 : 115	$\sigma = 0.60$	1	2.1
$C^{ar}-C$: 151	$C^{ar}-C^*-O$: 112	ϕ^-_2 : −115 ϕ^t_3 : 180	$\sigma_1 = 1.0$	1	0
C^*-O : 135	$C^{ar}-O-C^*$: 116	ϕ^c_3 : 0 ϕ_4 : 180	$\omega' = 0.60$	1	2.1
$C^{ar}-O$: 144	$C^{ar}-O-C$: 112		$\omega = 0.13$	1	8.4
C-C : 153	O-C*-C : 112	For OC-CC: t : 180 g$^+$: 60			
C-O : 144	O-C-C : 112	g$^-$: −60 For CC-CC:			
	C-C-C : 112	t : 180 g$^+$: 67.5			
	C*-O-C : 114	g$^-$: −67.5			

V 081   (see also **V 080** and **V 082**)

thermotropic polyesters: polyester II ; $y = 4, 5$

Yoon, D. Y.; Bruckner, S. *Macromolecules* **1985**, *18*, 651.

Bond Length [pm]	Valence Angles [°]	Torsion Angles [°]	ξ (for 500 K)	ξ_0	E_ξ [kJ·mol⁻¹]
C^{ar}-C^{ar} : 139	C^{ar}-C^{ar}-C^{ar} : 120	ϕ^+_1 : 115 ϕ^-_1 : −115	$\sigma = 0.60$	1	2.1
C^{ar}-C : 151	C-C-C : 112	ϕ_2 : 180 ϕ^+_3 : 144	$\sigma_3 = 0.36$	1	4.27
C^{ar}-N : 149	C^{ar}-O-C* : 116	ϕ^-_3 : −144 For CC-CC:	$\sigma_4 = 0.67$	1	1.63
C^{ar}-O : 144	O-C*-C : 112	t : 180 g^+ : 67.5	$\omega = 0.13$	1	8.4
C-C : 153	C^{ar}-N(O)=N : 112	g^- : −67.5 For OC*-CC:	$\psi = 1.71$	1	−2.22
N=N : 122	C^{ar}-N=N(O) : 115	t : 180 g^+ : 60 g^- : −60 For C*C-CC:			
C*-O : 135		t : 180 g^+ : 70 g^- : −70			

V 082   (see also V 080 and V 081)

thermotropic polyesters: polyester III ; $y = 4, 5$

Yoon, D. Y.; Bruckner, S. *Macromolecules* **1985**, *18*, 651.

Bond Length [pm]	Valence Angles [°]	Torsion Angles [°]	ξ (for 500 K)	ξ_0	E_ξ [kJ·mol^{-1}]
C^{ar}-C^{ar} : 139	C^{ar}-C^{ar}-C^{ar}: 120	ϕ_1 : 180	$\sigma = 0.60$	1	2.1
		ϕ_2 : 180			
C^{ar}-C : 151	C^{ar}-C*-O : 112	ϕ^+_3 : 144	$\sigma_1 = 1$ or 1.51	1	0 or -1.7
		ϕ^-_3 : -144			
C^{ar}-N : 149	C*-O-C : 114	For OC-CC:	$\sigma_2 = 0.70$	1	1.47
		t : 180			
N=N : 122	O-C-C : 112	g^+ : 60	$\omega = 0.60$	1	2.1
		g^- : -60			
C-C : 153	C^{ar}-N(O)=N: 112	For CC-CC:	$\omega' = 0.13$	1	8.4
		t : 180			
C-O : 144	C^{ar}-N=N(O): 115	g^+ : 67.5			
		g^- : -67.5			
C*-O : 135	C-C-C : 112	For C*O-CC:			
		t : 180			
		g^+ : 76			
		g^- : -76			

Comments:
The distribution of chain sequence extension, calculated by using RIS models, is compared with isotropic-nematic transition characteristics for a number of thermotropic polymers comprising rigid groups connected by polymethylene spacers. The distribution depends strongly not only on the odd-even character of the number of methylene units of the spacers, but also on the specific groups (or atoms) connected at the ends of polymethylene spacers.

Calcd. quantities: Distribution of chain sequence extension   Internal energy as a function of extension
                        Orientational correlations of rigid units   Values of enthalpy and entropy change at isotropic-nematic transitions

The following model requires two pages.
Therefore, this side has been left blank

V 083

poly(oxymethylene-1,4-*trans*-cyclohexylenemethyleneoxysebacoyl), PTCS

Riande, E.; Guzmàn, J.; de la Campa, J. G.; de Abajo, J. *Macromolecules* **1985**, *18*, 1583.

Bond Length [pm]	Valence Angles [°]	Torsion Angles [°]	ξ (for 303 K)	ξ_0	E_ξ [kJ·mol^{-1}]
C-C : 153	C*-O-CH$_2$: 113	For bonds 1,4,7: t : 180	$\sigma = 0.43$	1	2.1
C-O : 143	O-CH$_2$-CH$_2$: 110		$\sigma' = \sigma'' = 1$	1	0
C*-O : 133	O-C*-CH$_2$: 114	For O-CH$_2$: t : 180 g$^+$: 75	$\sigma_A = 0.29$	1	3.1
l$_{vb}$: 297	CH$_2$-CH$_2$-CH$_2$: 110	g$^-$: −75	$\sigma_B = 2.49$	1	−2.3
In cyclohexyl: C-C : 154	In cyclohexyl: C-C-C : 111.5	For C-C: t : 180 g$^+$: 60 g$^-$: −60	$\omega = 0.1$	1	5.8
	θ_e : 148.5 θ_a : 101.7				

$$U_1 = U_4 = U_7 = \begin{bmatrix} 1 & 0 & 0 \\ 1 & 0 & 0 \\ 1 & 0 & 0 \end{bmatrix} \quad U_2 = U_6 = \begin{bmatrix} 1 & \sigma_A & \sigma_A \\ 1 & \sigma_A & 0 \\ 1 & 0 & \sigma_A \end{bmatrix} \quad U_3 = U_5 = \begin{bmatrix} 1 & \sigma_B & \sigma_B \\ 1 & \sigma_B & 0 \\ 1 & 0 & \sigma_B \end{bmatrix}$$

$$U_8 = \begin{bmatrix} 1 & \sigma' & \sigma' \\ 1 & \sigma' & 0 \\ 1 & 0 & \sigma' \end{bmatrix} \quad U_9 = U_{15} = \begin{bmatrix} 1 & \sigma'' & \sigma'' \\ 1 & \sigma'' & \sigma''\omega \\ 1 & \sigma''\omega & \sigma'' \end{bmatrix} \quad U_{10} = \begin{bmatrix} 1 & \sigma & \sigma \\ 1 & \sigma & \sigma\omega \\ 1 & \sigma\omega & \sigma \end{bmatrix}$$

$$U_{11} = \ldots = U_{14} = \begin{bmatrix} 1 & \sigma & \sigma \\ 1 & \sigma & 0 \\ 1 & 0 & \sigma \end{bmatrix} \quad U_{16} = \begin{bmatrix} 1 & \sigma' & \sigma' \\ 1 & \sigma' & \sigma'\omega \\ 1 & \sigma'\omega & \sigma' \end{bmatrix}$$

rows and columns: t, g$^+$, g$^-$

Comments:
PCCS and PTCS are synthesized by condensation of the corresponding *cis* and *trans* isomers of 1,4-cyclohexanedimethanol with sebacic acid. Values of the mean-square dipole moments of both polyesters are determined from dielectric constant measurements on dilute solutions of the polymers in benzene. Theoretical calculations carried out with the RIS model give values of $d (\ln <r^2>_0) / dT$ in very good agreement with the experimental results. Fair agreement between theory and experiment is also found in the case of the dipole moments of the chains.

Calcd. quantities: $<r^2>_0 / nl^2$ $d (\ln <r^2>_0) / dT$ $<\mu^2> / xm^2$ $d (\ln <\mu^2>) / dT$

V 084

poly(oxymethylene-1,4-*cis*-cyclohexylenemethyleneoxysebacoyl), PCCS

Riande, E.; Guzmàn, J.; de la Campa, J. G.; de Abajo, J. *Macromolecules* **1985**, *18*, 1583.

Bond Length [pm]	Valence Angles [°]	Torsion Angles [°]	ξ (for 303 K)	ξ_0	E_ξ [kJ·mol^{-1}]				
C-C : 153	C*-O-CH$_2$: 113	For bonds 1,7: t : 180	$\sigma = 0.43$	1	2.1	$U_1 = U_4 = U_7 = \begin{bmatrix} 1 & 0 & 0 \\ 1 & 0 & 0 \\ 1 & 0 & 0 \end{bmatrix}$	$U_2 = U_6 = \begin{bmatrix} 1 & \sigma_A & \sigma_A \\ 1 & \sigma_A & 0 \\ 1 & 0 & \sigma_A \end{bmatrix}$	$U_3 = \begin{bmatrix} 1 & \sigma_B & \sigma_B \\ 1 & \sigma_B & 0 \\ 1 & 0 & \sigma_B \end{bmatrix}$	
C-O : 143	O-CH$_2$-CH$_2$: 110	For O-CH$_2$: t : 180 g$^+$: 75	$\sigma' = \sigma'' = 1$	1	0				
C*-O : 133	O-C*-CH$_2$: 114	g$^-$: −75	$\sigma_A = 0.29$	1	3.1				
l_{vb} : 297	CH$_2$-CH$_2$-CH$_2$: 110		$\sigma_B = 2.49$	1	−2.3	$U_5 = \begin{bmatrix} 0 & 1 & 1 \\ 0 & 0 & 0 \\ 0 & 0 & 0 \end{bmatrix}$	$U_8 = \begin{bmatrix} 1 & \sigma' & \sigma' \\ 1 & \sigma' & 0 \\ 1 & 0 & \sigma' \end{bmatrix}$	$U_9 = U_{15} = \begin{bmatrix} 1 & \sigma'' & \sigma'' \\ 1 & \sigma'' & \sigma''\omega \\ 1 & \sigma''\omega & \sigma'' \end{bmatrix}$	
In cyclohexyl:	In cyclohexyl:	For C-C: t : 180 g$^+$: 60 g$^-$: −60	$\omega = 0.1$	1	5.8				
C-C : 154	C-C-C : 111.5 θ_e : 148.5 θ_a : 101.7	For bond 4: c : 0				$U_{10} = \begin{bmatrix} 1 & \sigma & \sigma \\ 1 & \sigma & \sigma\omega \\ 1 & \sigma\omega & \sigma \end{bmatrix}$	$U_{11} = \ldots = U_{14} = \begin{bmatrix} 1 & \sigma & \sigma \\ 1 & \sigma & 0 \\ 1 & 0 & \sigma \end{bmatrix}$	$U_{16} = \begin{bmatrix} 1 & \sigma' & \sigma' \\ 1 & \sigma' & \sigma'\omega \\ 1 & \sigma'\omega & \sigma' \end{bmatrix}$	General matrix indices: rows and columns: t, g$^+$, g$^-$ For U_4 and U_5, rows and columns are, respectively: c, g$^+$, g$^-$

Comments:
PCCS and PTCS are synthesized by condensation of the corresponding *cis* and *trans* isomers of 1,4-cyclohexanedimethanol with sebacic acid. Values of the mean-square dipole moments of both polyesters are determined from dielectric constant measurements on dilute solutions of the polymers in benzene. Theoretical calculations carried out with the RIS model give values of $d(\ln \langle r^2 \rangle_0)/dT$ in very good agreement with the experimental results. Fair agreement between theory and experiment is also found in the case of the dipole moments of the chains.

Calcd. quantities: $\langle r^2 \rangle_0 / nl^2$ $d(\ln \langle r^2 \rangle_0)/dT$ $\langle \mu^2 \rangle / xm^2$ $d(\ln \langle \mu^2 \rangle)/dT$

poly(cis/trans-1,4-cyclohexane dimethanol-alt-formaldehyde), PCDO

Riande, E.; Guzmàn, J. *J. Polym. Sci., Polym. Phys. Ed.* **1985**, *23*, 1031.

Bond Length [pm]	Valence Angles [°]	Torsion Angles a) [°]	ξ (for 298 K)	ξ_o	E_ξ [kJ·mol^{-1}]
C-C (CH): 154	C-C-C(CH):111.5	t : 180	σ_e = 1.96	1	−1.67
C-C : 153	C-O-C : 110	g$^+$: 60	σ_a = 862.92	1	−16.75
C-O : 143	O-C-O : 110	g$^-$: −60	σ' = 0.22	1	3.77
l_{vb} : 297	θ_e : 148.5	or: ϕ : 0, 180	σ'' = 6.43	1	−4.61
	θ_a : 101.7		ω = 0.51	1	1.67

a) For the pair of bonds i-2, i-3 the situation depends on the two substitutions of the ring: if they are equivalent (eq-eq or ax-ax), $\phi = 0°$, whereas if they are either eq-ax or ax-eq, $\phi = 180°$.

Comments:
Theoretical analyses, performed using the RIS model, prove that the only parameters that have an appreciable effect on the calculated values of Δa are those concerning the cyclohexane ring. Conformational energies, geometrical parameters, and contributions to the optical anisotropies from the oxymethylene oxide have no noticable effect on the value of Δa calculated for the polymer. The theoretical values of Δa are roughly one order of magnitude lower than the experimental results.

ether type dimer liquid crystals [a]

Abe, A.; Furuya, H.; Yoon, D. Y. *Mol. Cryst. Liq. Cryst.* **1988**, *159*, 151.

Bond Length [pm]	Valence Angles [°]	Torsion Angles [°]	ξ (for 409 K)	ξ_0	E_ξ [kJ·mol^{-1}]
C-C : 153	C^{ar}-O-C$_1$: 120.0	For bond 1:	σ_1 = 0.692	1	1.25
O-C^1 : 140	O-C$_1$-C$_2$: 112.0	t : 180	σ_2 = 0.259	1	4.53
	O-C-C : 112.0	For bond 2:			
		t : 180	σ_3 = 0.541	1	2.1
		g$^+$: 63			
	O-C$_1$-D: 109.2	g$^-$: −63	σ_4 = 0.335	1	3.67
	C-C-D : 109.2	For bond 3-9:	σ_5 = 0.541	1	2.1
		t : 180			
		g$^+$: 67.5	$\omega = \omega' = 0$	1	∞
		g$^-$: −67.5			

[a] The bond conformations in the isotropic state may be estimated by the conventional RIS model. Conformational energy parameters required in the Boltzmann expression are taken from literature: $E(\sigma_1)$ = 0, $E(\sigma_i)$ = 2.1 (for i = 2-5), $E(\omega)$ = 8.4, $E(\omega')$ = 2.1; all values are given in kJ·mol^{-1}.

Comments:
An attempt is made to elucidate molecular conformations of ether- and ester type liquid crystals (DLC) carrying deuterated soft-spacers by utilizing the information provided by the D-NMR method. The analysis is carried out according to the following steps: 1. All possible configurations are enumerated for a free molecule within the framework of the RIS approximation. 2. Configurations which do not conform to the nematic ordering are discarded. 3. For the nematic ensemble thus selected, conformational statistical weight factors assigned to the individual bond rotations are adjusted so as to reproduce the observed profile of the D-NMR spectrum. The molecular axis of a conformer is defined along the line connecting the centers of the terminal mesogenic cores.

Calcd. quantities: NMR quadrupolar splittings   Bond conformations of flexible spacers   *trans*-Fraction f_t $\Delta <E>_{NI,conf}$ $\Delta <S>_{NI,conf}$ Order parameter S_{ZZ}

V 087

ester type dimer liquid crystals [a]

Abe, A.; Furuya, H.; Yoon, D. Y. *Mol. Cryst. Liq. Cryst.* **1988**, *159*, 151.

Bond Length [pm]	Valence Angles [°]	Torsion Angles [°]	ξ (for 367 K)	ξ_o	E_ξ [kJ · mol^{-1}]
O-C* : 135	C^{ar}-O-C* : 116.7	For bond 1: t : 180	ζ = 0.259	1	4.22
C*-C$_1$: 153	O-C*-C$_1$: 111.4	g$^+$: 57.3 g$^-$: -57.3	σ_1 = 0.445	1	2.47
	C*-C$_1$-C$_2$: 112.0	For internal C-C: t : 180	σ_2 = 0.509	1	2.06
	C*-C$_1$-D : 109.2	g$^+$: 67.5 g$^-$: -67.5	σ_3 = 0.220	1	4.62
	D-C-D : 107.9	For bond a: $\phi_1^\pm$: ± 58.0 $\phi_2^\pm$: ± 112.0	σ_4 = 0.509	1	2.06
			σ_5 = 0.227	1	4.52
		For bond b: t : 180	ω = 0	1	∞

[a] The bond conformations in the isotropic state may be estimated by the conventional RIS model. Conformational energy parameters required in the Boltzmann expression are taken from literature: E(ζ) = 4.2, E(σ_i) = 2.1 (i = 1-5), E(ω) = 8.4; all values are given in kJ · mol^{-1}.

Comments:
An attempt is made to elucidate molecular conformations of ether- and ester type liquid crystals (DLC) carrying deuterated soft-spacers by utilizing the information provided by the D-NMR method. The analysis is carried out according to the following steps: 1. All possible configurations are enumerated for a free molecule within the framework of the RIS approximation. 2. Configurations which do not conform to the nematic ordering are discarded. 3. For the nematic ensemble thus selected, conformational statistical weight factors assigned to the individual bond rotations are adjusted so as to reproduce the observed profile of the D-NMR spectrum. The molecular axis of a conformer is defined along the line connecting the centers of the terminal mesogenic cores.

Calcd. quantities: NMR quadrupolar splittings   Bond conformations of flexible spacers   *trans*-Fraction f_t $\Delta<E>_{Nl,conf}$ $\Delta<S>_{Nl,conf}$ Order parameter S_{ZZ}

4-ethoxy-4'-biphenylyl cyanide, 2OCB

Abe, A.; Kimura, N.; Nakamura, M. *Makromol. Chem., Theory Simul.* **1992**, *1*, 401.

Bond Length [pm]	Valence Angles [°]	Torsion Angles [°]	ξ a) (for 363 K)	ξ_o	E_ξ [kJ · mol^{-1}]
C^{ar}-C^{ar} : 148	C^{ar}-C^{ar}-C^{ar} : 120	For C^{ar}-O bond: t : 180 c : 0	σ = 0.14	1	5.9 (± 0.4)
C^{ar}-C : 144	C^{ar}-O-C* : 120				
C-O : 142	O-CH$_2$-CH$_2$: 110.0	For O-C bond: t : 180 g$^+$: 80			
C-C : 153	O-C-D : 109.2	g$^-$: − 80 For C-C bond:			
C-D : 110	C-C-D : 111.0	t : 180 g$^+$: 60			
C^{ar}-O : 136		g$^-$: − 60			

a) The given value corresponds to the isotropic solution.

Comments:
The conformation of the ethoxy tail flanking the cyanobiphenyl core is studied in the liquid-crystalline phase as well as in the isotropic solution. The conventional RIS analysis yields the energy difference between the *gauche* and the *trans* state to be of the order of 6.7 to 5.9 kJ · mol^{-1}. The deuterated compound, 4-ethoxy-4'-biphenylyl-d_7 cyanide, is prepared and studied in the liquid-crystalline state by ^{2}H-NMR. The orientational order parameters of the mesogenic core are determined by the analysis of dipolar and quadrupolar splitting data due to aromatic deuterons. The quadrupolar splitting data are also obtained for the deuterated ethoxy tail. The conformation of the tail is elucidated according to the RIS simulation scheme. The fraction of the *trans* conformation f_t is estimated to be 0.79 around the nematic-isotropic transition temperature (T_{NI} = 91.1 °C). The *trans* fraction increases moderately toward an asymptotic value as temperature decreases.

V 089 (see also V 090)

4-n-pentyl-4'-cyanobiphenyls, 5CB.

Abe, A.; Furuya, H. *Mol. Cryst. Liq. Cryst.* **1988**, *159*, 99

Bond Length [pm]	Valence Angles [a] [°]	Torsion Angles [°]	ξ [a] (for 308 K)	ξ_0	E_ξ [kJ·mol^{-1}]
C-C : 153	C^{ar}-C-C : 112.0	For C-C bonds: t : 180	σ_1 = 0.129 / 0.118	1	5.24 / 5.47
C-D : 110	C-C-C : 112.0	g$^+$: 58.3 g$^-$: −58.3	σ_2 = 0.441 / 0.387	1	2.10 / 2.43
l_{vb} : 1000	C-C-D (CD$_2$):109.2		σ_3 = 0.119 / 0.118	1	5.45 / 5.47
	C-C-D (CD$_3$):116.0 / 111.0	For C^{ar}-C bond: t : 180			
	D-C-D : 107.9				

[a] Numerical values correspond to the nematic state (ω = 0). They are given as model I / model II, if necessary (see: *comments*).

Comments:
Deuterium NMR studies are carried out for 4-cyanobiphenyls carrying deuterated n-alkyl tails such as 4-n-pentyl-4'-cyanobiphenyl (5CB) and 4-n-octyl-4'-cyanobiphenyl (8CB). A shape analysis of D-NMR spectra is performed within the framework of the RIS approximation, statistical weight parameters assigned to the individual bond rotations being treated as empirical variables. Values of ξ thus deduced are used in the estimation of the distribution of conformers permitted in the nematic phase. Two models are examined for the definition of the molecular axis: Model I: The para-axis of the cyanobiphenyl group is defined as the Z axis of the molecular frame X,Y,Z. The same rule applies to all the conformers. Model II: The molecular axis is assumed to lie along the direction connecting the N atom of the cyano group and the terminal carbon of the hydrocarbon tail. Model II differs from model I in that the orientation of the molecular axis is variable depending on the configuration of the tail when observed within the coordinate system fixed to the mesogenic core.

V 090 (see also V 089)

4-n-octyl-4'-cyanobiphenyls, 8CB

Abe, A.; Furuya, H. *Mol. Cryst. Liq. Cryst.* **1988**, *159*, 99

Bond Length [pm]	Valence Angles [a] [°]	Torsion Angles [°]	ξ [a] (for 306 K)	ξ_0	E_ξ [kJ·mol^{-1}]
C-C : 153	C^{ar}-C-C : 112.0	For C-C bonds: t : 180	σ_1 = 0.089 / 0.184	1	6.15 / 4.31
C-D : 110	C-C-C : 112.0	g$^+$: 58.3 g$^-$: − 58.3	σ_2 = 0.446 / 0.298	1	2.05 / 3.08
l$_{vb}$: 1000	C-C-D (CD$_2$):109.2		σ_3 = 0.064 / 0.144	1	6.99 / 4.93
		For C^{ar}-C bond:			
	C-C-D (CD$_3$):124.0 / 111.0	t : 180	σ_4 = 0.323 / 0.275	1	2.88 / 3.28
	D-C-D : 107.9		σ_5 = 0.089 / 0.169	1	6.15 / 4.52
			σ_6 = 0.446 / 0.234	1	2.05 / 3.70

[a] Numerical values correspond to the nematic state (ω = 0). They are given as model I / model II, if necessary (see: *comments*).

Comments:
Deuterium NMR studies are carried out for 4-cyanobiphenyls carrying deuterated *n*-alkyl tails such as 4-n-pentyl-4'-cyanobiphenyl (5CB) and 4-n-octyl-4'-cyanobiphenyl (8CB). A shape analysis of D-NMR spectra is performed within the framework of the RIS approximation, statistical weight parameters assigned to the individual bond rotations being treated as empirical variables. Values of ξ thus deduced are used in the estimation of the distribution of conformers permitted in the nematic phase. Two models are examined for the definition of the molecular axis: Model I: The para-axis of the cyanobiphenyl group is defined as the Z axis of the molecular frame X,Y,Z. The same rule applies to all the conformers. Model II: The molecular axis is assumed to lie along the direction connecting the N atom of the cyano group and the terminal carbon of the hydrocarbon tail. Model II differs from model I in that the orientation of the molecular axis is variable depending on the configuration of the tail when observed within the coordinate system fixed to the mesogenic core.

Further calculations on **aromatic polyesters**, **polyamides** and **liquid crystalline systems**:

V 091   Saiz, E.; Hummel, J. P.; Flory, P. J.; Plavšić, M.. *J. Phys. Chem.* **1981**, *85*, 3211.

The directions of the dipole moments in esters RCOOR' in which R and R' are alkyl or aryl groups are deduced by analysis of experimental values of dipole moments of six compounds containing two polar groups, one or both of which are esters.

V 092   Samulski, E. T.; Dong, R. Y. *J. Chem. Phys.* **1982**, *77*, 5090.

The scheme for computing deuterium quadrupolar splittings of a flexible solute in a nematic solvent is extended to study chain ordering in the 4-*n*-alkyl-4'-cyanobiphenyl liquid crystals (*n*-CB). All possible configurations {ϕ} of the *n*-CB molecule are generated by the RIS model of Flory. Each {ϕ} has a separate ordering matrix whose diagonal elements are assumed to scale with the shape anisotropy of the molecule as prescribed by the principal moments of inertia of the configuration I{ϕ}. The variation of deuterium quadrupolar splittings along the alkyl chain can be satisfactorily explained by the calculations, however, the detailed behavior of the computed splittings are sensitive to the assumptions about the placement of the molecular fixed frame.

V 093   Samulski, E. T.; Toriumi, H. *J. Chem. Phys.* **1983**, *79*, 5194.

A steric model with a single adjustable parameter is employed to compute configurationally averaged quadrupolar splittings of deuterium labeled alkyloxy chains in a homologous series of hexaalkyloxytriphenylenes (THE*n*). A comparison with experimental data permits inferences about the alkyl chain flexibility in discotic mesogens. The parameter, a control of the constraints on the chain mobility, is first optimized by fitting deuterium NMR data in a homologous series of nematogens with labeled alkyloxy chains. The same optimum value of the parameter is subsequently used to model the quadrupolar splittings in the columnar phase of THE*n* discotigens and succeeds in qualitatively accounting for the experimental observations. This result indicates that alkyl chain mobility is quite comparable in these two classes of liquid crystals. The distinctly different features of the quadrupolar splitting patterns observed for nematogens and columnar discotigens are shown to originate from the different orientation of the molecular order tensor relative to the mean direction of chain propagation in the two types of mesogens.

Samulski, E. T. *Israel J. Chem.* **1983**, *23*, 329.

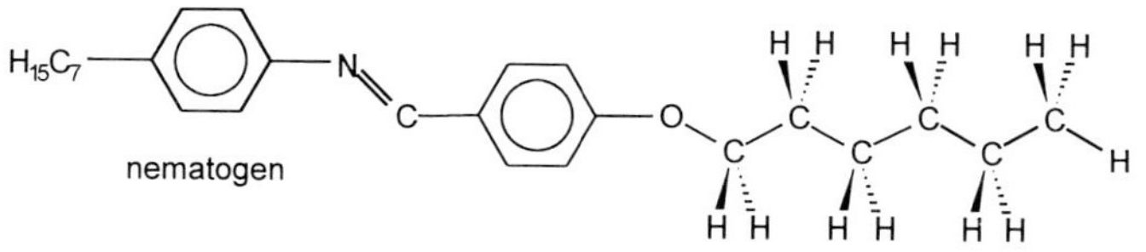

nematogen

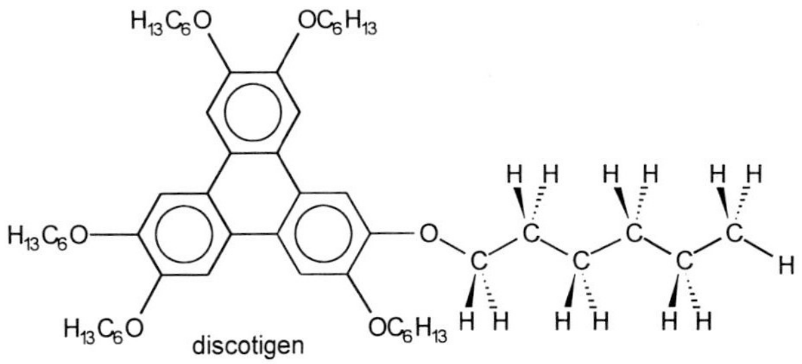

discotigen

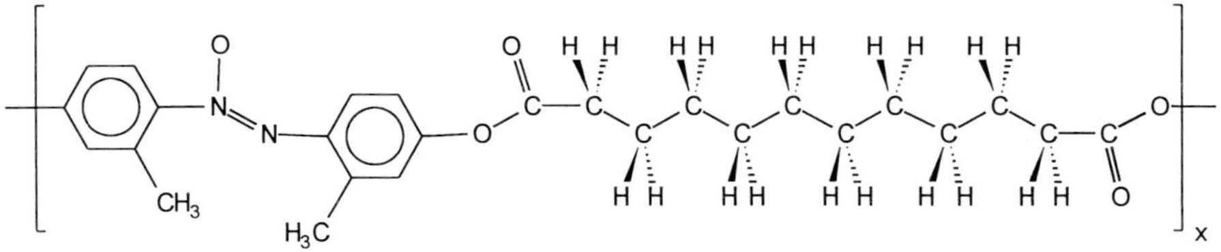

polymeric nematogen

Deuterium NMR results from labeled chains appended to nematogens and discotigens, chains linking mesogenic units in linear polymeric nematics and simple *n*-alkanes and lipids solubilized in a nematic solvent are examined. A phenomenological single chain model is used as a vehicle for contrasting alkyl chain flexibility in various ordered phases. Agreement between the model and experiment suggests a commonality in the behavior of all these types of alkyl chains originating from a repulsive (steric) constraint that is imposed on the isomerization of the chain by the local ordered environment.

V 095   Samulski, E. T.; Gauthier, M. M.; Blumstein, R. B.; Blumstein, A. *Macromolecules* **1984**, *17*, 479.

The order and mobility of a labeled flexible alkyl spacer in the linear thermotropic polymeric nematic liquid crystal poly(2,2'-dimethyl-4,4'-dioxyazoxybenzenedodecanedioyl-d_{20}) (poly[oxy(3-methyl-1,4-phenylene)azoxy(2-methyl-1,4-phenylene)oxy(1,12-dioxo-1,12-dodecanediyl-d_{20})]) is explored with deuterium NMR. The quadrupol splittings of the spacer methylene segments in the nematic melt of the polymer are reported as a function of the temperature and are contrasted with observations on model compounds solubilized in a nematic solvent.

V 096   Yoon, D. Y.; Bruckner, S.; Volksen, W.; Scott, J. C.; Griffin, A. C. *Faraday Discuss. Chem. Soc.* **1985**, *79*, 41.

The stability and the molecular order of nematic states of thermotropic polymers comprising rigid groups connected by flexible spacers are found to be dominated primarily by the characteristics (configurational partition function) of highly extended conformers, which are favoured for packing. Moreover, both the macroscopic consideration of the enthalpies and the entropies of isotropic-nematic transitions and the microscopic probe by deuterium NMR of labelled chains lead to the conclusion that the highly extended conformers are selected preferentially in polymeric nematogens. The findings on conformational order and orientational order of polymeric liquid crystals are compared with theoretical predicitions based on ideal lattice chains and worm-like chains.

V 097   Abe, A.; Furuya, H. *Polmer Bull.* **1988**, *19*, 403.

Deuterium NMR measurements are performed for dimer liquid crystals (DLC) having structures such as NC—Ph—Ph—O—$(CH_2)_n$—O—Ph—Ph—CN (CBA) with n = 9, 10. Fully deuterated CBAs with n = 9 and 10 exhibit, respectively, three and four splittings in the D-NMR spectra. By using partially deuterated samples, the signals corresponding to the largest splittings are found to include contributions from the α- and β-CD_2 groups. The origins of the rest of the signals are elucidated by the RIS method. Characteristic properties of the nematic mesophase are estimated for CBA-10.

V 098   Furuya, H.; Abe, A. *Polmer Bull.* **1988**, *20*, 467.

Deuterium NMR measurements are performed for main-chain polymer liquid crystals having structures such as –[Ph—O—CO—Ph—O—$(CH_2)_n$—O—Ph—CO—O—Ph—O—$(CH_2)_n$—O]$_x$– with n = 9 and 10. The general profiles of the spectra are found to be similar to those of the corresponding dimer. The results of the RIS analysis suggest that the spatial configurations of the spacer are nearly identical between the dimer and the polymer.

Deuterium NMR measurements are performed to elucidate orientational characteristics of ether-type main-chain liquid crystals such as DLC *I-n* and PLC *II-n* with $n = 9$ and 10. An RIS analysis of the D-NMR spectra is presented. The bond conformation probabilities of the flexible spacer in the nematic phase are estimated for DLC *I-n* and PLC *II-n* with $n = 9$ and 10. The RIS calculations are first performed for molecules in the unconstrained isotropic state. Conventional values of the conformational energy parameters are adopted in these calculations: $E(\sigma_1) = 0$, $E(\sigma_i) = 2.1$ with $i = 2\text{-}5$, $E(\omega) = 8.4$, and $E(\omega') = 2.1$, all energies being in kJ·mol^{-1}. For each conformation of the spacer, the inclination angles ψ_1 and ψ_2 are calculated. The distributions are bimodal, being consistent with the results of the previous RIS analysis where distributions are studied as a function of the angle θ between two successive mesogenic core axis (Abe, A. *Macromolecules* 1984, *17*, 2280). Under a nematic environment, distribution of the conformer should be restricted within certain ranges of the ($\psi_1\psi_2$) map. Here, configurations that do not conform to the nematic ordering are disregarded. For the nematic conformation, the second-order interactions, ω, are assumed to be entirely suppressed. In the simulation, statistical weight parameters σ_i ($i = 1$ to $n\text{-}1$) are varied in an uncorrelated manner within the range 0-1. A unique set of statistical weight factors is successfully selected for each molecular system by this technique, but is not given explicitly in the paper. From these values of σ_i, bond conformations expressed in terms of the *trans* fraction, $f_t = 1 - f_g$, are evaluated. The statistical weight factor σ_i is assigned to the *gauche* state of the *i*th C—C bond, the weight of unity being given to the corresponding *trans* state; temperature is 500 K throughout. The results indicate that the nematic conformation of the flexible segment is amazingly similar between the dimer and polymer with the same *n*. The difference between these two arises mostly from the orientation of the molecular axis in the liquid-crystalline domain. The orientational order parameters of the mesogenic core axis estimated from these analyses are found to be quite consistent with those observed directly by using mesogen-deuterated samples.

V 100   Mendicuti, F.; Patel, B.; Mattice, W. L. *Polymer* **1990**, *31*, 453.

Fluorescence is measured in dilute solution of model compounds for polymers of 2,6-naphthalene dicarboxylic acid and eight different glycols. The ratio of excimer to monomer emission depends on the glycol used. Studies as functions of temperature and solvent show that, in contrast with the analogous polyesters in which the naphthalene moiety is replaced with a benzene ring, there can be a substantial dynamic component to the excimer emission. Extrapolation to media of infinite viscosity shows that in the absence of rotational isomerism during the lifetime of the singlet excited state, there is an odd-even effect in the series in which the flexible spacers differ in the number of methylene units, but not in the series in which the flexible spacers differ in the number of oxyethylene units.

V 101   Mendicuti, F.; Patel, B.; Mattice, W. L. *Polymer* **1990**, *31*, 1877.

The fluorescence of the α,ω-diesters of $HO(CH_2)_mOH$, m = 2-6, and 1-naphthoic acid is characterized in dilute solution in solvents of different viscosity. An equilibrium RIS analysis can rationalize the observation of the odd-even effect and the occurrence of the maximum at $m = 5$.

V 102   Furuya, H.; Dries, T.; Fuhrmann, K.; Abe, A.; Ballauff, M.; Fischer, E. W. *Macromolecules* **1990**, *23*, 4122.

The magnetic susceptibilities of dimer liquid crystals such as $NC-Ph-Ph-O-(CH_2)_n-Ph-Ph-CN$ (n = 9, 10) are measured by a SQUID magnetometer. The results obtained are interpreted within the framework of the RIS approximation, the effect arising from the conformational anisotropy of the flexible spacer being strictly taken into account. The order parameters of the mesogenic core axis thus estimated are found to be consistent with those directly observed at just below T_{NI} by the ^2H NMR technique using mesogen-deuterated samples.

V 103   Furuya, H.; Abe, A.; Fuhrmann, K.; Ballauff, M.; Fischer, E. W. *Macromolecules* **1991**, *24*, 2999.

The magnetic susceptibilities of some ether-type liquid-crystalline polymers are measured by a SQUID magnetometer. The $\Delta\chi$ values estimated for the stable nematic state are analyzed according to a known RIS scheme.

V 104   Abe, A. *Makromol. Chem., Makromol. Symp.* **1992**, *53*, 13.

Deuterium NMR studies are performed to elucidate orientational characteristics of some ether-type main-chain liquid crystals. Spacers used are of the type $-O-(CH_2)_n-O-$ with n = 9, 10. Dimers, homopolymers as well as copolymers in which spacers n = 9 and 10 are arranged in an alternative fashion are investigated. The quadrupolar splitting data obtained from the deuterium-labeled mesogenic core and spacer are studied within the RIS approximation. The ordering characteristics thus estimated are found to be consistent with the magnetic susceptibility data obtained by using a SQUID magnetometer for the same polymers.

V 105   Mendicuti, F.; Mattice, W. L. *Polymer* **1992**, *33*, 4180.

The fluorescence in dilute solution is measured for five polyesters with terephthalate as the rigid aromatic unit and diols derived from cyclohexane as the flexible spacer. A conformational analysis concludes that the spacers most conducive to excimer formation are the 1,3-*cis*-cyclohexanediol and 1,4-*cis*-cyclohexanedimethanol. This result from calculations is compatible with experimental results.

V 106   Lakowicz, J. R.; Wiczk, W.; Gryczynski, I.; Fishman, M.; Johnson, M. L. *Macromolecules* **1993**, *26*, 349.

Frequency-domain measurements of fluorescence energy transfer are used to determine the end-to-end distance distribution of donor-acceptor (D-A) pairs linked by flexible alkyl chains. The length of the linker is varied from 11 to 28 atoms, and two different D-A pairs are used. In each case the D-A distributions are recovered from global analysis of measurements with different values for the Förster distance, which are obtained by collisional quenching of the donors. In all cases essentially the same distance distribution is recovered from the frequency-domain data for each value of the Förster distance. The experimentally recovered distance distributions are compared with those calculated from the RIS model. The experimentally recovered distance distributions for the largest chain molecules are in agreement with the predictions of the RIS model. However, the experimental and RIS distributions are distinct for the shorter D-A pairs.

V 107   Abe, A.; Iizumi, E.; Kimura, N. *Liq. Cryst.* **1994**, *16*, 655.

The conformations of flexible chain molecules incorporated in a nematic environment are investigated. Proton-proton and carbon-carbon dipolar coupling constant measurements are attempted for 1,2-dimethoxyethane and 1,2-diphenoxyethane, in addition to ^{2}H NMR observations of quadrupolar splittings. These conformation-dependent properties are analyzed according to the RIS scheme. Studies are further extended to a mixture of 1,2-diphenoxyethane with a nematic liquid crystal, 4,4'-azoxyanisole.

V 108   Abe, A.; Furuya, H.; Shimizu, R. N.; Nam, S. Y. *Macromolecules* **1995**, *28*, 96.

RIS analysis of the deuterium quadrupolar splitting data is performed for α,ω-bis[(4,4'-cyanobiphenyl)oxy]alkane dimer liquid crystals having $-O-(CH_2)_n-O-$ flexible spacers (n = 9 (CBA-9) and n = 10 (CBA-10)) between the 4,4'-cyanobiphenylyl ends according to the scheme previously established. The RIS analysis indicates that most of the conformers involved in the range $0 < \psi_1,\psi_2 < 45°$ adopt spatial configurations reasonably consistent with the nematic arrangement of mesogenic cores in both dimer LC systems, where ψ_1 and ψ_2 denote the inclination angles of the terminal mesogenic cores with respect to the molecular axis.

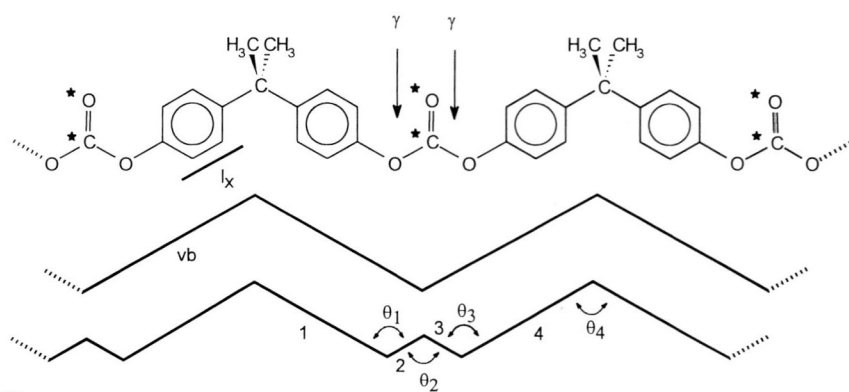

polycarbonate from 2,2-bis(4-hydroxyphenyl)propane, bisphenol A polycarbonate, PC

Williams, A. D.; Flory, P. J. *J. Polym. Sci.: Part A-2* **1968**, *6*, 1945.

Bond Length [pm]	Valence Angles [°]	Torsion Angles [°]	ξ a) (for 298 K)	ξ_0	E_ξ [kJ · mol^{-1}]				
C-C^{ar} : 152	C^{ar}-C-C^{ar} : 112	For bonds 1,4:	$\gamma \approx 0.11$	1	≈ 5.4				
C*-O : 134	C^{ar}-O-C : 113	ϕ : b)							
C^{ar}-O : 136	O-C*-O : 114					$U_1 = [1]$	$U_2 = [1\ \gamma]$	$U_3 = \begin{bmatrix} 1 & \gamma \\ 1 & 0 \end{bmatrix}$	$U_4 = \begin{bmatrix} 1 \\ 1 \end{bmatrix}$
C*=O* : 122	$\theta_1 = \theta_3$: 113								
l_x : 279	θ_2 : 114	For bonds 2,3:			rows / columns:	ϕ / ϕ	$\phi / t,c$	$t,c / t,c$	$t,c / \phi$
$l_1 = l_4$: 567	θ_4 : 112	t : 180							
$l_2 = l_3$: 134		c : 0							
l_{vb} : 700									

a) The parameter γ represents the statistical weight of the *cis*-configuration relative to a weight of unity for *trans*; *cis,cis* pairs are excluded.
b) Bonds 1 and 4 show a twofold symmetric potential. Here, they are treated as if bonds were permitted to undergo free rotation.

Comments:
The structure of the polycarbonate chain is analyzed from the point of view of the spatial configurations. It follows that the molecule can be treated as a freely rotating chain consisting of a succession of virtual bonds (l_{vb}) 7.0 Å in length, jointed at angles (θ) of ca. 112°.

Calcd. quantities: $(<r^2>_0 / M)_\infty$ = 0.78 (for γ = 0.5) to 0.85 (for γ = 0)   (*Exptl.*: $\approx$ 0.85)   (in: [Å^2 · mol · g^{-1}])

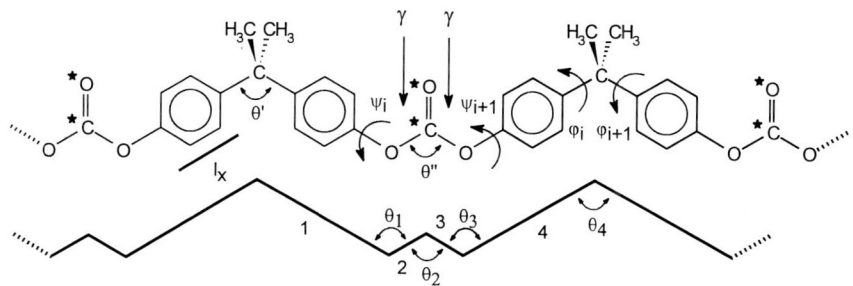

polycarbonate from 2,2-bis(4-hydroxyphenyl)propane, bisphenol A polycarbonate, PC

Laskowski, B. C.; Yoon, D. Y.; McLean, D.; Jaffe, R. L. *Macromolecules* **1988**, *21*, 1629.

Bond Length [pm]	Valence Angles [°]	Torsion Angles [°]	ξ a) (for 300 K)	ξ_o	E_ξ [kJ·mol^{-1}]			
C–C^{ar} : 152	C^{ar}–C–C^{ar} : 109.5	For bonds 1,4:	γ = 0.01	1	11.5			
C*–O : 134	C^{ar}–O–C : 113	ϕ : b)						
C^{ar}–O : 136	O–C*–O : 114							
C*=O* : 122								
	$\theta_1 = \theta_3$: 120.5 (for *trans*)	For bonds 2,3:						
l_x : 279	123.2 (for *cis*)	t : 180			$U_1 = [1]$	$U_2 = [1\ \gamma]$	$U_3 = \begin{bmatrix} 1 & \gamma \\ 1 & 0 \end{bmatrix}$	$U_4 = \begin{bmatrix} 1 \\ 1 \end{bmatrix}$
$l_1 = l_4$: 573	θ_2 : 106.5 (for *trans*)	c : 0						
$l_2 = l_3$: 132	111.7 (for *cis*)		rows / columns:		ϕ / ϕ	ϕ / t,c	t,c / t,c	t,c / ϕ
	θ_4 : 109.6							

a) The parameter γ represents the statistical weight of the *cis*-configuration relative to a weight of unity for *trans*; *cis,cis* pairs are excluded.
b) Bonds 1 and 4 show a twofold symmetric potential. Here, they are treated as if bonds were permitted to undergo free rotation.

Comments:
Ab initio calculations with full geometry optimization on diphenyl carbonate and diphenylpropane are carried out to determine the bond geometries and the conformational energies and then to compute the unperturbed chain dimensions of the bisphenol A polycarbonate. Application of these results to the RIS model of the PC chain leads to the prediction of the unperturbed chain dimension.

Calcd. quantities: $(\langle r^2 \rangle_o / M)_\infty$ = 1.1   (*Exptl.*: 0.87 to 1.28)   (in: [Å^2·mol·g^{-1}])

polycarbonate from 2,2-bis(4-hydroxyphenyl)propane, bisphenol A polycarbonate, PC

Hutnik, M.; Argon, A. S.; Suter, U. W. *Macromolecules* **1991**, *24*, 5956.
Hutnik, M.; Suter, U. W. *Polym. Prepr.* **1987**, *28*, 293.

Bond Length [pm]	Valence Angles [°]	Torsion Angles [°]	ξ (for 298 K)	ξ_0	E_ξ [kJ·mol^{-1}]				
C^{ar}-C^{ar} : 138	θ_2 : 109.8	For bonds 1,2,3,6:	γ = 0.05	1	7.4				
C-CH$_3$: 153		t$^+$: 135				$U_1 = \begin{bmatrix} 1 & 1 & 1 & 1 \\ 1 & 1 & 1 & 1 \\ 1 & 1 & 1 & 1 \\ 1 & 1 & 1 & 1 \end{bmatrix}$	$U_2 = \begin{bmatrix} 1 & 0 & 1 & 0 \\ 0 & 1 & 0 & 1 \\ 1 & 0 & 1 & 0 \\ 0 & 1 & 0 & 1 \end{bmatrix}$	$U_3 = \begin{bmatrix} 1 & 1 & 1 & 1 \\ 1 & 1 & 1 & 1 \\ 1 & 1 & 1 & 1 \\ 1 & 1 & 1 & 1 \end{bmatrix}$	For 4 × 4 matrices: rows and columns: t$^+$, c$^+$, c$^-$, t$^-$
$C^*=O^*$: 121	θ_4 : 124.0	c$^+$: 45							
C^{ar}-H : 110		c$^-$: – 45							
I* : 276	θ_5 : 109.0	t$^-$: – 135	rows / columns:						
$l_1 = l_2$: 154									
$l_3 = l_6$: 141	θ_6 : 124.0	For bonds 4,5:	U_4 : t$^+$, c$^+$, c$^-$, t$^-$ / t, c						
$l_4 = l_5$: 133		t : 180	U_5 : t, c / t, c						
	C^{ar}-C^{ar}-C^{ar} : 120	c : 0	U_6 : t, c / t$^+$, c$^+$, c$^-$, t$^-$			$U_4 = \begin{bmatrix} 1 & \gamma \\ 1 & \gamma \\ 1 & \gamma \\ 1 & \gamma \end{bmatrix}$	$U_5 = \begin{bmatrix} 1 & \gamma \\ 1 & 0 \end{bmatrix}$	$U_6 = \begin{bmatrix} 1 & 1 & 1 & 1 \\ 1 & 1 & 1 & 1 \end{bmatrix}$	
	O-C*-C* : 125.5								
	C^{ar}-C^{ar}-CH$_3$: 109								

Comments:
A classical force-field is developed to represent the conformational characteristics of PC based upon recent experimental and quantum mechanical data. This force field is an improvement upon previously published molecular mechanics force fields because it allows for rotation about all the single bonds in the PC repeat unit. An RIS model of PC is obtained using the force field results.

Calcd. quantities: $<r^2>_0 / M$ = 1.03   (*Exptl.:* 0.87 *to* 1.28)   (in: [Å^2·mol·g^{-1}])

*Further calculations on **polycarbonate** chains:*

V 112   Tonelli, A. E. *Macromolecules* **1972**, *5*, 558.

The conformational energies per independent repeat unit of poly(2,6-dimethyl-1,4-phenylene oxide) and the polycarbonate of bisphenol A are evaluated. Energetically allowed conformations are found to span the entire range of the rotation angle about the virtual bonds connecting neighboring ether oxygen atoms in the phenylene oxide polymer. Rotation about the virtual bonds in the polycarbonate chain is found to be similarly free of significant constraints. Consequently, both classes of polymers exhibit freely rotating chain statistics. Rotation about the virtual bonds in the phenylene oxide polymers and in polycarbonate is nearly truly free (each rotational angle is appreciably populated) and not just restricted to two symmetrically located rotational states of equal energy at 90° and -90°.

V 113   Sundararajan, P. R. *Can. J. Chem.* **1985**, *63*, 103.

Conformational energies are estimated for the segments of the bisphenol A polycarbonate chain, using the Lennard-Jones and Hill's empirical force field type of functions. It is found that the conformation of the carbonate group, defined by the torsion angle ζ, is restricted to the range of 45-65°. The rotations χ and χ' of the methyl groups also show similar limited flexibility. However, accessible conformations of the diphenyl propane segment, defined by torsion angles ϕ and ψ, span a wide area of the (ϕ,ψ) surface, with the restriction that the rotations of ϕ and ψ be synchronized such that $\phi + \psi \approx 90$ or $270°$. These features explain the slow thermal crystallization behaviour of the polycarbonate chains. The variability of the conformations of the repeat units is illustrated with a series of figures.

V 114 Sundararajan, P. R. *Macromolecules* **1987**, *20*, 1534.

The helical parameters corresponding to the various skeletal conformations of the bisphenol A polycarbonate chain are calculated. Combining these results with the conformational energy calculations shows that flat-helical and extended conformations are of equal energy for this chain. In addition, cyclic structures are also found to be stereochemically possible. The small values of the characteristic ratio of the unperturbed end-to-end distance and its temperature coefficient are attributed to the equal energy of the flat-helical and extended-helical, as well as the nonhelical, conformers.

V 115 Sundarajan, P. R. *Macromolecules* **1989**, *22*, 2149.

A comparison of the conformational freedom of rotation of the contiguous phenyl groups in polycarbonates, with various substituents at the C_α atom, is presented. Conformational maps are calculated for the polymer shown above. Synchronous rotation of the phenyls with a low-energy barrier is possible for **1**, **4**, **5**, and **8**. Although the extent of freedom of rotation depends on the nature of the substituent, there is very little difference in the characteristic ratio of the unperturbed end-to-end distance for these polycarbonates, and the temperature coefficient of the characteristic ratio is extremely small. In spite of the limited conformational freedom, it is shown that the steric symmetry and the geometric asymmetry of the chain segments enable the treatment of these chains in the framework of the freely rotating chain.

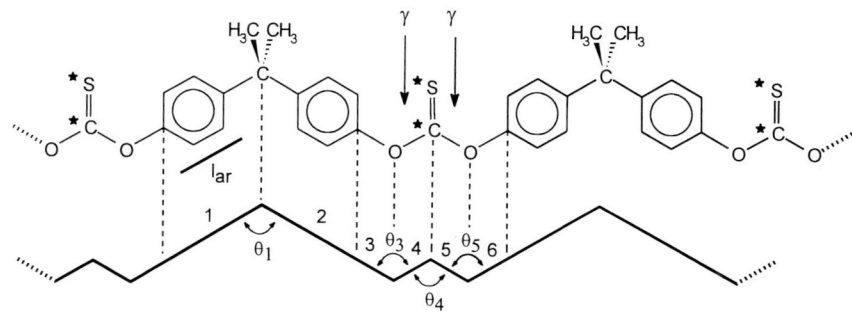

polythiocarbonate from 2,2-bis(4-hydroxyphenyl)propane, bisphenol A polythiocarbonate, PMTC

Saiz, E.; Fabre, M. J.; Gargallo, L.; Radić, D.; Hernández-Fuentes, I. *Macromolecules* **1989**, *22*, 3660.

Bond Length [pm]	Valence Angles [°]	Torsion Angles [°]	ξ (for 298 K)	ξ_0	E_ξ [kJ · mol^{-1}]
C^{ar}-C : 154	θ_1 : 109.8	For bonds 1,2,3,6:	$\gamma = 0.18$	1	4.2
C^{ar}-O : 141		t^+ : 135			
O-C* : 133	θ_3 : 117.7	c^+ : 45			
		c^- : -45			
l_{ar} : 276	θ_4 : 105.8	t^- : -135	rows / columns:		
$l_1 = l_2$: 430					
$l_3 = l_6$: 141	θ_5 : 117.7	For bonds 4,5:	U_4 : $t^+, c^+, c^-, t^- / t, c$		
$l_4 = l_5$: 133		t : 180	U_5 : t, c / t, c		
		c : 0	U_6 : t, c / t^+, c^+, c^-, t^-		

$$U_1 = \begin{bmatrix} 1 & 1 & 1 & 1 \\ 1 & 1 & 1 & 1 \\ 1 & 1 & 1 & 1 \\ 1 & 1 & 1 & 1 \end{bmatrix} \quad U_2 = \begin{bmatrix} 1 & 0 & 1 & 0 \\ 0 & 1 & 0 & 1 \\ 1 & 0 & 1 & 0 \\ 0 & 1 & 0 & 1 \end{bmatrix} \quad U_3 = \begin{bmatrix} 1 & 1 & 1 & 1 \\ 1 & 1 & 1 & 1 \\ 1 & 1 & 1 & 1 \\ 1 & 1 & 1 & 1 \end{bmatrix}$$

For 4 × 4 matrices: rows and columns: t^+, c^+, c^-, t^-

$$U_4 = \begin{bmatrix} 1 & \gamma \\ 1 & \gamma \\ 1 & \gamma \\ 1 & \gamma \end{bmatrix} \quad U_5 = \begin{bmatrix} 1 & \gamma \\ 1 & 0 \end{bmatrix} \quad U_6 = \begin{bmatrix} 1 & 1 & 1 & 1 \\ 1 & 1 & 1 & 1 \end{bmatrix}$$

Comments:
The dipole moment of PMTC is determined. Theoretical calculations are performed using the RIS model with the scheme developed for polycarbonates. The calculated values of μ_{eff} are almost insensitive to the conformational parameters used in the calculations.

Calcd. quantities: $<\mu^2>_0$ $\qquad \mu_{eff} = (<\mu^2>_0 / x)^{1/2}$

polythiocarbonates with chlorophenyl or dichlorophenyl side groups

Saiz, E.; Abradelo, C.; Mogín, J.; Tagle, L. H.; Hernández-Fuentes, I. *Macromolecules* **1991**, *24*, 5594.

Bond Length [pm]	Valence Angles [°]	Torsion Angles [°]	ξ (for 308 K)	ξ_0	E_ξ [kJ·mol^{-1}]				
C^{ar}-C^{ar} : 138	θ_1 : 109.8	For bonds 3,6:	$\gamma = 0.19$	1	4.2				
C-CH$_3$: 153		t^+ : 135							
C^*=O^* : 121	θ_3 : 117.7	c^+ : 45				$U_1 = \begin{bmatrix} 1 & 1 & 1 & 1 \\ 1 & 1 & 1 & 1 \\ 1 & 1 & 1 & 1 \\ 1 & 1 & 1 & 1 \end{bmatrix}$	$U_2 = \begin{bmatrix} 1 & 0 & 1 & 0 \\ 0 & 1 & 0 & 1 \\ 1 & 0 & 1 & 0 \\ 0 & 1 & 0 & 1 \end{bmatrix}$	$U_3 = \begin{bmatrix} 1 & 1 & 1 & 1 \\ 1 & 1 & 1 & 1 \\ 1 & 1 & 1 & 1 \\ 1 & 1 & 1 & 1 \end{bmatrix}$	For 4 × 4 matrices: rows and columns: t^+, c^+, c^-, t^-
C^{ar}-H : 110		c^- : –45							
l^* : 276	θ_4 : 105.8	t^- : –135		rows / columns:					
$l_1 = l_2$: 154									
$l_3 = l_6$: 141	θ_5 : 117.7	For bonds 4,5:		U_4 : t^+, c^+, c^-, t^- / t, c					
$l_4 = l_5$: 133		t : 180		U_5 : t, c / t, c		$U_4 = \begin{bmatrix} 1 & \gamma \\ 1 & \gamma \\ 1 & \gamma \\ 1 & \gamma \end{bmatrix}$	$U_5 = \begin{bmatrix} 1 & \gamma \\ 1 & 0 \end{bmatrix}$	$U_6 = \begin{bmatrix} 1 & 1 & 1 & 1 \\ 1 & 1 & 1 & 1 \end{bmatrix}$	
		c : 0		U_6 : t, c / t^+, c^+, c^-, t^-					
	C^{ar}-C^{ar}-C^{ar} : 120								
	O-C^*-C^* : 125.5								
	C^{ar}-C^{ar}-CH$_3$: 109	For bonds 1 / 2:							
		t^+ : 100 / 160							
		c^+ : 20 / 80							
		c^- : –80 / –20							
		t^- : –160 / –100							

Comments:
The dipole moments of three asymmetric polythiocarbonates derived from bisphenol A are determined in benzene solution at 298 K. Good agreement between theoretical and experimental results can be obtained assuming that the direction of the dipole moment of the thiocarbonate group is opposite to that in carbonate residues.

Calcd. quantities: $\langle \mu^2 \rangle_0$ $\mu_{eff} = (\langle \mu^2 \rangle_0 / x)^{1/2}$

poly(4-oxy-1-benzoate), PHBA [a]

Rutledge, G. C. *Macromolecules* **1992**, *25*, 3984.

Bond Length [pm]	Valence Angles [°]	Torsion Angles [°]	ζ (for 300 K)	ζ_0	E_ζ [kJ·mol^{-1}]	(see: [b])
C^*-O : 137	C^{ar}-C^*-O : 118.3	For bond 1: t : 180 c : 0	$\zeta = 0.05$	1	7.47	*Conventional RIS model:*
C^{ar}-C^{ar} : 139	C^{ar}-O-C^* : 110.9					
C^{ar}-O : 141	θ_1 : 118.3	For bond 3: t$^+$: 135 c$^+$: 45 c$^-$: –45 t$^-$: –135				$U_4' = \begin{bmatrix} 1 \\ 1 \\ 1 \\ 1 \end{bmatrix}$ $U_1 = \begin{bmatrix} 1 & 1 \end{bmatrix}$ $U_3 = \begin{bmatrix} 1 & 1 & 1 & 1 \\ 1 & 1 & 1 & 1 \end{bmatrix}$
C^{ar}-C : 149	$\theta_2 = \theta_3 = 90$					rows / columns: U_4' : t$^+$, c$^+$, c$^-$, t$^-$ / t U_1 : t / t, c U_3 : t, c / t$^+$, c$^+$, c$^-$, t$^-$
C^{ar}-C^{ar} : 145	θ_4 : 110.9					
l_1 : 288		For bond 4 (conv. model): t : 180				*Variance RIS model:*
l_3 : 280		or (variance model): t : 180 t$^+$: 127 t$^-$: –127				$U_4' = \begin{bmatrix} 1 & \zeta & \zeta \\ 1 & \zeta & \zeta \\ 1 & \zeta & \zeta \\ 1 & \zeta & \zeta \end{bmatrix}$ $U_1 = \begin{bmatrix} 1 & 1 \\ 1 & 1 \\ 1 & 1 \end{bmatrix}$ $U_3 = \begin{bmatrix} 1 & 1 & 1 & 1 \\ 1 & 1 & 1 & 1 \end{bmatrix}$
l_4 : 137						rows / columns: U_4' : t$^+$, c$^+$, c$^-$, t$^-$ / t, t$^+$, t$^-$ U_1 : t, t$^+$, t$^-$ / t, c U_3 : t, c / t$^+$, c$^+$, c$^-$, t$^-$

[a] The models shown here are just a few examples of the general RIS scheme developed.
[b] U_2 is the identity matrix E_s of the order s, where s is the number of columns in U_1 or the number of rows in U_3.

Comments:
A statistical mechanical analysis based on the RIS theory is applied to the analysis of single-chain behaviour in a family of aromatic polyester compositions, including homopolymers and nonregular copolymers, and a series of nonregular terpolymers containing the *m*-phenylene aromatic moiety. Averages of chain rigidity and shape are determined over all allowed conformations and constitutions of the chains. Calculated values of chain persistence length a ranging from 200 down to 20 Å for a variety of compositions are in good agreement with the available experimental data.

Calcd. quantities: $[|<r>|/l^*]_\infty$ $(<r^2>/nl^{*2})_\infty$ Persistence length *a* $[<s^2>/nl^{*2}]_\infty$ (l^* : scaling length)

V 119   (see also **V 118, V120, V 121**)

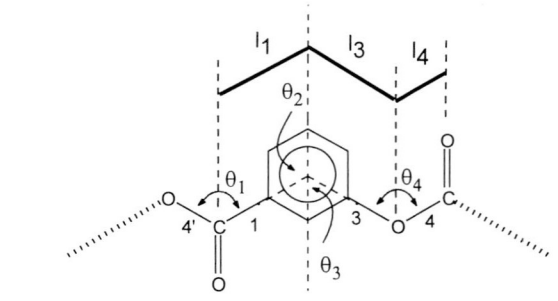

poly(3-oxy-1-benzoate) [a]

Rutledge, G. C. *Macromolecules* **1992**, *25*, 3984.

Bond Length [pm]	Valence Angles [°]	Torsion Angles [°]	ξ (for 300 K)	ξ_0	E_ξ [kJ·mol^{-1}]	(see: [b])
C*-O : 137	C^{ar}-C*-O : 118.3	For bond 1: t : 180 c : 0	$\zeta = 0.05$	1	7.47	*Conventional RIS model:*
C^{ar}-C^{ar} : 139	C^{ar}-O-C* : 110.9					
C^{ar}-O : 141	θ_1 : 118.3	For bond 3: t$^+$: 135 c$^+$: 45 c$^-$: – 45 t$^-$: – 135				$U_4' = \begin{bmatrix} 1 \\ 1 \\ 1 \\ 1 \end{bmatrix}$ $U_1 = \begin{bmatrix} 1 & 1 \end{bmatrix}$ $U_3 = \begin{bmatrix} 1 & 1 & 1 & 1 \\ 1 & 1 & 1 & 1 \end{bmatrix}$
C^{ar}-C : 149	θ_2 : 120					
C^{ar}-C^{ar} : 145	θ_3 : 60					
l_1 : 288	θ_4 : 110.9	For bond 4 (conv. model): t : 180 or (variance model): t : 180 t$^+$: 127 t$^-$: – 127				*Variance RIS model:* $U_4' = \begin{bmatrix} 1 & \zeta & \zeta \\ 1 & \zeta & \zeta \\ 1 & \zeta & \zeta \\ 1 & \zeta & \zeta \end{bmatrix}$ $U_1 = \begin{bmatrix} 1 & 1 \\ 1 & 1 \\ 1 & 1 \end{bmatrix}$ $U_3 = \begin{bmatrix} 1 & 1 & 1 & 1 \\ 1 & 1 & 1 & 1 \end{bmatrix}$
l_3 : 280						
l_4 : 137						

rows / columns:
U_4' : t$^+$, c$^+$, c$^-$, t$^-$ / t
U_1 : t, t$^+$, t$^-$ / t, c
U_3 : t, c / t$^+$, c$^+$, c$^-$, t$^-$

rows / columns:
U_4' : t$^+$, c$^+$, c$^-$, t$^-$ / t, t$^+$, t$^-$
U_1 : t, t$^+$, t$^-$ / t, c
U_3 : t, c / t$^+$, c$^+$, c$^-$, t$^-$

[a] The models shown here are just a few examples of the general RIS scheme developed.
[b] U_2 is the identity matrix E_s of the order s, where s is the number of columns in U_1 or the number of rows in U_3.

Comments:
A statistical mechanical analysis based on the RIS theory is applied to the analysis of single-chain behaviour in a family of aromatic polyester compositions, including homopolymers and nonregular copolymers, and a series of nonregular terpolymers containing the *m*-phenylene aromatic moiety. Averages of chain rigidity and shape are determined over all allowed conformations and constitutions of the chains. Calculated values of chain persistence length a ranging from 200 down to 20 Å for a variety of compositions are in good agreement with the available experimental data.

Calcd. quantities: $[|\langle r \rangle|/l^*]_\infty$ $[\langle r^2 \rangle / nl^{*2}]_\infty$ Persistence length *a* $[\langle s^2 \rangle / nl^{*2}]_\infty$ (l* : scaling length)

V 120 (see also V 118, V 119, V 121)

poly(4'-oxybiphenyl-4-carboxylate) [a]

Rutledge, G. C. *Macromolecules* **1992**, *25*, 3984.

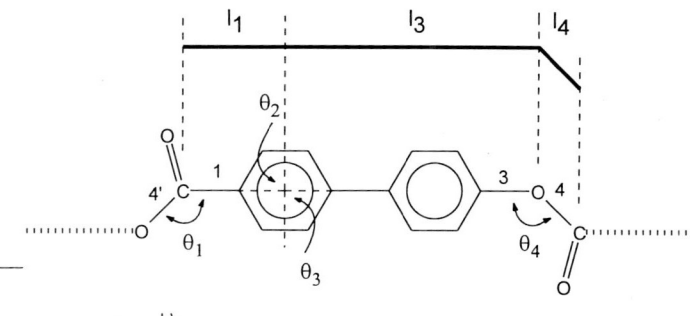

Bond Length [pm]	Valence Angles [°]	Torsion Angles [°]	ξ (for 300 K)	ξ_0	E_ξ [kJ·mol^{-1}]	(see: [b])		
C^*-O : 137	C^{ar}-C^*-O : 118.3	For bond 1: t : 180 c : 0	$\zeta = 0.05$	1	7.47	*Conventional RIS model:*		rows / columns:
C^{ar}-C^{ar} : 139	C^{ar}-O-C^* : 110.9							U_4' : t^+, c^+, c^-, t^- / t
C^{ar}-O : 141	θ_1 : 118.3	For bond 3: t^+ : 135 c^+ : 45 c^- : −45 t^- : −135				$U_4' = \begin{bmatrix} 1 \\ 1 \\ 1 \\ 1 \end{bmatrix}$	$U_1 = \begin{bmatrix} 1 & 1 \end{bmatrix}$ $U_3 = \begin{bmatrix} 1 & 1 & 1 & 1 \\ 1 & 1 & 1 & 1 \end{bmatrix}$	U_1 : t, t^+, t^- / t, c U_3 : t, c / t^+, c^+, c^-, t^-
C^{ar}-C : 149	θ_2 : 90							
C^{ar}-C^{ar} : 145	θ_3 : 90					*Variance RIS model:*		rows / columns:
l_1 : 288 [c]	θ_4 : 110.9	For bond 4 (conv. model): t : 180				$U_4' = \begin{bmatrix} 1 & \zeta & \zeta \\ 1 & \zeta & \zeta \\ 1 & \zeta & \zeta \\ 1 & \zeta & \zeta \end{bmatrix}$	$U_1 = \begin{bmatrix} 1 & 1 \\ 1 & 1 \\ 1 & 1 \\ 1 & 1 \end{bmatrix}$ $U_3 = \begin{bmatrix} 1 & 1 & 1 & 1 \\ 1 & 1 & 1 & 1 \end{bmatrix}$	U_4' : t^+, c^+, c^-, t^- / t, t^+, t^- U_1 : t, t^+, t^- / t, c U_3 : t, c / t^+, c^+, c^-, t^-
l_3 : 703		or (variance model): t : 180 t^+ : 127 t^- : −127						
l_4 : 137								

[a] The models shown here are just a few examples of the general RIS scheme developed.
[b] U_2 is the identity matrix E_s of the order s, where s is the number of columns in U_1 or the number of rows in U_3.
[c] There seems to be a misprint in the original paper: l_1 is given as 711 pm.

Comments:
A statistical mechanical analysis based on the RIS theory is applied to the analysis of single-chain behaviour in a family of aromatic polyester compositions, including homopolymers and nonregular copolymers, and a series of nonregular terpolymers containing the *m*-phenylene aromatic moiety. Averages of chain rigidity and shape are determined over all allowed conformations and constitutions of the chains. Calculated values of chain persistence length a ranging from 200 down to 20 Å for a variety of compositions are in good agreement with the available experimental data.

Calcd. quantities: $[|<r>|/l^*]_\infty$ $(<r^2>/nl^{*2})_\infty$ Persistence length a $[<s^2>/nl^{*2}]_\infty$ (l^* : scaling length)

V 121   (see also **V 118 - V 120**)

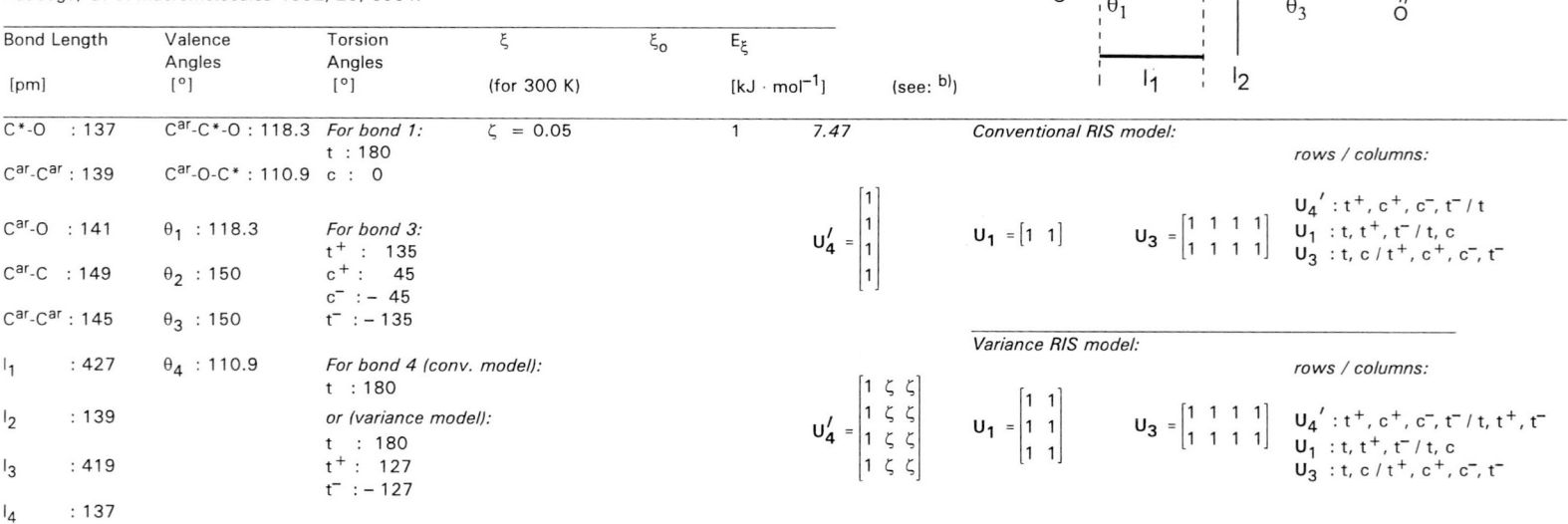

poly(6-oxy-2-naphthoate), PHNA [a]

Rutledge, G. C. *Macromolecules* **1992**, *25*, 3984.

Bond Length [pm]	Valence Angles [°]	Torsion Angles [°]	ξ (for 300 K)	ξ_o	E_ξ [kJ·mol^{-1}]	(see: [b])
C^*-O : 137	C^{ar}-C^*-O : 118.3	For bond 1: t : 180 c : 0	ζ = 0.05	1	7.47	*Conventional RIS model:*
C^{ar}-C^{ar} : 139	C^{ar}-O-C^* : 110.9					
C^{ar}-O : 141	θ_1 : 118.3	For bond 3: t^+ : 135 c^+ : 45 c^- : −45 t^- : −135				
C^{ar}-C : 149	θ_2 : 150					
C^{ar}-C^{ar} : 145	θ_3 : 150					
l_1 : 427	θ_4 : 110.9	For bond 4 (conv. model): t : 180 or (variance model): t : 180 t^+ : 127 t^- : −127				*Variance RIS model:*
l_2 : 139						
l_3 : 419						
l_4 : 137						

Conventional RIS model:

$$U_4' = \begin{bmatrix} 1 \\ 1 \\ 1 \\ 1 \end{bmatrix} \quad U_1 = \begin{bmatrix} 1 & 1 \end{bmatrix} \quad U_3 = \begin{bmatrix} 1 & 1 & 1 & 1 \\ 1 & 1 & 1 & 1 \end{bmatrix}$$

rows / columns:
U_4' : t^+, c^+, c^-, t^- / t
U_1 : t, t^+, t^- / t, c
U_3 : t, c / t^+, c^+, c^-, t^-

Variance RIS model:

$$U_4' = \begin{bmatrix} 1 & \zeta & \zeta \\ 1 & \zeta & \zeta \\ 1 & \zeta & \zeta \\ 1 & \zeta & \zeta \end{bmatrix} \quad U_1 = \begin{bmatrix} 1 & 1 \\ 1 & 1 \\ 1 & 1 \end{bmatrix} \quad U_3 = \begin{bmatrix} 1 & 1 & 1 & 1 \\ 1 & 1 & 1 & 1 \end{bmatrix}$$

rows / columns:
U_4' : t^+, c^+, c^-, t^- / t, t^+, t^-
U_1 : t, t^+, t^- / t, c
U_3 : t, c / t^+, c^+, c^-, t^-

[a] The models shown here are just a few examples of the general RIS scheme developed.
[b] U_2 is the identity matrix E_s of the order s, where s is the number of columns in U_1 or the number of rows in U_3.

Comments:
A statistical mechanical analysis based on the RIS theory is applied to the analysis of single-chain behaviour in a family of aromatic polyester compositions, including homopolymers and nonregular copolymers, and a series of nonregular terpolymers containing the *m*-phenylene aromatic moiety. Averages of chain rigidity and shape are determined over all allowed conformations and constitutions of the chains. Calculated values of chain persistence length a ranging from 200 down to 20 Å for a variety of compositions are in good agreement with the available experimental data.

Calcd. quantities: $[|<r>|//^*]_\infty$ $(<r^2>/nl^{*2})_\infty$ Persistence length *a* $[<s^2>/nl^{*2}]_\infty$ (l^* : scaling length)

Further calculations on **aromatic** chains:

V 122 Tonelli, A. E. *Macromolecules* **1973**, *6*, 503.

The intramolecular flexibilities of poly(1,4-phenylene oxide), poly(2,6-dimethyl-1,4-phenylene oxide), poly(2-methyl-6-phenyl-1,4-phenylene oxide), and poly(2,6-diphenyl-1,4-phenylene oxide) are evaluated through estimation of the resistance to rotation about the $C_{1,4}-O$ bonds in their backbones. A 6-12 potential is used to account for the van der Waals interactions between nonbonded atoms and groups encountered during the backbone rotations, while the twofold intrinsic potential to rotation about the $C_{1,4}-O$ bonds resulting from the π-electron delocalization is also included.

V 123 Hummel, J. P.; Flory, P. J. *Macromolecules* **1980**, *13*, 479.

Structural data from X-ray crystallographic investigations on aromatic amides and esters are examined for the purpose of deducing bond length and bond angles appropriate for poly(*p*-benzamide), poly(*p*-phenyleneterephthalamide), and the corresponding polyesters. Conformational energies are calculated for acetanilide, *N*-methylbenzamide, phenyl acetate, and methyl benzoate as functions of torsion angles about the phenylene axis. An empirical force field (6-exp type) supplemented by terms for frame distorsion and electron delocalization is used for this purpose. Bond angles and bond lengths are adjusted to values that minimize the total energy at each value of the torsion angle. The empirical torsional energies are compared with ab initio molecular orbital calculations.

V 124 Erman, B.; Flory, P. J.; Hummel, J. P. *Macromolecules* **1980**, *13*, 484.

Moments of rank 1-4 formed from the components x,y, and z of the chain vector **r** and expressed in the coordinate system affixed to the first unit are calculated as functions of chain length *n* for *p*-phenylene polyamides of type I, $-(NH-C_6H_4-CO)_n-$, of type II, $-(NH-C_6H_4-NH-CO-C_6H_4-CO)_{n/2}-$, and of the corresponding polyesters.

V 125 Jung, B.; Schürmann, B. L. *Macromolecules* **1989**, *22*, 477.

A theoretical approach is applied to elucidate the molecular conformations, associated flexibility, and dynamics of poly(*p*-hydroxybenzoic acid) esters, pHB. Properties such as the radius of gyration and persistence length which are characteristic for the stiffness of a macromolecule are calculated on the basis of two different theoretical methods: (a) Molecular dynamics and (b) the RIS model augmented by the more recent scheme for the matrix computations. The analysis of the results obtained by the latter method reflects a strong dependence on the choice of the structural parameters of the system.

V 126   Jung, B.; Schürmann, B. L. *Makromol. Chem., Rapid Commun.* **1989**, *10*, 419.

The RIS model and the Porod-Kratky model are discussed with regard to their respective merits to predict the persistence length of macromolecules on the basis of a molecular dynamics trajectory. Three different polyesters are compared: poly(p-hydroxybenzoate) [poly(oxy-1,4-phenylenecarbonyl)], poly(ethylene terephthalate) [poly(oxyethyleneoxyterephthaloyl)], and poly(4-hydroxybicyclo[2.2.2]octane-1-carboxylate) [poly(oxybicyclo[2.2.2]octane-1,4-diylcarbonyl)].

V 127   Depner, M.; Schürmann, B. L. *Polymer* **1992**, *33*, 398.

A theoretical analysis of the possible conformations of poly(p-phenylene terephthalate) (PPTA) and poly(p-phenylene isophthalate) (PPIA) is performed on the basis of molecular mechanics and molecular dynamics trajectories. The dependence of the persistence length on the fluctuations of the torsional angle around the ester bond is discussed for PPTA in the frame of the RIS model. Realistic parameters like bond length and bond angles are provided by computer simulations using MD.

V 128   Charati, S. G.; Vetrivel, R.; Kulkarni, M. G.; Kulkarni, S. S. *Macromolecules* **1992**, *25*, 2215.

The paper describes the use of molecular modeling to study the flexibility and conformation of single-chain sections of three polyarylates based on terephthalic acid and (i) bisphenol A, (ii) bisphenol based on methyl isobutyl ketone, and (iii) phenolphthalein. Configurational entropies are calculated based on the ease of rotation of various bonds. Entropies calculated for the cooperative rotation of the bisphenol phenyl rings do not correlate with the experimentally determined glass transition temperatures. However, a correlation is found using entropies determined for the independent rotation of various bonds keeping others in their minimum-energy conformation.

V 129   Fusco, R.; Longo, L.; Caccianotti, L.; Aratari, C.; Allegra, G. *Makromol. Chem., Theory Simul.* **1993**, *2*, 685.

The chain rigidity of poly(p-hydroxybenzoate) is estimated through the theoretical evaluation of its persistence length. A non-Brownian molecular dynamics simulation of an isolated chain with 20 monomeric units is performed. The sampled conformational population is analyzed and the orientational correlation function between monomeric units along the chain is calculated. An algorithm based on the worm-like chain model is applied to evaluate the persistence length. The results are compared with those obtained from equilibrium models like the freely-rotating-chain and the rotational-matrix method.

*Calculations on **ribbon-like polyaromatic** chains:*

V 130 Welsh, W. J.; Bhaumik, D.; Mark, J. E. *Macromolecules* **1981**, *14*, 947.

Planarity, or departures therefrom, can be of considerable importance in the formation of either crystalline or liquid-crystalline phases in rigid-rod polymeric systems. Of the four aromatic heterocyclic polymers *cis*- and *trans*-poly(benzobisoxazoles) (PBO) and -poly(benzobisthiazoles) (PBT), the *cis*-PBO and *trans*-PBT are known to form such phases. The present investigation involves energy calculations carried out to characterize any deviations from planarity arising from *p*-phenylene group rotations along these chain backbones. Intramolecular (conformational) energy calculations correctly indicate that the two PBO polymers should be planar, and similar calculations on the PBT polymers correctly predict nonplanarity.

V 131 Bhaumik, D.; Welsh, W. J.; Jaffe, H. H.; Mark, J. E. *Macromolecules* **1981**, *14*, 951.

Interaction energies between relatively rigid benzobisoxazole and benzobisthiazole polymers are calculated in an attempt to gain insight into the very high mechanical strength and unusual solvent resistance of these materials. The predicted details of the chain packing and the corresponding densities are found to be in good agreement with experimental results obtained on model compounds in the crystalline state.

V 132 Welsh, W. J.; Bhaumik, D.; Mark, J. E. *J. Macromol. Sci. - Phys.* **1981**, *B 20*, 59.

Some aromatic heterocyclic polymers are rigid or rodlike and are very nearly intractable unless some atoms or groups of high flexibility are introduced along the chains. This theoretical investigation employs semiempirical methods to calculate intramolecular energies of various conformations about such "molecular swivels" in order to characterize their flexibility. In addition, simple geometric arguments are used to determine which swivels have one or more conformations exhibiting the parallel or colinear chain continuation.

V 133 Welsh, W. J.; Jaffé, H. H.; Kondo, N.; Mark, J. E. *Makromol. Chem* **1982**, *183*, 801.

Aromatic heterocyclic polymers are very difficult to process unless "swivel" atoms or groups are inserted along the chains to increase conformational flexibility. The present theoretical investigation employs the CNDO/2 method with direct geometry optimization to calculate such flexibility for the wholly aromatic swivels biphenyl, 2,2´-bipyridyl, 2-phenylpyridine, 2,2´-bipyrimidyl, and 2-phenylpyrimidine. The most important result is the prediction that both flexibility and accessability of coplanar conformations should increase significantly with the number of *ortho*-CH groups replaced by N-atoms. The calculations also provide information on other conformation-dependent properties such as optimized geometries, charge distributions, and dipole moments.

V 134 Welsh, W. J.; Mark, J. E. *J. Mater. Sci.* **1983**, *18*, 1119.

While essentially rigid in the axial direction, the rod-like polymers *cis*- and *trans*-polybenzbisoxazole (PBO) and polybenzobisthiazole (PBT) do exhibit conformational flexibility with respect to rotations about the bonds between alternating phenylenes and heterocyclic groups. Since preparation of high-strength materials from these polymers required a high degree of alignment, flexibility should be important in this regard. CNDO/2 molecular orbital calculations are therefore carried out to obtain the conformational-energy profiles of related model compounds.

V 135   Welsh, W. J.; Mark, J. E. *Polym. Engin. Sci.* **1983**, *23*, 140.

Geometry-optimized CNDO/2 calculations are carried out in an attempt to predict the effect of protonation on the conformational characteristics and geometry of PBO model compounds. Values of the conformational energy *vs.* rotation of the end-phenylenes about the heterocyclic group are calculated for *cis*-PBO model compounds in the unprotonated form and as $2H^+$ and $4H^+$ ions.

V 136   Welsh, W. J.; Mark, J. E. *Polym. Engin. Sci.* **1985**, *25*, 965.

Geometry-optimized CNDO/2 molecular orbital calculations are carried out on poly(5,5´-bibenzoxazole-2,2´-diyl-1,4-phenylene)- and poly(2,5-benzoxazole)-model compounds to determine conformational energies as a function of rotation about each type of rotatable bond within the repeat units.

V 137   Viswanadhan, V. N.; Bergmann, W. R.; Mattice, W. L. *Macromolecules* **1987**, *20*, 1539.

Configuration-dependent physical properties of homopolymers of C(4)-C(8) linked (+)-catechin or (−)-epicatechin, linked from C(4) of the one monomer to C(8) of the next, are investigated by using RIS theory. The characteristic ratios and components of the persistence vector are evaluated for these polymers using structural information obtained from MM2 calculations for dimers of (+)-catechin and (−)-epicatechin. There are two rotational isomers at the interflavan bond between monomer units. Helices are formed if one rotational isomer is populated to the exclusion of the other. The chirality of the helix is determined solely by the selection of the rotational isomer. It is independent of the stereochemistry at C(3) and C(4). The polymers form random coils with unperturbed dimensions smaller than those of atactic polystyrene with the same molecular weight if the relative population of the two rotational isomers is assigned in the manner required by recent time-resolved fluorescence measurements and the rotational isomers are randomly distributed along the chain.

V 138   Zhang, R.; Mattice, W. L. *Macromolecules* **1992**, *25*, 4937.

RIS theory provides a relatively simle formalism for the evaluation of the persistence vector, **a**, for a chain that can be represented by a repeating sequence of independent virtual bonds such as polybenzobisoxazole (PBO) and polybenzobisthiazole (PBT). The present study combines RIS theory with long molecular dynamics simulations for small fragments in order to evaluate the limiting length of **a** for very stiff chains. The approach can be applied to other stiff chain polymers.

cis-PBO:		*trans*-PBO:		*cis*-PBT:		*trans*-PBT:	
$\langle l_l \rangle = 648.5$	$\langle \theta_l \rangle = 174.90$	$\langle l_l \rangle = 648.4$	$\langle \theta_l \rangle = 174.82$	$\langle l_l \rangle = 681.3$	$\langle \theta_l \rangle = 165.12$	$\langle l_l \rangle = 683.5$	$\langle \theta_l \rangle = 169.20$
$\langle l_s \rangle = 286.1$	$\langle \theta_s \rangle = 175.44$	$\langle l_s \rangle = 286.1$	$\langle \theta_s \rangle = 175.42$	$\langle l_s \rangle = 288.0$	$\langle \theta_s \rangle = 175.61$	$\langle l_s \rangle = 287.9$	$\langle \theta_s \rangle = 175.64$
$\langle l_c \rangle = 141.0$		$\langle l_c \rangle = 141.0$		$\langle l_c \rangle = 141.5$		$\langle l_c \rangle = 141.5$	

Bond length are given in [pm], angles in degrees.

With regard to this paper, see also the controversial discussion in: V 138a   Roitman, D. B.; McAdon, M. *Macromolecules* **1993**, *26*, 4381, and
   V 138b   Zhang, R.; Mattice, W. L. *Macromolecules* **1993**, *26*, 4384.

V 139   Zhang, R.; Mattice, W. L. *Macromolecules* **1993**, *26*, 6100.

The flexibility of a thermoplastic polyimide, PI-2, synthesized from 3,3',4,4'-benzophenonetetracarboxylic dianhydride (BTDA) and 2,2-dimethyl-1,3-bis(4-aminophenoxy)propane (DMDA), is studied with atom-based molecular modeling. By molecular dynamics simulations and conformational grid search it can be shown that there is a substantial amount of flexibility within BTDA, the so-called rigid part of PI-2. Quantitative characterization of the flexibility of a PI-2 single chain is achieved by studying the persistence length and the characteristic ratio, using the transformation matrix and generator matrix formalism.

poly(cyclohexene sulphone)

Fawcett, A. H.; Ivin, K. J. *Polymer* **1975**, *16*, 569, 573.

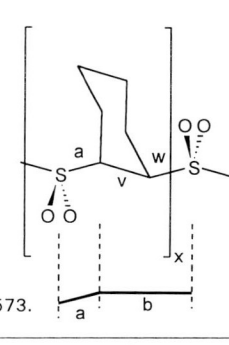

ξ	(E_ξ [kJ·mol^{-1}]) a)
(for 298 K)	
For atactic chains	
f = 0	(∞)
σ = 1	(0)
ω = 0	(∞)
α = 8.42	(−5.28)
γ = 2.33	(−2.1)

	w ←	Bond a Chirality of ring adj. to bond				Bond w Chirality of central ring		
		L		D		L		D
Bond w		t g$^+$ g$^-$		t g$^+$ g$^-$	Bond a	t g$^+$ g$^-$		t g$^+$ g$^-$
L	t	1 0 f		f 1 0	LLL t	1 0 α		1 α 0
	g$^+$	0 0 0		0 0 0	DDD g$^+$	0 0 0		α ωγ 0
	g$^-$	f 0 σ		1 f 0	g$^-$	α 0 ωγ		0 0 0
D	t	f 0 1		1 f 0	DLL,LLD t	f 0 α		1 α 0
	g$^+$	1 0 f		f σ 0	DLD,LDD g$^+$	0 0 0		α γ 0
	g$^-$	0 0 0		0 0 0	DDL,LDL g$^-$	α 0 γ		0 0 0

rows and columns: t, g$^+$, g$^-$

Isotactic: −[S−(C)−C(D)−(S)]−

Syndiotactic: −[(S)−C−(C)(L)−S−(C)−C(D)−(S)]−

Isotactic matrices:
$$\begin{bmatrix} 1 & f & 0 \\ f & \sigma & 0 \\ 0 & 0 & 0 \end{bmatrix} \begin{bmatrix} 1 & \alpha & 0 \\ \alpha & \gamma\omega & 0 \\ 0 & 0 & 0 \end{bmatrix}$$

Syndiotactic matrices:
$$\begin{bmatrix} f & 0 & 1 \\ 1 & 0 & f \\ 0 & 0 & 0 \end{bmatrix} \begin{bmatrix} 1 & 0 & \alpha \\ 0 & 0 & 0 \\ \alpha & 0 & \gamma \end{bmatrix} \begin{bmatrix} f & 1 & 0 \\ 0 & 0 & 0 \\ 1 & f & 0 \end{bmatrix} \begin{bmatrix} 1 & \alpha & 0 \\ \alpha & \gamma & 0 \\ 0 & 0 & 0 \end{bmatrix}$$

a) For more details, see original paper.

Comments:
The RIS model, coupled with the Flory matrix method, is applied to the calculation of the unperturbed mean-square end-to-end distance in poly(cyclohexene sulphone) as a function of several parameters. The calculations are performed for atactic, isotactic and syndiotactic chains; the tacticity arises from the two possible ways, D and L, in which the rings can be attached to the main chain, assuming that the C−C bonds are all in the *trans* conformation, as indicated by dielectric measurements.

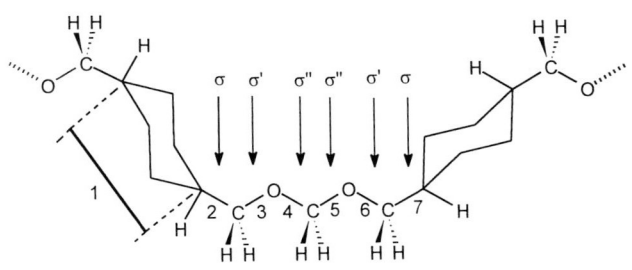

poly(*trans*-1,4-cyclohexylene-dimethylene-oxymethylene oxide), PTCDM

Riande, E.; Guzmán, J.; Saiz, E. *Polymer* **1981**, *22*, 465.

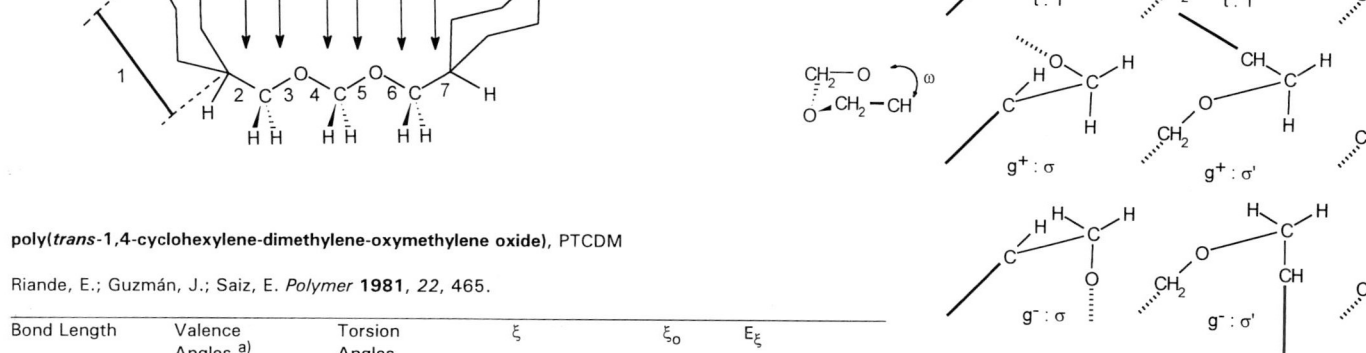

Bond Length [pm]	Valence Angles [°] [a)]	Torsion Angles [°]	ξ (for 298 K)	ξ_0	E_ξ [kJ · mol^{-1}]				
C-C : 154	C-C-C : 111.5	For bonds 2-7:	$\sigma = 0.51$	1	1.67				
		t : 180							
		g$^+$: 60	$\sigma' = 0.51$	1	1.67	$U_2 = \begin{bmatrix} 1 & \sigma & \sigma \end{bmatrix}$	$U_3 = \begin{bmatrix} 1 & \sigma' & \sigma' \\ 1 & \sigma' & 0 \\ 1 & 0 & \sigma' \end{bmatrix}$	$U_4 = \begin{bmatrix} 1 & \sigma'' & \sigma'' \\ 1 & \sigma'' & \sigma''\omega \\ 1 & \sigma''\omega & \sigma'' \end{bmatrix}$	For matrix U_2: rows / columns:
		g$^-$: -60							
			$\sigma'' = 6.4$	1	-4.6 ± 0.2				ϕ / t, g$^+$, g$^-$
		For virtual bond 1:							
		ϕ [b)]	$\omega = 0.51$	1	1.67	$U_5 = \begin{bmatrix} 1 & \sigma'' & \sigma'' \\ 1 & \sigma'' & 0 \\ 1 & 0 & \sigma'' \end{bmatrix}$	$U_6 = \begin{bmatrix} 1 & \sigma' & \sigma' \\ 1 & \sigma' & \sigma'\omega \\ 1 & \sigma'\omega & \sigma' \end{bmatrix}$	$U_7 = \begin{bmatrix} 1 & \sigma & \sigma \\ 1 & \sigma & 0 \\ 1 & 0 & \sigma \end{bmatrix}$	For 3 × 3 matrices: rows and columns: t, g$^+$, g$^-$

a) Since the energy of the boat form is 23 kJ · mol^{-1} above the chair form, the chair conformation is the only one that is considered in these calculations. Moreover, the equatorial-equatorial conformation is the most favoured one in *trans*-1,4-cyclohexane dimethanol.

b) Virtual bonds are fixed at the position shown in the figure without allowing rotational freedom.

Calcd. quantities: $\langle\mu^2\rangle / xm^2$ = 0.17 ± 0.01 (*Exptl.*: 0.17 to 0.21)

$d(\ln \langle\mu^2\rangle)/dT$ = 5.0 × 10^{-3} K^{-1} (*Exptl.*: 5.5 × 10^{-3} K^{-1})

poly(thiomethylene-1,4-*trans*-cyclohexylenemethylenethiomethylene), PTCMT

de la Peña, J. L.; Riande, E.; Guzmán, J. *Macromolecules* **1985**, *18*, 2739.

Bond Length [pm]	Valence Angles [°] [a]	Torsion Angles [°]	ξ (for 300 K)	ξ_0	E_ξ [kJ·mol^{-1}]
C-C : 153	C-C-C : 111.5	For bonds 2-7: t : 180	$\sigma = 1.95$	1	−1.67
C-S : 181.5	C-S-C : 100	g^+ : 60 g^- : −60	$\sigma' = 7.48$	1	−5.02
C-H : 109	C-C-S : 114				
l_{vb} : 297	C-C-H : 110	For virtual bond 1: ϕ [b]			

$$U_1 = \begin{bmatrix} 1 & 0 & 0 \\ 1 & 0 & 0 \\ 1 & 0 & 0 \end{bmatrix} \quad U_2 = \begin{bmatrix} 1 & \sigma & \sigma \\ 0 & 0 & 0 \\ 0 & 0 & 0 \end{bmatrix} \quad U_3 = U_6 = \begin{bmatrix} 1 & 0 & 0 \\ 1 & 0 & \sigma \\ 1 & \sigma & 0 \end{bmatrix}$$

$$U_4 = U_5 = \begin{bmatrix} 1 & \sigma' & \sigma' \\ 1 & \sigma' & 0 \\ 1 & 0 & \sigma' \end{bmatrix} \quad U_7 = \begin{bmatrix} 1 & \sigma & \sigma \\ 0 & 0 & \sigma \\ 0 & \sigma & 0 \end{bmatrix}$$

rows and columns: t, g^+, g^-

[a] Since the energy of the boat form is 23 kJ·mol^{-1} above the chair form, the chair conformation is the only one that is considered in these calculations. Moreover, the equatorial-equatorial conformation is the most favoured one in *trans*-1,4-cyclohexane dimethanol.

[b] Virtual bonds are fixed at the position shown in the figure without allowing rotational freedom.

Calcd. quantities:

$<\mu^2> / xm^2$	= 0.25	(Exptl.: 0.265)
$d(\ln <\mu^2>)/dT$	= 3.7×10^{-3} K^{-1}	(Exptl.: 4.3×10^{-3} K^{-1})

Configurational entropy S_c
Configurational optical parameter Δa

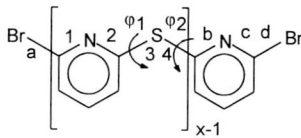

poly(2,6-pyridinediyl sulfide), PPγS

Tarazona, M. P.; Boileau, S.; de Leuze, A.; Saiz, E.; Sanchez, E.; Días-Calleja, R.; Riande, E. *Macromolecules* **1992**, *25*, 5020.

Bond Length [pm]		Valence Angles [°]		Torsion Angles [°]		ξ (for 318 K)	ξ_0	E_ξ [kJ·mol^{-1}]
C^{ar}-H	: 108	H-C^{ar}-N	: 119.4	For bonds 3,4:	$\gamma = 0.79$	1	0.63	
		Br-C^{ar}-N	: 119.4	g^+	: 80			
C^{ar}-Br	: 185	S-C^{ar}-N	: 119.4	g^-	: −80			
		C^{ar}-N-C^{ar}	: 121.5					
C^{ar}-N	: 134.9	N-C^{ar}-C^{ar}	: 120.4					
		H-C^{ar}-C^{ar}	: 120.4					
C^{ar}-S	: 177	C^{ar}-C^{ar}-C^{ar}	: 119.2					
		C^{ar}-S-C^{ar}	: 97.4					
C^{ar}-C^{ar}	: 139.8							

$$U_a = \begin{bmatrix} 1 & 0 \end{bmatrix} \quad U_1 = \begin{bmatrix} 1 & 0 \\ 0 & 0 \end{bmatrix} \quad U_2 = \begin{bmatrix} 1 & 0 \\ 0 & 0 \end{bmatrix} \quad U_3 = \begin{bmatrix} 1 & 1 \\ 0 & 0 \end{bmatrix}$$

$$U_4 = \begin{bmatrix} 1 & \gamma \\ \gamma & 1 \end{bmatrix} \quad U_b = \begin{bmatrix} 1 & 0 \\ 1 & 0 \end{bmatrix} \quad U_c = \begin{bmatrix} 1 & 0 \\ 0 & 0 \end{bmatrix} \quad U_d = \begin{bmatrix} 1 \\ 1 \end{bmatrix}$$

For U_4: rows and columns: g^+, g^-; other matrix indices are assigned accordingly

Calcd. quantities: $\langle \mu^2 \rangle^{1/2}$ = 3.52 D (Exptl.: 3.53 D)

polypyrrole

Yurtsever, E.; Erman, B. *Polymer* **1993**, *34*, 3887.

Bond Length [pm]	Valence Angles [°]	Torsion Angles [°]	ξ (for 300 K)	ξ_o	E_ξ [kJ·mol^{-1}]		
l_{vb} : 388	θ : 43.6	t : 180	α = 0.266	1	3.31		
		g^+ : 30	β = 0.121	1	5.28	$U = \begin{bmatrix} 1 & \alpha & \alpha \\ 1 & \beta & \beta \\ 1 & \beta & \beta \end{bmatrix}$	rows and columns: t, g^+, g^-
		g^- : −30					

Comments:
Conformational energy surfaces are determined as functions of torsional angles of bonds between rings for the α-α pyrrole trimer by using semiempirical quantum-mechanical calculations (AM1). Isomeric states for torsional angles are identified as positions of minimum energy. Statistical weights determined on this basis are used in calculating the average dimensions of polypyrrole chains by the RIS model and the matrix multiplication scheme. Results of calculations indicate that the chains have unusually large characteristic ratios and persistence length with very strong temperature dependence. Characteristic ratios obtained by the RIS model show perfect agreement, in a wide temperature range, with the predictions of the worm-like chain model.

Calcd. quantities:

$\langle r^2 \rangle_o / nl^2$ = 53.5

Persistence vector $\langle r \rangle$ ≈ 112 Å

$d (\ln \langle r^2 \rangle_o) / dT$ = −3.69 × 10^{-3} K^{-1}

$\langle r \rangle / l_{vb}$ ≈ 29.0

poly(tetramethyl-*p*-silphenylene-siloxane), PTMPS

Wang, S.; Mark, J. E. *Comput. Polym. Sci.* **1993**, *3*, 33.

Bond Length [pm]	Valence Angles [°]	Torsion Angles [°]	ξ	ξ_0	E_ξ [kJ·mol^{-1}]			
Si-O : 164	Si-O-Si : 142.5	t : 180	σ = 0 to 1	1	∞ to 0	$U_1 = \begin{bmatrix} 1 & \sigma & \sigma \\ 1 & \alpha\sigma & \beta\sigma \\ 1 & \beta\sigma & \alpha\sigma \end{bmatrix}$	$U_2 = \begin{bmatrix} 1 & \sigma & \sigma \\ 1 & \gamma\sigma & 0 \\ 1 & 0 & \gamma\sigma \end{bmatrix}$	$U_3 = \begin{bmatrix} 1 & \sigma' & \sigma' \\ 1 & \alpha\sigma' & \beta\sigma' \\ 1 & \beta\sigma' & \alpha\sigma' \end{bmatrix}$
Si-C^{ar} : 190	O-Si-C^{ar} : 104	g$^+$: 60	σ' = 1	1	0			
C^{ar}-C^{ar} : 139		g$^-$: –60	α = 1	1	0			
l$_{vb}$: 658 a)			β = 1	1	0	rows and columns: t, g$^+$, g$^-$		
			γ = 0 to 1	1	∞ to 0			

Comments:
The unperturbed dimensions and characteristic ratios of PTMPS chains are calculated using the RIS approximation. The low experimental value of the characteristic ratio reported for this polymer is successfully interpreted by these computations. The stiffening effect expected from introduction of the *p*-phenylene group into the chain backbone is apparently offset by its reducing the repulsive interactions that would otherwise occur among the atoms or groups before and after it along the chain.

Calcd. quantities: $<r^2>_0 / nl^2$ = 1.60 or 4.22 a)

Entropy change at constant volume, $(\Delta S_u)_V$
Configurational entropy, S_{conf}

a) The characteristic ratio changes from 1.3 to 2.8 with σ changing from 0 to 1, when the virtual bond is used. On the other hand, when each bond of the phenylene group is taken into account individually, the two extreme values are 3.41 and 7.40. By assuming all the statistical weight factors to be unity, which corresponds to the freely-rotating chain, the characteristic ratio is 1.60 when the virtual bond is used, and 4.22 if it is not.

*Further calculations on **polyaromatic** chains:*

V 146   Orchard, B. J.; Freidenreich, B.; Tripathy, S. K. *Polymer* **1986**, *27*, 1533.

Structural modeling of two relatively stable conducting polymers, polythiophene and polypyrrole, is carried out using a number of theoretical modeling tools. These have included determination of optimized valence molecular geometry: conformational properties of single isolated chains and their crystalline packing arrangements.

V 147   Ito, Y.; Ihara, E.; Murakami, M.; Sisido, M. *Macromolecules* **1992**, *25*, 6810.

Empirical conformational energy calculations are performed on helical poly(2,3-quinoxaline)s to predict stable conformations. Two energy minimum conformations are found by varying the dihedral angle, ψ, between two adjacent quinoxaline units from 5 to 180°. Circular dichroism spectra are calculated for the two stable conformations (ψ = 45 and 135°) on the basis of exciton theory.

ethylene – carbon monoxide copolymers

Wittwer, H.; Pino, P.; Suter, U. W. *Macromolecules* **1988**, *21*, 1262.

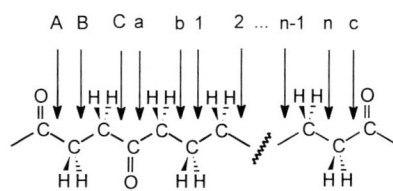

ξ a)	ξ_0	E_ξ [kJ·mol^{-1}]	ξ a)	ξ_0	E_ξ [kJ·mol^{-1}]
(for 400 K)			(for 400 K)		
$\sigma_b = 0.29$	0.7	2.9	$\sigma_A = 0.07$	1.6	10.5
$\sigma_a = 0.54$	1.7	3.8	$\omega_{AB} = 1.70$	0.8	−2.5
$\omega_{ab} = 0.90$	0.9	0.0	$\psi = 0.48$	1.3	3.3
$\kappa_{ab} = 0.63$	0.8	0.8	$\kappa_{AB} = 0.33$	0.8	2.9
$\kappa'_{ab} = 1.50$	1.5	0.0	$\kappa'_{AB} = 0.54$	0.9	1.7
$\kappa''_{ab} = 0.59$	1.1	2.1	$\kappa''_{AB} = 0.20$	0.8	4.6
$\lambda = 0.12$	1.0	0.4	$\sigma_{PE} = 0.53$	1.0	2.1
$\sigma_B = 0.54$	0.8	1.3	$\omega_{PE} = 0.08$	1.0	8.4

rows and columns: t^-, t or t^0, t^+, g^+, g^-

$$U_x = \begin{bmatrix} 1 & \sigma_{PE} & \sigma_{PE} \\ 1 & \sigma_{PE} & \sigma_{PE}\omega_{PE} \\ 1 & \sigma_{PE}\omega_{PE} & \sigma_{PE} \end{bmatrix}$$

$$U_A = \begin{bmatrix} 1 & 1 & 1 & \sigma_A & \sigma_A \\ 1 & 1 & 1 & \sigma_A & \sigma_A \\ 1 & 1 & 1 & \sigma_A & \sigma_A \\ 1 & 1 & 1 & \sigma_A & \sigma_A \\ 1 & 1 & 1 & \sigma_A & \sigma_A \end{bmatrix}$$

$$U_a = \begin{bmatrix} \lambda & 1 & \lambda & \sigma_a & \sigma_a \\ \lambda & 1 & \lambda & \sigma_a & \sigma_a \\ \lambda & 1 & \lambda & \sigma_a & \sigma_a \\ \lambda & 1 & \lambda & \sigma_a & \sigma_a \\ \lambda & 1 & \lambda & \sigma_a & \sigma_a \end{bmatrix}$$

$$U_B = \begin{bmatrix} \kappa_{AB} & \sigma_B\kappa'_{AB} & \sigma_B\kappa''_{AB} \\ 1 & \sigma_B & \sigma_B \\ \kappa_{AB} & \sigma_B\kappa''_{AB} & \sigma_B\kappa'_{AB} \\ 1 & \sigma_B\psi & \sigma_B\omega_{AB} \\ 1 & \sigma_B\omega_B & \sigma_B\psi \end{bmatrix}$$

$$U_b = \begin{bmatrix} 1 & \sigma_b & \sigma_b \\ 1 & \sigma_b & \sigma_b \\ 1 & \sigma_b & \sigma_b \\ 1 & \sigma_b & \sigma_b\omega_{ab} \\ 1 & \sigma_b\omega_{ab} & \sigma_b \end{bmatrix}$$

$$U_C = \begin{bmatrix} \kappa_{AB} & 1 & \kappa_{AB} & \sigma_A & \sigma_A \\ \kappa'_{AB} & 1 & \kappa''_{AB} & \sigma_A\psi & \sigma_A\omega_{AB} \\ \kappa''_{AB} & 1 & \kappa'_{AB} & \sigma_A\omega_{AB} & \sigma_A\psi \end{bmatrix}$$

$$U_c = \begin{bmatrix} \kappa_{ab} & 1 & \kappa_{ab} & \sigma_a & \sigma_a \\ \kappa_{ab} & 1 & \kappa_{ab} & \sigma_a & \sigma_a \\ \kappa_{ab} & 1 & \kappa_{ab} & \sigma_a\omega_{ab} & \sigma_a \end{bmatrix}$$

a) For details, see the original publication.

Comments:
Bond length and bond angles are kept fixed at values used earlier (Suter, U. W. *J. Am. Chem. Soc.* **1979**, *101*, 6481). For the CH_2–CH_2 bonds three approximately staggered conformations are appropriate. At first, rotation about CO–CH_2 bond is subdivided into the same three isomeric states, but in the course of the calculations it becomes clear that the *trans* region cannot be satisfactorily modeled by a single conformation. The region is consequently split into three equal domains, so that five rotational isomeric states result: g^-: 0° to −120°, t^-: −120° to −160°, t^0: −160° to 160°, t^+: 160° to 120°, g^+: 120° to 0°.

ethylene – vinyl chloride copolymers

Mark, J. E. *Polymer* **1973**, *14*, 553.

Bond Length [pm]	Valence Angles [°]	Torsion Angles [a)] [°]	ξ (for 298 K)	ξ_o	E_ξ [kJ · mol^{-1}]	For the assignment of the statistical weight parameters, ξ, see the models of the corresponding homopolymers
C-C : 153	C-C-C : 112	t : 180	η = 4.2	1	– 3.556	
		g$^+$: 60	$\omega = \omega'' = 0.032$	1	8.528	
		g$^-$: –60	$\omega' = 0.071$	1	6.554	
			τ = 0.45	1	1.978	
			τ^* = 1.0	1	0	

$$U_e = \begin{bmatrix} 1 & \tau/\eta & \tau/\eta \\ 1 & \tau/\eta & \tau\omega/\eta \\ 1 & \tau\omega/\eta & \tau/\eta \end{bmatrix} \quad U_d = \begin{bmatrix} \eta\tau^* & 1 & \tau \\ \eta & 1 & \tau\omega \\ \eta & \omega & \tau \end{bmatrix} \quad U_{dd} = \begin{bmatrix} \eta'\omega'' & \tau\omega' & 1 \\ \eta & \tau\omega' & \omega \\ \eta\omega' & \tau\omega'' & \omega' \end{bmatrix}$$

$$U_{dl} = \begin{bmatrix} \eta & \omega' & \tau\omega'' \\ \eta\omega' & 1 & \tau\omega \\ \eta\omega'' & \omega & \tau\omega' \end{bmatrix}^2 \quad U_{de} = \begin{bmatrix} \eta/\tau & \omega' & 1 \\ \eta/\tau & 1 & \omega \\ \eta/\tau & \omega & \omega' \end{bmatrix} \quad U_{ed} = \begin{bmatrix} \eta & \tau & 1 \\ \eta & \tau\omega' & \omega \\ \eta\omega' & \tau\omega & 1 \end{bmatrix}$$

rows and columns: t, g$^+$, g$^-$ (see: b))

a) Since intramolecular steric interactions may be relatively large within ethylene units, the model allows for the displacement of these rotational states by an amount of $\Delta\phi$ = 0 and 10° from their symmetric locations. Specifically, these states are located at 180°, 60° + $\Delta\phi$, and – 60° – $\Delta\phi$, respectively. However, it can be shown that the location of the rotational states is not of crucial importance in the present analysis. Thus, $\Delta\phi$ = 0° can be considered for all calculations performed here.

b) For CH$_2$–CH$_2$–CH$_2$ (ethylene) bond pairs, the statistical weight matrix is designated U_e.

Comments:
RIS theory is used to calculate mean-square unperturbed dimensions $<r^2>_o$ and dipole moments $<\mu^2>$ of ethylene-vinyl chloride copolymers as a function of chemical composition, chemical sequence distribution, and stereochemical composition of the vinyl chloride sequences. As was previously found for several other copolymeric chains, $<\mu^2>$ is much more sensitive to chemical composition and chemical sequence distribution than is $<r^2>_o$. The present calculations also indicate that both $<r^2>_o$ and $<\mu^2>$ are most strongly dependent on chemical sequence distribution for ethylene-vinyl chloride chains having vinyl chloride sequences which are significantly syndiotactic in structure.

Calcd. quantites: $<r^2_o> / nl^2$ $<\mu^2> / x$ (x = n/2 = 100)

ethylene – vinyl bromide copolymers, PEVB

Tonelli, A. E. *Macromolecules* **1982**, *15*, 290.

Bond Length [pm]	Valence Angles [°]	Torsion Angles [°]	(for 323 K)
C-C : 153	C-C-C : 112	t : 180	
C-H : 110	H-C-H : 109	g^+ : 60	
C-Br: *191*	H-C-Br : 109	g^- : –60	

rows and columns: t, g^+, g^-

$$U_{dd(CHBr-CH_2-CHBr)} = \begin{bmatrix} 0.005 & 0.006 & 0.370 \\ 0.370 & 0.010 & 0.224 \\ 0.006 & 0.000 & 0.010 \end{bmatrix} \quad U_{dl(CHBr-CH_2-CHBr)} = \begin{bmatrix} 0.184 & 0.061 & 0.001 \\ 0.061 & 0.638 & 0.028 \\ 0.001 & 0.028 & 0.000 \end{bmatrix}$$

$$U_{d(CH_2-CHBr-CH_2)} = \begin{bmatrix} 0.143 & 0.143 & 0.143 \\ 0.143 & 0.143 & 0.000 \\ 0.143 & 0.000 & 0.143 \end{bmatrix} \quad U_{de(CHBr-CH_2-CH_2)} = \begin{bmatrix} 0.180 & 0.017 & 0.086 \\ 0.339 & 0.204 & 0.075 \\ 0.086 & 0.009 & 0.004 \end{bmatrix}$$

$$U_{ed(CH_2-CH_2-CHBr)} = \begin{bmatrix} 0.180 & 0.086 & 0.339 \\ 0.086 & 0.004 & 0.075 \\ 0.017 & 0.009 & 0.204 \end{bmatrix} \quad U_{e(CH_2-CH_2-CH_2)} = \begin{bmatrix} 0.205 & 0.094 & 0.094 \\ 0.205 & 0.094 & 0.004 \\ 0.205 & 0.004 & 0.094 \end{bmatrix}$$

Comments:
Conformational energies are calculated for chain segments in poly(vinyl bromide) (PVB) homopolymer and the copolymers of vinyl bromide (VB) and ethylene (E), PEVB. Semiempirical potential functions are used to account for the nonbonded van der Waals and electrostatic interactions. RIS models are developed for PVB and PEVB from the calculated conformational energies. Dimensions and dipole moments are calculated for PVB and PEVB using their RIS models, where the effects of stereosequence and comonomer sequence are explicitly considered. It is concluded from the calculated dimensions and dipole moments that the dipole moments are most sensitive to the microstructure of PVB homopolymers and PEVB copolymers and may provide an experimental means for their structural characterization.

Calcd. quantites: $\quad <r^2_0>/nl^2 \quad\quad C_m \quad\quad C_n \quad\quad d(\ln<r^2_0>)/dT \quad\quad d(\ln<\mu^2>)/dT$

propylene – vinyl chloride copolymers

Mark, J. E. *J. Polym. Sci.: Polym. Phys. Ed.* **1973**, *11*, 1375.

Bond Length [pm]	Valence Angles [°]	Torsion Angles [a] [°]	ξ (for 298 K)	ξ_0	E_ξ [kJ · mol^{-1}]	
C-C : 153	C-C-C : 112	t : 180	τ = 0.45	1	1.98	For the assignment of the statistical weight parameters, ξ, see the models of the corresponding homopolymers
						For a bond-pair meeting at a $CHCH_3$ group:
		g$^+$: 60	$\omega = \omega'' = 0.032$	1	8.528	
		g$^-$: −60	$\omega' = 0.071$	1	6.554	$U_d = \begin{bmatrix} \eta & 1 & \tau \\ \eta & 1 & \tau\omega \\ \eta & \omega & \tau \end{bmatrix}$ rows and columns: t, g$^+$, g$^-$
			η (PP) = 1.0	1	0	
			η' (VC) = 4.2	1	−3.56	For bond pairs separating two CHCl groups:

$$U_{dd} = \begin{bmatrix} \eta'\omega'' & \tau\omega' & 1 \\ \eta' & \tau\omega' & \omega \\ \eta'\omega' & \tau\omega\omega'' & \omega' \end{bmatrix} \quad U_{dl} = \begin{bmatrix} \eta' & \omega' & \tau\omega'' \\ \eta'\omega' & 1 & \tau\omega \\ \eta'\omega'' & \omega & \tau\omega'^2 \end{bmatrix}$$

For $CHCH_3$–CH_2–$CHCl$ bond pairs: For rotational states about $CHCl$–CH_2–$CHCH_3$ bond pairs:

$$U_{dd} = \begin{bmatrix} \eta'\omega' & \tau\omega & 1 \\ \eta' & \tau\omega' & \omega \\ \eta'\omega' & \tau\omega\omega' & \omega \end{bmatrix} \quad U_{dl} = \begin{bmatrix} \eta' & \omega & \tau\omega' \\ \eta'\omega' & 1 & \tau\omega \\ \eta'\omega' & \omega & \tau\omega\omega' \end{bmatrix} \qquad U_{dd} = \begin{bmatrix} \eta\omega' & \tau\omega' & 1 \\ \eta & \tau\omega & \omega \\ \eta\omega & \tau\omega\omega' & \omega' \end{bmatrix} \quad U_{dl} = \begin{bmatrix} \eta & \omega' & \tau\omega' \\ \eta\omega & 1 & \tau\omega \\ \eta\omega' & \omega & \tau\omega\omega' \end{bmatrix}$$

[a] Since intramolecular steric interactions may be relatively large within propylene units, the model allows for the displacement of these rotational states by an amount of 10° from their symmetric locations. Specifically, these states are located at 180° + Δϕ / 60° / −60° − Δϕ and 180° − Δϕ / 60° + Δϕ / −60° for the two skeletal bonds leading, respectively, into and out of a $CHCH_3$ group of d atomic configuration. The same two sets of rotational angles pertain to the two skeletal bonds leading, respectively, out of and into a $CHCH_3$ group of l configuration.

Comments:
Mean-square unperturbed dimensions and dipole moments are calculated for propylene-vinyl chloride copolymers by means of RIS theory. The calculations indicate that for these chain molecules $<\mu^2>$ is much more sensitive to chemical sequence distribution than is $<r^2>_0$, a conclusion in agreement with results of previous studies of ethylene-propylene copolymers and styrene-substituted styrene copolymers. In the case of propylene-vinyl chloride chains, both $<r^2>_0$ and $<\mu^2>$ are most strongly dependent on chemical sequence distribution in the case of copolymers which are significantly syndiotactic in stereochemical structure.

Calcd. quantities: $<r^2>_0 / nl^2$ $<\mu^2> / x$ ($x = n/2 = 100$)

propylene – 1-pentene copolymers

Biskup, U.; Cantow, H.-J. *Makromol. Chem.* **1973**, *168*, 329.

Bond Length [pm]	Valence Angles [°]	Torsion Angles [a)] [°]	ξ [PP/PPE] (for 403 K)	ξ_0	E_ξ [PP/PPE] [kJ · mol^{-1}]		For the assignment of the statistical weight parameters, ξ, see the models of the corresponding homopolymers
C-C : 153	C-C-C: 112/112	t : $180 - \Delta\phi$	*Homopolymer parameters of the calculations:*				
		g^+ : $60 + \Delta\phi$	$\eta = 0.829/1$	1	0.63 / 0.0		
			$\tau = 0.571/0.368$	1	1.88 / 3.35	$U'_A = \begin{bmatrix} \eta_A\tau*_A & 1 & \tau_A \\ \eta_A & \omega_A & \tau_A \\ \eta_A & 1 & \tau_A\omega_A \end{bmatrix}$	rows and columns: t, g^+, g^-
		g^- : -60	$\omega = 0.105/0.056$	1	7.54 / 9.63		
			$\omega' = 0.056/0.024$	1	9.63 / 12.56		
		$\Delta\phi = 2.5°/10°$	$\omega'' = 0.056/0.024$	1	9.63 / 12.56		
			$\tau^* = 1/0.731$	1	0 / 1.05		
			Additional copolymer-parameter:			$U''_{iAB} = \begin{bmatrix} \eta_A\omega''_{AB} & 1 & \tau_A\omega'_A \\ \eta_A & \omega_A & \tau_A\omega'_A \\ \eta_A\omega'_A & \omega'_A\tau_A\omega_A\omega''_{AB} \end{bmatrix}$	$U''_{AB} = \begin{bmatrix} \eta_A & \omega'_A & \tau_A\omega''_{AA} \\ \eta_A\omega'_A & 1 & \tau_A\omega_A \\ \eta_A\omega''_{AB} & \omega_A & \tau_A\omega'^2 \end{bmatrix}$
			$\omega''_{AB} = 0.032$	1	11.30		

a) The corresponding values for the homopolymers are used here throughout.

Comments:
For atactic as well as for isotactic copolymers of propylene with 1-pentene, styrene, or vinyl chloride the unperturbed dimensions, $<r^2>_0 / nl^2$, are evaluated using the RIS model. The chemical composition is varied over the whole range. $<r^2>_0 / nl^2$ turns out to be lower than an average of the corresponding homopolymer values for all copolymers, if the mole fractions of the components are weighted. The sequence length distribution – being characterized by the product $r_A r_B$ of the copolymerization parameters – has only little influence on $<r^2>_0 / nl^2$, especially for $r_A r_B < 1$.

propylene – styrene copolymers

Biskup, U.; Cantow, H.-J. *Makromol. Chem.* **1973**, *168*, 329.

Bond Length [pm]	Valence Angles [°]	Torsion Angles [a)] [°]	ξ [PP/PS] (for 403 K)	ξ_0	E_ξ [PP/PS] [kJ · mol^{-1}]	For the assignment of the statistical weight parameters, ξ, see the models of the corresponding homopolymers
C-C : *153*	C-C-C : *112/114*	t : $180 - \Delta\phi$	*Homopolymer parameters of the calculations:*			
		g^+ : $60 + \Delta\phi$	$\eta = 0.829/1.456$	1	0.63 / -1.26	
			$\tau = 0.571/0.473$	1	1.88 / 2.51	
		g^- : -60	$\omega = 0.105/0.082$	1	7.54 / 8.37	$U'_A = \begin{bmatrix} \eta_A \tau^*_A & 1 & \tau_A \\ \eta_A & \omega_A & \tau_A \\ \eta_A & 1 & \tau_A \omega_A \end{bmatrix}$ rows and columns: t, g^+, g^-
			$\omega' = 0.056/0.034$	1	9.63 / 11.30	
			$\omega'' = 0.056/0.034$	1	9.63 / 11.30	
		$\Delta\phi = 2.5°/5°$	$\tau^* = 1/1$	1	0 / 0	
			Additional copolymer-parameter:			$U''_{iAB} = \begin{bmatrix} \eta_A \omega''_{AB} & 1 & \tau_A \omega'_A \\ \eta_A & \omega_A & \tau_A \omega'_A \\ \eta_A \omega'_A & \omega'_A & \tau_A \omega_A \omega''_{AB} \end{bmatrix}$ $\quad U''_{AB} = \begin{bmatrix} \eta_A & \omega'_A & \tau_A \omega''_{AA} \\ \eta_A \omega'_A & 1 & \tau_A \omega_A \\ \eta_A \omega''_{AB} & \omega_A & \tau_A \omega'^2 \end{bmatrix}$
			$\omega''_{AB} = 0.044(0.024)$	1	10.47 (12.56)	

a) The corresponding values for the homopolymers are used here throughout.

Comments:
For atactic as well as for isotactic copolymers of propylene with 1-pentene, styrene, or vinyl chloride the unperturbed dimensions, $<r^2>_o / nl^2$, are evaluated using the RIS model. The chemical composition is varied over the whole range. $<r^2>_o / nl^2$ turns out to be lower than an average of the corresponding homopolymer values for all copolymers, if the mole fractions of the components are weighted. The sequence length distribution — being characterized by the product $r_A r_B$ of the copolymerization parameters — has only little influence on $<r^2>_o / nl^2$, especially for $r_A r_B < 1$.

propylene – vinyl chloride copolymers

Biskup, U.; Cantow, H.-J. *Makromol. Chem.* **1973**, *168*, 329.

Bond Length [pm]	Valence Angles [°]	Torsion Angles [°] a)	ξ [PP/PVC] (for 403 K)	ξ_o	E_ξ [PP/PVC] [kJ · mol^{-1}]	For the assignment of the statistical weight parameters, ξ, see the models of the corresponding homopolymers
C-C : 153	C-C-C : 112/112	t : 180 – $\Delta\phi$	*Homopolymer parameters of the calculations:*			
		g$^+$: 60 + $\Delta\phi$	η = 0.829/2.893	1	0.63 / -3.56	$U'_A = \begin{bmatrix} \eta_A \tau^*_A & 1 & \tau_A \\ \eta_A & \omega_A & \tau_A \\ \eta_A & 1 & \tau_A \omega_A \end{bmatrix}$ rows and columns: t, g$^+$, g$^-$
			τ = 0.571/0.645	1	1.88 / 1.47	
		g$^-$: – 60	ω = 0.105/0.105	1	7.54 / 7.54	
			ω' = 0.056/0.153	1	9.63 / 6.28	
			ω'' = 0.056/0.105	1	9.63 / 7.54	
			τ^* = 1 / 1	1	0.0 / 0.0	
		$\Delta\phi$: 2.5°/0°				
			Additional copolymer-parameter:			$U''_{iAB} = \begin{bmatrix} \eta_A \omega''_{AB} & 1 & \tau_A \omega'_A \\ \eta_A & \omega_A & \tau_A \omega'_A \\ \eta_A \omega'_A & \omega'_A & \tau_A \omega_A \omega''_{AB} \end{bmatrix}$ $U''_{AB} = \begin{bmatrix} \eta_A & \omega'_A & \tau_A \omega''_{AA} \\ \eta_A \omega'_A & 1 & \tau_A \omega_A \\ \eta_A \omega''_{AB} & \omega_A & \tau_A \omega/2 \end{bmatrix}$
			ω''_{AB} = 0.082	1	8.37	

a) The corresponding values for the homopolymers are used here throughout.

Comments:
For atactic as well as for isotactic copolymers of propylene with 1-pentene, styrene, or vinyl chloride the unperturbed dimensions, $<r^2>_o / nl^2$, are evaluated using the RIS model. The chemical composition is varied over the whole range. $<r^2>_o / nl^2$ turns out to be lower than an average of the corresponding homopolymer values for all copolymers, if the mole fractions of the components are weighted. The sequence length distribution – being characterized by the product $r_A r_B$ of the copolymerization parameters – has only little influence on $<r^2>_o / nl^2$, especially for $r_A r_B < 1$.

Bond rotations considered in the matrices U_{E1}, U_{E2}, U_{E3}.

ethylene – vinyl acetate copolymers, short-branch formation

Viswanadhan, V. N.; Mattice, W. L. *J. Polym. Sci.; Polym. Phys. Ed.* **1985**, *23*, 1957.

Bond Length [pm]	Valence Angles [°]	Torsion Angles [°] a)	ξ (for 573 K)	ξ_0	E_ξ [kJ · mol^{-1}]	For the assignment of the statistical weight parameters, ξ, see the models of the corresponding homopolymers
		t : 180	σ = 0.44	1	2.09	
		g$^+$: 60 – $\Delta\phi$	σ_E = 5.79	1	– 8.37	$U_{E1} = \begin{bmatrix} \sigma_E & \sigma'_E\sigma & \sigma \\ \sigma_E & \sigma'_E\sigma\omega'_E & \sigma\omega \\ \sigma_E\omega_E & \sigma'_E\sigma\omega & \sigma \end{bmatrix}$ $U_{E2} = \begin{bmatrix} \sigma_E & \sigma & \sigma'_E\sigma \\ \sigma_E & \sigma & \sigma'_E\sigma\omega \\ \sigma_E & \sigma\omega & \sigma'_E\sigma \end{bmatrix}$ $U_{E3} = \begin{bmatrix} 1 & \sigma & \sigma\omega_E \\ 1 & \sigma\omega'_E & \sigma\omega \\ 1 & \sigma\omega & \sigma \end{bmatrix}$
		g$^-$: – 60 – $\Delta\phi$	σ'_E = 0.77	1	1.26	
			ω = 0.14	1	9.21	
			ω_E = 0.38	1	4.61	
			ω'_E = 0.25	1	6.70	rows and columns: t, g$^+$, g$^-$

a) Dihedral angles for the *gauche* states are located $\pm(2\pi/3 – \Delta\phi)$ from the dihedral angle for the *trans* states. The value of $\Delta\phi$ is 0.13 for all bonds except those involving the carbon atom bearing the acetate group. For these bonds, a larger value of $\Delta\phi$, 0.35, is suggested by the locations of the minima in the conformational energy surface.

Comments:
A realistic RIS model is used to estimate the relative probabilities of the formation of various types of short branches in ethylene-vinyl acetate copolymers that are rich in ethylene. Butyl is predicted to be the most common short branch in all of the copolymers examined, although it is less common in the copolymers than in low-density PE. The major factor responsible for the suppression of the R_{04} backbiting intrachain radical transfer is the increased preference for *trans* states at the main chain bonds flanking the attachment site for an isolated acetoxy side chain.

*Further calculations on **copolymer** chains:*

C 009   Biskup, U.; Cantow, H.-J. *Makromol. Chem.* **1971**, *148*, 31.

Using an Ising model for rotational isomeric states the unperturbed dimensions, $<r^2>_0/nl^2$, are evaluated for copolymers of propylene and pentene-1. Chemical composition, tacticity, and sequence length distribution are varied. It is found that only for atactic copolymers $<r^2>_0/nl^2$ depends linear on the chemical composition. Deviations from linearity cannot be attributed to the influence of diads.

C 010   Mark, J. E. *Polym. Prepr.* **1972**, *13*, 135.

Dipole moments are used to characterize the chemical sequence distributions in vinyl copolymers. The RIS model used follows very closely that developed for poly(*p*-chlorostyrene). Rotational states for the two skeletal bonds leading into and out of C^α atoms of *d* configuration are located at $180° + \Delta\phi$, $60°$, and $-60° - \Delta\phi$, and $180° - \Delta\phi$, $60° + \Delta\phi$, and $-60°$ respectively. The same two sets of rotational angles pertain to the two skeletal bonds leading, respectively, out of and into C^α atoms of *l* configuration. The statistical weight factors used to calculate the configuration-dependent properties of the *p*-chlorostyrene copolymers are simply those used in the treatment of the corresponding homopolymer. Calculations are carried out assuming (i) skeletal bond angles of $112°$, (ii) tetrahedral orientation of the group dipoles with respect to the adjoining skeletal bonds, (iii) $\Delta\phi = 10°$, the average value of the range 0-20° assumed appropriate for polystyrene chains, (iv) bond dipole moments m_1 (*p*-chlorostyrene units) = 1.68 D, (v) bond dipole moments m_2 (*p*-methylstyrene units) = 0.0 D, (vi) bond dipole moments m_2 (styrene units) = 0.36 D, (vii) reactivity ratio product $r_1 r_2 = 0.75$ for the comonomer pair *p*-chlorostyrene, *p*-methylstyrene, and (viii) $r_1 r_2 = 0.85$ for the pair *p*-chlorostyrene, styrene.

C 011   Mark, J. E. *J. Am. Chem. Soc.* **1972**, *94*, 6645.

An RIS scheme is developed in order to investigate the feasibility of using dipole moments to characterize chemical sequence distribution in vinyl copolymers. By means of this scheme, dipole moments of such chains may be calculated as a function of chain length, temperature, chemical sequence distribution, and stereochemical composition. The method is illustrated by application to the simplest case, vinyl copolymers in which the pendant groups have dipoles of different magnitude, but are of sufficient similar chemical structure that the conformation energy of the chain is independent of the chemical composition. The copolymers poly(*p*-chlorostyrene-*p*-methylstyrene) and poly(*p*-chlorostyrene-styrene) have this structural feature. Dipole moments are calculated for these two copolymers, using chemical sequence distributions characteristic of the reactivity ratios determined for these two sets of comonomers and random stereochemical sequence distribution. The results obtained are in good agreement with experimental results.

C 012   Mark, J. E. *J. Chem. Phys.* **1972**, *57*, 2541.

Mean-square unperturbed dimensions $<r^2>_0$ and their temperature coefficient, $d(\ln <r^2>_0)/dT$, are calculated for ethylene-propylene copolymers by means of the RIS theory. Conformational energies required in the analysis are shown to be readily obtained from previous analyses of PE and PP, without additional approximations. Results thus calculated are reported as a function of chemical composition, chemical sequence distribution, and stereochemical composition of the PP sequences. Calculations of $<r^2>_0/nl^2$ are carried out using (i) $r_1 r_2 = 0.01$, 1.0, 10.0, and 100.0, (ii) $p_r = 0.95, 0.50$, and 0.05, (iii) bond length of 153 pm and bond angles of 112° for all skeletal bonds, (iv) $\Delta\phi = 0$ and 10°, and (v) statistical weight factors appropriate for temperatures of 248, 298, and 348 K. Matrices used are:

For CH_2-CH_2-CH_2 (ethylene) bond pairs:

$$U_e = \begin{bmatrix} 1 & \tau/\eta & \tau/\eta \\ 1 & \tau/\eta & \tau\omega/\eta \\ 1 & \tau\omega/\eta & \tau/\eta \end{bmatrix}$$

For propylene units:

$$U_d = \begin{bmatrix} \eta & 1 & \tau \\ \eta & 1 & \tau\omega \\ \eta & \omega & \tau \end{bmatrix} \quad U_{dd} = \begin{bmatrix} \eta\omega & \tau\omega & 1 \\ \eta & \tau\omega & \omega \\ \eta\omega & \tau\omega^2 & \omega \end{bmatrix} \quad U_{dl} = \begin{bmatrix} \eta & \omega & \tau\omega \\ \eta\omega & 1 & \tau\omega \\ \eta\omega & \omega & \tau\omega^2 \end{bmatrix}$$

For mixed units: (rows and columns: t, g^+, g^-)

$$U_{de} = \begin{bmatrix} \eta/\tau & \omega & 1 \\ \eta/\tau & 1 & \omega \\ \eta/\tau & \omega & \omega \end{bmatrix} \quad U_{ed} = \begin{bmatrix} \eta & \tau & 1 \\ \eta & \tau\omega & \omega \\ \eta/\omega & \tau\omega & 1 \end{bmatrix}$$

C 013   Tonelli, A. E. *Macromolecules* **1974**, *7*, 632.

The conformational entropies of the various diads, *i.e.*, A-A, B-B, and A-B, present in a copolymer of the monomer units A and B, are calculated through the utilization of semiempirical potential energy functions. It is assumed that the glass transition temperature, T_g, is inversely related to the intramolecular equilibrium flexibility of a polymer chain as manifested by its conformational entropy. This approach is applied to the alternating vinyl copolymer composed of styrene (S) and acrylonitrile (AN) monomer units, where the stereoregularity of the various diads is explicitly considered. Based on the calculated conformational energies of the S-S, AN-S, AN-AN, and S-AN diads of various stereoregularity, in addition to the total conformational entropies of various copolymers chains estimated through adoption of the RIS model, the following conclusions may be drawn: 1) T_g for the isotactic (syndiotactic) copolymers should be higher (lower) than T_g(Fox), and 2) the T_g's of the regularly alternating isotactic (syndiotactic) copolymers should be higher (lower) than the T_g's of the random copolymers with the same overall monomer composition.

C 014   Tonelli, A. E. *Macromolecules* **1975**, *8*, 544.

The conformational entropies of copolymer chains are calculated through utilization of semiempirical potential energy functions and adoption of the RIS model of polymers. It is assumed that the glass transition temperature, T_g, is inversely related to the intramolecular, equilibrium flexibility of a copolymer chain as manifested by its conformational entropy. This approach is applied to the vinyl copolymers of vinyl chloride and vinylidene chloride with methyl acrylate, where the stereoregularity of each copolymer is explicitly considered, and correctly predicts the observed deviations from the Fox relation when they occur. It therefore appears that the sequence distribution – T_g effects observed in many copolymers may have an intramolecular origin in the form of specific molecular interactions between adjacent monomer units, which can be characterized by estimating the resultant conformational entropy.

C 015   Tonelli, A. E. *Macromolecules* **1977**, *10*, 633.

Conformational energies are estimated for the diads composed of two styrene (S) units, two methyl methacrylate (MMA) units, and neighboring S and MMA units which constitute the homo- and copolymers of styrene and methyl methacrylate. These diad energies are used to evaluate the isolated chain conformational energies and entropies of S, MMA, and S – MMA homo- and copolymers, through adoption of the RIS model. Independent of sequence and stereoregularity, the conformational entropies calculated for S – MMA copolymers are found to exceed the weighted sum of S and MMA homopolymer entropies indicating an increase in equilibrium chain flexibility for the S – MMA copolymers.

C 016 Oka, M.; Nakajima, A. *Polymer J.* **1978**, *4*, 465.

The equation for the unperturbed mean-square radius of gyration, $<s^2>_0$, of copolymers is obtained for two cases by using the RIS method. For one case it is assumed that the total mass of each structural unit of the chain is situated on the skeletal atom. For the other case the deviation is considered of the center of mass of each structural unit from each skeletal atom.

C 017 Tonelli, A. E. *Macromolecules* **1978**, *11*, 634.

Ethylene-propylene (E-P) copolymers of low ethylene content, which insures a predominance of isolated ethylene units, are studied to determine the ^{13}C NMR chemical shifts expected at each of the methylene carbons in $-CH_2-CH_2-CH_2-$ as a function of the stereosequence of the surrounding polypropylene chain segments. A slightly modified version of *Mark´s* RIS model for E-P copolymers (**C 012**) is used to determine the average number of three-bond carbon-carbon *gauche* or γ-interactions involving the above methylene groups as a function of the surrounding polypropylene stereoregularity. Each γ-interaction is assigned a shielding effect of – 5.3 ppm.

C 018 Tonelli, A. E.; Schilling, F. C. *Macromolecules* **1981**, *14*, 74.

^{13}C NMR chemical shifts expected for the carbon atoms in ethylene – vinyl chloride (E – VC) copolymers are calculated as functions of the E and VC monomer sequence distribution and the stereoregularity of the VC sequences. The γ gauche effect, which results in upfield chemical shifts for those carbon atoms in a *gauche* arrangement with carbon or chlorine substituents in the γ position, is utilized, together with the predicted bond rotational probabilities obtained from *Mark´s* conformational model of E – VC copolymers (**C 002**), to calculate the ^{13}C NMR chemical shifts for carbons in the various microstructural environments possible in E – VC copolymers. The methine and methylene carbon chemical shifts are predicted to occur over ≈ 6 and ≈ 30 ppm ranges, respectively, solely as a result of different monomer sequence distributions. The several-fold greater range of methylene carbon chemical shifts is due to the downfield shift produced by the presence of different numbers of deshielding chlorine atoms in the β position.

C 019 Tanaka, N. *Polymer* **1981**, *22*, 647.

The composition dependence of T_m is investigated for RIS models of *atactic* PP and random ethylene – propylene copolymers using a modified *Flory´s* equation.

C 020 Khanarian, G.; Cais, R. E.; Kometani, J. M.; Tonelli, A. E. *Macromolecules* **1982**, *15*, 866.

Experimental values are presented of the molar Kerr constants $_mK/x$ and dipole moments squared, $<\mu^2>/x$, for the copolymers poly(styrene-*co*-*p*-bromostyrene), where x is the degree of polymerization. Some results are also presented for poly(styrene-*co*-*p*-chlorostyrene) and related polymers. The RIS model of *Yoon et al.* (Yoon, D. Y.; Sundararajan, P. R.; Flory, P. J. *Macromolecules* **1975**, *8*, 776) is used to calculate $_mK/x$ and $<\mu^2>/x$ values as a function of tacticity and composition. The statistical weight matrices are identical with those used by *Saiz et al.* (Saiz, E.; Mark, J. E.; Flory, P. J. *Macromolecules* **1977**, *10*, 967), with the following parameters:

$\eta = 0.8 \exp\{397/RT\}$, $\omega = \omega' = 1.3 \exp\{-(1987/RT)\}$ and $\omega'' = 1.8 \exp\{-(2186/RT)\}$, where T = 298 K is the temperature.

C 021   Tonelli, A. E.; Schilling, F. C. *Macromolecules* **1984**, *17*, 1946.

^{13}C NMR chemical shifts expected for the carbon atoms in propylene – vinyl chloride (P – VC) copolymers of low propylene content are calculated as a function of copolymer stereosequence. Mark´s conformational model of P – VC copolymers (**C 004**) is coupled with the γ *gauche* effect, which results in upfield chemical shifts for those carbon atoms in a *gauche* arrangement with carbon or chlorine substituents in the γ position, to calculate the ^{13}C NMR chemical shifts of the carbon atoms in the vicinity of a propylene unit surrounded by vinyl chloride units. Agreement of the calculated chemical shifts and those which are observed is excellent.

C 022   Saiz, E.; Mark, J. E. *Makromol. Chem.* **1987**, *188*, 2185.

RIS theory is used to predict values of the optical-configuration parameter Δa for ethylene – propylene copolymers as a function of chemical composition, chemical sequence distribution, and stereochemical structure of the propylene sequences. The calculations are based on information available for ethylene and propylene homopolymers, and on the model used to interpret the unperturbed dimensions of these copolymers. Values of Δa are generally found to decrease significantly with increase in the fraction of propene units, but to be relatively insensitive to chemical sequence distribution and stereochemical structure. Geometries and conformational energies are the same as those used for the interpretation of the unperturbed dimensions of these chains. The conformational energies used are $E(\eta) = 0$, $E(\tau) = 2.09$, and $E(\omega) = 0.37$ kJ · mol^{-1}.

C 023   Bahar, I.; Erman, B.; Haliloglu, T. *Macromolecules* **1994**, *27*, 1703.

Effects of intrinsic structural and conformational properties on segmental orientation in uniaxially deformed copolymers are considered. Dependence of segmental orientation on equilibrium values of bond angles, torsional states, and probability distribution of rotameric states is studied. Calculations are carried out for chains with independent as well as pairwise interdependent rotameric states for neighboring bonds using the matrix generation technique of RIS formalism. Results are interpreted with reference to polarized Fourier transform infrared spectroscopy measurements in which the orientation of transition moment vectors is detected. Calculations carried out independently by Monte Carlo simulations show that this method yields an adequate qualitative description of the orientational behavior of chain segments while precise quantitative determination requires the use of the exact matrix generation technique.

X

The following model requires two pages.
Therefore, this side has been left blank

Molecule with $n_1 = 2$, $n_2 = 4$, and $n_3 = 3$

Interactions for $_1U_{n(1)}$

Interactions for $_2U_1$

Interactions for $_3U_1$

Interactions for $_2U_2$

Interactions for $_3U_2$

branched macromolecules in general, and polyethylene containing *n*-butyl groups

Mattice, W. L. *Macromolecules* **1975**, *8*, 644.

Procedures are presented for computing the configuration partition function for branched macromolecules which are subject to the RIS approximation. The assumptions required are (1) the short-range interations can be adequately represented by three- and four-bond interactions, (2) bond length and bond angles are constant, and (3) there are a minimum of two bonds between each branching point. A molecule with any number of branches can be treated. Procedures are also presented for the computation of a priori and conditional bond probabilities. Illustrative calculations are carried out for a low-density polyethylene containing n-butyl groups (σ = 0.54, ψ = 1.00, ω = 0.088 at 298 K). The appropriate statistical weight matrix for those bonds sufficiently remote from the branching point is U_{chain}. Other matrices are:

$$U_{chain} = \begin{bmatrix} 1 & \sigma & \sigma \\ 1 & \sigma\psi & \sigma\omega \\ 1 & \sigma\omega & \sigma\psi \end{bmatrix} \quad {}_1U_{n_1} = \sigma\begin{bmatrix} 1 & 1 & \sigma \\ \omega & \psi & \sigma\psi\omega \\ \psi & \omega & \sigma\psi\omega \end{bmatrix} \quad {}_2U_1 = \sigma\begin{bmatrix} 1 & \sigma & 1 \\ 1 & \sigma\psi & \omega \\ 1 & \sigma\omega & \psi \end{bmatrix} \quad {}_3U_1 = \sigma\begin{bmatrix} \omega & \omega & \sigma\psi^2 \\ \omega & 1 & \sigma\psi \\ \omega & \psi & \sigma\psi\omega \end{bmatrix} \text{ (first layer)} \quad {}_3U_1 = \sigma\begin{bmatrix} \psi & \omega & \sigma\psi\omega \\ \psi & 1 & \sigma\omega \\ \psi & \psi & \sigma\omega^2 \end{bmatrix} \text{ (second layer)} \quad {}_3U_1 = \sigma\begin{bmatrix} 1 & \omega & \sigma\psi \\ 1 & 1 & \sigma \\ 1 & \psi & \sigma\omega \end{bmatrix} \text{ (third layer)}$$

$$_2U_2 = \begin{bmatrix} 1 & \sigma\psi & \sigma\omega \\ 1 & \sigma\psi\omega & \sigma\psi\omega \\ 1 & \sigma\omega & \sigma\psi \end{bmatrix} \quad {}_3U_2 = \begin{bmatrix} 1 & \sigma\omega & \sigma\psi \\ 1 & \sigma\psi & \sigma\omega \\ 1 & \sigma\psi\omega & \sigma\psi\omega \end{bmatrix}$$

In the cases where n_2 = 2, it is $_2U_2 = {}_2U_{n_2}$, and the appropriate expressions are here:

$$_2U_2 = {}_2U_{n_2} = \begin{bmatrix} \omega & \psi & \sigma\psi\omega \\ \psi\omega & \psi\omega & \sigma\psi^2\omega^2 \\ \psi & \omega & \sigma\psi\omega \end{bmatrix} \text{ (ll pair)} \quad {}_2U_2 = {}_2U_{n_2} = \begin{bmatrix} \psi & \omega & \sigma\psi\omega \\ \omega & \psi & \sigma\psi\omega \\ \psi\omega & \psi\omega & \sigma\psi^2\omega^2 \end{bmatrix} \text{ (dl pair)}$$

If the branches are numbered so that $_3A_1$ is behind the plane of the paper, the following substitutions are required:

$$_1U_{n_1} = \sigma\begin{bmatrix} 1 & \sigma & 1 \\ \psi & \sigma\psi\omega & \omega \\ \omega & \sigma\psi\omega & \psi \end{bmatrix} \quad {}_2U_1 = \sigma\begin{bmatrix} 1 & 1 & \sigma \\ 1 & \psi & \sigma\omega \\ 1 & \omega & \sigma\psi \end{bmatrix} \quad {}_2U_2 = \begin{bmatrix} 1 & \sigma\psi & \sigma\omega \\ 1 & \sigma\psi\omega & \sigma\psi\omega \\ 1 & \sigma\omega & \sigma\psi \end{bmatrix} \quad {}_3U_2 = \begin{bmatrix} 1 & \sigma\omega & \sigma\psi \\ 1 & \sigma\psi & \sigma\omega \\ 1 & \sigma\psi\omega & \sigma\psi\omega \end{bmatrix}$$

$$_3U_1 = \sigma\begin{bmatrix} \omega & \sigma\psi^2 & \omega \\ \omega & \sigma\psi\omega & \psi \\ \omega & \sigma\psi & 1 \end{bmatrix} \text{ (first layer)} \quad {}_3U_1 = \sigma\begin{bmatrix} \psi & \omega & \sigma\psi\omega \\ \psi & 1 & \sigma\omega \\ \psi & \psi & \sigma\omega^2 \end{bmatrix} \text{ (second layer)} \quad {}_3U_1 = \sigma\begin{bmatrix} 1 & \omega & \sigma\psi \\ 1 & 1 & \sigma \\ 1 & \psi & \sigma\omega \end{bmatrix} \text{ (third layer)} \qquad \text{rows and columns: t, } g^+, g^-$$

X 002

branched macromolecules in general

Mattice, W. L. *Macromolecules* **1976**, *9*, 48.

Expressions for the unperturbed mean-square radius of gyration, unperturbed mean-square dipole moment, and unperturbed mean-square optical anisotropy are obtained for a molecule consisting of three branches which emenate from a common atom, using RIS theory. The results are equally valid for homopolymers and copolymers, and for molecules in which the branches are of identical or different length. With the exception of a matrix occurring at the branch point, the dimensions of the generator matrices are no larger than those required for a linear molecule. The procedure adopted can be extended to treat more highly branched molecules.

molecules containing a single trifunctional or tetrafunctional branch point

Mattice, W. L.; Carpenter, D. K. *Macromolecules* **1976**, *9*, 53.

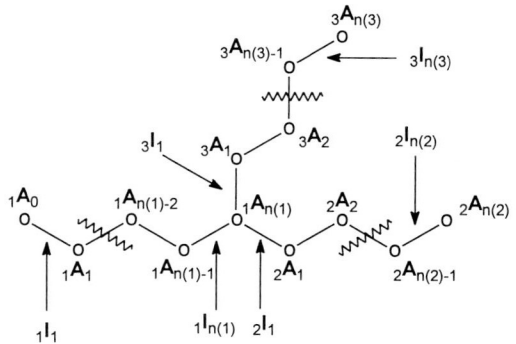

The mean-square unperturbed radius of gyration, $<s^2>_o$, for a molecule containing a single trifunctional or tetrafunctional branch point is computed using RIS theory. Various symmetric threefold rotation potentials are used. Bond length are all assumed to be identical, and bond angles are all tetrahedral. Mean-square unperturbed radii of gyration for the corresponding linear molecule, $<s_{lin}^2>_o$, are also computed. The asymptotic limit for $g = <s^2>_o/<s_{lin}^2>_o$ as the number of bonds increases is in agreement with results obtained using random flight statistics. However, the number of bonds required for g to obtain the asymptotic limit, as well as the nature of the approach to that limit, are extremely sensitive to the three-bond and four-bond interactions present in the branched molecule. Statistical weights appropriate for polyethylene cause g to approach the asymptotic limit from below. The statistical weight matrix for the nonterminal bonds which are sufficiently remote from a branch point is U_{chain}.

Near a ***trifunctional*** branch point the following statistical weight matrices are required (the term σ is factored out of $_1U_{n(1)}$, and σ^2 is factored out of $_2U_1 \ominus {_3U_1}$):

$$U_{chain} = \begin{bmatrix} 1 & \sigma & \sigma \\ 1 & \sigma\psi & \sigma\omega \\ 1 & \sigma\omega & \sigma\psi \end{bmatrix} \quad {_1U_{n_1}} = \begin{bmatrix} 1 & 1 & \sigma \\ \omega & \psi & \sigma\psi\omega \\ \psi & \omega & \sigma\psi\omega \end{bmatrix} \quad {_2U_1} \ominus {_3U_1} = \begin{bmatrix} \omega & \omega & \sigma\psi^2 & \sigma\psi & \sigma\omega & \sigma^2\psi\omega & 1 & \omega & \sigma\psi \\ \omega & 1 & \sigma\psi & \sigma\psi^2 & \sigma\psi & \sigma^2\psi\omega & \omega & \omega & \sigma\omega \\ \omega & \psi & \sigma\psi\omega & \sigma\psi\omega & \sigma\psi\omega & \sigma^2\omega^3 & \psi & \psi^2 & \sigma\psi\omega \end{bmatrix}$$

$$_2U_2 = \begin{bmatrix} 1 & \sigma\psi & \sigma\omega \\ 1 & \sigma\psi\omega & \sigma\psi\omega \\ 1 & \sigma\omega & \sigma\psi \end{bmatrix} \quad {_3U_2} = \begin{bmatrix} 1 & \sigma\omega & \sigma\psi \\ 1 & \sigma\psi & \sigma\omega \\ 1 & \sigma\psi\omega & \sigma\psi\omega \end{bmatrix}$$

The following matrices are required near a ***tetrafunctional*** branch point:

$$U_{chain} = \begin{bmatrix} 1 & \sigma & \sigma \\ 1 & \sigma\psi & \sigma\omega \\ 1 & \sigma\omega & \sigma\psi \end{bmatrix} \quad {_1U_{n_1}} = \begin{bmatrix} 1 & 1 & 1 \\ \psi\omega & \psi\omega & \psi\omega \\ \psi\omega & \psi\omega & \psi\omega \end{bmatrix} \quad {_2U_2} = {_3U_2} = {_4U_2} = \begin{bmatrix} 1 & \sigma\psi\omega & \sigma\psi\omega \\ 1 & \sigma\psi\omega & \sigma\psi\omega \\ 1 & \sigma\psi\omega & \sigma\psi\omega \end{bmatrix} \quad (_2U_1 \ominus {_3U_1}) \ominus {_4U_1} \text{ is given in the paper.}$$

A *gauche* three-bond interaction is assigned the statistical weight σ, while ψ and ω are the statistical weights for four-bond interactions generated by two successive *gauche* placements of the same and opposite sign, respectively. Bond angles are fixed at 100° for the molecules with a trifunctional branch point, and at 109.5° for those with a tetrafunctional branch point. The rotational angles are t: 180°, g$^+$: 60°, g$^-$: –60°.

X 004

alkanes containing a single trifunctional or tetrafunctional branch point

Mattice, W. L. *J. Am. Chem. Soc.* **1976**, *98*, 3466.

RIS theory is used to assess the effects of a single trifunctional or tetrafunctional branch point on the optical anisotropy exhibited by alkanes.

When a ***trifunctional*** branch point is present, the configuration partition function is written as $Z = {}_1U_1^{(n(1))}({}_2U_1 \ominus {}_3U_1)\{({}_2U_1^{(n(2)-1)}) \otimes {}_3U_1^{(n(3)-1)}\}$. Statistical weight matrices used are:

$$U_{chain} = \begin{bmatrix} 1 & \sigma & \sigma \\ 1 & \sigma\psi & \sigma\omega \\ 1 & \sigma\omega & \sigma\psi \end{bmatrix} \quad {}_1U_{n_1} = \begin{bmatrix} 1 & 1 & \tau \\ \omega & \psi & \tau\psi\omega \\ \psi & \omega & \tau\psi\omega \end{bmatrix} \quad {}_2U_1 \ominus {}_3U_1 = \begin{bmatrix} \omega & \omega & \tau\psi^2 & \tau\psi & \tau\omega & \tau^2\psi\omega & 1 & \omega & \tau\psi \\ \omega & 1 & \tau\psi & \tau\psi^2 & \tau\psi & \tau^2\psi\omega & \omega & \omega & \tau\omega \\ \omega & \psi & \tau\psi\omega & \tau\psi\omega & \tau\psi\omega & \tau^2\omega^3 & \psi & \psi^2 & \tau\psi\omega \end{bmatrix} \quad {}_2U_2 = \begin{bmatrix} 1 & \sigma\psi & \sigma\omega \\ 1 & \sigma\psi\omega & \sigma\psi\omega \\ 1 & \sigma\omega & \sigma\psi \end{bmatrix} \quad {}_3U_2 = \begin{bmatrix} 1 & \sigma\omega & \sigma\psi \\ 1 & \sigma\psi & \sigma\omega \\ 1 & \sigma\psi\omega & \sigma\psi\omega \end{bmatrix}$$

When a **tetrafunctional** branch point is present, the configuration partition function is written as

$Z = {}_1U_1^{(n(1))}\{({}_2U_1 \ominus {}_3U_1) \ominus {}_4U_1\}\{({}_2U_2^{(n(2)-1)}) \otimes {}_3U_2^{(n(3)-1)}) \otimes ({}_4U_2^{(n(4)-1)})\}$; ${}_1U_1 = [1\ 0\ 0]$, ${}_2U_{n2} = {}_3U_{n3} = \text{col}\,[1\ 1\ 1]$. Statistical weight matrices used are:

$$U_{chain} = \begin{bmatrix} 1 & \sigma & \sigma \\ 1 & \sigma\psi & \sigma\omega \\ 1 & \sigma\omega & \sigma\psi \end{bmatrix} \quad {}_1U_{n_1} = \begin{bmatrix} 1 & 1 & 1 \\ \psi\omega & \psi\omega & \psi\omega \\ \psi\omega & \psi\omega & \psi\omega \end{bmatrix} \quad {}_2U_2 = {}_3U_2 = {}_4U_2 = \begin{bmatrix} 1 & \sigma\psi\omega & \sigma\psi\omega \\ 1 & \sigma\psi\omega & \sigma\psi\omega \\ 1 & \sigma\psi\omega & \sigma\psi\omega \end{bmatrix} \quad ({}_2U_1 \ominus {}_3U_1) \ominus {}_4U_1 \text{ is given in the paper.}$$

As in the case of the treatment of *n*-alkanes, the geometry about methyl groups is tetrahedral and methylene groups have <CCC and <HCH of 112° and 109°, respectively. The geometry at trifunctional branch points is that of isobutane, i.e. <CCC = 111° and <CCH = 107.9°. Symmetry dictates that the geometry about the tetrafunctional branch point must be tetrahedral. The dihedral angles for the *trans* and *gauche* states for bonds sufficiently remote from a trifunctional branch point are 180°, 60°, and −60°. Bonds involving the atom at a trifunctional branch point have the rotational state assigned where ${}_1C_{(n(1)-2}-{}_1C_{n(1)-1}-{}_1C_{n(1)}-H$, ${}_2C_2-{}_2C_1-{}_1C_{n(1)}-H$, or ${}_3C_2-{}_3C_1-{}_1C_{n(1)}-H$ is in the *trans* conformation. The other two rotational states are assigned at $\pm(60° + \Delta\phi)$ from the above angle.

X 005

flexible molecules containing a trifunctional branch point

Mattice, W. L. *Macromolecules* **1977**, *10*, 1177.

Expressions are developed for the center of mass vectors and moment of inertia tensors for a molecule which contains a trifunctional branch point. Evaluation is achieved using RIS theory. Configurational averages are obtained for unperturbed molecules, the averaging being accomplished in the internal coordinate system defined by the first two bonds in a branch. Under certain limiting conditions simple relationships exist between persistence vectors and average center of mass vectors for the branched molecule.

X 006

branched polyethylene

Mattice, W. L. *Macromolecules* **1977**, *10*, 1182.

RIS theory is used to compute the persistence vectors, average center of mass vectors, and average moment of inertia tensors for unperturbed polyethylenes containing a trifunctional branch point. In each case averaging occurs in an internal coordinate system defined by the first two bonds in a particular branch. Results may depend upon the selection of this branch if the three branches contain different numbers of bonds. Spherical symmetry is always attained at sufficiently high molecule weights. In the calculations, all <CCC are 112° except angles at the branch point, which are assigned 111°. Dihedral angles for rotational states occur at 180° and ± 60° for bonds which do not involve the carbon atom at the branch point. Rotational states for bonds involving the branch point carbon atom have their dihedral angles separated by 120°. Statistical weight matrices for bonds of the chain, and near the trifunctional branch point are, respectively:

$$U_{chain} = \begin{bmatrix} 1 & \sigma & \sigma \\ 1 & \sigma & \sigma\omega \\ 1 & \sigma\omega & \sigma \end{bmatrix} \quad {}_1U_{n_1} = \begin{bmatrix} 1 & 1 & \tau \\ \omega & 1 & \tau\omega \\ 1 & \omega & \tau\omega \end{bmatrix} \quad {}_2U_1 \ominus {}_3U_1 = \begin{bmatrix} \omega & \omega & \tau & \tau & \tau\omega & \tau^2\omega & 1 & \omega & \tau \\ \omega & 1 & \tau & \tau & \tau & \tau^2\omega & \omega & \omega & \tau\omega \\ \omega & 1 & \tau\omega & \tau\omega & \tau\omega & \tau^2\omega 3 & 1 & 1 & \tau\omega \end{bmatrix} \quad {}_2U_2 = \begin{bmatrix} 1 & \sigma & \sigma\omega \\ 1 & \sigma\omega & \sigma\omega \\ 1 & \sigma\omega & \sigma \end{bmatrix} \quad {}_3U_2 = \begin{bmatrix} 1 & \sigma\omega & \sigma \\ 1 & \sigma & \sigma\omega \\ 1 & \sigma\omega & \sigma\omega \end{bmatrix}$$

Rows and columns: t, g^+, g^- and/or

$tt, tg^+, tg^-, g^+t, g^+g^+, g^+g^-, g^-t, g^-g^+, g^-g^-$

A value of unity is assigned to ψ. Numerical values for σ and ω of 0.54 and 0.088, respectively, reproduce the experimental characteristics of unperturbed linear polyethylene chains at 413 K. The appropriate value for τ is uncertain; calculations are performed for $\tau = \sigma$ and $\tau = \sigma/10$. Calculations are also carried out using $\sigma = 0.43$ and $\omega = 0.034$, values appropriate for an unperturbed polymethylene at 298 K.

branched polyethylene, low density polyethylene, LDPE

Mattice, W. L.; Stehling, F. C. *Macromolecules* **1981**, *14*, 1479.

Bond Length [pm]	Valence Angles [°]	Torsion Angles [°]	ξ (for 413 K)	ξ_o	E_ξ [kJ·mol^{-1}]
C-C : 153	C-C-C : 112	t : 180	$\sigma = 0.54$	1	2.093
C-H : 110		g^+ : 68.5	$\omega = 0.09$	1	8.374
		g^- : −68.5	$\tau \leq 0.54$	1	$E_\tau \geq E_\sigma$

$$U_i = \begin{bmatrix} 1 & \sigma & \sigma \\ 1 & \sigma & \sigma\omega \\ 1 & \sigma\omega & \sigma \end{bmatrix} \qquad U_4 = \begin{bmatrix} 1 & 1 & \tau \\ \omega & 1 & \tau\omega \\ 1 & \omega & \tau\omega \end{bmatrix} \qquad U_6 = \begin{bmatrix} 1 & \sigma & \sigma\omega \\ 1 & \sigma\omega & \sigma\omega \\ 1 & \sigma\omega & \sigma \end{bmatrix}$$

$$U \ominus U = \begin{bmatrix} \omega & \omega & \tau & \tau & \tau\omega & \tau^2\omega & 1 & \omega & \tau \\ \omega & 1 & \tau & \tau & \tau & \tau^2\omega & \omega & \omega & \tau\omega \\ \omega & 1 & \tau\omega & \tau\omega & \tau\omega & \tau^2\omega^3 & 1 & 1 & \tau\omega \end{bmatrix}$$

rows and/or columns:
t, g^+, g^- or
tt, tg^+, tg^-, g^+t, g^+g^+,
g^+g^-, g^-t, g^-g^+, g^-g^-

Comments:
Formation of short branches in LDPE is investigated. It is assumed that the probability of an intramolecular rearrangement of the *Roedel* type is proportional to the probability that the reacting groups are separated by a distance r* ± Δr and adhere sufficiently closely to "three in a line" geometry. Excluded volume effects are ignored. The calculations rationalize many of the structural features observed in LDPE. For the present purpose, probabilities must be evaluated by using branched-molecule RIS theory.

branched polyethylene, low density polyethylene, LDPE

Mattice, W. L. *Macromolecules* **1983**, *16*, 487.

Bond Length [pm]	Valence Angles [°]	Torsion Angles [°]	ξ (for 473 K)	ξ_0	E_ξ [kJ·mol^{-1}]
C-C : 153	C-C-C : 112	t : 180	σ = 0.54	1	2.093
C-H : 110		g$^+$: 68.5	ω = 0.09	1	8.374
		g$^-$: −68.5	$\tau \leq 0.54$	1	$E_\tau \geq E_\sigma$

$$U = \begin{bmatrix} 1 & \sigma & \sigma \\ 1 & \sigma & \sigma\omega \\ 1 & \sigma\omega & \sigma \end{bmatrix} \quad U_4 = \begin{bmatrix} 1 & 1 & \tau \\ \omega & 1 & \tau\omega \\ 1 & \omega & \tau\omega \end{bmatrix} \quad U_6 = \begin{bmatrix} 1 & \sigma & \sigma\omega \\ 1 & \sigma\omega & \sigma\omega \\ 1 & \sigma\omega & \sigma \end{bmatrix} \quad U_7 = \begin{bmatrix} 1 & \sigma\omega & \sigma \\ 1 & \sigma & \sigma\omega \\ 1 & \sigma\omega & \sigma\omega \end{bmatrix}$$

$$U \ominus U = \begin{bmatrix} \omega & \omega & \tau & \tau & \tau\omega & \tau^2\omega & 1 & \omega & \tau \\ \omega & 1 & \tau & \tau & \tau & \tau^2\omega & \omega & \omega & \tau\omega \\ \omega & 1 & \tau\omega & \tau\omega & \tau\omega & \tau^2\omega^3 & 1 & 1 & \tau\omega \end{bmatrix}$$

rows and/or columns:
t, g$^+$, g$^-$ or
tt, tg$^+$, tg$^-$, g$^+$t, g$^+$g$^+$,
g$^+$g$^-$, g$^-$t, g$^-$g$^+$, g$^-$g$^-$

Comments:

Formation of complex branches, which might arise from two intramolecular rearrangements of the *Roedel* type, are investigated in polyethylene. An RIS model is used for the polyethylene chain statistics, with inclusion of the effect of a trifunctional branch point on weighting of all configurations.

X 009

triacetin

Mattice, W. L.; Saiz, E. *J. Am. Chem. Soc.* **1978**, *100*, 6308.

Bond Length [pm]	Valence Angles [°]	Torsion Angles [a)] [°]	$\xi^{b,c)}$ (for 298 K)	ξ_0	E_ξ [kJ·mol^{-1}]
C-C : 151	C-C'-O' : 126.3		$\sigma_1 = 2.33 / 0.18$ 1		$-2.1 / 4.19$
C-C' : 150	O-C'-O' : 122.3		$\sigma_2 = 0.15 / 0.28$ 1		$4.61 / 3.14 (\pm 1.88)$
C-O : 144	C-O-C' : 116.7		$\sigma_3 = 0.71 / 0.31$ 1		$0.84 / 2.93 (\pm 1.26)$
C'-O : 135	C-C-C : 114.7		$\omega_1 = 0.03 / 0.11$ 1		$8.4 / 5.44$ <
C'-O' : 120	O-C-C : 106.4		$\omega_2 = 1.00 / 0.43$ 1		$0 / 2.1 (\pm 2.1)$
C-H : 100			$\omega_3 = 0.03 / 0.60$ 1		$8.4 / 1.26$ <
			$\omega_4 = 0.03 / 0.03$ 1		$8.4 / 8.4$ <

$$_1U_3 = \begin{bmatrix} 1 & 1 \end{bmatrix} \qquad _2U_1 = \begin{bmatrix} \sigma_1 & \sigma_2 & 1 \\ \sigma_1 & \sigma_2 & \omega_1 \end{bmatrix} \qquad _2U_2 = \begin{bmatrix} 1 & \sigma_3 & \sigma_3\omega_4 \\ 1 & \sigma_3\omega_4 & \sigma_3\omega_4 \\ 1 & \sigma_3\omega_4 & \sigma_3 \end{bmatrix}$$

$$_2U_1 \ominus {_3U_1} = \begin{bmatrix} \sigma_1^2\omega_2 & \sigma_1\omega_1 & \sigma_1\sigma_2 & \sigma_1\sigma_2 & \sigma_2\omega_1 & \sigma_2^2\omega_3 & \sigma_1 & \omega_1 & \sigma_2 \\ \sigma_1^2\omega_2 & \sigma_1 & \sigma_1\sigma_2 & \sigma_1\sigma_2 & \sigma_2 & \sigma_2^2\omega_3 & \sigma_1\omega_1 & \omega_1 & \sigma_2\omega_1 \end{bmatrix}$$

Remaining statistical weight matrices for triacetin are
$_2U_3 = {_3U_3} = \text{col}(1,1,1)$ and $_1U_2 = {_3U_4} = 1$

$$_3U_2 = \begin{bmatrix} 1 & \sigma_3\omega_4 & \sigma_3 \\ 1 & \sigma_3 & \sigma_3\omega_4 \\ 1 & \sigma_3\omega_4 & \sigma_3\omega_4 \end{bmatrix}$$

rows and columns: t, g$^+$, g$^-$

---

a) Ester groups are maintained in the planar *trans* conformation ($_1\phi_2 = {_2\phi_3} = {_3\phi_3} = 180°$). A detailed discussion of the rotational angles is provided in the original paper.
b) The configuration partition function, Z, is generated in the manner appropriate for a molecule containing a single trifunctional branch point: $Z = {_1U_1}^{(3)}({_2U_1} \ominus {_3U_1})[({_2U_2}^{(3)}) \otimes ({_3U_2}^{(3)})]$.
c) A summary of the 16 most prelevant configurations is given together with the optimized values of the ratio of the (combined) statistical weights to the configuration partition function, using a given reference set of conformational energies and $T = 298$ K (reference set / probable range).

Comments:

An RIS treatment of unperturbed triacetin is presented. The objective is to determine conformational preferences of the glycerol moiety in unperturbed triglycerides. Calculations based on the model provide excellent agreement with the experimental dipole moment, optical anisotropy, and molar Kerr constant. The orientation of the ester group dipole moment and composition of the anisotropic part of the ester group polarizability tensor are among the critical parameters in the calculation. Statistical weights deduced from the analysis reveal that a variety of configurations are accessible to the glycerol moiety in an unperturbed triglyceride.

Calcd. quantities: $\langle\mu^2\rangle$ $\langle\gamma^2\rangle$ $\langle_mK\rangle$

The following model requires two pages.
Therefore, this side has been left blank

X 010

unperturbed triglycerides

Mattice, W. L. *J. Am. Chem. Soc.* **1979**, *101*, 732.

A triglyceride in the interior of a chylomicron or very low density lipoprotein is essentially unperturbed by long-range interactions. Configuration-dependent properties for unperturbed triglycerides are obtained from a representative sample generated by Monte-Carlo methods. Necessary *a priori* probabilities and conditional probabilities are obtained from an RIS treatment which incorporates first- and second-order interaction. Triglycerides studies have 1-22 carbon atoms in each acyl group. Unperturbed radii of gyration, $<s^2>_o^{1/2}$, are in the range 8-10 Å for those triglycerides.

The statistical weight matrix for bond i in branch j is denoted by $_jU_i$. Matrices are given as $(_2U_3 = {_3U_3} = \text{col}(1,1,1);\ _1U_{n(1)-1} = {_2U_{n(2)}} = {_3U_{n(3)}} = \text{col}(1,1,1)$:

$$_1U_{n(1)} = \begin{bmatrix} 1 & 1 \end{bmatrix}$$

$$_2U_1 = \begin{bmatrix} \sigma_1 & \sigma_2 & 1 \\ \sigma_1 & \sigma_2 & \omega_1 \end{bmatrix}$$

$$_2U_2 = \begin{bmatrix} 1 & \sigma_3 & \sigma_3\omega_4 \\ 1 & \sigma_3\omega_4 & \sigma_3\omega_4 \\ 1 & \sigma_3\omega_4 & \sigma_3 \end{bmatrix}$$

$$_2U_1 \ominus {_3U_1} = \begin{bmatrix} \sigma_1^2\omega_2 & \sigma_1\omega_1 & \sigma_1\sigma_2 & \sigma_1\sigma_2 & \sigma_2\omega_1 & \sigma_2^2\sigma_3 & \sigma_1 & \omega_1 & \sigma_2 \\ \sigma_1^2\omega_2 & \sigma_1 & \sigma_1\sigma_2 & \sigma_1\sigma_2 & \sigma_2 & \sigma_2^2\sigma_3 & \sigma_1\omega_1 & \omega_1 & \sigma_2\omega_1 \end{bmatrix}$$

rows:
columns: $t(160°),\ g^+(77.3°)$

$t,\ g^+$
$t(185°),\ g^+(53.6°),\ g^-(-62.7°)$

$t,\ g^+, g^-$
$t(180°),\ g^+(77.3°),\ g^-(-77.3°)$

t, g^+
$tt(175°),\ tg^+(62.7°),\ tg^-(-53.6°),\ g^+t,\ g^+g^+,\ g^+g^-,\ g^-t,\ g^-g^+,\ g^-g^-$

$$_3U_2 = \begin{bmatrix} 1 & \sigma_3\omega_4 & \sigma_3 \\ 1 & \sigma_3 & \sigma_3\omega_4 \\ 1 & \sigma_3\omega_4 & \sigma_3\omega_4 \end{bmatrix}$$

$$_1U_{n(1)-3} = \begin{bmatrix} 1 & 1 & 1 \\ 1 & 1 & \omega \\ 1 & \omega & 1 \end{bmatrix}$$

$$_1U_{n(1)-2} = \begin{bmatrix} 1 & 1 & 1 \\ \omega_a & 1 & \omega_a \\ \omega_a & \omega_a & 1 \end{bmatrix}$$

$$_2U_4 = {_3U_4} = \begin{bmatrix} 1 & 1 & 1 \end{bmatrix}$$

rows:
columns:

t,g^+,g^-
$t(180°),\ g^+(77.3°),\ g^-(-77.3°)$

t,g^+,g^-
$t(180°),\ g^+(65°),\ g^-(-65°)$

$t,\ g^+,g^-$
$t(180°),\ g^+(65°),\ g^-(-65°)$

$t(180°),\ g^+(65°),\ g^-(-65°)$

$$_2U_5 = {_3U_5} = \begin{bmatrix} 1 & \omega_a & \omega_a \\ 1 & 1 & \omega_a \\ 1 & \omega_a & 1 \end{bmatrix}$$

$$U_{C-C_{chain}} = \begin{bmatrix} 1 & \sigma & \sigma \\ 1 & \sigma & \sigma\omega \\ 1 & \sigma\omega & \sigma \end{bmatrix}$$

rows:
columns:

t,g^+,g^-
$t(180°),\ g^+(65°),\ g^-(-65°)$

t,g^+,g^-
$t(180°),\ g^+(65°),\ g^-(-65°)$

Dihedral angles are given in the column headings. Statistical weights at 298 K are $\sigma_1 = 2.3$, $\sigma_2 = 0.16$, $\sigma_3 = 0.71$, $\omega_1 = \omega_3 = \omega_4 = 0.034$, and $\omega_2 = 1$. Values for σ and ω at 298 K are 0.43 and 0.034, respectively. Planar ester groups are assumed.

Additional statistical weight matrices are required to describe interactions which occur near the *cis* carbon-carbon double bond in oleic acid. If the *i*th bond in branch *j* is a *cis* carbon-carbon double bond, interactions can be taken into account using the following matrices:

$$_jU_{i-2} = \begin{bmatrix} 1 & \sigma' & \sigma' \\ 1 & \sigma' & \sigma'\omega \\ 1 & \sigma'\omega & \sigma' \end{bmatrix} \quad _jU_{i-1} = \begin{bmatrix} 1 & 1 & 1 \\ 1 & 1 & 1 \\ 1 & 1 & 1 \end{bmatrix} \quad _jU_i = \begin{bmatrix} 1 & 0 & 0 \\ 0 & 1 & 0 \\ 0 & 0 & 1 \end{bmatrix} \quad _jU_{i+1} = \begin{bmatrix} 0 & \zeta & \zeta \\ \zeta & 1 & 1 \\ \zeta & 1 & 1 \end{bmatrix} \quad _jU_{i+2} = \begin{bmatrix} 1 & \sigma' & \sigma' \\ 1 & \sigma' & \sigma' \\ 1 & \sigma' & \sigma' \end{bmatrix}$$

rows: t, g^+, g^- t, g^+, g^- t, g^+, g^- t cis, g^+ cis, g^- cis t, g^+, g^-

columns: $t(180°), g^+(65°), g^-(-65°)$ $t(180°), g^+(120°), g^-(-120°)$ t cis(0°), g^+ cis(0°), g^- cis(0°) $t(180°), g^+(120°), g^-(-120°)$ $t(180°), g^+(65°), g^-(-65°)$

Appropriate values for ζ and σ' are 0.1 and 1.4, respectively.

Bond Length [pm]		Valence Angles [°]		Bond Length [pm]		Valence Angles [°]	
C-C'	: 150	C-C'-O'	: 126.3	C=O	: 134	O-C-C	: 106.4
C-O	: 144	O-C'-O'	: 122.3	C-C (glycerol)	: 151	C-C=C	: 125.0
C'-O	: 135	C-O-C'	: 116.7	C-C (acyl)	: 153	C-C-C (glycerol)	: 114.7
C'-O'	: 120	C-C-C'	: 113.7			C-C-C (acyl)	: 112.0

Calcd. quantities: $<s^2>_0 / \Sigma l^2$, $<r^2>_0 / \Sigma l^2$

vinyl polymers with flexible side chains; application to poly(alkyl vinyl ethers)

Abe, A. *J. Polym. Sci.: Symp.* **1976**, *54*, 135.

The general method for treating the configurational statistics of vinyl polymer chains of *Flory* is extended to include vinyl chains with flexible side chains. Elements of the generator matrix **G** are modified so that statistical mechanical averages include all configurations of the skeletal backbone as well as side chains. An expression is derived for the mean-square of the group dipole moment attached to these side chains in the RIS approximation. The mean-square dipole moment $<\mu^2>/x$ of poly(alkyl vinyl ethers) are treated according to this scheme.

The expressions generally applicable to any vinyl chains such as $H-(CH_2-CHX)_x-CH_3$ is:

$$U = \begin{bmatrix} 1 & 1 & 1 \\ 1 & \omega & 1 \\ 1 & 1 & 0 \end{bmatrix} \quad U''_m = \begin{bmatrix} \eta^2\omega'' & \eta & \eta\tau\omega' \\ \eta & \omega & \tau\omega' \\ \eta\tau\omega' & \tau\omega' & \tau^2\omega\omega'' \end{bmatrix} \quad U''_r = \begin{bmatrix} \eta^2 & \eta\omega' & \eta\tau\omega'' \\ \eta\omega' & 1 & \tau\omega \\ \eta\tau\omega'' & \tau\omega & \tau^2\omega'^2 \end{bmatrix}$$

Application to poly(alkyl vinyl ethers):

$$U' = \begin{bmatrix} 0 & 1 & 1 \\ 1 & 0.050 & 2 \\ 1 & 2 & 0 \end{bmatrix} \quad U''_m = \begin{bmatrix} 0.678 & 2.100 & 0.215 \\ 2.100 & 0.025 & 0.027 \\ 0.215 & 0.027 & 0.000 \end{bmatrix} \quad U''_r = \begin{bmatrix} 4.339 & 0.328 & 0.134 \\ 0.328 & 1.000 & 0.019 \\ 0.134 & 0.019 & 0.000 \end{bmatrix}$$

The rows and columns of these matrices are indexed in the order t, g, g⁻. The parameters η and τ represent statistical weights for the first-order interactions associated with t and g⁻ conformations, and ω, ω', and ω'' are those for the second order interactions between groups $CH_2\cdots CH_2$, $CH_2\cdots X$, and $X\cdots X$, respectively. Each element (ξ,ζ) {ξ and ζ denote each of the rotational states about the skeletal bonds (i-1 and i)} of the **U'** matrix given above is now multiplied by the corresponding statistical weight factor $\beta(\xi,\zeta)$ to complete the expressions required for the vinyl polymer system with flexible side chains. Thus, $U' = [u'_{\xi\zeta}\beta(\xi,\zeta)]$. As is shown by inspection of a model statistical weight matrix, **U''** should not be affected by the side-chain conformations under the assumption that interactions of higher orders are negligible.

The mean-square dipole moments $<\mu^2>/x$ are calculated for chains of 100 units whose stereochemical configurations are generated by Monte Carlo methods. Geometrical parameters used are as follows: bond length C-C: 153 pm, C-O: 143 pm; bond angles $<CCC = 112.0°$, $<COC = 111.5°$, $<CCO = 100°$. Possible displacements of rotational states are taken into account by adopting the expression: $\phi(t) = 180° - \Delta\phi$, $\phi(g^+) = 60° + \Delta\phi_g$, and $\phi(g^-) = -60°$; $\Delta\phi_g = 0°, 10°, 20°$.

X 012 (see also X 013 and X 014)

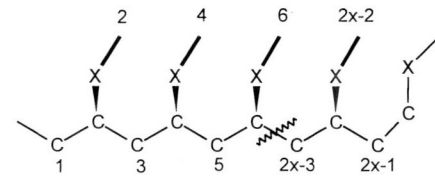

vinyl polymers containing articulated side chains with threefold rotational potentials

Mattice, W. L. *Macromolecules* **1977**, *10*, 1171.

The configuration partition function, Z, for vinyl polymers containing articulated side chains with threefold rotational potentials is evaluated using RIS theory in the form appropriate for branched molecules. Of particular interest is the form adopted by Z when it is to be utilized for the computation of an average property of the main chain, such as its mean-square unperturbed end-to-end distance, $<r^2>_0$, or the average end-to-end vector, $<\mathbf{r}>$. The second and succeeding atoms of the articulated side chain then give rise to five factors which appear in seven of the nine elements in statistical weight matrices for alternate bonds of the main chain. Precise definitions of these factors are presented for articulated side chains containing two or three bonds, threefold rotational potentials being assumed. Their effect on the conformations adopted by bonds in the main chain, on $<r^2>_0$, and on $d(\ln <r^2>_0)/dT$ is evaluated.

Side chains containing two bonds; treatment as meso and racemic diads:

$$_1U_1 = [1] \qquad _2U_1 \ominus {_3U_1} = [\eta\omega_3 \; 1 \; \tau \; \eta \; 1 \; \tau \; \eta\tau_1 \; \tau_1 \; \tau\tau_1\omega_3]$$

In general rows and columns are indexed by t, g, g⁻ and/or tt, tg, tg⁻, gt, gg, gg⁻, g⁻t, g⁻g, g⁻g⁻

$$(_3U_2)_m = (_5U_2)_m = \ldots = (_{2x-3}U_2)_m = \begin{bmatrix} \eta\omega_2 & 1 & \tau\omega_1 \\ \eta & \omega & \tau\omega_1 \\ \eta\omega_1 & \omega_1 & \tau\omega\omega_2 \end{bmatrix}$$

$$(_3U_2)_r = (_5U_2)_r = \ldots = (_{2x-3}U_2)_r = \begin{bmatrix} \eta & \omega_1 & \tau\omega_2 \\ \eta\omega_1 & 1 & \tau\omega \\ \eta\omega_2 & \sigma & \tau\omega_1^2 \end{bmatrix}$$

$$_2U_2 = {_4U_2} = \ldots = {_{2x-2}U_2} = {_{2x-1}U_4} = \begin{bmatrix} 1 \\ 1 \\ 1 \end{bmatrix}$$

$$_4U_1 \ominus {_5U_1} = {_6U_1} \ominus {_7U_1} = \ldots = {_{2x-2}U_1} \ominus {_{2x-1}U_1} = \begin{bmatrix} \eta\omega_3 & 1 & \tau & \eta\omega_3 & \omega_3 & \tau_3 & \eta\tau_1 & \tau_1 & \tau\tau_1\omega_3 \\ \eta\omega_3 & \omega & \tau & \eta & \omega & \tau & \eta\tau_1 & \tau_1 & \tau\tau_1\omega_3 \\ \eta\omega_3 & 1 & \tau\omega & \eta & 1 & \tau\omega & \eta\tau_1\omega & \tau_1\omega_3 & \tau\tau_1\omega\omega_3^2 \end{bmatrix}$$

$$_{2x-1}U_2 = \begin{bmatrix} 1 & \sigma & \sigma\omega_1 \\ 1 & \sigma\omega & \sigma \\ 1 & \sigma\omega_1 & \sigma\omega \end{bmatrix} \qquad _{2x-1}U_3 = \begin{bmatrix} 1 & \sigma & \sigma \\ 1 & \sigma\omega & \sigma \\ 1 & \sigma & \sigma\omega \end{bmatrix}$$

$$_jU'_m = \begin{bmatrix} \eta\tau^* & 1 & \tau f_1 \\ \eta & \omega(2-\tau^*) & \tau f_2 \\ \eta f_1 & f_2 & \tau\omega f_4 \end{bmatrix} \begin{bmatrix} \eta\omega_2 & 1 & \tau\omega_1 \\ \eta & \omega & \tau\omega_1 \\ \eta\omega_1 & \omega_1 & \tau\omega\omega_2 \end{bmatrix}$$

$$_jU'_r = \begin{bmatrix} \eta\tau^* & 1 & \tau f_1 \\ \eta & \omega(2-\tau^*) & \tau f_2 \\ \eta f_1 & f_2 & \tau\omega f_4 \end{bmatrix} \begin{bmatrix} \eta & \omega_1 & \tau\omega_2 \\ \eta\omega_1 & 1 & \tau\omega \\ \eta\omega_2 & \omega & \tau\omega_1^2 \end{bmatrix}$$

$\tau^* = (\tau_1 + 2\omega_3)(1 + \tau_1 + \omega_3)^{-1}$, $f_1 = (1 + \omega_3 + \tau_1\omega_3)(1 + \tau_1 + \omega_3)^{-1}$, $f_2 = (2 + \tau_1\omega_3)(1 + \tau_1 + \omega_3)^{-1}$, $f_3 = (2 + \tau_1)(1 + \tau_1 + \omega_3)^{-1} = 2 - \tau^*$, $f_4 = (2 + \tau_1\omega_3^2)(1 + \tau_1 + \omega_3)^{-1}$.

X 013   (see also X 012 and X 014)

vinyl polymers containing articulated side chains with threefold rotational potentials

Mattice, W. L. *Macromolecules* **1977**, *10*, 1171.

Side chains containing two bonds; treatment as d and l centers:

$$_jU'_{dd} = \begin{bmatrix} \eta\tau^* & 1 & \tau f_1 \\ \eta f_1 & f_2 & \tau\omega f_4 \\ \eta & \omega(2-\tau^*) & \tau f_2 \end{bmatrix} \begin{bmatrix} \eta\omega_2 & \tau\omega_1 & 1 \\ \eta & \tau\omega_1 & \omega \\ \eta\omega_1 & \tau\omega\omega_2 & \omega_1 \end{bmatrix} \qquad _jU'_{dl} = \begin{bmatrix} \eta\tau^* & 1 & \tau f_1 \\ \eta f_1 & f_2 & \tau\omega f_4 \\ \eta & \omega(2-\tau^*) & \tau f_2 \end{bmatrix} \begin{bmatrix} \eta & \omega_1 & \tau\omega_2 \\ \eta\omega_1 & 1 & \tau\omega \\ \eta\omega_2 & \omega & \tau\omega_1^2 \end{bmatrix}$$

$$_jU'_{ld} = \begin{bmatrix} \eta\tau^* & \tau f_1 & 1 \\ \eta & \tau f_2 & \omega(2-\tau^*) \\ \eta f_1 & \tau\omega f_4 & f_2 \end{bmatrix} \begin{bmatrix} \eta & \tau\omega_2 & \omega_1 \\ \eta\omega_2 & \tau\omega_1^2 & \omega \\ \eta\omega_1 & \tau\omega & 1 \end{bmatrix} \qquad _jU'_{ll} = \begin{bmatrix} \eta\tau^* & \tau f_1 & 1 \\ \eta & \tau f_2 & \omega(2-\tau^*) \\ \eta f_1 & \tau\omega f_4 & f_2 \end{bmatrix} \begin{bmatrix} \eta\omega_2 & 1 & \tau\omega_1 \\ \eta\omega_1 & \omega_1 & \tau\omega\omega_2 \\ \eta & \omega & \tau\omega_1 \end{bmatrix}$$

$\tau^* = (\tau_1 + 2\omega_3)(1 + \tau_1 + \omega_3)^{-1}$, $f_1 = (1 + \omega_3 + \tau_1\omega_3)(1 + \tau_1 + \omega_3)^{-1}$, $f_2 = (2 + \tau_1\omega_3)(1 + \tau_1 + \omega_3)^{-1}$, $f_3 = (2 + \tau_1)(1 + \tau_1 + \omega_3)^{-1} = 2 - \tau^*$, $f_4 = (2 + \tau_1\omega_3^2)(1 + \tau_1 + \omega_3)^{-1}$.

Side chain	τ^*	f_1	f_2	f_3	f_4	$E(\tau)$	$E(\omega_3)$	$E(\sigma_1)$	$E(\omega_4)$
CH_3	1.00	1.00	1.00	1.00	1.00				
CH_2CH_3	0.21	0.85	1.65	1.79	1.64	4.187	8.374		
$CH_2CH_2CH_3$	0.17	0.89	1.72	1.83	1.72	4.187	8.374	2.093	8.374
OCH_3	0.09	0.91	1.83	1.91	1.83	5.862	∞		
OCH_2CH_3	0.07	0.93	1.86	1.93	1.86	5.862	∞	3.768	∞
$OOCCH_3$	0.07	1.00	1.93	1.93	1.93	∞	8.374		
CH_2OCH_3	0.79	0.89	1.11	1.21	1.05	1.256	1.424	3.768	∞

X 014   (see also X 012 and X 013)

vinyl polymers containing articulated side chains with threefold rotational potentials

Mattice, W. L. *Macromolecules* **1977**, *10*, 1171.

Extension to longer side chains; side chains containing three bonds:

The preceeding treatment is readily applied to longer side chains. Necessary procedures are illustrated by considering a molecule in which the side chain contains three bonds. Some equations are replaced; here, σ_1 and ω_4 denote first- and second-order interactions of the third atom in the side chain with atoms in the main chain. Previous expressions for $_jU'_m$ and $_jU'_r$ apply without modification provided each τ_1 in the definition for τ^* and the f_i is replaced by $c\tau_1$, where $c = (1 + 2\sigma_1\omega_4)(1 + \sigma_1 + \sigma_1\omega_4)^{-1}$.

$$_1U_1 = [1]$$

$$_2U_1 \ominus {_3U_1} = [\eta\omega_3 \ 1 \ \tau \ \eta \ 1 \ \tau \ \eta\tau_1 \ \tau_1 \ \tau\tau_1\omega_3]$$

$$_2U_2 = {_4U_2} = \ldots = {_{2x-2}U_2} = \begin{bmatrix} 1 & \sigma_1 & \sigma_1\omega_1 \\ 1 & \sigma_1\omega_4 & \sigma_1 \\ 1 & \sigma_1\omega_4 & \sigma_1\omega_4 \end{bmatrix}$$

$$_2U_3 = {_4U_3} = \ldots = {_{2x-2}U_3} = {_{2x-1}U_4} = \begin{bmatrix} 1 \\ 1 \\ 1 \end{bmatrix}$$

$$_4U_1 \ominus {_5U_1} = {_6U_1} \ominus {_7U_1} = \ldots = {_{2x-2}U_1} \ominus {_{2x-1}U_1} = \begin{bmatrix} \eta\omega_3 & 1 & \tau & \eta\omega_3 & \omega_3 & \tau_3 & \eta\tau_1 & \tau_1 & \tau\tau_1\omega_3 \\ \eta\omega_3 & \omega & \tau & \eta & \omega & \tau & \eta\tau_1 & \tau_1 & \tau\tau_1\omega_3 \\ \eta\omega_3 & 1 & \tau\omega & \eta & 1 & \tau\omega & \eta\tau_1\omega & \tau_1\omega_3 & \tau\tau_1\omega\omega_3^2 \end{bmatrix}$$

$$_{2x-1}U_2 = \begin{bmatrix} 1 & \sigma & \sigma\omega_1 \\ 1 & \sigma\omega & \sigma \\ 1 & \sigma\omega_1 & \sigma\omega \end{bmatrix} \quad _{2x-1}U_3 = \begin{bmatrix} 1 & \sigma & \sigma \\ 1 & \sigma\omega & \sigma \\ 1 & \sigma & \sigma\omega \end{bmatrix}$$

$$_jU'_m = \begin{bmatrix} \eta\tau^* & 1 & \tau f_1 \\ \eta & \omega(2-\tau^*) & \tau f_2 \\ \eta f_1 & f_2 & \tau\omega f_4 \end{bmatrix} \begin{bmatrix} \eta\omega_2 & 1 & \tau\omega_1 \\ \eta & \omega & \tau\omega_1 \\ \eta\omega_1 & \omega_1 & \tau\omega\omega_2 \end{bmatrix}$$

$$_jU'_r = \begin{bmatrix} \eta\tau^* & 1 & \tau f_1 \\ \eta & \omega(2-\tau^*) & \tau f_2 \\ \eta f_1 & f_2 & \tau\omega f_4 \end{bmatrix} \begin{bmatrix} \eta & \omega_1 & \tau\omega_2 \\ \eta\omega_1 & 1 & \tau\omega \\ \eta\omega_2 & \omega & \tau\omega_1^2 \end{bmatrix}$$

$\tau^* = (\tau_1 + 2\omega_3)(1 + \tau_1 + \omega_3)^{-1}$, $f_1 = (1 + \omega_3 + \tau_1\omega_3)(1 + \tau_1 + \omega_3)^{-1}$, $f_2 = (2 + \tau_1\omega_3)(1 + \tau_1 + \omega_3)^{-1}$, $f_3 = (2 + \tau_1)(1 + \tau_1 + \omega_3)^{-1} = 2 - \tau^*$, $f_4 = (2 + \tau_1\omega_3^2)(1 + \tau_1 + \omega_3)^{-1}$.

X 015

poly(alkyl vinyl ether)s

Abe, A. *Macromolecules* **1977**, *10*, 34.

Bond Length [pm]	Valence Angles [°]	Torsion Angles [°]	ξ (for 298 K)	ξ_0	E_ξ [kJ·mol^{-1}]
C-C : 153	C-C-C : 112	*For meso (dd):*	$\eta = 2.10$	1	−1.84
C-O : 143	C-O-C : 111.5	tt : 7, 7	$\tau = 0.43$	1	2.09
		tg : 0, 115			
C-H : 110	C-C-O : 110	gt : 115, 0	$\omega'' = 0.16$	1	4.61
		t$\bar{g}$: 23, −118			
		$\bar{g}$t : −118, 23	$\omega' = 0.22\text{-}0.07$	1	3.77 - 6.70
	C-C-H : 110	gg : 116, 79; 79, 116			
		g$\bar{g}$: 90, −109	$\omega = 0.04\text{-}0.02$	1	7.95 - 9.21
		$\bar{g}$,g : −109, 90			

$$U''_m = \begin{bmatrix} \eta^2\omega'' & \eta & \eta\tau\omega' \\ \eta & \omega & \tau\omega' \\ \eta\tau\omega' & \tau\omega' & 0 \end{bmatrix} \quad U''_r = \begin{bmatrix} \eta^2 & \eta\omega' & \eta\tau\omega'' \\ \eta\omega' & 1 & \tau\omega \\ \eta\tau\omega'' & \tau\omega & 0 \end{bmatrix} \quad U'_a = \begin{bmatrix} 1 & 1 & 1 \\ 1 & \omega & 1 \\ 1 & 1 & 0 \end{bmatrix}$$

$$U' = [u'_a(\xi\zeta) \times \beta(\xi\zeta)]$$

rows and columns: t, g, $\bar{g}$

For racemic (dl):

tt : 3, 3
tg : 23, 113
gt : 113, 23
t$\bar{g}$: 13, −113
$\bar{g}$t : −113, 13
gg : 115, 115
g$\bar{g}$: 77, −118
$\bar{g}$g : −118, 77

$E(\eta) \approx -2.1$, $E(\tau) \approx 2.1$ kJ·mol^{-1}

With R = C*H(CH$_3$)CH$_2$CH$_3$ (S), a and b may be identified as:

$a = \eta + 1 + \tau$, $b = 1 + \tau$

With R = CH$_2$C*H(CH$_3$)CH$_2$CH$_3$ (S), a and b are given by:
$E(\sigma) = 3.77$, $E(\tau') = 2.1$, $E(\tau'') = 6.28$, $E(\omega') = 2.93$ kJ·mol^{-1}

$a = 3 + 2\sigma + 2\tau' + 2\tau'' + \sigma\tau'' + \omega'$
$b = 3 + \sigma + 2\tau' + 2\tau'' + \sigma\tau'' + \omega' + \sigma\omega'$

For symmetric chains:

$$\beta = \begin{bmatrix} 0 & 1 & 1 \\ 1 & 1 & 2 \\ 1 & 2 & 2 \end{bmatrix}$$

For asymmetric chains:

$$\beta_d = \begin{bmatrix} 0 & b & b \\ a & a+b & a+b \\ a & a+b & a+b \end{bmatrix} \quad \beta_l = \begin{bmatrix} 0 & a & a \\ b & a+b & a+b \\ b & a+b & a+b \end{bmatrix}$$

For symmetric chains, furthermore:

$$U_1 = \begin{bmatrix} 1 & 2 & 2 \\ 0 & 0 & 0 \\ 0 & 0 & 0 \end{bmatrix} \quad U_x = \begin{bmatrix} 1 \\ 2 \\ 2 \end{bmatrix}$$

For asymmetric chains, furthermore:

$$U_{d;1} = \begin{bmatrix} a & a+b & a+b \\ 0 & 0 & 0 \\ 0 & 0 & 0 \end{bmatrix} \quad U_{l;1} = \begin{bmatrix} b & a+b & a+b \\ 0 & 0 & 0 \\ 0 & 0 & 0 \end{bmatrix} \quad U_{d;x} = \begin{bmatrix} b \\ a+b \\ a+b \end{bmatrix} \quad U_{l;x} = \begin{bmatrix} a \\ a+b \\ a+b \end{bmatrix}$$

Comments:
Conformational energies are estimated for the poly(alkyl vinyl ether) chains using semiempirical energy expressions. The results are tested for NMR data on 2,4-dimethoxypentane. The 3 × 3 statistical weight matrices derived therefrom are applied to the analysis of various configuration dependent properties of these polymers: $H-(CH_2-CH(OR))-CH_3$, R = methyl, ethyl, isopropyl, (S)-1-methylpropyl, and (S)-2-methylbutyl groups. The characteristic ratio of the unperturbed dimensions (C_∞ = 6.1 - 8.0) estimated from fractionated samples of poly(methyl vinyl ether) in a good solvent is reproduced in a moderately isotactic region. Experimental values of the mean-square dipole moment per repeat unit $<\mu^2>/x$ for isotactic samples of poly(isopropyl vinyl ether) (0.67) and poly(isobutyl vinyl ether) (0.97 - 1.35) are also found to be in agreement with those calculated in a reasonable range of tacticity.

poly(oxy(1-alkyl ethylenes)) with side chains such as ethyl, isopropyl, and *tert.*-butyl

Abe, A.; Hirano, T.; Tsuji, K.; Tsuruta, T. *Macromolecules* **1979**, *12*, 1100.

X 016   poly(oxy(1-ethyl ethylene)), R = CH$_2$-CH$_3$

Bond Length [pm]	Valence Angles [°]	Torsion Angles [°]	ξ (for 303 K)	ξ_0	E_ξ [kJ·mol^{-1}]
C-C : 153	O-C-O : 111.5	t	α = 1.65	1	−1.256
			β = 0.56 / 0.44	1	1.47 / 2.09
C-O : 143	C-C-O : 111.5	g$^+$	γ = 0.44	1	2.09 (± 1.05)
			δ = 0.19	1	4.19 (± 2.09)
C-H : 110		g$^-$	σ = 0.12 / 0.22	1	5.44 / 3.77
			ω = 0.52	1	1.67
			ω' = 0.31	1	2.93

$$V_R = \begin{bmatrix} \gamma\omega+\delta & 1+\gamma\omega' & 1+\gamma\omega'+\delta \\ \gamma+\delta & 1+\gamma & 1+\gamma+\delta \\ \gamma & 1+\gamma & 1+\gamma \end{bmatrix}$$

rows and columns: t, g$^+$, g$^-$

X 017   poly(oxy(1-isopropyl ethylene)), R = CH(CH$_3$)$_2$

Bond Length [pm]	Valence Angles [°]	Torsion Angles [°]	ξ (for 303 K)	ξ_0	E_ξ [kJ·mol^{-1}]
C-C : 153	O-C-O : 111.5	t	ξ = 0.19 / 0.04	1	4.19 / 8.37
C-O : 143	C-C-O : 111.5	g$^+$	ν = 0.61 - 0.14	1	1.26 - 5.02
C-H : 110		g$^-$	ω' = 0.19 / 0.31	1	4.19 / 2.93

$$V_R = \begin{bmatrix} \xi\omega' & \nu\omega' & 1+\nu\omega'+\xi\omega' \\ \xi & \nu & 1+\nu+\xi \\ 0 & \nu & \nu \end{bmatrix}$$

rows and columns: t, g$^+$, g$^-$

X 018   poly(oxy(1-tert.-butyl ethylene)), R = C(CH$_3$)$_3$

C-C : 153	O-C-O : 111.5	$\omega' = 0.19$	1	4.19	
C-O : 143	C-C-O : 111.5	$\omega'' = 0.007$	1	12.56	$V_R = \begin{bmatrix} 0 & 0 & \omega' \\ 0 & 0 & 1 \\ 0 & 0 & \omega'' \end{bmatrix}$
C-H : 110					

rows and columns: t, g$^+$, g$^-$

Comments:
Configuration-dependent properties of a series of poly(oxy(1-alkyl ethylene)) chains having various side chains differing in size and shape are examined. Conformational energy parameters established for polyoxypropylene (**A 101**) are adopted in common to the skeletal configuration, steric interaction imposed by larger substituents being taken into account separately. Each element (η,ζ) of statistical weight matrices V_R represents the effect attributable to the side chain when skeletal bonds i and $i+1$ are in the rotational states η and ζ, respectively. Each element $u_{\eta\zeta}$ of the $U_{b,R}$ matrix given in **A 101** is multiplied by the corresponding statisical weight factor $V_R(\eta\zeta)$ to complete a revised expression for the polymer system under consideration. Conformational energy calculations carried out for a portion of the PEO chain (**S 140**) give energy minima for the g$^\pm$g$^\mp$ states at ϕ(C-C) = $\pm$ 70° and ϕ(C-O) = $\mp$ 95°. Within a reasonable range of conformational energies, observed values of the characteristic ratio and the dipole moment ratio for isotactic polymers are reproduced.

Calcd. quantities: $<r^2>_0 / nl^2$ $<\mu^2>_0 / xm^2$

X 019

poly(1-alkenes)

Wittwer, H.; Suter, U. W. *Macromolecules* **1985**, *18*, 403.

Bond Length [pm]	Valence Angles [°]	Torsion Angles a) [°]	ξ (for 300 K)	ξ_o	E_ξ [kJ·mol^{-1}]
C - C : 153	C-CH$_2$-C : 114.0	t	σ = 0.53	1.0	2.08
C - H : 110	C-CH-H : 108.5	g^{+*}	η = 0.92	1.0	0.29
	H-C-H : 108.5	g$^+$	τ = 0.27	0.5	2.08
	C-CH-C : 112.0	g$^-$	ω = 0.09	0.9	7.48
	C-C-H : 106.8	g^{-*}	ω^* = 0.09	0.9	7.48

$$U'_{PP} = \begin{bmatrix} 1 & 1 & 1 & 1 & 1 \\ 1 & 1 & 1 & 1 & 1 \\ 1 & 1 & 0 & 0 & 1 \\ 1 & 1 & 0 & 0 & 1 \\ 1 & 1 & 1 & 1 & 0 \end{bmatrix} \Rightarrow U' = \begin{bmatrix} a & a & b & b & c \\ a & a & b & b & c \\ b' & b' & 0 & 0 & d \\ b' & b' & 0 & 0 & d \\ c' & c' & d & d & 0 \end{bmatrix}$$

rows and columns: t, g^{+*}, g$^+$, g$^-$, g^{-*}

$$U''_m = \begin{bmatrix} 0 & \eta\omega^* & 0 & \eta & 0 \\ \eta\omega^* & 0 & 0 & 0 & \tau\omega^* \\ 0 & 0 & 0 & \omega^* & \tau\omega^* \\ \eta & 0 & \omega^* & 0 & 0 \\ 0 & \tau\omega^* & \tau\omega^* & 0 & 0 \end{bmatrix} \quad U''_r = \begin{bmatrix} \eta^2 & 0 & \eta\omega^* & 0 & 0 \\ 0 & 0 & 0 & \omega^* & \tau\omega^* \\ \eta\omega^* & 0 & 0 & 0 & \tau\omega^* \\ 0 & \omega^* & 0 & 1 & 0 \\ 0 & \tau\omega^* & \tau\omega^* & 0 & 0 \end{bmatrix}$$

For P1B:

$a = \tau + 2$
$b = b' = 1 + \omega + \tau\omega$
$c = c' = 1 + \omega + \tau\omega$
$d = 2 + \tau\omega$

For P1P:

$a = \tau + 2\omega + 2\sigma\omega + 2\sigma\tau\omega$
$b = b' = 1 + \sigma + \tau + \omega + 2\sigma\omega + \sigma\tau\omega$
$c = c' = 1 + \sigma + \omega + 2\sigma\omega + \tau\omega$
$d = 2 + 2\sigma + 2\sigma\omega + \tau\omega$

For P4MP:

$a = 2\omega + 2\tau\omega$
$b = b' = 1 + 2\omega + 3\tau\omega$
$c = c' = 1 + 2\omega + \tau\omega$
$d = 2 + 2\omega + 2\tau\omega$

For PS4MH

$a = \omega + 2\eta\omega + 4\tau\omega + \eta\tau\omega + 2\tau^2\omega$
$b = 1 + 3\omega + 2\eta\omega + \eta + 6\tau\omega + \eta\tau\omega + 2\tau^2\omega$
$b' = 2\eta + 2\omega + \eta\tau + 7\tau\omega + 2\tau^2\omega$
$c = 1 + 3\omega + 2\eta\omega + \tau + 3\tau\omega + \eta\tau\omega$
$c' = 2\eta + 2\omega + \eta\tau + 4\tau\omega$
$d = 1 + 2\eta + 4\omega + \tau + \eta\tau + 6\tau\omega$

a) For explicit values of the side-chain torsion angle, χ, see the original reference.

Comments:
RIS models are constructed, based on the model for polypropylene (PP) with five states per bond, and assuming mutually independent side groups, for poly(1-butene) (P1B), poly(1-pentene) (P1P), poly(4-methyl-1-pentene) (P4MP), and poly[(S)-4-methyl-1-hexene] (PS4MH). Values for the unperturbed dimensions are calculated and found to be in good agreement with experiment for P1P, and poor agreement with experiment for P4MP and PS4MH. The reason for the poor agreement is found to lie in the mutual interdependence between side groups. A remedy is offered by direct numerical integration; this approach, however, is limited by the amount of computational effort necessary.

poly(2-chlorocyclohexyl acrylate), PCCHA

Díaz-Calleja, R.; Riande, E.; San Román, J.; Compañ, V. *Macromolecules* **1994**, *27*, 2092.

Bond Length [pm]	Valence Angles [°]	Torsion Angles [°]	ξ (for 373 K)	ξ_o	E_ξ [kJ·mol^{-1}]
	C-C-C : 112.5	$(tt)_m$: 164, 164	β = 1.79	1	1.8
		$(tg)_m$: 177, −66			
		$(gt)_m$: 66, 177	τ = 1.1	1	0.3
		$(tt)_r$: 177, 177			
			γ = 1.79	1	1.8
		χ : 0 or 180			
			γ_1 = 1.4	1	1.0
			γ_2 = 1	1	0

$$U' = \begin{bmatrix} 1 & 0 & \rho & 0 \\ 0 & \rho & 0 & \rho \\ 1 & 0 & 0 & 0 \\ 0 & \rho & 0 & 0 \end{bmatrix} \quad U_r = \begin{bmatrix} 1 & \gamma_1 & 0 & 0 \\ \gamma_1 & \gamma_2 & 0 & 0 \\ 0 & 0 & 0 & 0 \\ 0 & 0 & 0 & 0 \end{bmatrix} \quad U_m = \begin{bmatrix} 1 & \gamma & \beta & \beta \\ \gamma & 1 & \beta & \beta \\ \beta & \beta & 0 & 0 \\ \beta & \beta & 0 & 0 \end{bmatrix}$$

rows and columns: t (χ=0°), t (χ = 180°), g (χ=0°), g (χ=180°)

Comments:
The temperature and frequency dependence of the complex dielectric permittivity ε^* for both 2-chlorocyclohexyl isobutyrate (CCHI) and poly(2-chlorocyclohexyl acrylate) (PCCHA) is reported. The analysis of the dielectric results in terms of the electric modulus suggests that whereas the conductive processes in CCHI are produced only by free charges, the conductivity observed in PCCHA involves both free charges and interfacial phenomena. The 4×4 RIS scheme presented which accounts for two rotational states about the CH−CO bonds of the side group reproduces the intramolecular correlation coefficient of the polymer.

Calcd. quantities: $\langle \mu^2 \rangle / x$ $\quad\quad$ $d(\ln \langle \mu^2 \rangle)/dT$

X 021

itaconate polymers

Saiz, E.; Horta, A.; Gargallo, L.; Hernández-Fuentes, I.; Radić, D. *Macromolecules* **1988**, *21*, 1736.

Bond Length [pm]	Valence Angles [°]	Torsion Angles [°]	ξ (for 298 K)	ξ_o	E_ξ [kJ·mol^{-1}]			
	θ' : 112	$(tt)_m$: 178, −155	α = 0.09	1	5.9 (± 0.8)	$U' = \begin{bmatrix} 1 & 1 \\ 1 & 0 \end{bmatrix}$	$U''_m = \begin{bmatrix} 1 & \alpha \\ \alpha & \alpha^2/\beta \end{bmatrix}$	$U''_r = \begin{bmatrix} \beta & \alpha \\ \alpha & \alpha^2/\beta \end{bmatrix}$
		$(tg)_m$: 178, 66						
	θ'' : 110	$(gg)_m$: 66, 66	β = 0.36	1	2.5 (± 0.8)			
		$(tt)_r$: 178, −154						
		$(tg)_r$: 181, 60						
		$(gg)_r$: 60, 60						
		χ_1 : 0, 180				rows and columns: t, g$^+$		
		χ_2 : 180, ± 60						
		χ_3 : 120, 110, −70, −60						

Comments:
The dipole moments (μ) of poly(monobenzyl itaconate), PMBzl, and of poly(dibenzyl itaconate), PDBzl, of known tacticity are determined in dioxane at 298 K. A two state scheme (t, g$^+$) is used, with backbone bond angles close to each other. 2,2-Dimethylsuccinate is used as a model compound to obtain the modulus and orientation of the dipole moment of the repeat unit. The polymer μ calculated is scarcely sensitive to tacticity. Comparison with experimental μ shows excellent agreement in the case of PDBzl. For PMBzl, the experimental μ is higher than calculated, and agreement requires modification of the conformational parameters. According to the results shown, the result of conformational energy calculations, namely, E(α) = 5.9 kJ·mol^{-1} and E(β) = 2.5 kJ·mol^{-1} leads to μ_{eff} = 1.54 and 1.62, for w_m = 0.25 and 0.5, respectively. The unperturbed dimensions and the dipole moments of PMBzl and PDBzl can be reproduced theoretically with a two state RIS model analogous to that used for PMMA.

Calcd. quantities: $<r^2>_o / nl^2$ μ_{eff}

X 022

poly(alkyl methacrylates)

Sasaki, T.; Arisawa, H.; Yamamoto, M. *Polymer J.* **1991**, *23*, 108.

Bond Length [pm]	Valence Angles [°]	Torsion Angles (ϕ_i/ϕ_{i+1}) [°]	ξ (for 300 K)	ξ_0	E_ξ [kJ·mol^{-1}]
		MMA:	*PMMA:*		
		meso: 20(−15) / −15(20)	α = 0.232	1.6	4.8
		rac: 10 / 10	β = 3.83	1.4	− 2.5
	Ethyl groups:	*EMA:*	*PEMA:*		
	O-C-C : 110	meso: 20(−15) / −15(20)	α = 0.389	2.4	4.56
		rac: 10 / 10	β = 3.83	1.9	− 2.5
	Isopropyl groups:	*iPMA:*	*PiPMA:*		
	C-O-C : 117.3	meso: 105 (10)/ 10 (105)	α = 70		
	O-C-C : 108.6	rac: 5 / 5	β = 3000		
	(COC)(OCC) : 104,133				

$$U' = \begin{bmatrix} 1 & 1 \\ 1 & 0 \end{bmatrix} \qquad U''_m = \begin{bmatrix} 1 & \alpha \\ \alpha & \alpha^2/\beta \end{bmatrix} \qquad U''_r = \begin{bmatrix} \beta & \alpha \\ \alpha & \alpha^2/\beta \end{bmatrix}$$

rows and columns: t, g$^+$

poly(isopropyl methacrylate)

Comments:
Conformational energy calculations and estimation of the static chain stiffness are carried out for a number of poly(alkyl methacrylates). The intramolecular energies of the diad model compounds of MMA, EMA, and *i*-PMA are calculated.

X 023

R' = -COOCH$_2$C$_6$H$_{11}$
R'' = -CH$_2$COOCH$_2$C$_6$H$_{11}$

diester of 2,2-dimethylsuccinic acid, used as a model compound

poly(dicyclohexylmethylene itaconate), PDCMI

Díaz-Calleja, R.; Saiz, E.; Riande, E.; Gargallo, L.; Radić, D. *Macromolecules* **1993**, *26*, 3795.

Bond Length [pm]	Valence Angles [°]	Torsion Angles [°]	ξ (for 300 K)	ξ_0	E_ξ [kJ · mol^{-1}]			
	θ_1 : 112	For main chain: t : 180	$\alpha = 0.10$	1	5.86	$U_1 = \begin{bmatrix} 1 & 1 \\ 1 & 0 \end{bmatrix}$	$U_2 = \begin{bmatrix} 1 & \alpha \\ \alpha & \alpha^2/\beta \end{bmatrix}$	$U_3 = \begin{bmatrix} \beta & \alpha \\ \alpha & \alpha^2/\beta \end{bmatrix}$
	θ_2 : 110	g : 60	$\beta = 0.37$	1	2.51			
		For side chain: $\chi_1 = 0, 180$ $\chi_2 = \pm 60, 180$ $\chi_3 = 60, 70, 290, 300$					rows and columns: t, g	

Comments:
The relaxation behavior of amorphous rigid polymers is studied by using the model PDCMI. Moreover, the conformational characteristics of the polymer are investigated by critically analyzing the dipole moments of the chains.

Calcd. quantities: *A priori* probabilities $<\mu^2>$ μ_{eff}

X 024

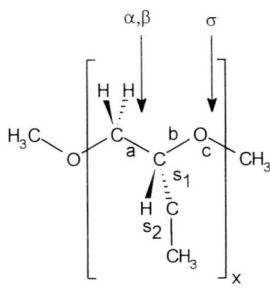

poly[oxy(1-ethylethylene)]

Abe, A.; Hirano, T.; Tsuji, K.; Tsuruta, T. *Macromolecules* **1979**, *12*, 1100.

Bond Length [pm]	Valence Angles [°]	Torsion Angles[a] [°]	ξ [a] (for 303 K)	ξ_0	E_ξ [kJ·mol^{-1}]
C-C : 153	C-C-O : 111.5	For C-C bonds:	γ = 0.44	1	2.09 (± 1.05)
C-O : 143	C-O-C : 111.5	t : 180 both			
		g^+ : 60 or 70	δ = 0.19	1	4.19 (± 2.09)
		g^- : −60 or −70			
		For C-O bonds:	ω' = 0.31	1	2.93 (± 0.42)
		t : −160 or −170			
		g^+ : 60 both	α = 1.65	1	−1.26 both
		g^- : −80 or −70			
		For O-C bonds:	β = 0.56 or 0.44	1	1.47 or 2.09
		t : 180 both			
		g^+ : 80 or 70	σ = 0.11 or 0.22	1	5.44 or 3.77
		g^- : −80 or −70			
			ω = 0.53	1	1.67 both

Matrix Basis Set:

$$U_a^R = \begin{bmatrix} 1 & \alpha & \beta \\ 0 & \alpha & \beta\omega \\ 1 & \alpha\omega & 0 \end{bmatrix} \quad U_b^R = \begin{bmatrix} 1 & 0 & 1 \\ 1 & 0 & \omega \\ 1 & 0 & 1 \end{bmatrix} \quad U_c^R = \begin{bmatrix} 1 & \sigma & 0 \\ 1 & 0 & 0 \\ 1 & 0 & \sigma \end{bmatrix}$$

For Side Chains:

$$V^R = \begin{bmatrix} \gamma\omega'+\delta & 1+\gamma\omega' & 1+\gamma\omega'+\delta \\ \gamma+\delta & 1+\gamma & 1+\gamma+\delta \\ \gamma & 1+\gamma & 1+\gamma \end{bmatrix}$$

$U_b^R(E) = [u_{\eta,\zeta} V^R(\eta,\zeta)]$

rows and columns: t, g^+, g^-

[a] Two parameter sets are given in the original paper.

Comments:
Configuration dependent properties of a series of poly[oxy(1-alkylethylene)] chains having various side chains differing in size and shape are examined. Conformational energy parameters established for POP are adopted in common to the skeletal configuration, steric interactions imposed by larger substituents being taken into account separately. Within a reasonable range of conformational energies, observed values of the characteristic ratio and the dipole moment ratio for isotactic polymeres are reproduced. Fractions of the conformation about internal C−C bonds calculated by using the same parameter set compare favorably with those estimated from NMR data. Calculations are extended to evaluate $<r^2>_0 / nl^2$ and $<\mu^2> / xm^2$ for atactic and syndiotactic chains.

Calcd. quantities: (for x = 200)		
$<r^2>_0 / nl^2$	= 6.0 or 5.7	(Exptl.: 5.4 or 6.3)
$d(\ln <r^2>_0)/dT$	= −1.4 (or −2.0) × 10^{-3} K^{-1}	
$<\mu^2> / xm^2$	= 0.77 or 0.72	
$d(\ln <\mu^2>)/dT$	= −1.0 (or −1.1) × 10^{-3} K^{-1}	

X 025

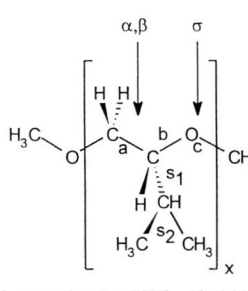

poly[oxy(1-isopropylethylene)]

Abe, A.; Hirano, T.; Tsuji, K.; Tsuruta, T. *Macromolecules* **1979**, *12*, 1100.

Bond Length [pm]	Valence Angles [°]	Torsion Angles[a] [°]	ξ [a] (for 298 K)	ξ_0	E_ξ [kJ·mol^{-1}]
C-C : 153	C-C-O : 111.5	For C-C bonds:	ν = 0.60 to 0.36	1	1.26 to 2.51
C-O : 143	C-O-C : 111.5	t : 180 both			
		g^+ : 110 both	ξ = 0.18 or 0.034	1	4.19 or 8.37
		g^- : −60 or −70			
		For C-O bonds:	ω' = 0.18 or 0.31	1	4.19 or 2.93
		t : −160 or −170			
		g^+ : 60 both	α = 1.66	1	−1.26 both
		g^- : −85 both			
		For O-C bonds:	β = 0.55 or 0.43	1	1.47 or 2.09
		t : 180 both			
		g^+ : 80 or 70	σ = 0.22 or 0.11	1	3.77 or 5.44
		g^- : −80 or −70			
			ω = 0.51	1	1.67

Matrix Basis Set:

$$U_a^R = \begin{bmatrix} 1 & \alpha & \beta \\ 0 & \alpha & \beta\omega \\ 1 & \alpha\omega & 0 \end{bmatrix} \quad U_b^R = \begin{bmatrix} 1 & 0 & 1 \\ 1 & 0 & \omega \\ 1 & 0 & 1 \end{bmatrix} \quad U_c^R = \begin{bmatrix} 1 & \sigma & 0 \\ 1 & 0 & 0 \\ 1 & 0 & \sigma \end{bmatrix}$$

For Side Chains:

$$V^R = \begin{bmatrix} \xi\omega' & \nu\omega' & 1+\nu\omega'+\xi\omega' \\ \xi & \nu & 1+\nu+\xi \\ 0 & \nu & \nu \end{bmatrix}$$

U_b^R (IP) = $[u_{\eta,\zeta} V^R (\eta,\zeta)]$

rows and columns: t, g^+, g^-

[a] Two parameter sets are given in the original paper.

Comments:
Configuration dependent properties of a series of poly[oxy(1-alkylethylene)] chains having various side chains differing in size and shape are examined. Conformational energy parameters established for POP are adopted in common to the skeletal configuration, steric interactions imposed by larger substituents being taken into account separately. Within a reasonable range of conformational energies, observed values of the characteristic ratio and the dipole moment ratio for isotactic polymeres are reproduced. Fractions of the conformation about internal C−C bonds calculated by using the same parameter set compare favorably with those estimated from NMR data. Calculations are extended to evaluate $<r^2>_0/nl^2$ and $<\mu^2>/xm^2$ for atactic and syndiotactic chains.

Calcd. quantities:
(for x = 200)

$<r^2>_0/nl^2$	= 7.4 or 6.9 or 6.8 or 6.3	
$d(\ln <r^2>_0)/dT$	= −0.78 × 10^{-3} K^{-1}	
$<\mu^2>/xm^2$	= 0.83 or 0.82 or 0.81 or 0.82	(Exptl.: 0.83)
$d(\ln <\mu^2>)/dT$	= −1.61 × 10^{-3} K^{-1}	

X 026

poly[oxy(1-*tert.*-butylethylene)]

Abe, A.; Hirano, T.; Tsuji, K.; Tsuruta, T. *Macromolecules* **1979**, *12*, 1100.

Bond Length [pm]	Valence Angles [°]	Torsion Angles[a] [°]	ξ [a] (for 353 K)	ξ_0	E_ξ [kJ·mol^{-1}]
C-C : 153	C-C-O : 111.5	For C-C bonds: t : 180 *both*	$\omega' = 0.24$	1	4.19
C-O : 143	C-O-C : 111.5	g^+ : 110 *both* g^- : -60 or -70 For C-O bonds: t : -160 or -170	$\omega'' = 0.058$ or 0.014 $\alpha = 1.54$	1 1	8.37 or 12.56 -1.26
		g^+ : 60 *both* g^- : -85 *both* For O-C bonds: t : 180 *both*	$\beta = 0.61$ or 0.49 $\sigma = 0.28$ or 0.16	1 1	1.47 or 2.09 3.77 or 5.44
		g^+ : 80 or 70 g^- : -80 or -70	$\omega = 0.57$	1	1.67

Matrix Basis Set:

$$U_a^R = \begin{bmatrix} 1 & \alpha & \beta \\ 0 & \alpha & \beta\omega \\ 1 & \alpha\omega & 0 \end{bmatrix} \quad U_b^R = \begin{bmatrix} 1 & 0 & 1 \\ 1 & 0 & \omega \\ 1 & 0 & 1 \end{bmatrix} \quad U_c^R = \begin{bmatrix} 1 & \sigma & 0 \\ 1 & 0 & 0 \\ 1 & 0 & \sigma \end{bmatrix}$$

For Side Chains:

$$V^R = \begin{bmatrix} 0 & 0 & \omega' \\ 0 & 0 & 1 \\ 0 & 0 & \omega'' \end{bmatrix}$$

U_b^R (TB) = $[u_{\eta,\zeta} V^R (\eta,\zeta)]$

rows and columns: t, g^+, g^-

[a] Two parameter sets are given in the original paper.

Comments:
Configuration dependent properties of a series of poly[oxy(1-alkylethylene)] chains having various side chains differing in size and shape are examined. Conformational energy parameters established for POP are adopted in common to the skeletal configuration, steric interactions imposed by larger substituents being taken into account separately. Within a reasonable range of conformational energies, observed values of the characteristic ratio and the dipole moment ratio for isotactic polymeres are reproduced. Fractions of the conformation about internal C—C bonds calculated by using the same parameter set compare favorably with those estimated from NMR data. Calculations are extended to evaluate $<r^2>_0 / nl^2$ and $<\mu^2> / xm^2$ for atactic and syndiotactic chains.

Calcd. quantities: (for x = 200)

$<r^2>_0 / nl^2$	= 16.0 or 10.7	(*Exptl.*: ~ 16)
$d(\ln <r^2>_0) / dT$	= -4.1×10^{-3} K^{-1}	
$<\mu^2> / xm^2$	= 1.57 or 0.99	
$d(\ln <\mu^2>) / dT$	= -5.6×10^{-3} K^{-1}	

X 027   (see also X 028 – X 030)

polydialkylsiloxanes, PDAS: **polydimethylsiloxane**, PDMS

Mattice, W. L. *Macromolecules* **1978**, *11*, 517.

Unperturbed dimensions and dipole moments of polydialkylsiloxanes (alkyl = methyl, ethyl, *n*-propyl, isopropyl) are investigated using RIS theory. Polymers are treated as branched molecules in which each silicon atom constitutes a tetrafunctional branch point. All significant first- and second-order interactions are included in the configuration partition function. Higher order interactions not suppressed by second-order interactions are also evaluated and accountered for in the statistical weights used. PDAS can rigorously be treated by incorporating only two additional parameters (τ^* and f_2) into statistical weight matrices used previously for PDMS. These two parameters are unity for PDMS but may differ from unity if side chains are articulated.

Bond Length [pm]	Valence Angles [°]	Torsion Angles [°]	ξ (for 298 K)	ξ_0	E_ξ [kJ · mol^{-1}]		
Si-O : 164	O-Si-O : 109.5	t : 180	$\tau^* = 1$	/	0		
Si-C : 190	Si-O-Si : 143	g$^+$: 60	$f_2 = 1$	/	0	$U_a = \begin{bmatrix} \tau^* & \sigma_1 & \sigma_1 \\ 1 & \sigma_1 f_2 & 0 \\ 1 & 0 & \sigma_1 f_2 \end{bmatrix}$	$U_b = \begin{bmatrix} 1 & \sigma_1 & \sigma_1 \\ 1 & \sigma_1 & \sigma_1\omega_1 \\ 1 & \sigma_1\omega_1 & \sigma \end{bmatrix}$
		g$^-$: −60	$\sigma_1 = 0.238$	1	3.56		
			$\omega_1 = 0.156$	1	4.61	rows and columns: t, g$^+$, g$^-$	

Calcd. quantities: $(\langle r^2 \rangle_0 / nl^2)_\infty$ $(\langle \mu^2 \rangle_0 / xm^2)_\infty$

X 028 (see also X 027, X 029 and X 030)

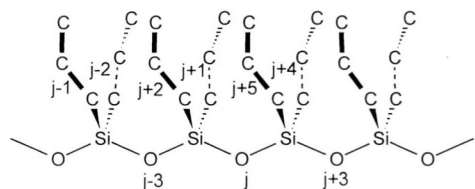

polydialkylsiloxanes, PDAS: polydiethylsiloxane, PDES

Mattice, W. L. *Macromolecules* **1978**, *11*, 517.

Unperturbed dimensions and dipole moments of polydialkylsiloxanes (alkyl = methyl, ethyl, n-propyl, isopropyl) are investigated using RIS theory. Polymers are treated as branched molecules in which each silicon atom constitutes a tetrafunctional branch point. All significant first- and second-order interactions are included in the configuration partition function. Higher order interactions not suppressed by second-order interactions are also evaluated and accountered for in the statistical weights used. PDAS can rigorously be treated by incorporating only two additional parameters (τ^* and f_2) into statistical weight matrices used previously for PDMS. These two parameters are unity for PDMS but may differ from unity if side chains are articulated.

Bond Length [pm]	Valence Angles [°]	Torsion Angles [°]	ξ (for 298 K)	ξ_0	E_ξ [kJ·mol^{-1}]		
Si-O : 164	O-Si-O : 109.5	t : 180	τ^* : see equation				
Si-C : 190	Si-O-Si : 143	g^+ : 60	f_2 : see equation			$U_a = \begin{bmatrix} \tau^* & \sigma_1 & \sigma_1 \\ 1 & \sigma_1 f_2 & 0 \\ 1 & 0 & \sigma_1 f_2 \end{bmatrix}$	$U_b = \begin{bmatrix} 1 & \sigma_1 & \sigma_1 \\ 1 & \sigma_1 & \sigma_1\omega_1 \\ 1 & \sigma_1\omega_1 & \sigma \end{bmatrix}$
C-C : 153	Si-C-C : 112	g^- : -60	$\sigma_1 = 0.238$	1	3.56		
	C-C-C : 112		$\omega_1 = 0.156$	1	4.61	rows and columns: t, g^+, g^-	
			$\sigma_2 \sim 1$	1	"small"	$\tau^* = [\sigma_2(\sigma_2 + 4\omega_2) + 2\omega_2^2(1+\omega_3)]\{\sigma_2[1+\omega_2]^2 + \sigma_2\omega_2] + 2\omega_2 + \omega_3(1+\omega_2^2)\}^{-1}$	
			$\omega_2 = 0.71 - 0.36$	1	0.84 - 2.51		
			$\omega_3 = 0.43$	1	2.09	$f_2 = [\sigma_2\omega_2(4+\sigma_2\omega_2) + 2(1+\omega_3)]\{\sigma_2[(1+\omega_2)^2 + \sigma_2\omega_2] + 2\omega_2 + \omega_3(1+\omega_2^2)\}^{-1}$	

Calcd. quantities: $(<r^2>_0 / nl^2)_\infty$ $(<\mu^2>_0 / xm^2)_\infty$

X 029   (see also X 027, X 028 and X 030)

polydialkylsiloxanes, PDAS: poly(di-*n*-propyl siloxane)

Mattice, W. L. *Macromolecules* **1978**, *11*, 517.

Unperturbed dimensions and dipole moments of polydialkylsiloxanes (alkyl = methyl, ethyl, *n*-propyl, isopropyl) are investigated using RIS theory. Polymers are treated as branched molecules in which each silicon atom constitutes a tetrafunctional branch point. All significant first- and second-order interactions are included in the configuration partition function. Higher order interactions not suppressed by second-order interactions are also evaluated and accounterred for in the statistical weights used. PDAS can rigorously be treated by incorporating only two additional parameters (τ^* and f_2) into statistical weight matrices used previously for PDMS. These two parameters are unity for PDMS but may differ from unity if side chains are articulated.

Bond Length [pm]	Valence Angles [°]	Torsion Angles [°]	ξ (for 298 K)		ξ_0	E_ξ [kJ·mol^{-1}]
Si-O : 164	O-Si-O : 109.5	t : 180	τ^* : see equation			
Si-C : 190	Si-O-Si : 143	g$^+$: 60	f_2 : see equation			
C-C : 153	Si-C-C : 112	g$^-$: −60	$\sigma_1 = 0.238$	1		3.56
	C-C-C : 112		$\omega_1 = 0.156$	1		4.61
			$\sigma_2 \approx 1$	1		"small"
			$\sigma_3 \leq 0.71$	1		≥ 0.84
			$\omega_2 = 0.71 - 0.36$	1		0.84 - 2.51
			$\omega_3 = 0.43$	1		2.09

$$U_a = \begin{bmatrix} \tau^* & \sigma_1 & \sigma_1 \\ 1 & \sigma_1 f_2 & 0 \\ 1 & 0 & \sigma_1 f_2 \end{bmatrix} \qquad U_b = \begin{bmatrix} 1 & \sigma_1 & \sigma_1 \\ 1 & \sigma_1 & \sigma_1 \omega_1 \\ 1 & \sigma_1 \omega_1 & \sigma \end{bmatrix}$$

rows and columns: t, g$^+$, g$^-$

Extension of side chains to n-propyl can be accomodated by multiplying each σ_2 by c, where c = $(1 + 2\sigma_3 \omega_4)[1 + \sigma_3(\omega_4 + \omega_5)]^{-1}$

$\tau^* = [\sigma_2(\sigma_2 + 4\omega_2) + 2\omega_2^2(1 + \omega_3)] \{\sigma_2[1 + \omega_2]^2 + \sigma_2\omega_2] + 2\omega_2 + \omega_3(1 + \omega_2^2)\}^{-1}$

$f_2 = [\sigma_2 \omega_2(4 + \sigma_2 \omega_2) + 2(1 + \omega_3)] \{\sigma_2[(1 + \omega_2)^2 + \sigma_2 \omega_2] + 2\omega_2 + \omega_3(1 + \omega_2^2)\}^{-1}$

Calcd. quantities: $(<r^2>_0 / nl^2)_\infty$ $(<\mu^2>_0 / xm^2)_\infty$

X 030   (see also X 027 – X 029)

polydialkylsiloxanes, PDAS: poly(di-isopropyl siloxane)

Mattice, W. L. *Macromolecules* **1978**, *11*, 517.

Unperturbed dimensions and dipole moments of polydialkylsiloxanes (alkyl = methyl, ethyl, n-propyl, isopropyl) are investigated using RIS theory. Polymers are treated as branched molecules in which each silicon atom constitutes a tetrafunctional branch point. All significant first- and second-order interactions are included in the configuration partition function. Higher order interactions not suppressed by second-order interactions are also evaluated and accountered for in the statistical weights used. PDAS can rigorously be treated by incorporating only two additional parameters (τ^* and f_2) into statistical weight matrices used previously for PDMS. These two parameters are unity for PDMS but may differ from unity if side chains are articulated.

Bond Length [pm]	Valence Angles [°]	Torsion Angles [°]	ξ (for 298 K)	ξ_0	E_ξ [kJ·mol^{-1}]		
Si-O : 164	O-Si-O : 109.5	t : 180	τ^* : see equation				
Si-C : 190	Si-O-Si : 143	g$^+$: 60	f_2 : see equation			$U_a = \begin{bmatrix} \tau^* & \sigma_1 & \sigma_1 \\ 1 & \sigma_1 f_2 & 0 \\ 1 & 0 & \sigma_1 f_2 \end{bmatrix}$	$U_b = \begin{bmatrix} 1 & \sigma_1 & \sigma_1 \\ 1 & \sigma_1 & \sigma_1\omega_1 \\ 1 & \sigma_1\omega_1 & \sigma \end{bmatrix}$
C-C : 153	Si-C-C : 112	g$^-$: –60	$\sigma_1 = 0.238$	1	3.56		
	C-C-C : 112		$\omega_1 = 0.156$	1	4.61	rows and columns: t, g$^+$, g$^-$	
			$\sigma_2 \sim 1$	1	"small"	$\tau^* = [2\sigma_2^2(1 + \omega_3) + \omega_2\omega_3(4\sigma_2 + \omega_2\omega_3)] \, [\sigma_2(\omega_3/\omega_2)(\sigma_2 + 1)(\omega_2^2 + 1) +$	
			$\sigma_3 \leq 0.71$	1	≥ 0.84	$+ \, 2\sigma_2(\sigma_2 + \omega_3^2)]^{-1}$	
			$\omega_2 = 0.71 - 0.36$	1	0.84 - 2.51	$f_2 = \{2\sigma_2[\sigma_2\omega_2(1 + \omega_3) + 2\omega_3] + \omega_3^2/\omega_2\} \, [\sigma_2(\omega_3/\omega_2)(\sigma_2 + 1)(\omega_2^2 + 1) +$	
			$\omega_3 = 0.43$	1	2.09	$+ \, 2\sigma_2(\sigma_2 + \omega_3) + \omega_3^2]^{-1}$	

Calcd. quantities: $(\langle r^2 \rangle_0 / nl^2)_\infty$ $(\langle \mu^2 \rangle_0 / xm^2)_\infty$

polyethylene and polydimethylsiloxane short-chain unimodal networks

Mark, J. E.; Curro, J. G. *J. Chem. Phys.* **1983**, *79*, 5705.

Bond Length [pm]	Valence Angles [°]	Torsion Angles [°]	ξ	ξ_0	E_ξ [kJ·mol^{-1}]			
polyethylene:			(for 473 K)					
C-C : 153	C-C-C : 112	t : 180 g$^+$: 60 g$^-$: −60	$\sigma = 0.59$ $\omega = 0.15$	1 1.3	2.09 8.38	$U = \begin{bmatrix} 1 & \sigma & \sigma \\ 1 & \sigma & \sigma\omega \\ 1 & \sigma\omega & \sigma \end{bmatrix}$		rows and columns: t, g$^+$, g$^-$
polydimethylsiloxane:			(for 443.5K)					
Si-O : 164	Si-O-Si : 143 O-Si-O : 110	t : 180 g$^+$: 60 g$^-$: −60	$\sigma' = 0.38$ $\omega' = 0.37$	1 1.3	3.559 4.605	$U_a = \begin{bmatrix} 1 & \sigma' & \sigma' \\ 1 & \sigma' & 0 \\ 1 & 0 & \sigma' \end{bmatrix}$	$U_b = \begin{bmatrix} 1 & \sigma' & \sigma' \\ 1 & \sigma' & \sigma'\omega' \\ 1 & \sigma'\omega' & \sigma' \end{bmatrix}$	

Comments:
The present theoretical approach to rubberlike elasticity is novel in that it utilizes the wealth of information which RIS theory provides on the spatial configurations of chain molecules. Specifically, Monte Carlo calculations based on the RIS approximation are used to simulate spatial configurations, and thus distribution functions for end-to-end separation *r* of the chains. Results are presented for polyethylene and polydimethylsiloxane chains most of which are quite short, in order to elucidate non-Gaussian effects due to limited chain extensibility.

polyurethane networks from polyoxypropylene tetrols and hexamethylene diisocyanate

Lee, K.-J.; Eichinger, B. E. *Polymer* **1990**, *31*, 406.

Bond Length [pm]	Valence Angles [°]	Torsion Angles [°]	ξ (for 353 K)	ξ_0	E_ξ [kJ·mol^{-1}]				
C-O : 143	O-C-C : 110	t : 180	α = 1.54	1	−1,26				
C-C : 153	C-O-C : 110	g$^+$: 60	β = 0.49	1	2.09	$U_{a,R} = \begin{bmatrix} 1 & \alpha & \beta \\ 0 & \alpha & \beta\omega \\ 1 & \alpha\omega & 0 \end{bmatrix}$	$U_{b,R} = \begin{bmatrix} 1 & 0 & 1 \\ 1 & 0 & \omega \\ 1 & 0 & 1 \end{bmatrix}$	$U_{c,R} = \begin{bmatrix} 1 & \sigma & 0 \\ 1 & 0 & 0 \\ 1 & 0 & \sigma \end{bmatrix}$	$Q = \begin{bmatrix} 1 & 0 & 0 \\ 0 & 0 & 1 \\ 0 & 1 & 0 \end{bmatrix}$
		g$^-$: −60	σ = 0.28	1	3.77				
			ω = 0.57	1	1.67	rows and columns: t, g$^+$, g$^-$		$U_S = Q\,U_R\,Q$	(see: a))

a) The given statistical weight matrices are applied to the repeating units indicated in the above figure. The remaining bonds, which are not repeating, may be safely taken to be *trans*.

Comments:
Model urethane networks prepared from polyoxypropylene tetrols and hexamethylene diisocyanate are studied with the aid of the computer, and good agreement of theory with experiment is found for gel points when the crosslinkers are treated as sticks.

The following model requires two pages.
Therefore, this side has been left blank

X 033   (see also X 034)

LHT-240, x = 3, n ca. 4

LG-56, x = 0, n ca. 15

three conformations for LG-56:
(**a**) arm 1 is treated as side chain
(**b**) arm 3 is treated as side chain
(**c**) arm 2 is treated as side chain

polyurethane networks from polyoxypropylene triols and 4,4'-diphenylmethane diisocyanate: Niax triol LG-56

Lee, K.-J.; Eichinger, B. E. *Polymer* **1990**, *31*, 414.

Bond Length [pm]	Valence Angles [°]	Torsion Angles [°]	ξ (for 298 K)	ξ_0	E_ξ [kJ · mol^{-1}]	(see:a))
C-O : 143	O-C-C : 110	t : 180	α = 0.51	1	1.67	
			β = 0.36	1	2.51	
C-C : 153	C-O-C : 110	g$^+$: 60	σ = 0.43	1	2.10	
			ω = 0.03	1	8.40	
		g$^-$: −60	σ'_2 = 0.22	1	3.77	
			σ_3 = 1.96	1	−1.67	
			σ'_3 = 0.22	1	3.77	

rows and columns: t, g$^+$, g$^-$

$$U_{a,R} = \begin{bmatrix} 1 & \alpha & \beta \\ 0 & \alpha & \beta\omega \\ 1 & \alpha\omega & 0 \end{bmatrix} \quad U_{b,R} = \begin{bmatrix} 1 & 0 & 1 \\ 1 & 0 & \omega \\ 1 & 0 & 1 \end{bmatrix} \quad U_{c,R} = \begin{bmatrix} 1 & \sigma & 0 \\ 1 & 0 & 0 \\ 1 & 0 & \sigma \end{bmatrix}$$

$$U_{b1} = \begin{bmatrix} \omega_3 + \alpha\omega_2 + \beta & 0 & \omega_3 + \alpha + \beta \\ 1 + \alpha\omega_2 + \beta & 0 & \omega(1 + \alpha + \beta) \\ 1 + \alpha\omega_2 + \beta\omega_3 & 0 & 1 + \alpha + \beta\omega_3 \end{bmatrix} \quad U_{c1} = \begin{bmatrix} 1 & 0 & \sigma'_2 \\ 1 & \sigma'_2 & \sigma'_2\omega_2 \\ 1 & \sigma'_2\omega_2 & 0 \end{bmatrix}$$

$$U_{c2} = \begin{bmatrix} 1 & \sigma'_2 & \sigma'_2 \\ 1 & \sigma'_2 & 0 \\ 1 & 0 & \sigma'_2 \end{bmatrix} \quad U_{d1} = \begin{bmatrix} 1 & \beta & \sigma_3 \\ 1 & \beta\omega_2 & 0 \\ \omega_2 & 0 & \sigma_3\omega_2 \end{bmatrix} \quad U'_{d1} = \begin{bmatrix} \omega_2\sigma_2(1+\sigma'_2) & \sigma_3(\omega_2+\sigma'_2) & \beta(\omega_2+\sigma'_2) \\ \sigma_2(1+\sigma'_2\omega_2) & \sigma_3(1+\sigma'_2) & \beta\omega_3(1+\sigma'_2) \\ \sigma_2(1+\sigma'_2\omega_2) & \sigma_3\omega_3(1+\sigma'_2) & \beta(1+\sigma'_2) \end{bmatrix} \quad U_{c3} = \begin{bmatrix} 1 & \sigma'_3\omega_2 & \sigma'_3 \\ 1 & \sigma'_3 & 0 \\ 1 & 0 & \sigma'_3\omega_2 \end{bmatrix} \quad U_{c4} = Q\, U_{c3}\, Q$$

a) The statistical weight matrices for the bonds a', b', c' and c'1 are obtained by treating $U_{a,R}$, U_{b1}, $U_{c,R}$ and U_{c1} with equation (2) given in the original reference of **X 032** with $U_{2a,R}$ and $U_{c,R}$ given there in equations (1a) and (1c).

X 034 (see also **X 033**)

LHT-240, x = 3, n *ca.* 4

LG-56, x = 0, n *ca.* 15

three conformations for LHT-240:
(**a**) arm 1 is treated as side chain
(**b**) arm 3 is treated as side chain
(**c**) arm 2 is treated as side chain

polyurethane networks from polyoxypropylene triols and 4,4'-diphenylmethane diisocyanate: Niax triol LHT-240

Lee, K.-J.; Eichinger, B. E. *Polymer* **1990**, *31*, 414.

The chemical structure of *LHT-240* is similar to that of *LG-56*, except for $x = 3$. Thus, the procedure used to obtain the statistical weight matrices for *LG-56* is applicable here to *LHT-240*. The three conformations based on the central carbon (*C) are summarized in the Figure. In the following, only those matrices that differ from those of *LG-56* are written:

Bond Length [pm]	Valence Angles [°]	Torsion Angles [°]	ξ (for 298 K)	ξ_0	E_ξ [kJ·mol^{-1}]
C-O : 143	O-C-C : 110	t : 180	$\alpha = 0.51$	1	1.67
			$\beta = 0.36$	1	2.51
C-C : 153	C-O-C : 110	g$^+$: 60	$\sigma = 0.43$	1	2.10
			$\omega = 0.03$	1	8.40
		g$^-$: −60	$\sigma_4 = 1.40$	1	−0.84
			$\sigma'_4 = 0.43$	1	2.10
			$\sigma''_4 = 0.22$	1	3.77
			$\omega_4 = 0.36$	1	2.51
			$\gamma = 0.43$	1	2.10
			$\omega' = 0.31$	1	2.93
			$\delta = 0.19$	1	4.18

$$U_{e1} = \begin{bmatrix} 1 & 0 & \sigma'_4 \\ 1 & \sigma'_4 & \sigma'_4\omega_4 \\ 1 & \sigma'_4\omega_4 & 0 \end{bmatrix} \quad U_{e3} = \begin{bmatrix} 1 & \sigma_4 & \sigma_4 \\ 1 & \sigma_4 & \sigma_4\omega_4 \\ 1 & \sigma_4\omega_4 & \sigma_4 \end{bmatrix} \quad U_{c11} = \begin{bmatrix} 1 & \sigma''_4 & \sigma''_4 \\ 1 & \sigma''_4 & 0 \\ 1 & 0 & \sigma''_4 \end{bmatrix}$$

$$U_{e2} = \begin{bmatrix} 1 & \sigma & \sigma \\ 1 & \sigma & \sigma\omega \\ 1 & \sigma\omega & \sigma \end{bmatrix} \quad U_{b2} = \begin{bmatrix} \gamma\omega'+\delta & 1+\gamma\omega'+\delta & 0 \\ \gamma & 1+\gamma & 0 \\ \gamma+\delta & \omega(1+\gamma+\delta) & 0 \end{bmatrix} \quad U_{e2'} = \begin{bmatrix} 1 & \sigma & \sigma \\ 1 & \sigma\omega_4 & 0 \\ 1 & 0 & \sigma \end{bmatrix}$$

$$U_{e3'} = \begin{bmatrix} \sigma_2\sigma'_2\omega_2 & \sigma'_2\sigma_4 & \sigma'_2\beta \\ \sigma_2(1+\sigma'_2\omega_2) & \sigma_4(1+\sigma'_2) & \beta\omega_4(1+\sigma'_2) \\ \sigma_2(1+\sigma'_2\omega_2) & \sigma_4\omega_4(1+\sigma'_2) & \beta(1+\sigma'_2) \end{bmatrix}$$

rows and columns: t, g$^+$, g$^-$

Comments:

Polyoxypropylene triol-based urethane networks are simulated. Simulations show that cyclic molecules are present in substantial amounts in the sol fraction when the stoichometric ratio $r_s \approx 1.0$. Simulations also show that the fractions of loops are much higher than those obtained from modified cascade theory.

*Further calculations on **stars, grafts, networks,** and other **systems with articulated side chains**:*

X 035   Zimm, B. H.; Stockmayer, W. H. *J. Chem. Phys.* **1949**, *17*, 1301.

Formulas for the mean-square radii of various branched and ringed polymer molecules are developed under the usual assumptions regarding the statistics of chain configuration. For branched molecules, it is shown that the mean square radii vary less rapidly with molecular weight than for strictly linear molecules, while for systems containing only rings and unbranched chains the variation is more rapid than for the linear case.

X 036   Nagai, K.; Ishikawa, T. *J. Chem. Phys.* **1966**, *45*, 3128.

Stress-optical data of polydimethylsiloxane networks are analyzed by assuming rotational isomeric states to be three in number, *i.e.*, t, g^+, and g^-. An expression in matrix form for the optical anisotropy $\Delta\Gamma$ of *Kuhn's* random link (an equivalent to the stress-optical coefficient) of two-repeat polymers is derived, based on the additivity principle of bond polarizabilities and the RIS approximation. The interdependency of neighboring bonds is taken into account. It is found that semiquantitative agreement is achieved by assuming $E(g) > 0$, and $\Delta a_e > 0$. Here, $E(g)$ is the energy of the *gauche* conformation relative to the *trans*, and $\Delta \alpha_e = (\alpha_1 - \alpha_2)_{Si-O} - (\alpha_1 - \alpha_2)_{C-Si} + (\alpha_1 - \alpha_2)_{C-H}$ is the "effective" anisotropy of the Si–O bond. A harmonized choice of values of the internal-rotational engle for the *gauche* conformation, and ω_1, the supplement to < SiOSi, is needed for interpretation. Definite conclusion can not be obtained about to what extent the g^+g^- conformations for neighboring bonds are suppressed.

X 037   Abe, A. *J. Am. Chem. Soc.* **1968**, *90*, 2205.

The conformational analysis of optically active poly(α-olefins) such as poly[(S)-3-methyl-1-pentene], poly[(R)-3,7-dimethyl-1-octene], poly[(S)-4-methyl-1-hexene], poly[(S)-5-methyl-1-heptene], and poly[(S)-6-methyl-1-octene] is presented. Polymer chains of various tacticities are treated in the RIS approximation, in consideration of the steric interactions between nonbonded groups separated by three and by four C–C bonds. Proper account of the side-chain conformations is shown to be essential in each instance. Optical activities of these polymers are evaluated within the framework of the revised version of the *Whiffen-Brewster* empirical rule. It is shown that the optical rotatory power of a polymer having asymmetric sites in the α or β position with respect to the main chain is large and nearly insensitive to the stereochemical arrangements of the chain units. When the asymmetric center in the side chain is situated at the γ or a farther position, however, quite a remarkable dependence of the rotatory power on stereoregularity is anticipated. Within the reasonable range of conformational energies, the calculated temperature coefficients of the optical rotations show satisfactory agreement with observations. Asymptotic behavior of the optical rotation with degree of polymerization is also discussed.

X 038   Abe, A. *Polymer J.* **1970**, *1*, 232.

The unperturbed dimensions of poly(α-olefins) are treated in consideration of the side chain configuration. Within the reasonable range of conformational energies, the root mean square end-to-end distance, $<r^2>_0 / nl^2$, and its temperature coefficient $d (\ln <r^2>_0) / dT$ are evaluated and compared with those observed. The presence of several percent of heterotactic units in an allegedly isotactic poly(1-butene) or poly(1-pentene) is the primary requirement in finding the agreement between theory and experiments. When the pendant group is methyl, on the other hand, the experimental value of $<r^2>_0 / nl^2$ is found to be consistent with those calculated over a wide range of isotacticity (0.92-0.98), provided that the thermal coefficient is negative and large. The treatment is further extended to the polymer systems with bulky substituents. It is pointed out, finally, that the magnitude of $<r^2>_0 / nl^2$ cannot provide a good criterion to define the stiffness of the vinyl chain system except when the polymer possesses a stereoregular configuration.

X 039   Mattice, W. L. *Macromolecules* **1981**, *14*, 143.

A comparison is presented between the behavior of unperturbed stars of finite size whose configurational statistics are evaluated by RIS theory and the Kratky-Porod wormlike chain model. Emphasis is placed on the initial slopes of the characteristic ratio, C, or g when plotted as a function of the reciprocal of the number of bonds, n.

X 040   Mattice, W. L.; Skolnick, J. *Macromolecules* **1981**, *14*, 1463.

Average configuration-dependent physical properties are evaluated for tri- tetra-, and hexafunctional polyethylene stars perturbed by electrostatic repulsion of charges placed at the free chain ends. Configuration-dependent properties evaluated are the probability for a *trans* placement, expansion of $<s^2>$, the mean-square radius of gyration, asymmetry of the distribution of the chain atoms, and asymmetry of the distribution described by the atoms considered to bear the charges.

X 041   Mattice, W. L. *Macromolecules* **1983**, *16*, 1623.

Expansion is considered for finite, regular polyethylene stars perturbed by the excluded volume effect. An RIS model is used for the chain statistics. The number of bonds in each branch ranges up to 10 240, and the functionality of the branch point ranges up to 20. The form of the calculation employed here provides a lower bound for the expansion. If the number, n, of bonds in the polymers is held constant, expansion is found to decrease with increasing branch point functionality. Two factors dictate the manner in which finite stars approach the limiting behavior expected for very large stars. These two factors are the chain length dependence at small n of the characteristic ratio and of $(\alpha^5-\alpha^3)/n^{1/2}$.

X 042   Tonelli, A. E.; Schilling, F. C.; Bovey, F. A. *J. Am. Chem. Soc.* **1984**, *106*, 1157.

From *Mark*'s RIS model for ethylene-propylene copolymers (*J. Chem. Phys.* **1972**, *57*, 2541) it is determined that $P(t) = 0.380$, $P(g^+) = 0.014$, and $P(g^-) = 0.606$ in 2,4-dimethylhexane (2,4-DMH). Using this RIS model, furthermore, for all the branched alkanes considered whose isopropyl groups are separated by at least one methylene carbon from the next substituted carbon and the RIS model developed by Asakura et al. (*Makromol. Chem.* **1976**, *177*, 1493) for head-to-head polypropylene to treat 2,3-dimethyl pentane, $\Delta\delta$'s are calculated for a large number of branched alkanes. The agreement between the observed and the calculated nonequivalent ^{13}C NMR chemical shifts is quite good, including the prediction that separation of the isopropyl group from the next substituted carbon by four or more methylene carbons removes the nonequivalence.

X 043   Curro, J. G.; Mark, J. E. *J. Chem. Phys.* **1984**, *80*, 4521.

Bimodal polydimethylsiloxane (PDMS) networks containing a large mole fraction of very short chains are known to be unusually tough elastomers. The purpose of this investigation is to understand the rubber elasticity behavior of these bimodal networks. As a first approach, it is assumed that the *average* chain deformation is affine. This deformation, however, is partitioned nonaffinely between the long and short chains so that the free energy is minimized. Gaussian statistics is used for long chains. The distribution function for the short chains is found from Monte Carlo calculations. Monte Carlo calculations are performed on PDMS chains having $n = 10, 20, 30$, and 40 skeletal bonds using a method described in by *Mark* and *Curro* (*J. Chem. Phys.* **1983**, *79*, 5705). The statistical weight matrices and weighting factors are also described in this paper.

X 044   Mattice, W. L. *Macromolecules* **1984**, *17*, 415.

Expansion factors are evaluated for tri- and tetrafunctional polyethylene stars and for selected fragments. A detailed RIS model is used for unperturbed stars. Expansion is produced by forcing atoms participating in long-range interactions to behave as hard spheres. The longest chains in stars experience a greater expansion than do linear chains containing the same number of bonds. Examination of subchains shows that the extra expansion occurs primarily in that portion of the chain near the branch point. While branching increases expansion of individual chains, overall expansion of the entire star is smaller than that of the linear chain of the same number of bonds, n. The reduced expansion of stars arises because the longest chain in a molecule of specified n contains fewer bonds as the functionality of the branch point increases. This latter effect is of greater consequence than the extra expansion of individual chains in the star.

X 045   Mattice, W. L.; Scheraga, H. A. *Macromolecules* **1984**, *17*, 2690.

A tractable matrix formulation is developed for the formulation of the conformational partition function for a chain undergoing formation of intramolecular antiparallel sheets with tight bends. The distribution function for the number of residues per strand is among the physical properties which can be extracted from this conformational partition function. The behavior at the high end of the distribution function determines the dimensions of the smallest statistical weight matrix that may be used in the characterization of the transition. A simple relationship, which incorporates the degree of polymerization and the two statistical weights arising from end effects, provides a rapid estimate of the dimensions of the matrix. The validity of the approximation is demonstrated by a consideration of several different chains. Some of the chains considered are assigned end-effect parameters that are in qualitative agreement with those expected for linear polyethylene.

X 046   Damewood, J. R., Jr. *Macromolecules* **1985**, *18*, 1793.

The present calculations on poly(di-*n*-hexylsilylene) are consistent with the hypothesis that the origin of thermochromic behavior in selected polysilane polymers resides in a change in population of conformational states along the silicon backbone with temperature.

X 047   Mattice, W. L. *Macromolecules* **1986**, *19*, 2303.

Reduction of the unperturbed dimensions of the main chain is calculated when ethyl groups are attached to a polyethylene backbone. Values of most of the paramters are taken from the well-known RIS model for unperturbed polyethylene (Abe, A.; Jernigan, R. L.; Flory, P. J. *J. Am. Chem. Soc.* **1966**, *88*, 631): the bond angle is 112°, *gauche* states are located at ± 60° (*trans:* 180°). First- and second-order interactions are weighted by using $\sigma = 0.43$ and $\omega = 0.034$ (for 300 K). An additional statistical weight, denoted by τ, is required at each bond to an atom that constitutes a trifunctional branch point (Flory, P. J.; Sundararajan, P. R.; DeBolt, L. C. *J. Am. Chem. Soc.* **1974**, *96*, 5015). Calculations are performed with $\tau = 0$ and $\tau = \sigma$.

X 048   Huber, K.; Burchard, W.; Bantle, S.; Fetters, L. J. *Polymer* **1987**, *28*, 1990.

Light and neutron scattering studies are carried out on 12-arm star polystyrene molecules in toluene (a good solvent) and in cyclohexane (a ϑ solvent). The global dimensions of all samples under ϑ dimensions are larger than those predicted by the existing theories. Monte Carlo simulations of 12-arm star chains as a function of chain length are performed using *Flory´s* RIS model. Two different types of star chains, a pure combinatorial star and a star with a specific centre are used. The procedure was first applied by *Yoon et al.* (*J. Chem. Phys.* **1974**, *61*, 5366) and starts from the matrix **P**, containing the *a priori* probabilities p for the rotational angle of a bond, being in the η state, supposing that the preceding bond is in the ξ state. 180° (*trans*) and ± 60° (for *gauche*) are used as rotational isomeric states. The C−C bond length is 153 pm, and the valence angle is 112°.

$$\mathbf{P} = \begin{bmatrix} 0.321 & 0.138 & 0.138 \\ 0.138 & 0.059 & 0.00052 \\ 0.138 & 0.00052 & 0.059 \end{bmatrix}$$

X 049 Farmer, B. L.; Rabolt, J. F.; Miller, R. D. *Macromolecules* **1987**, *20*, 1167.

Conformational calculations are carried out on poly(di-*n*-hexylsilanes). The most significant finding from the energy calculations is that the all-*trans* conformation is not the lowest energy structure for the symmetrically alkyl-substituted silane polymers. A helical structure is preferred for the isolated molecule.

X 050 Mathur, S. C.; Mattice, W. L. *Macromolecules* **1987**, *20*, 2165.

The characteristic ratio, C, is evaluated by RIS theory for polyethylene chains that contain randomly placed short branches. The size of the short branches ranges from methyl to octyl. Calculations are restricted to chains with few branches, where C is nearly a linear function of branch content. The calculations provide numerical values for $(\partial \ln C / \partial P_b)_o$, where P_b is the probability that a repeat unit has a short branch and zero as a subscript denotes the initial slope.

X 051 Mathur, S. C.; Mattice, W. L. *Makromol. Chem.* **1988**, *189*, 2893.

The characteristic ratio and its temperature coefficient are examined for alternating copolymers of ethylene and propylene. The stereochemical composition covers the entire range *racemic* through *meso*. The energies used in these calculations are 2.1 kJ · mol^{-1} for σ, 8.3 kJ · mol^{-1} for ω, and 0 or 2.1 kJ · mol^{-1} for τ. The bond angle is 112°, torsion angles are t (180°), and g$^{\pm}$ ($\pm$60, $\pm$55, or $\pm$50). Satisfactory agreement between theory and experiment is obtained when the first-order interactions are assumed to be pair-wise additive, causing $\tau = \sigma$. At 290.4 K, the comparison of theory and experiment yields 6.9 vs. 6.5 ($\pm$ 0.2) for C, and -0.0011 K^{-1} vs. -0.0010 ($\pm$ 0.0003) K^{-1} for $d (\ln C) / d T$.

X 052 Vega, S.; Riande, E.; Guzmán, J. *Macromolecules* **1990**, *23*, 3573.

Poly(*trans*-7-oxabicyclo[4.3.0]nonane) (PTOXN) is synthesized by cationic polymerization. Rotations within this molecule about the C_{Cy}—O and C_{Cy}—CH$_2$ bonds, flanking the cyclohexylene group, give rise to severe interactions between neighbor methylene groups. Semiempirical potential calculations indicate that the C_{Cy}—O bond is restricted to a single rotational state (either g$^-$ or g$^+$), this rotational state being dependent on the location of the neighbor C_{Cy}—CH$_2$ bond. In general, the RIS model gives a good account of the dielectric conformation-dependent properties of the polymer.

X 053 Neuburger, N. A.; Mattice, W. L. *Polym. Prepr. (Div. Am. Chem. Soc.)* **1991**, *32(3)*, 279.

Molecular dynamic simulations are performed on a series of siloxane chains of the form polydialkylsiloxane. Analysis of the trajectories are performed by generating pair-wise probability maps. Using various minimization techniques, a system of equations is evaluated to yield a set of statistical weights for each of the systems.

X 054 Auriemma, F.; Corradini, P.; Vacatello, M. *Macromolecules* **1992**, *25*, 2955.

Theory and comparison with experiments on the mesophasic behavior of rodlike polymers with flexible side chains is described. A statistical-thermodynamical lattice treatment of systems of chain molecules, previously developed to investigate the nematic/isotropic phase transition of polymers with rigid and flexible parts alternating along the main chain, is modified to study the mesophasic behavior of these polymers. Two series of poly(1,4-phenylene terephthalate) and of poly(1,4-oxybenzoyl), with alkyl groups of various length attached to the aromatic rings, are modeled as proper sequences of isodiametric units placed at the nodes of the face-centered cubic lattice. In the hypothesis that the behavior of these polymers is mainly determined by the high rigidity of the main chains, rather than by anisotropic attractive interactions, and that the side chains can be treated as completely flexible, the model requires a single adjustable parameter, related to the inflexibility of the main chain. Using the same value of this parameter for all the members of each series, the existence of a stable anisotropic phase is predicted for the investigated systems, in good semiquantitative agreement with experiments.

X 055   Neuburger, N.; Bahar, I.; Mattice, W. L. *Macromolecules* **1992**, *25*, 2447.

Molecular dynamics trajectories are computed for isolated polydialkylsiloxane chains in which the alkyl groups are methyl, ethyl, propyl, isopropyl, butyl, isobutyl, or *tert*-butyl. The joint probability profiles for the pair of Si—O bonds flanking a silicon atom are insensitive to the changes in the alkyl substituents. In contrast, the joint probability profiles for the pair of bonds flanking an oxygen atom change systematically as the size of alkyl group increases. There is a strong correlation in the dihedral angles at these two bonds when the alkyl groups are large. The features in these probability profiles are incorporated into an RIS analysis that permits predictions of the characteristic ratio, C_∞. If the alkyl groups are not branched, the values of C_∞ increase with the size of the alkyl group and reach a limit of about 8.3. Branching of the side chain reduces the characteristic ratio.

X 056   Bahar, I.; Neuburger, N.; Mattice, W. L. *Macromolecules* **1992**, *25*, 4619.

Results from molecular dynamics simultions of isolated polydialkylsiloxane chains are analyzed with emphasis on the conformational dynamics of a fragment of five repeat units, in which the alkyl group is *tert*-butyl. This fragment furnishes a typical example of a chain in which the bulky substituents restrict to a considerable extent the rotational motions of the backbone, which otherwise enjoys a high degree of flexibility. The rotational motions of the bonds flanking the oxygen atom are observed to be strongly correlated. The orientational motion of the pair of bonds flanking the oxygen atom is analyzed within the framework of the dynamic RIS formalism.

X 057   Aoki, A.; Hayashi, T.; Asakura, T. *Macromolecules* **1992**, *25*, 155

^{13}C NMR chemical shift assignments of comonomer sequences in a 1-butene-propylene copolymer are obtained from the ^{13}C two-dimensional INADEQUATE NMR experiment and from the calculated chemical shifts due to the γ effect. By tracing carbon-carbon connectivities in the 2D-INADEQUATE spectrum, the validity of previous assignments of triad and tetrad sequences is confirmed. Referring to the confirmed assignments, the chemical shift differences among comonomer sequences longer than pentad are predicted by the chemical shift calculation (the γ-effect method) based on the γ-effect of the ^{13}C NMR chemical shift and Mark´s RIS model (*J. Chem. Phys.* **1972**, *57*, 2541) modified by considering the side chain conformation in a 1-butene unit. Assignments provided in this study agree well with *Cheng*´s assignments by a reaction probability model (*J. Polym. Sci., Polym. Phys. Ed.* **1983**, *21*, 573). Since the bond conformational probability of the main chain should be influenced by the side chain conformation, Mark´s RIS model for an ethylene – propylene copolymer was modified for the application of a 1-butene-propylene copolymer considering the rotational isomeric states of bond *j* in the side chain. Modified matrices for the interdiad bond pair (about *i-1* and *i*) meeting at CH—CH$_3$ and CH—CH$_2$—CH$_3$ groups are $U_{p(d)}$ and $U_{b(d)}$, respectively:

$$U_{p(d)} = \begin{bmatrix} \eta & 1 & \tau \\ \eta & 1 & \tau\omega \\ \eta & \omega & \tau \end{bmatrix} \quad \beta_{kl} = \begin{bmatrix} a & b & c \\ c & d & e \\ b & f & d \end{bmatrix} \quad U_{b(d)} = \begin{bmatrix} \eta \times a & b & \tau \times c \\ \eta \times c & d & \tau\omega \times e \\ \eta \times b & \omega \times f & \tau \times d \end{bmatrix} \quad \begin{array}{ll} a = \tau + 2\omega & d = \tau\omega + 2 \\ b = \tau + \omega + 1 & e = \tau\omega^2 + 2 \\ c = \tau\omega + \omega + 1 & f = \tau + 2 \end{array}$$

The matrix $U_{b(d)}$ for a 1-butene unit is obtained by multiplying every element in $U_{p(d)}$ with the appropriate β_{kl}. Elements in β_{kl} denote statistical weights corresponding to three conformations (t, g$^+$, g$^-$) of the side chain, determined by combination of rotational isomeric states of bonds *i-1* and *i* in the main chain. Matrices $U_{p(d)}$ and $U_{b(d)}$ correspond to the dextro configuration of propylene and 1-butene units. Matrices $U_{p(l)}$ and $U_{b(l)}$ for the levo configuration can be obtained in a similar manner. The values of the statistical weight, η, was taken to be 1.0, and the values of four-bond pentane interference (ω) and three-bond *gauche* interaction (τ) are characterized with E(ω) = 6.3 kJ · mol^{-1} and E(τ) = 2.1 kJ · mol^{-1}, respectively.

Polypeptide chains and related compounds:

N 001   "Theory of the One-Dimensional Phase Transition in Polypeptide Chains"
Zimm, B. H.; Bragg, J. K. *J. Chem. Phys.* **1958**, *28*, 1246.

In the helical form of a polypeptide chain the hydrogen of each amide group forms a hydrogen bond to the oxygen of the third group preceding it in the chain. In the presented model, a hydrogen-bonded oxygen is represented by the digit 1 and an unbonded one by 0. Thus, a configuration of the chain is represented by a sequence of digits. It is assumed that the statistical weight of any configuration is given by the product of the following factors: *(1)* a quantity α for every 0 that appears, *(2)* a quantity αs for every 1 that follows a 1, *(3)* a quantity $\alpha \sigma s$ for every 1 that follows a 0. Further, the sequence 101 is prohibited as being sterically impossible. The factor α is a measure of the freedom available to each segment of the chain in its own configuration space when it is not hydrogen-bonded into the helix. The factor *s* measures the statistical weight of the bonded form relative to the unbonded form. The factor $\alpha \sigma s$ is introduced for the first bond to follow a series of unbonded configurations; σ is expected to be much less than unity because the freedom of the two segments between the segments bearing the oxygen and hydrogen is also restricted. The matrix developed is:

$$U = \begin{bmatrix} 1 & \sigma s & 0 & 0 \\ 0 & 0 & 1 & s \\ 1 & 0 & 0 & 0 \\ 0 & 0 & 1 & s \end{bmatrix}$$

According to this model, the observed sharpness of the helix-coil transition in polypeptide chains is due to the following circumstance: the formation of the first turn of the helix is difficult because of the large reduction of entropy, but, once formed, the first turn acts as a nucleus to stabilize the formation of further turns by hydrogen bonding.

N 002   "Theory of the Phase Transition between Helix and Random Coil in Polypeptide Chains"
Zimm, B. H.; Bragg, J. K. *J. Chem. Phys.* **1959**, *31*, 526.

The transition between the helical and randomly coiled forms of a polypeptide chain is discussed by reference to a model that allows hydrogen bonding only between each group and the third preceding. The partition function for this model is handeled in two alternative ways, either as a summation suitable for short chains, or in terms of the eigenvalues and eigenvectors of a characteristic matrix; the latter is more suitable for long chains. A transition from the random to the helical form is encountered as either the bonding parameter or the chain length is increased. It is assumed that a given state of the chain can be completely described by the state of the oxygen atoms alone, *i.e.*, by a statement as to whether or not each is bonded to the hydrogen of the third preceding segment. If the digit 1 represents a bonded oxygen atom and 0 an unbonded atom, then a state of the chain is described by a sequence such as 0000111000011.... The statistical weight of a given state of the chain is assumed to be the product of the following factors: *(1)* the quantity unity for every 0 that appears (unbonded segment); *(2)* the quantity *s* for every 1 that follows a 1 (bonded segment); *(3)* the quantity σs for every 1 that follows μ or more 0's (boundary between bonded and unbonded regions); *(4)* the quantity 0 for every 1 that follows a number of 0's less than μ. The effect of assumption *(4)* is that sequences of less than μ zeros do not appear. For the α helix, μ is usually considered to be about three. The theory is compared with published data on polybenzylglutamate with fair agreement.

$$U_{\mu=1} = \begin{bmatrix} 1 & \sigma s \\ 1 & s \end{bmatrix}$$

$$U_{\mu=3} = \begin{bmatrix} 1 & \sigma s & 0 & 0 & 0 & 0 & 0 & 0 \\ 0 & 0 & 1 & s & 0 & 0 & 0 & 0 \\ 0 & 0 & 0 & 0 & 1 & 0 & 0 & 0 \\ 0 & 0 & 0 & 0 & 0 & 0 & 1 & s \\ 1 & 0 & 0 & 0 & 0 & 0 & 0 & 0 \\ 0 & 0 & 1 & s & 0 & 0 & 0 & 0 \\ 0 & 0 & 0 & 0 & 1 & 0 & 0 & 0 \\ 0 & 0 & 0 & 0 & 0 & 0 & 1 & s \end{bmatrix}$$

The matrix is of the order $\rho \times \rho$, where $\rho = 2^{\mu}$.

The assumptions made give matrices as given here for $\mu = 1$ and 3, respectively.

N 003   "The Helix-Coil Transition in Charged Macromolecules"
Zimm, B. H.; Rice, S. A. *Mol. Phys.* **1960**, *3*, 391.

Starting from the Grand Partition Function and using matrix methods introduced previously, a formulation of the theory of helix-coil transitions in charged macromolecules is given. The theory is compared with experimental data for the system polyglutamic acid – dioxane – water – NaCl. If the parameters describing the extra difficulty associated with initiating a helical configuration is assigned the same value as for polybenzylglutamate, the calculations are in excellent agreement with experiment. An analysis of the physical significance of the Grand Partition Function leads to methods for determining the fraction of polymer in helical form, the equilibrium constant for the addition of a residue to an already formed helix, as well as the initiation parameter. The value of the initiation parameter deduced from titration data on polyglutamic acid is in rough agreement with the value for polybenzylglutamate. The nature of the approximations used is discussed briefly.

N 004   "On the Theory of the Helix-Coil Transition in Polypeptides"
Lifson, S.; Roig, A. *J. Chem. Phys.* **1961**, *34*, 1963.

The evaluation of the configurational partition function of a polypeptide molecule, with the internal rotation angles as variables, leads to an improved treatment of the phenomenon of helix-coil transition in polypeptide molecules. The conditional probabilities of occurrence of helical and coiled states of the peptide units are obtained in the form of a 3 × 3 matrix. The order of this matrix is the lowest possible for the model employed, and is derived by a logical procedure which served to eliminate redundancies in the enumeration of states. The eigenvalues of this matrix yield the various molecular averages as functions of the degree of polymerization, temperature, and molecular constants. Explicit formulas are given for the degree of intramolecular hydrogen bonding, average number of helical sequences, and the distribution of their length, as well as the number average and the weight average of these lengths.

$$U = \begin{bmatrix} w & v & 0 \\ 0 & 0 & u \\ v & v & u \end{bmatrix}$$

The rules in constructing the matrix are:
    (1) a peptide unit in the coil state always contributes a factor u;
    (2) a peptide unit at the beginning or at the end of an uninterrupted sequence of helical states contributes a factor v;
    (3) a peptide unit at the interior of an uninterrupted sequence of helical states contributes a factor of w.

Rows (corresponding to the value of $\rho_{i-1}\rho_i$):   hh, hc, c {h∪c}
Columns (corresponding to the value of $\rho_i\rho_{i+1}$):   hh, hc, c {h∪c}   [h ≡ helical, c ≡ coil]

"The Configuration of Random Polypeptide Chains"

Brant, D. A.; Flory, P. J. *J. Am. Chem. Soc.* **1965**, *87*, 2788, 2791.

Bond Length [pm]		Valence Angles [°]		Torsion Angles [°]			
C^α-C	: 153	C^α-C-N	: 114	For bond 1:		For bond 3:	
C-N	: 132	C-N-C^α	: 123	t	: 180	t	: 180
N-C^α	: 147	N-C^α-C	: 110	g^+	: 60	g^+	: 60
C=O	: 124	C^α-C=O	: 121	g^-	: –60	g^-	: –60
N-H	: 100	C-N-H	: 123	For bond 2:		or:	
C^α-C^β	: 154	N-C^α-C^β	: 110	t	: 180	c	: 0
C^α-H^α	: 100	N-C^α-H^α	: 110			s^+	: 120
						s^-	: –120

Comments:

The experimentally measured dimensions of polypeptide chains in the unperturbed, random coil form are correlated with the chain structure using the RIS model and statistical mechanical methods applicable to linear systems of interacting subunits. The polypeptide chain with all its amide groups in the *trans* conformation may be treated as a sequence of virtual bonds of fixed length connecting successive α-carbons; the mutual orientation of a pair of adjoining virtual bonds is determined by the angles of rotation about the single bonds at the intervening α-carbon atom. Values of the characteristic ratio in the limit of large n_p are calculated. A temperature $T = 310$ K is used throughout. Because of uncertainties in several of the parameters entering into the rotational potential function, these are varied over the reasonable ranges. Those varied are the torsional barrier heights, the dielectric constant, and the van der Waals radius of the β-methylene group. The effect of replacing torsional potential function $V_{t,a}^{(3)}$ (minima at 180°, 60°, and –60°) by $V_{t,b}^{(3)}$ (minima at 0°, 120°, and –120°) is also investigated.

Calcd. quantities: $<r^2>_0 / n_p l_p^2$ $<s^2>_0 / n_p l_p^2$

N 006 "Statistical Mechanics of Noncovalent Bonds in Polyamino Acids. III. Interhelical Hydrophobic Bonds in Short Chains"
Poland, D. C.; Scheraga, H. A. *Biopolymers* **1965**, *3*, 305.

A partition function is derived for a simple model of interacting helices in a short (20 residues) chain of poly-L-alanine. It is found that interhelical hydrophobic bonds effect a marked stabilization of helical forms, and give rise to a sharp transition of the type found in many polymers.

N 007 "Statistical Mechanics of Noncovalent Bonds in Polyamino Acids. IV. Matrix Treatment of Hydrophobic Bonds in the Random Coil and of the Helix-Coil Transition for Chains of Arbitrary Length"
Poland, D. C.; Scheraga, H. A. *Biopolymers* **1965**, *3*, 315.

The *Lifson-Roig* matrix theory of the helix-coil transition in polyglycine is extended to situations where side-chain interactions (hydrophobic bonds) are present both in the helix and in the random coil. It is shown that the conditional probabilities of the occurrence of any number and size of hydrophobic pockets in the random coil can be adequately described by a 2×2 matrix. This is combined with the *Lifson-Roig* 3×3 matrix to produce a 4×4 matrix which represents all possible combinations of any amount and size sequence of α-helix with random coil containing all possible types of hydrophobic pockets in molecules of any given chain length. The total set of rules is *(1)* a state h preceded and followed by states h contributes a factor $w\sigma$ to the partition function; *(2)* a state h preceded and followed by states c contributes a factor v' to the partition function; *(3)* a state h preceded or followed by one state c contributes a factor v^* to the partition function; *(4)* a state c' contributes a factor u to the partition function; *(5)* a state d preceded by a state other than d contributes a factor s to the partition function; *(6)* a state d preceded by a state d contributes a factor r to the partition function.

States under discussion are:
helical (h)
random coil (c):   f ≡ free to realize positions of internal rotation
b ≡ near neighbor R-R hydrophobic bond
d ≡ a pocket of hydrophobic bonding
Thus, c = b ∪ f ∪ d = c' ∪ d, where c' = b ∪ f.

The appropriate matrix is:

$$U_i = \begin{bmatrix} w\sigma & v^* & 0 & 0 \\ 0 & 0 & u & s \\ v^* & v' & u & s \\ v^* & v' & u & r \end{bmatrix}$$

Rows (i-1,i) and columns (i,i+1):
hh, hc, c' {h ∪ c}, d {h ∪ c}

N 008 "Statistical Mechanics of Noncovalent Bonds in Polyamino Acids. V. Treatment of Long Chains by the Method of Sequence-Generating Functions: Hydrophobic Bonding in Random Coil, and Interactions Between Helical Segments"
Poland, D. C.; Scheraga, H. A. *Biopolymers* **1965**, *3*, 335.

The method of sequence-generating functions is applied to long polypeptide chains to describe various types of hydrophobic bonding in the random coil. These results are then combined with a similar treatment of the α-helix in order to discuss the helix-coil transition in single helices and in molecules whose helical segments interact by side-chain hydrophobic bonding. Numerical calculations are presented to show the relative probabilities of occurrence in the random coil of neighbor-neighbor hydrophobic bonds and pockets of hydrophobic bonding, and the relative probabilities of occurrence of the various states in a system of interacting helices. A discussion is also presented of the dependence of the helix-coil nucleation and growth parameters on solvent and side chain.

"A General Treatment of Helix-Coil Equilibria in Macromolecular Systems"

Flory, P. J.; Miller, W. G. *J. Mol. Biol.* **1966**, *15*, 284.

Statistical mechanical methods are applied to helix-coil equilibria in systems of single-strand polypeptide and double-strand polynucleotide helices. Chains of any finite length and copolymeric chains comprising units having different helix-forming tendencies can be treated rigorously by these methods. The partition function, the average degree of helicity and related quantities are formulated in terms of products of factors (matrices). Execution of numerical calculations requires specification of the sequential order of the various units in the macromolecular chain. In the case of α-helices, in accordance with the notation of *Zimm* and *Bragg*, s represents the statistical weight for a helical unit situated within a very long helical sequence relative to a statistical weight of unity for the random coil state. That is, $-RT \ln s$ will represent the free energy change per mol for transfer of a unit from the random coil to the interior of a helical sequence remote from end effects. In further compliance with the notation of *Zimm & Bragg*, σ represents the statistical weight factor to be associated with the two ends of a polypeptide helical sequence. That is, $-RT \ln \sigma$ is taken to be the change in free energy for the process whereby a very long sequence is fractured within its mid portion and the two long fragments are separated by a sequence of units in the random coil state, the total number of helical units being conserved in the process. An analogous procedure is discussed for the two-strand helices as well.

For molecules *limited to one helical sequence*, the basic statistical weight matrix used is given by:

$$U = \begin{bmatrix} 1 & \sigma s & 0 \\ 0 & s & 1 \\ 0 & 0 & 1 \end{bmatrix}$$

"Dimensions of Polypeptide Chains in Helicogenic Solvents"

Miller, W. G.; Flory, P. J. *J. Mol. Biol.* **1966**, *15*, 298.

Bond Length [pm]	Valence Angles [°]	Torsion Angles [°]	σ a,b) (for 310 K)			
C^α-C : 153	θ' : 114	For bond 1:	$\sigma = 0.003$: agreement is achieved for:			
C-N : 132	θ'' : 123	t : 180	s = 2.5 to 3.5			
N-C^α : 147	θ''' : 110	g^+ : 60		$\langle T_c \rangle = \begin{bmatrix} 0.54 & 0.18 & 0.62 \\ 0.06 & -0.74 & 0.08 \\ 0.59 & -0.12 & -0.43 \end{bmatrix}$	$U_i^{methodA} = \begin{bmatrix} 1 & \sigma s \\ 1 & s \end{bmatrix}$	$U_i^{methodB} = \begin{bmatrix} 1 & \sigma s & 0 & 0 & 0 & 0 & 0 & 0 \\ 0 & 0 & 1 & s & 0 & 0 & 0 & 0 \\ 0 & 0 & 0 & 0 & 1 & 0 & 0 & 0 \\ 0 & 0 & 0 & 0 & 0 & 0 & 1 & s \\ 1 & 0 & 0 & 0 & 0 & 0 & 0 & 0 \\ 0 & 0 & 1 & s & 0 & 0 & 0 & 0 \\ 0 & 0 & 0 & 0 & 1 & 0 & 0 & 0 \\ 0 & 0 & 0 & 0 & 0 & 0 & 1 & s \end{bmatrix}$
C=O : 124	θ^O : 121	g^- : -60	$\sigma = 0.0005$: agreement is achieved for:			
N-H : 100	θ^H : 123	For bond 2:	s = 1.3 to 1.4			
C^α-C^β : 154	θ^β : 110	t : 180				
C^α-H^α : 100	θ^α : 110	For bond 3:	$\sigma = 0.0005$: agreement is achieved for:	matrix **U**, method A:		
		t : 180	s = 1.1 to 1.2	rows and columns: c, h		
l_p : 380	ξ : 13.2	g^+ : 60				
	η : 22.2	g^- : -60	$\sigma = 0.0001$: agreement is achieved for:	matrix **U**, method B:		
			s < 1.1	rows: ccc, cch, chc, chh, hcc, hch, hhc, hhh		
In the helical conformation:		ψ : 46.8		columns: ccc, hcc, chc, hhc, cch, hch, chh, hhh		
		ϕ : 58.1				

a) σ is the helix initiation parameter.
b) Inasmuch as neither σ nor s could be estimated from data available, the most that could be done was to indicate combinations of them which describe the observations presented in Figures of the paper.

Comments:
Observed departures from the mean square end-to-end distance $\langle r^2 \rangle$ and radius of gyration $\langle s^2 \rangle$ of polypeptides of high molecular weight in helicogenic solvents such as DMF from the values of these quantities expected for rigid rods may be attributed either to (i) random deviations of the bond rotation angles from their preferred (α-helical) values, or to (ii) occasional interruptions of the helix by sequences of random coil units. Methods are presented for exact calculation of $\langle r^2 \rangle$ and $\langle s^2 \rangle$ for chains of any average degree of helicity, and it is shown that agreement with experiment on the basis of the second explanation can be achieved by suitable choice of parameters.

Calcd. quantites: $\langle r^2 \rangle / n_p l_p^2$ $\langle s^2 \rangle / n_p l_p^2$

N 011   "Computation of the Sterically Allowed Conformations of Peptides"
         Leach, S. J.; Némethy, G.; Scheraga, H. A. *Biopolymers* **1966**, *4*, 369.

The sterically permitted conformations for various di- and tripeptides are described using mathematical and computer methods. The effects of variations in the size and shape of the side-chain groups on the allowed conformations are assessed. Other factors which are investigated are the effects of possible variations in the geometry of the planar amide backbone and in the van der Waal's contact distances between atoms on the sterically permitted backbone conformations. The evaluation of the steric restrictions emphasizes their important role as a determinant in protein conformation.

N 012   "Conformational Analysis of Macromolecules.
         III. Helical Structures of Polyglycine and Poly-L-alanine"
         Scott, R. A.; Scheraga, H. A. *J. Chem. Phys.* **1966**, *45*, 2091.

polyglycine   poly(L-alanine)

A theoretical study of the regular conformations of isolated helices (*i.e.*, with no intermolecular interactions) of polyglycine and poly-L-alanine is carried out. The energy of each helical conformation is calculated by using semiempirical potential energy functions for the barriers to internal rotation about single bonds, nonbonded interactions, dipole-dipole interactions between the amide groups, and hydrogen bonding between backbone NH and CO groups. All bond length and bond angles are held fixed, and the amide group is fixed in the planar *trans* conformation. The effect of varrying the parameters in the semiempirical potential functions is studied.

N 013   "Kinetics of the Helix-Coil Transition in Polyamino Acids"
         Poland, D.; Scheraga, H. A. *J. Chem. Phys.* **1966**, *45*, 2071.

The matrix theory of the helix-coil transition in polyamino acids is extended to give the initial rate of change of helical content when the system is suddenly perturbed from equilibrium. Using the matrix to treat chains of arbitrarily length, it is possible to assign not only the equilibrium statistical weights for the initial probability of occurrence of each species but also the rate constants for all possible initial reactions involving the formation and breakdown of helical states. It is shown that all the rate constants can be related to the equilibrium statistical weights and to only one rate constant. The technique used to compute the initial rate can also be applied to calculate the moments of the equilibrium distribution of length of helical sequences.

N 014   "Conformational Energy Estimates for Statistically Coiling Polypeptide Chains"
Brant, D. A.; Miller, W. G.; Flory, P. J. *J. Mol. Biol.* **1967**, *23*, 47.

Conformational energies appropriate for an internal segment of an unperturbed, randomly coiling poly(α-amino acid) chain are calculated on the basis of several alternative approximate conformational potential energy functions. Calculations applicable to homopolymers having a substituent -CH_2-R' at the α-carbon in the L-configuration and to polyglycine are carried out. On the basis of the conformational energy estimates, statistical mechanical average properties of unperturbed randomly coiling polypeptide chains are evaluated. The important conclusions are insensitive to the details of the potential functions although, as previously reported, terms accounting for electrostatic interaction of neighboring amide groups must be included in order to obtain agreement with experimental values for the characteristic ratio of the unperturbed mean square end-to-end distance in poly(L-amino acid) chains possessing a substituent -CH_2-R' at the α-carbon. This value is $<r^2>_0 / n_p \, l_p^2 \approx 9$. On the other hand, the calculated dimensions of random polyglycine chains are found to be little influenced by such electrostatic terms, the configuration being influenced predominantly by the symmetry of the residue. The characteristic ratio calculated for polyglycine is $<r^2>_0 / n_p \, l_p^2 \approx 2$.

Bond Length [pm]		Valence Angles [°]	
C^α-C	: 153	θ'	: 114
C-N	: 132	θ''	: 123
N-C^α	: 147	θ'''	: 110
C=O	: 124	θ^O	: 121
N-H	: 100	θ^H	: 123
C^α-C^β	: 154	θ^β	: 110
C^α-H^α	: 100	θ^α	: 110
l_p	: 380 (for free rotation: l_p : 193)		

$$<T_{L\text{-Ala}}> = \begin{bmatrix} 0.51 & 0.20 & 0.59 \\ -0.046 & -0.61 & 0.21 \\ 0.65 & -0.23 & -0.30 \end{bmatrix}$$

$$<T_{Gly}> = \begin{bmatrix} 0.36 & -0.077 & 0.00 \\ -0.092 & -0.37 & 0.0 \\ 0.0 & 0.0 & -0.12 \end{bmatrix}$$

poly(L-alanine)

polyglycine

N 015   "Random Coil Configuration of Polypeptide Copolymers"
Miller, W. G.; Brant, D. A.; Flory, P. J. *J. Mol. Biol.* **1967**, *23*, 67.

The average dimensions of unperturbed, random-coil polypeptide copolymers composed either of L-alanine (or other residues having -CH_2-R' side chains) and glycine or of D- and L-alanine are treated theoretically, and results of numerical calculations are presented. The effect of amino acid sequence as well as over-all composition is investigated. Preliminary experimental results are presented for random D-L-polyglutamates and glycine-glutamic acid copolymers for comparison with the theoretical predictions. The calculation of $<r^2>_0$ is greatly simplified by the fact that bond rotations within a given amino acid residue are sensibly independent of rotations in neighboring residues. This circumstance permits the matrices T_i to be averaged independently from one another. Thus, the averaged matrix $<T_i>$ is determined by contributions to the conformational energy which depend on ϕ_i and ψ_i; these contributions are taken to be independent of neighboring bond rotation angles. The elements of $<T_i>$ are obtained therefore as their statistical mechanical averages over ϕ_i and ψ_i using the relevant contributions to the conformational energy.

a) random glycine-alanine-type copolymers:

$$<T_{L\text{-Ala}}> = \begin{bmatrix} 0.49 & 0.19 & 0.60 \\ -0.10 & -0.58 & 0.22 \\ 0.66 & -0.27 & -0.27 \end{bmatrix} \quad <T_{Gly}> = \begin{bmatrix} 0.33 & -0.11 & 0.0 \\ -0.16 & -0.34 & 0.0 \\ 0.0 & 0.0 & -0.11 \end{bmatrix}$$

b) D,L-copolymers having alanine-type side chains:

$$<T_{D\text{-Ala}}> = \begin{bmatrix} 0.49 & 0.19 & -0.60 \\ -0.10 & -0.58 & -0.22 \\ -0.66 & 0.27 & -0.27 \end{bmatrix}$$

N 016   "Dipole Moments in Relation to Configuration of Polypeptide Chains"
Flory, P. J.; Schimmel, P. R. *J. Am. Chem. Soc.* **1967**, *89*, 6807.

Methods are presented for calculating mean-square dipole moments, $<\mu^2>$, of polypeptide chains, averaged over all configurations of the chain skeleton. They are applicable to chains of any number $(x + 1)$ of residues, the residues being in any specified sequence. Dipole moments of glycine peptides are calculated and compared with experimental determinations. The effects of stereosequence on the dipole moment are well reproduced by the calculations. l_p is taken to be 380 pm.

$$<T_{i,gly}> = \begin{bmatrix} 0.36 & -0.077 & 0.0 \\ -0.092 & -0.37 & 0.0 \\ 0.0 & 0.0 & -0.12 \end{bmatrix} \quad <T_{0,gly}> = \begin{bmatrix} 0.36 & -0.027 & 0.0 \\ -0.086 & -0.39 & 0.0 \\ 0.0 & 0.0 & -0.12 \end{bmatrix} \quad <T_{x,gly}> = \begin{bmatrix} 0.53 & -0.84 & 0.0 \\ 0.76 & -0.52 & 0.0 \\ 0.0 & 0.0 & -0.92 \end{bmatrix}$$

$$<T_{i,L\text{-ala}}> = \begin{bmatrix} 0.51 & 0.20 & 0.59 \\ -0.046 & -0.61 & 0.21 \\ 0.65 & -0.23 & -0.30 \end{bmatrix} \quad <T_{0,L\text{-ala}}> = \begin{bmatrix} 0.36 & -0.015 & 0.93 \\ 0.85 & -0.40 & -0.34 \\ 0.37 & 0.92 & -0.13 \end{bmatrix} \quad <T_{x,L\text{-ala}}> = \begin{bmatrix} 0.52 & 0.85 & 0.077 \\ 0.72 & -0.50 & 0.33 \\ 0.32 & -0.12 & -0.87 \end{bmatrix}$$

$$<T_{i,D\text{-ala}}> = \begin{bmatrix} 0.51 & 0.20 & -0.59 \\ -0.046 & -0.61 & -0.21 \\ -0.65 & 0.23 & -0.30 \end{bmatrix}$$

The matrices $<T_{o/x,\text{ D-ala}}>$ are obtained by reversing the signs of the 13, 23, 31, and 32 elements of $<T_{o/x,\text{ L-ala}}>$
$0 < i < x$

polyglycine

poly(L-alanine)

N 017   "Conformational Analysis of Macromolecules. IV. Helical Structures of Poly(L-alanine), Poly(L-valine), Poly(β-methyl-L-aspartate), Poly(γ-methyl-L-glutamate), and Poly(L-tyrosine)"
Ooi, T.; Scott, R. A.; Vanderkooi, G.; Scheraga, H. A. *J. Chem. Phys.* **1967**, *46*, 4410.

Energy calculations are carried out for several isolated (single-stranded) homopolymer polyamino acids in order to find the most stable regular (helical) conformation. The energy is expressed as a function of the dihedral angles ϕ, ψ, and the set of χ_i's for rotation about the $N-C^\alpha$ and $C^\alpha - C^*$ bonds of the backbone, and the j single bonds of the side chains, respectively. Torsional, nonbonded, hydrogen-bonded, and dipole-dipole interaction energy contributions are included. In all cases examined, the nonbonded interaction energy would favour the right-handed α-helix over the left-handed one. However, the dipole-dipole interaction energy between the side chains and the backbone is apparently of sufficient importance, in the case of the asparate polymer, to reverse the screw sense.

N 018   "Conformational Energy and Configurational Statistics of Poly(L-proline)"
Schimmel, P. R.; Flory, P. J. *Procl. Nat. Acad. Sci. US* **1967**, *58*, 52.

Two distinctly different forms of poly(L-proline) designated *I* and *II*, respectively, are investigated in the solid state and in solution. In form *I* the imide group is in the *cis* configuration, in form *II* it is *trans*. Hydrogen bonds of the kind usually occurring in polypeptides are precluded by the absence of an amino hydrogen. All bond length and bond angles are taken from *Sasisekharan's* investigation of poly(L-proline) *II* (*Acta Cryst.* **1959**, *12*, 897). The imide group is assigned to the planar *trans* configuration. The distance between consecutive α-carbon atoms is then fixed at l_p = 380 pm. Except in the range ψ = $-75°$ to $170°$, steric repulsions give rise to energies more than 28 kJ·mol^{-1} above the minimum which occurs at $-124°$. With ψ restricted to this range, rotations about each $C^\alpha-C$ bonds are sensibly independent of rotations of adjacent bonds. The transformation matrix T_i is a function of rotation angles ϕ_i and ψ_i and of the angles θ, η and ξ. These latter angles are fixed by the structure at $110.5°$, $22.2°$, and $13.2°$, respectively, and the pyrrolidine ring fixes ϕ_i at *ca.* $78°$. Hence, T_i is a function only of ψ_i apart from fixed geometrical parameters.

$$\langle T \rangle = \begin{bmatrix} 0.44 & 0.30 & 0.81 \\ -0.54 & -0.62 & 0.52 \\ 0.71 & -0.69 & -0.11 \end{bmatrix}$$

Calcd. quantities:

$\langle r^2 \rangle_0 / n_p l_p^2$

N 019   "Conformational Energy Estimates for Helical Polypeptide Molecules"
Brant, D. A. *Macromolecules* **1968**, *1*, 291.

Conformational energy functions used to account for the mean-square unperturbed dimensions and dipole moments of randomly coiling polypeptides are used, after modification to account for the possibility of intramolecular hydrogen bonding, to evaluate the conformational energies of polypeptide helices. It is concluded that van der Waals energies are predominantly responsible for dictating the helical residue conformations of least energy. The relative stability of these conformations is greatly enhanced by the possibility for formation in the corresponding helical structures of intramolecular hydrogen bonds and to a lesser extent by favorable dipole interactions which occur in addition to the hydrogen-bonded interaction.

N 020   "Conformational Energies and Configurational Statistics of Copolypeptides containing L-Proline"
Schimmel, P. R.; Flory, P. J. *J. Mol. Biol.* **1968**, *34*, 105.

The conformational energy calculated for a *trans*-L-prolyl residue when isolated from other prolyl residues in a polypeptide chain is characterized by two minima of comparable energy occurring near ψ = $125°$ (A) and ψ = $325°$ (C), where ψ is the angle of rotation about the $C^\alpha-C$ bond; the angle ϕ of rotation about the $N-C^\alpha$ bond is fixed at approximately $122°$ by the proline ring.

$$\langle T_{L\text{-ala(P)}} \rangle = \begin{bmatrix} 0.50 & 0.14 & 0.81 \\ -0.031 & -0.73 & 0.098 \\ 0.67 & -0.084 & -0.32 \end{bmatrix}$$

$$\langle T_{L\text{-pro(Y)}} \rangle = \begin{bmatrix} 0.33 & 0.18 & 0.34 \\ -0.61 & -0.013 & 0.30 \\ 0.64 & -0.54 & 0.11 \end{bmatrix}$$

$$\langle T_{L\text{-gly(P)}} \rangle = \begin{bmatrix} 0.47 & 0.20 & 0.0 \\ 0.019 & -0.0021 & 0.0 \\ 0.0 & 0.0 & 0.038 \end{bmatrix}$$

N 021 "Dimensions of Protein Random Coils"
Miller, W. G.; Goebel, C. V. *Biochemistry* **1968**, *7*, 3925.

The dimensions of protein random coils are calculated for a variety of proteins of known amino acid sequence. Glycine and proline contribute to reducing the dimensions of random coil proteins. Branched side chains expand the chain only slightly more than unbranched side chains. Side chains represented as structured to the γ position are compared with structureless representations. It is demonstrated that the two approaches give comparable chain dimensions. The effect of sequence is investigated.

N 022 "Conformational Analysis of Macromolecules. V. Helical Structures of Poly(L-aspartic acid) and Poly(L-glutamic acid), and Related Compounds"
Yan, J. F.; Vanderkooi, G.; Scheraga, H. A. *J. Chem. Phys.* **1968**, *49*, 2713.

Previous energy calculations for isolated (single-stranded) homopolymer poly(amino acids), to find the most stable regular (helical) conformation, are extended to poly(L-aspartic acid), poly(L-glutamic acid), and several related compounds. Since rotation about all single bonds (including the acid and ester groups of the side chains) is taken into account, it is necessary to obtain expressions for the torsional potentials for these internal rotations. The computed helix sense is found to agree with the experimental one in all cases except one, using a single set of potential functions and parameters.

N 023 "Theoretical Conformation of Proline Oligomers"
Hopfinger, A. J.; Walton, A. G. *J. Macromol. Sci.-Phys.* **1969**, *B3(1)*, 171.

Minimum energy conformations of proline oligomers and poly(L-proline) are calculated. The left-hand helix of *trans*-polyproline *II* becomes stable at the tetramer, whereas the right-hand helix of the *cis*-polyproline *I* is not established until at least the pentamer. The potential minima include values of ψ (ψ = 163° *trans*, ψ = 56° *cis*) which yield forms of the polymer that are virtually identical with polyproline *I* and *II* in the solid state.

N 024 "Cooperativity in Poly(L-proline) *I* → *II* Transitions"
Holzwarth, G.; Chandrasekaran, R. *Macromolecules* **1969**, *2*, 245.

The poly(L-proline) *I* → *II* transition is known to exhibit positive cooperativity. In the present study, the contributions of intramolecular steric and electrostatic interactions to this phenomenon are computed. The calculations suggest that the total end effects are 48 kJ · mol^{-1} for form *I* but only 14 kJ · mol^{-1} for form *II*.

N 025 "Solution Properties of Synthetic Polypeptides. V. Helix-Coil Transition in Poly(β-benzyl-L-asparate)"
Hayashi, Y.; Teramoto, A.; Kawahara, K.; Fujita, H. *Biopolymers* **1969**, *8*, 403.

Simple approximate expressions are derived from the theory of *Zimm* and *Bragg* for use in the analysis of experimental data on the helix-coil transition in polypeptide. On the basis of the resulting expressions practical procedures are proposed to determine two basic parameters characterizing a thermally induced transition, *i.e.*, helix initiation parameter σ and enthalpy change for the helix formation, ΔH. They are applied to the data for poly(β-benzyl-L-aspartate) (PBLA) with the result σ = 1.6 × 10^{-4} and ΔH = − 1.9 kJ · mol^{-1} for PBLA in *m*-cresol; σ = 0.6 × 10^{-4} and ΔH = 1.09 kJ · mol^{-1} for PBLA in chloroform containing 5.7 vol-% of dichloroacetic acid. This result gives evidence that σ may change not only from one polypeptide to another but also for a given polypeptide in different solvents.

N 026 "Conformational Characteristics of L-Proline Oligomers"
Tonelli, A. E. *J. Am. Chem. Soc.* **1970**, *92*, 6187.

The conformational characteristics of L-proline oligomers are studied by calculating the intramolecular energies of all conformers distinguishable from each other by the *cis* or *trans* character of each of their imide bonds. Intrinsic torsional and nonbonded van der Waals potentials and dipolar electrostatic interactions which depend on the *cis* or *trans* character of the amide groups and the conformation about the $C^\alpha - C$ bonds are considered. Thus, the intramolecular conformational energy of the poly(L-proline) chain is a function of the angles of rotation ψ and ω. Beginning with the trimer, all nonterminal residues are predicted to be exclusively *trans*, while the N- and C-terminal residues in all the oligomers are found to be mixtures of *cis* and *trans*. However, according to the present calculations the reported onset of the form *II* helical conformation abruptly at the pentamer has no intramolecular origin.

N 027 "The Conformation about the Nitrogen – α-Carbon Bond in Random-Coil Polypeptides"
Tonelli, A. E.; Bovey, F. A. *Macromolecules* **1970**, *3*, 410.

The conformation about the $N-C^\alpha$ bond in random-coil polypeptides is studied by computing the average vicinal coupling $J_{N\alpha}$ of the peptide NH and C^α proton. Calculated conformational energies of random-coil polypeptides make possible the computation of the vicinal proton coupling by taking an average over the calculated conformations and assuming a reasonable Karplus-like dependence of $J_{N\alpha}$ on the dihedral angle ϕ'. The calculated coupling agrees within the probable experimental error with the values observed in solution for three different random-coil polypeptides, poly(L-alanine), poly(L-methionine), and poly(β-benzyl-L-aspartate).

N 028 "The Conformational Characteristics of Dipeptides"
Tonelli, A. E.; Brewster, A. I.; Bovey, F. A. *Macromolecules* **1970**, *3*, 412.

The conformational energies of the diastereoisomeric dipeptides L,L- and L,D-alanylalanine are calculated for the zwitterionic state, taking into account torsional and van der Waals potentials, monopole-monopole electrostatic interactions, and coulombic forces between the charged ends. The vicinal couplings $J_{N\alpha}$ of the peptide NH and C-terminal α-proton is computed for each isomer by taking an average over the calculated conformations and assuming a reasonable Karplus-like dependence of $J_{N\alpha}$ on the dihedral angle ϕ'. The calculated couplings approximate the values observed in aqueous solution.

N 029 "Ring Closure and Local Conformational Deformations of Chain Molecules"
Gō, N.; Scheraga, H. A. *Macromolecules* **1970**, *3*, 178.

A mathematical method is developed to provide a solution to two problems hitherto arising in conformational energy calculations of oligomers and polymers, when bond length and bond angles are maintained fixed. The two problems are the calculation of the sets of dihedral angles which lead to *(a)* exact ring closure in cyclic molecules and *(b)* local conformational deformations of linear or cyclic molecules. Most of the emphasis is placed on polypeptide chain molecules.

N 030 "Calculation of the Conformation of the Pentapeptide
cyclo-(glycyl glycyl glycyl prolyl prolyl). I. A Complete Energy Map"
Gō, N.; Scheraga, H. A. *Macromolecules* **1970**, *3*, 188.

The complete energy surface of cyclo-(glycyl glycyl glycyl prolyl prolyl) is computed. With rigid geometry (fixed bond length and bond angles, and planar amide groups), the condition of exact ring closure makes two dihedral angles (taken as the angles ψ between Pro-Pro and Pro-Gly) independent; the remaining six dihedral angles are computed as a function of the two independent ones.

N 031 "Location of Proline Derivatives in Conformational Space. I. Conformational Calculations, Optical Activity and NMR Experiments"
Madison, V.; Schellman, J. *Biopolymers* **1970**, *9*, 511.

In order to develop methods of analysis applicable to the determination of the conformation of biological polymers in solution, a series of proline derivatives is studied. The steric constraints of the pyrrolidine ring limit these compounds to a relatively small set of conformations. This set is further reduced by eliminating conformations whith large computed conformational energy. Computations reveal that the conformational energy of the proline derivatives fits into one of three classes, depending on the bulk and the polarity of the C-terminal group. Three classes of optical activity are observed. The optical activity data are analyzed in terms of conformations computed to be of low energy. Nuclear magnetic resonance provides an experimental measure of the fraction of molecules which have *cis* unsymmetrically-substituted tertiary amide groups. This information aids and confirms the other measures of molecular conformation.

N 032 "Conformational Properties of Poly(L-proline) Form *II* in Dilute Solution"
Mattice, W. L.; Mandelkern, L. *J. Am. Chem. Soc.* **1971**, *93*, 1769.

The intrinsic viscosity of poly(L-proline) is studied as a function of molecular weight and temperature in five commonly used solvents: water, trifluoroethanol, acetic acid, propionic acid, and benzyl alcohol. The characteristic ratio is 14 in water and 18-20 in the organic solvents at 303 K, and $d (\ln <r^2>_0) / dT$ is negative. The theoretical rotational potential function obtained by *Hopfinger* and *Walton* for L-prolyl-L-prolyl-L-prolyl-L-proline (*J. Macromol. Sci. Phys.* **1969**, *3*, 171) correctly predicts the characteristic ratio at 303 K but predicts the wrong sign for $d (\ln <r^2>_0) / dT$.

N 033   "Unperturbed Dimensions of Sequential Copolypeptides Containing Glycine, L-Alanine, L-Proline, and γ-Hydroxy-L-proline"
Mattice, W. L.; Mandelkern, L. *Biochemistry* **1971**, *10*, 1934.

The characteristic ratios of poly(pro-gly), poly(hyp-gly), poly(gly-gly-pro-gly), poly(gly-gly-hyp-gly), and poly(pro-ala) are determined in water. The results confirm the main features of the theoretical conformational maps derived by *Flory* and coworkers for glycine followed by either L-proline or a nonproline residue. Small adjustments, well within the uncertainty described by *Schimmel* and *Flory*, are suggested in the conformational map for L-proline followed by glycine. The constants for the Lennard-Jones functions of *Scheraga* and coworkers, as used by *Madison* and *Schellman*, produce a conformational map for L-proline followed by a nonproline residue which is in somewhat poorer agreement with experiment. The two sets of modified constants introduced by *Madison* and *Schellman* fail to predict the conformational properties of these sequential copolypeptides.

N 034   "Helix-Coil Stability Constants for the Naturally Occurring Amino Acids in Water. I. Properties of Copolymers and Approximate Theories"
von Dreele, P. H.; Poland, D.; Scheraga, H. A. *Macromolecules* **1971**, *4*, 396.

Since the *Zimm-Bragg* parameters σ and s of the naturally occurring amino acids (in water) cannot be obtained from studies of the helix-coil transition in homopolymers, because of experimental difficulties, a technique is developed to circumvent these problems. It involves the study of the thermally induced transition curves for random copolymers of "guest" amino acid residues in a water-soluble "host" poly(amino acid). The data may be interpreted with the aid of suitable theories for the helix-coil transition in random copolymers to obtain σ and s for the "guest" residues. It is shown in this paper that, for the usual ranges of parameters found for poly(amino acids), one of the two lowest order approximations (corresponding to earlier treatments by *Lifson* and *Allegra*) is completely adequate. In essence, the low-order approximations hold if σ and s for the two constituents of the copolymer do not differ appreciably from each other.

N 035   "Helix-Coil Stability Constants for the Naturally Occurring Amino Acids in Water. II. Characterization of the Host Polymers and Application of the Host-Guest Technique to Random Poly(hydroxypropylglutamine-*co*-hydroxybutylglutamine)"
von Dreele, P. H.; Ananthanarayanan, V. S.; Andreatta, R. H.; Poland, D.; Scheraga, H. A. *Macromolecules* **1971**, *4*, 408.

The melting behavior of the homopolymers poly[N^5-(3-hydroxypropyl)-L-glutamine] and poly[N^5-(4-hydroxybutyl)-L-glutamine], and copolymers of two amino acids, is determined in water. The homopolymer data are treated by the *Zimm-Bragg* theory and the copolymer data are analyzed with the lowest order approximation of the theory discussed in the previous paper (**N 034**). Within the experimental error, it is not possible to detect any temperature dependence of the parameter σ or of the parameters ΔH and ΔS for the homopolymers. Using these temperature-independent parameters for one of the homopolymers, it is possible to compute those for the other homopolymer by applying the host-guest technique (and associated theory) to the copolymer data. Good agreement is obtained between the parameters computed directly from homopolymer data and those obtained by the host-guest technique.

N 036   "Helix-Coil Stability Constants for the Naturally Occurring Amino Acids in Water. III. Glycine Parameters from Random Poly(hydroxybutylglutamine-*co*-glycine)"
Ananthanarayanan, V. S.; Andreatta, R. H.; Poland, D.; Scheraga, H. A. *Macromolecules* **1971**, *4*, 417.

Water-soluble random copolymers, containing glycine and hydroxybutylglutamine, are synthesized, fractionated, and characterized. From an analysis of their thermally induced helix-coil transition curves, using an approximate theory for random copolymers and the host-guest technique, it is possible to obtain the *Zimm-Bragg* parameters σ and s which characterize the (hypothetical) helix-coil transition of polyglycine in water. The relatively low values of s in the temperature range of 273 to 343 K, in water, indicate that glycyl residues in polyglycine do not adopt the α-helical conformation under these conditions; when incorporated in a copolymer, they act as strong helix brakers. The values of the *Zimm-Bragg* parameter s for polyglycine in water are reported as follows (for σ = 1 × 10^{-5}): 0.510 (273 K), 0.550 (283 K), 0.591 (293 K), 0.615 (303 K), 0.625 (313 K), 0.631 (333 K), 0.610 (343 K).

N 037   "Further Comparison with Experiment of the Calculated Results Obtained from Semiempirical and Quantum Mechanical Conformational Energy Maps Appropriate to Random-Coil Polypeptides"

Tonelli, A. E. *Macromolecules* **1971**, *4*, 618.

The average vicinal coupling constants $J_{N\alpha}$ between amide and α-protons, the mean-square unperturbed end-to-end distance $<r^2>_0$ for randomly coiling polypeptides with side chains R of the type R = -CH_2—R′, and the mean-square unpertubed dipole moments $<\mu^2>_0$ of the zwitterionic diastereoisomers of the tri- and tetrapeptides of alanine are calculated from the conformational energy map reported to be appropriate to these random-coil polypeptides according to PCILO quantum mechanical molecular orbital calculations. These results are compared with experiment and with the corresponding quantities calculated from the conformational energy maps obtained by using semiempirical potential energy functions. Both conformational energy maps predict the correct vicinal coupling $J_{N\alpha}$ observed for three different random-coil polypeptides in solution. However, the PCILO map leads to dimensions $<r^2>_0$ almost identical with those calculated for a polypeptide chain with free rotation about the N—C^α and C^α—C backbone bonds, and predict the mean-square dipole moments of the diastereoisomeric tri- and tetrapeptides of alanine to increase with decreasing stereoregularity and to be virtually independent of stereosequence, respectively, in disagreement with the experimental observations. The semiempirical map yields calculated dimensions and predicts the effect of stereosequence on the dipole moments of the zwitterionic tri- and tetrapeptides in excellent agreement with experiment. On this basis, and because other quantum mechanical calculations conform very closely to the semiempirical energy maps, it appears that the PCILO map is in error. The transformation matrix for an interior L-alanine residue averaged over the PCILO map is (for 298 K):

$$<T_{L\text{-ala}}> = \begin{bmatrix} 0.066 & -0.68 & -0.13 \\ -0.34 & -0.42 & 0.32 \\ -0.29 & 0.064 & -0.39 \end{bmatrix}$$

where $<T_{D\text{-ala}}>$ is identical with $<T_{L\text{-ala}}>$ except for the signs of elements (1,3), (2,3), (3,1), and (3,2). The average transformation matrices for the terminal residues in the zwitterionic peptides are taken from *Flory* and *Schimmel* (*J. Am. Chem. Soc.* **1967**, *89*, 6807).

"A Model for Helix-Coil Transition in Specific-Sequence Copolymers of Amino Acids"

Gō, N.; Lewis, P. N.; Gō, M.; Scheraga, H. A. *Macromolecules* **1971**, *4*, 692.

State[a]	(ϕ/ψ)	Hydrogen bond on CO	Hydrogen bond on NH	Statistical weight	
1	c	No	No	1	p_1
2	c	No	Yes	$\sigma^{-1/4} s^{1/2}$	p_2
3	c	Yes	No	$\sigma^{-1/4} s^{1/2}$	p_3
4	c	Yes	Yes	$\sigma^{-1/2} s$	p_4
5	h	No	No	$\sigma^{-1/2}$	p_5
6	h	No	Yes	$\sigma^{1/4} s^{1/2}$	p_6
7	h	Yes	No	$\sigma^{1/4} s^{1/2}$	p_7
8	h	Yes	Yes	s	p_8

$$W = \begin{bmatrix} p_1 & p_2 & 0 & 0 & 0 & 0 & 0 & 0 & 0 & 0 \\ 0 & 0 & p_5 & 0 & 0 & 0 & 0 & p_7 & 0 & 0 \\ 0 & 0 & 0 & p_5 & 0 & 0 & 0 & 0 & 0 & 0 \\ 0 & 0 & 0 & 0 & 0 & p_5 & 0 & 0 & 0 & 0 \\ 0 & 0 & 0 & 0 & 0 & p_5 & 0 & 0 & 0 & 0 \\ p_2 & p_4 & 0 & 0 & 0 & 0 & 0 & 0 & 0 & 0 \\ 0 & 0 & 0 & 0 & 0 & p_6 & 0 & 0 & 0 & 0 \\ 0 & 0 & 0 & 0 & 0 & p_6 & 0 & 0 & 0 & 0 \\ 0 & 0 & 0 & 0 & p_5 & 0 & 0 & 0 & p_7 & 0 \\ 0 & 0 & 0 & 0 & 0 & 0 & p_5 & 0 & 0 & p_7 \\ 0 & 0 & 0 & 0 & 0 & 0 & p_6 & 0 & 0 & p_8 \end{bmatrix}$$

The Zimm-Bragg model:

$$W_{A(f)}^{Zimm/Bragg} = \begin{bmatrix} 1 & 0 & 0 & \sigma s_{A(f)} \\ 1 & 0 & 0 & 0 \\ 0 & 1 & 0 & 0 \\ 0 & 0 & 1 & s_{A(f)} \end{bmatrix}$$

$$W_{A(f)}^{Zimm/Bragg} = \begin{bmatrix} p_1 & p_2 \\ p_3 & p_4 \end{bmatrix}$$

$p_1 = 1$
$p_2 = \sigma s_{A(f)}$
$p_3 = 1$
$p_4 = s_{A(f)}$

For the Lifson-Roig model, see next page.

[a] An amino acid residue in a polypeptide chain can exist in any of eight distinct states according to three independent factors: (a) whether or not the values of the two dihedral angles ϕ and ψ are constrained to those characteristic of the right-handed α helix, α_R, designated "h" and "c", respectively; (b) whether or not a hydrogen bond is formed between the CO group of the *i*th residue in question and the NH group of the *(i+4)*th residue, when the three intermediate residues are all in h states; and (c) whether or not a hydrogen bond is formed between the NH group of the *i*th residue in question and the CO group of the *(i−4)*th residue, when the three intermediate residues are all in h states.

"A Model for Helix-Coil Transition in Specific-Sequence Copolymers of Amino Acids"

Gō, N.; Lewis, P. N.; Gō, M.; Scheraga, H. A. *Macromolecules* 1971, *4*, 692.

Comparison of the Zimm-Bragg and Lifson-Roig models for homopoly(amino acids):
Statistical weights for the Lifson-Roig model:

State [a]	(ϕ/ψ)	Hydrogen bond on CO	NH	Statistical weight [b]	
1	c	No	No	u	p_1
2	c	No	Yes	$v^{-1/2}w^{1/2}u$	p_2
3	c	Yes	No	$v^{-1/2}w^{1/2}u$	p_3
4	c	Yes	Yes	$v^{-1}wu$	p_4
5	h	No	No	v	p_5
6	h	No	Yes	$v^{1/2}w^{1/2}$	p_6
7	h	Yes	No	$v^{1/2}w^{1/2}$	p_7
8	h	Yes	Yes	w	p_8

$$W = \begin{bmatrix} p_1 & p_3 & p_5 & 0 & 0 & 0 & 0 & 0 & 0 & 0 & 0 \\ 0 & 0 & 0 & p_5 & 0 & 0 & 0 & 0 & p_7 & 0 & 0 \\ p_1 & p_3 & 0 & 0 & p_5 & 0 & 0 & 0 & 0 & 0 & 0 \\ 0 & 0 & 0 & 0 & 0 & p_5 & 0 & 0 & 0 & 0 & 0 \\ p_1 & p_3 & 0 & 0 & 0 & 0 & 0 & 0 & 0 & 0 & 0 \\ 0 & 0 & 0 & 0 & 0 & 0 & p_5 & 0 & 0 & 0 & 0 \\ p_2 & p_4 & 0 & 0 & 0 & 0 & 0 & 0 & 0 & 0 & 0 \\ 0 & 0 & 0 & 0 & 0 & 0 & 0 & p_5 & 0 & 0 & 0 \\ 0 & 0 & 0 & 0 & 0 & p_6 & 0 & 0 & 0 & 0 & 0 \\ 0 & 0 & 0 & 0 & 0 & 0 & 0 & p_6 & 0 & 0 & 0 \\ 0 & 0 & 0 & 0 & 0 & 0 & p_5 & 0 & 0 & p_7 & 0 \\ 0 & 0 & 0 & 0 & 0 & 0 & 0 & 0 & p_5 & 0 & p_7 \\ 0 & 0 & 0 & 0 & 0 & 0 & 0 & 0 & p_6 & 0 & p_8 \end{bmatrix}$$

The partition functions for the Zimm-Bragg (ZB) and Lifson-Roig (LR) are given by:

$$Z_{ZB}(\sigma,s,N) = (1,0,0,0) \begin{bmatrix} 1 & 0 & 0 & \sigma s \\ 1 & 0 & 0 & 0 \\ 0 & 1 & 0 & 0 \\ 0 & 0 & 1 & s \end{bmatrix}^{N-4} \begin{bmatrix} 1 \\ 1 \\ 1 \\ 1 \end{bmatrix}$$

$$Z_{LR}(u,v,w,N) = (0,0,1) \begin{bmatrix} w & v & 0 \\ 0 & 0 & u \\ v & v & u \end{bmatrix}^{N} \begin{bmatrix} 0 \\ 1 \\ 1 \end{bmatrix}$$

[a] An amino acid residue in a polypeptide chain can exist in any of eight distinct states according to three independent factors: *(a)* whether or not the values of the two dihedral angles ϕ and ψ are constrained to those characteristic of the right-handed α helix, α_R, designated "h" and "c", respectively; *(b)* whether or not a hydrogen bond is formed between the CO group of the *i*th residue in question and the NH group of the *(i+4)*th residue, when the three intermediate residues are all in h states; and *(c)* whether or not a hydrogen bond is formed between the NH group of the *i*th residue in question and the CO group of the *(i−4)*th residue, when the three intermediate residues are all in h states.
[b] In practice, the coil state in the *Lifson-Roig* model, which is characterized by nonhelical (ϕ,ψ) angles, is assigned a statistical weight u of unity so that the following relationships result:
$\sigma^{1/2} = v / (1 + v)$ *or* $v = 1 / (\sigma^{-1/2} - 1)$
$s = w / (1 + v)$ *or* $w = s / (1 - \sigma^{1/2})$, where the v and w are really v/u and w/u, since u is taken as the reference state.

Comments:
A model for the helix-coil transition in specific-sequence copolymers of amino acids is presented. While the statistical weight matrix, required for the correlation of the states of four consecutive residues in this model, is 44×44, it is shown that this can be concentrated to an 11×11 matrix, whose secular equation is the same as that for the Zimm-Bragg 8×8 matrix, which is contractable to a 4×4 matrix. The statistical weight of a residue in one of eight distinct states reflects the major intramolecular interactions responsible for helix stabilization. It is shown that, for copolymers, the form of the individual statistical weights is unique, rather than arbitrary as it is for homopolymers. While the 11×11 matrix is required for a copolymer, it reduces to a 4×4 matrix for a homopolymer. Numerical examples on a specific-sequence binary random copolymer indicate that the average helix contents, as determined by the 11×11 and by the *Zimm-Bragg* 2×2 matrix methods, differ in most cases by only a few per cent. Thus, the 2×2 matrix formulation can be used as a good approximation for the analysis of experimental data on polypeptide copolymers.

N 040 "Conformational Energy and Unperturbed Chain Dimensions of Polypeptide Homopolymers"
Tanaka, S.; Nakajima, A. *Polymer J.* **1971**, *2*, 717.

Conformational energy and characteristic ratio $<r^2>_0 / nl^2$ are theoretically investigated for polyglycine, poly(L-alanine), poly(*N*-methylglycine), and poly(*N*-methyl-L-alanine), assuming *trans* conformation for amide and imide bonds, and employing suitable molecular parameters and plausible potential functions. The chain is regarded as a sequence of virtual bonds whose length are 380 pm. The theoretically calculated characteristic ratios are, respectively, $<r^2>_0 / nl^2$ = 2.17, 8.38, 2.97, and 0.58. These results are discussed in connection with experimental and other theoretical results.

Low-energy conformations on the energy contour diagrams:

residue	rotational angles [°]	
	φ	ψ
ala	160	−160
	60	40
	−60	−60
gly	80 (−80)	−80 (80)
	60 (−60)	60 (−60)
nmg	80 (−80)	180 (180)
	120 (−120)	−80 (80)
	60 (−60)	80 (−80)
nma	120	−80
	120	−160
	−120	−100
	−120	−160

The averaged transformation matrices used (at 298 K) are:

$$<T>_{Ala} = \begin{bmatrix} 0.565 & 0.370 & 0.499 \\ -0.055 & -0.573 & 0.417 \\ 0.566 & -0.417 & 0.330 \end{bmatrix} \quad <T>_{Gly} = \begin{bmatrix} 0.368 & -0.004 & 0.000 \\ -0.163 & -0.392 & 0.000 \\ 0.000 & 0.000 & -0.140 \end{bmatrix}$$

$$<T>_{NMG} = \begin{bmatrix} 0.578 & 0.653 & 0.000 \\ -0.267 & -0.032 & 0.000 \\ 0.000 & 0.000 & 0.000 \end{bmatrix} \quad <T>_{NMA} = \begin{bmatrix} 0.332 & 0.132 & 0.835 \\ -0.166 & 0.552 & -0.222 \\ -0.557 & -0.027 & 0.223 \end{bmatrix}$$

N 041 "Conformational Energy and Unperturbed Chain Dimensions of Polypeptide Copolymers"
Tanaka, S.; Nakajima, A. *Polymer J.* **1971**, *2*, 725.

Copolypeptides composed of two components among L-alanine, glycine, *N*-methylglycine, and *N*-methyl-L-alanine, and D,L-copolypeptides composed either of D-alanine and L-alanine or of *N*-methyl-D-alanine, *N*-methyl-L-alanine are theoretically investigated for conformational energies and unperturbed average chain dimensions, assuming *trans* conformation both for amide and imide bonds. The characteristic ratios of copolymers are calculated as a function of sequence probability that a residue of the minor component is followed by a residue of the same kind at varied copolymer compositions. For equimolar copolymers, the effect of the degrees of polymerization is also investigated. In racemic D,L-copolymers, the chain dimensions are predicted as markedly dependent on the isotacticity of the polymer.

N 042 "Molecular Theory of the Helix-Coil Transition in Poly(amino acids). III. Evaluation and Analysis of s and σ for Polyglycine and Poly(L-alanine) in water"
Gō, M.; Gō, N.; Scheraga, H. A. *J. Chem. Phys.* **1971**, *54*, 4489.

Previously, the *Zimm-Bragg* parameters s and σ for the helix-coil transition in polyglycine and poly(L-alanine) are calculated in terms of molecular quantities (Gō, N.; Gō, M.; Scheraga, H. A. *Procl. Natl. Acad. Sci. U. S.* **1968**, *59*, 1030). These calculations are extended here to take into account the effects of water as the solvent and are analyzed to deduce the relative importance of the various interaction terms for the helix-coil transition. The screening effect of the electrostatic interactions by the water molecules which lie between or near two charged atoms is taken into account by using a dielectric constant $D = 4.0$ for short-range interactions, and by cutting off all interactions of longer range than about 600 pm. The calculated values of the parameter σ, which are very sensitive to the value of D, agree well with experimental ones. The origin of the greater stability of the α-helical conformation of poly(L-alanine) compared to that of polyglycine is analyzed. The effect on the value of the parameter s, arising from the binding of water molecules to free CO and NH groups of the residue in the coil state, is considered. Parameters which express the strength of the binding of water molecules are adjusted so as to make the calculated values of s for poly(L-alanine) in water as a function of temperature fit best with experimental values. Finally, a simple but realistic model of the helix-coil transition, deduced from the calculation of s and σ in terms of molecular quantities, is proposed as a first approximation not only for homopolymers but also for copolymers.

N 043 "Unperturbed Dimensions of Poly(N^5-ω-hydroxyethyl-L-glutamine) in Water"
Mattice, W. L.; Lo, J.-T. *Macromolecules* **1972**, *5*, 734.

The characteristic ratio of poly(N^5-ω-hydroxyethyl-L-glutamine) in water at 303 K is found to be 10 ± 1, in agreement with results obtained by *Brant* and *Flory* (*J. Am. Chem. Soc.* **1965**, *87*, 2791) for four other polypeptides with $-CH_2-R$ side chains. The circular dichroism of poly(N^5-ω-hydroxyethyl-L-glutamine) under these conditions, where the polypeptide is in a statistical conformation, exhibits a positive band at 216 nm.

N 044 "Helix-Coil Stability Constants for the Naturally Occurring Amino Acids in Water. IV. Alanine Parameters from Random Poly(hydroxypropylglutamine-*co*-L-alanine)"
Platzer, K. E. B.; Ananthanarayanan, V. S.; Andreatta, R. H.; Scheraga, H. A. *Macromolecules* **1972**, *5*, 177.

The synthesis and characterization of water-soluble random copolymers containing L-alanine and N^5-(3-hydroxypropyl)-L-glutamine are described, and the thermally induced helix-coil transition of these copolymers in water is studied. The incorporation of L-alanine is found to increase the helix-content of the polymer. The *Zimm-Bragg* parameters σ and s for the helix-coil transition in poly(L-alanine) in water are deduced from an analysis of the melting curves of the copolymers. The values of s, computed from both the *Lifson* and *Allegra* theories using the value $\sigma = 0.0008 \pm 0.0002$ are (*Lifson/Allegra*): s = 1.078/1.081 (273 K), 1.068/1.071 (293 K), 1.040/1.042 (313 K), 1.007/1.008 (333 K).

N 045 "Helix-Coil Stability Constants for the Naturally Occurring Amino Acids in Water. VI. Leucine Parameters from Random Poly(hydroxypropylglutamine-*co*-L-leucine) and Poly(hydroxybutylglutamine-*co*-L-leucine)"
Alter, J. E.; Taylor, G. T.; Scheraga, H. A. *Macromolecules* **1972**, *5*, 739.

The synthesis and characterization of water-soluble random copolymers containing L-leucine with either N^5-(3-hydroxypropyl)-L-glutamine or N^5-(4-hydroxybutyl)-L-glutamine are described, and the thermally induced helix-coil transition of these copolymers in water is studied. The incorporation of L-leucine is found to increase the helix-content of the polymer. The *Zimm-Bragg* parameters σ and s for the helix-coil transition in poly(L-leucine) in water are deduced from an analysis of the melting curves of the copolymers. The values of s, computed from both the *Lifson* and *Allegra* theories are (*Lifson*, $\sigma = 3 \times 10^{-3}$ / *Allegra*, $\sigma = 3.3 \times 10^{-3}$ / *Allegra*, $\sigma = 1.2 \times 10^{-3}$): s = 1.09/1.10/1.09 (273 K), 1.12/1.14/1.15 (293 K), 1.12/1.13/1.16 (313 K), 1.087/1.09/1.14 (333 K).

N 046   "Thermal and Charge-Induced Coil to α-Helix Transition of Poly(L-glutamic acid) and Random L-Glutamic Acid — L-Alanine Copolymers"
Warashina, A.; Ikegami, A. *Biopolymers* **1972**, *11*, 529.

Thermal and charge induced random coil to α-helix transitions of poly(L-glutamic acid) (PGA) are measured by optical rotatory dispersion in various solvents. The data of PGA in 0.1 M NaCl are analyzed by the *Zimm-Rice* theory. The initiation parameter, σ, of the *Zimm-Rice* theory is given by a value of 5 ($\pm$ 1) $\times$ 10^{-3}. Random copolymers of L-glutamic acid and L-alanine containing 10, 30, and 40 molar percents of alanyl residue are analyzed as well.

N 047   "Helix-Coil Stability Constants for the Naturally Occurring Amino Acids in Water. V. Serine Parameters from Random Poly(hydroxybutylglutamine-co-L-serine)"
Hughes, L. J.; Andreatta, R. H.; Scheraga, H. A. *Macromolecules* **1972**, *5*, 187.

Water-soluble random copolymers containing L-serine and N^5-(4-hydroxybutyl)-L-glutamine are prepared, and the thermally induced helix-coil transition of these copolymers in water is studied. The *Zimm-Bragg* parameters σ and s for the (hypothetical) helix-coil transition in poly(L-serine) in water are deduced. The values of s, computed from both the *Lifson* and *Allegra* theories are (*Lifson*, σ = 1 $\times$ 10^{-4} / *Allegra*, σ = 7.5 $\times$ 10^{-5}): s = 0.667/0.726 (273 K), 0.757/0.784 (293 K), 0.777/0.792 (313 K), 0.731/0.744 (333 K).

N 048   "Conformational Properties of Poly(L-proline) Containing a Flexible Pyrrolidine Ring"
Mattice, W. L.; Nishikawa, K.; Ooi, T. *Macromolecules* **1973**, *6*, 443.

Two conformational maps for the dipeptide unit in the interior of poly(L-proline) containing peptide bonds in the planar *trans* configuration are found to correctly predict the characteristic ratios experimentally observed for this polypeptide at 278 and 303 K. These two conformational maps differ from previous maps in allowing for flexibility in pyrrolidine rings. They suggest that there are two rotational isomeric states, separated by about 180° rotation about ψ, for poly(L-proline) containing peptide bonds in the planar *trans* conformation. While one of these rotational states is populated to only a minor extent at 303 K, it nevertheless exerts a significant effect upon the unperturbed dimensions of poly(L-proline). The averaged transformation matrices calculated from the two conformational maps, using a temperature of 303 K, are given. Thereby, two different positions are considered for C_i^γ: in the plane of C_i^α–N_i–C_i^δ (γ^1 position) and in the plane of C_i^β–C_i^α–N_i (γ^2 position). The unspecified bond angles in the pyrrolidine ring of residue *i* are functions of both ϕ_i and the location of C_i^γ. The characteristic ratios calculated from the conformational maps are 13.0 at 278 K and 12.4 at 303 K for the γ^1 position, and the γ^2 position leads to the slight higher characteristic ratios of 20.0 at 278 K and 17.0 at 303 K. These predictions are in excellent agreement with the experimentally observed values.

$$\langle T \rangle_{\gamma 1} = \begin{bmatrix} 0.423 & 0.474 & 0.548 \\ -0.589 & -0.221 & 0.582 \\ 0.640 & 0.667 & 0.186 \end{bmatrix} \qquad \langle T \rangle_{\gamma 2} = \begin{bmatrix} 0.444 & 0.523 & 0.589 \\ -0.589 & -0.272 & 0.644 \\ 0.641 & 0.686 & 0.175 \end{bmatrix}$$

N 049 "Comparison of the Conformational Map for Poly(L-proline) with Conformational Maps for Polysarcosine and Poly(N-methyl-L-alanine)"
Mattice, W. L. *Macromolecules* **1973**, *6*, 855.

The presence of the four minima in the conformational map reported by *Mark* and *Goodman* for poly(N-methyl-L-alanine) (*Biopolymers* **1967**, *5*, 809) is confirmed. The conformational map reported by *Tanaka* and *Nakajima* for polysarcosine (*Polymer J.* **1970**, *1*, 71; see also *Polymer J.* **1971**, *2*, 717), however, is found to be incorrect due to their faillure to consider several crucial interatomic contacts. The conformational maps for both polysarcosine and poly(N-methyl-L-alanine) are found to be sensitive to the orientation selected for the methyl groups.

poly(N-methyl-L-alanine)

N 050 "Helix-Coil Stability Constants for the Naturally Occurring Amino Acids in Water. VII. Phenylalanine Parameters from Random Poly(hydroxypropylglutamine-co-L-phenylalanine)"
Van Wart, H. E.; Taylor, G. T.; Scheraga, H. A. *Macromolecules* **1973**, *6*, 266.

Water-soluble random copolymers containing L-phenylalanine and N^5-(3-hydroxypropyl)-L-glutamine are prepared. The thermally induced helix-coil transition in these copolymers in water is studied. The incorporation of L-phenylalanine is found to increase the helix-content of the host polymer. The *Zimm-Bragg* parameters σ and s for the helix-coil transition in poly(L-phenylalanine) in water are deduced from an analysis of the melting curves of the copolymers. The values of s, computed from both the *Lifson* and *Allegra* theories are (*Lifson/Allegra*; σ = *0.0018*): s = 1.056/1.061 (273 K), 1.078/1.086 (293 K), 1.041/1.047 (313 K), 1.000/1.003 (333 K).

N 051 "Conformational Statistics of Short Chains of Poly(L-alanine) and Polyglycine Generated by Monte Carlo Method and the Partition Function of Chains with Constrained Ends"
Premilat, S.; Hermans, J., Jr. *J. Chem. Phys.* **1973**, *59*, 2602.

Properties are calculated of samples of short poly(L-alanine) and polyglycine chains generated by the Monte Carlo method. The average square of the end-to-end distance $<r^2>$ and the average end-to-end vector are very close to the values calculated according to the method of *Flory*. The average fourth power of the end-to-end distance $<r^4>$ is compared with that calculated for a Kratky-Porod type wormlike chain, and the agreement is found to be good for all chain length, *i.e.*, including those for which the distribution of r^2 is decidedly non-Gaussian.

N 052 "Helix-Coil Stability Constants for the Naturally Occurring Amino Acids in Water. VIII. Valine Parameters from Random Poly(hydroxypropylglutamine-co-L-valine) and Poly(hydroxybutylglutamine-co-L-valine)"
Alter, J. E.; Andreatta, R. H.; Taylor, G. T.; Scheraga, H. A. *Macromolecules* **1973**, *6*, 564.

The synthesis and characterization of water-soluble "random" copolymers containing L-valine with either N^5-(3-hydroxypropyl)-L-glutamine or N^5-(4-hydroxybutyl)-L-glutamine are described, and the thermally induced helix-coil transitions of these copolymers in water are studied. The incorporation of *L*-valine is found to decrease the helix content of the polymer at low temperatures and increase it at high temperatures. The *Zimm-Bragg* parameters σ and s for the helix-coil transition in poly(*L*-valine) in water are deduced from an analysis of the melting curves of the copolymers. The values of s, computed for σ = 1 × 10^{-4}, are: s = 0.85 (273 K), 0.93 (293 K), 1.00 (313 K), 1.06 (333 K).

N 053 "Conformational Properties of Poly(γ-hydroxy-L-proline) Based on
Rigid and Flexible Pyrrolidine Rings"
Ooi, T.; Clark, D. S.; Mattice, W. L. *Macromolecules* **1974**, *7*, 337.

Conformational energy maps are computed for the internal dipeptide unit in poly(γ-hydroxy-L-proline) containing planar *trans* peptide bonds. The conformational energy maps based on rigid pyrrolidine rings which have the conformation observed in the solid state exhibit one low-energy region at $\psi = 145° \pm 40°$ (using the convention in which $\phi,\psi = 180°,180°$ for the fully extended chain). The characteristic ratios for this geometry are much higher than the result obtained experimentally for poly(L-proline). Conformational energy maps based on flexible pyrrolidine rings contain a region of low energy near $\psi = -50°$ in addition to the low-energy region at $150° \pm 70°$. The characteristic ratio based on the opportunity for flexibility in the pyrrolidine rings of residues i and $i+1$ is close to the result obtained experimentally for poly(L-proline) in water. The hydroxyl group in the γ-hydroxy-L-proline residue decreases the flexibility of the pyrrolidine ring, leading to a smaller configurational entropy for the γ-hydroxy-L-proline residue than for the L-proline residue. This effect would lead to an increased thermal stability of the collagen triple helix when γ-hydroxy-L-proline is substituted for L-proline.

N 054 "Calculation of the Characteristic Ratio of Randomly Coiled Poly(L-proline)"
Tanaka, S.; Scheraga, H. A. *Macromolecules* **1975**, *8*, 623.

Allowing for rotation about the $C^\alpha - C$ bond (*i.e.*, variation of ψ) and for some degree of freedom about the peptide bond (*i.e.*, small variation of ω), the characteristic ratios of the form *I* (*cis*) and form *II* (*trans*) poly(L-proline) chain are calculated by a Monte Carlo method in which the conformational energies are used as weighting factors. The Monte Carlo method enabled short-range interactions (beyond those involved in a single residue) to be taken into account.

N 055 "Helix-Coil Stability Constants for the Naturally Occurring Amino Acids in Water. IX. Glutamic Acid Parameters from Random Poly(hydroxybutylglutamine-*co*-L-glutamic Acid)"
Maxfield, F. R.; Alter, J. E.; Taylor, J. E.; Scheraga, H. A. *Macromolecules* **1975**, *8*, 479.

The synthesis and characterization of water-soluble random copolymers containing L-glutamic acid with N^5-(4-hydroxybutyl)-L-glutamine and the thermally induced helix-coil transitions of these copolymers in water and in 0.1 N KCl are described. The incorporation of L-glutamic acid is found to increase the helix content of the polymer at low pH and to decrease it at high pH even though the presence of 0.1 N KCl effectively eliminated the difference between the electrostatic free energies of the helix and the coil. The *Zimm-Bragg* parameters σ and *s* for the helix-coil transition in poly(L-glutamic acid) in water and in 0.1 N KCl are deduced from analysis of the melting curves of the copolymers. The values of *s*, computed from the *Allegra* theory are (for: pH 2.3, $\sigma = 1.0 \times 10^{-2}$ / pH 8 (in 0.1 N KCl), $\sigma = 6 \times 10^{-4}$ / pH 8 (in water), $\sigma = 6 \cdot 10^{-4}$): *s* = 1.47/0.98 / 0.96 (273 K), 1.35/0.97/0.96 (293 K), 1.21/0.95/0.93 (313 K), 1.07/0.93/0.89 (333 K).

N 056 "The Effect of Neighboring Charges on the Helix Forming Ability of Charged Amino Acids in Proteins"
Maxfield, F. R.; Scheraga, H. A. *Macromolecules* **1975**, *8*, 491.

It has been found that the fraction of glutamic acid residues which are helical in proteins is larger than might be expected from the experimentally determined value of the helical stability constants of glutamic acid. In order to understand this difference, the effect of neighboring charged side chains on the glutamic acid residues in proteins of known structure is examined. It is found that a positively charged side chain four residues away from a glutamic acid greatly enhances its probability to be helical. Similar results are obtained for aspartic acid, lysine, arginine, and histidine.

N 057 "Investigation of the Conformation of Four Tetrapeptides by Nuclear Magnetic Resonance and Circular Dichroism Spectroscopy, and Conformational Energy Calculations"
Howard, J. C.; Ali, A.; Scheraga, H. A.; Momany, F. A. *Macromolecules* **1975**, *8*, 607.

Proton NMR and circular dichroism studies are carried out on aqueous solutions of the tetrapeptide Asp-Lys-Thr-Gly and of its sequence variants Gly-Thr-Asp-Lys, Asp-Lys-Gly-Thr, and Lys-Thr-Gly-Asp; the N and C termini of all tetrapeptides are blocked with CH_3CO and $NHCH_3$ groups, respectively. The spectroscopic data suggest that bend conformations may exist, to some extent, among the distributions of conformations in the first, third, and fourth, but not in the second, tetrapeptide. The result is consistent with empirical probabilities for the prediction of bend conformations in proteins. Conformational energy calculations on these four tetrapeptides support the indications from the experimental data. It thus appears that, because of short-range interactions, the tendency toward bend formation exists in short peptides, provided that both the composition and amino acid sequence are energetically favourable for bend formation.

N 058 "Moments and Distribution Functions for Polypeptide Chains. Poly(L-alanine)"
Conrad, J. C.; Flory, P. J. *Macromolecules* **1976**, *9*, 41.

Statistical mechanical averages of vectors and tensors characterizing the configurations of polypeptides are calculated for poly(L-alanines) of x_u = 2 to 400 peptide units. These quantities are expressed in the reference frame of the first peptide unit, the X axis being situated along the virtual bond, the Y axis in the plane of the peptide unit. The persistence vector $\mathbf{a} \equiv <\mathbf{r}>$ converges rapidly with the chain length to its limit $\mathbf{a}_\infty$ which lies virtually in the XZ plane. Configurational averages of Cartesian tensors up to the sixth rank formed from the displacement vector $\rho = \mathbf{r} - \mathbf{a}$ are computed. For $x_u > 50$ the even moments of fourth and sixth rank formed from the reduced vector ρ for the real chain are well represented by the freely jointed chain with 21.7 virtual bonds equivalent to one of the model. The angles η and ξ formed by the virtual bond vector and the $N-C^\alpha$ and $C^\alpha-C$ skeletal bonds are 22.2° and 13.2°, respectively. The length of the virtual bond vector l_p of the *trans* peptide unit is 380 pm. All calculations are performed for a temperature of 298 K. The averaged transformation matrix computed for the PLA chain is:

$$<T_{L\text{-Ala}}> = \begin{bmatrix} 0.50 & 0.18 & 0.61 \\ -0.035 & -0.62 & 0.19 \\ 0.66 & -0.21 & -0.32 \end{bmatrix}$$

N 059 "Conformational Properties of the Complexes Formed by Proteins and Sodium Dodecyl Sulfate"
Mattice, W. L.; Riser, J. M.; Clark, D. S. *Biochemistry* **1976**, *15*, 4264.

Circular dichroism spectra are obtainied for albumin, α-chymotrypsinogen, collagen, concanavalin A, elastase, hemoglobin, histone f_2b, α-lactalbumin, lactate dehydrogenase, β-lactoglobulin, lysozyme, myoglobin, papain, ribonuclease A, and thermolysin in the presence of sodium dodecyl sulfate and dithiothreitol. While all spectra have the shape anticipated for a mixture of random coil and α-helix, the intensities differ markedly. The variation in the circular dichroism can be quantitatively explained by a model which assumes that the arginyl, histidyl, and lysyl residues have an enhanced probability of propagating a helical segment in the presence of the detergent. The model also permits the computation of dimensional properties (unperturbed end-to-end distance and radius of gyration) for polypeptides of known amino acid sequence. Such computations are performed for 67 proteins. The computed dimensions are compatible with experimental values and with the molecular weight dependence of the transport properties of the complexes. Furthermore the model can account for the abnormal transport properties of the sodium dodecyl sulfate complexes formed by ribonuclease A, collagen fragments, and histones f_2a_1, f_2a_2, f_2b, and f_3. Even though some of the protein-sodium dodecyl sulfate complexes have helical contents as high as 50%, their overall conformation more closely approximates that of a random coil rather than a rod.

Initial values for σ and s:

Amino acid residue	σ	s
Pro	0	0
Gly	0.00001	0.60
Ser,Cys,Thr	0.0001	0.76
Val,Ilu	0.0001	0.95
Glu,Arg,Asn,Asp,Gln,Lys	0.0006	0.96
Ala	0.0008	1.06
Phe,His,Try,Tyr	0.0018	1.07
Leu,Met	0.003	1.12

Configuration partition function: $Z = U_i^n$

$$U_i = \begin{bmatrix} 1 & \sigma s \\ 1 & s \end{bmatrix}_j \quad U_1 = \begin{bmatrix} 1 & 0 \end{bmatrix} \quad U_n = \begin{bmatrix} 1 \\ 1 \end{bmatrix}$$

Fraction Helix: $f = z^{-1}(n-2)^{-1} U'^n_i$

$$U'_i = \begin{bmatrix} 1 & \sigma s & 0 & \sigma s \\ 1 & s & 0 & s \\ 0 & 0 & 1 & \sigma s \\ 0 & 0 & 1 & s \end{bmatrix}_j \quad U'_1 = \begin{bmatrix} 1 & 0 & 0 & 0 \end{bmatrix} \quad U'_n = \begin{bmatrix} 0 \\ 0 \\ 1 \\ 1 \end{bmatrix}$$

Transformation matrices:

$$\langle T^{coil}_{Ala} \rangle = \begin{bmatrix} 0.51 & 0.20 & 0.59 \\ -0.046 & -0.61 & 0.21 \\ 0.65 & -0.23 & -0.30 \end{bmatrix} \quad \langle T^{coil}_{Gly} \rangle = \begin{bmatrix} 0.36 & -0.077 & 0 \\ -0.092 & -0.037 & 0 \\ 0 & 0 & -0.012 \end{bmatrix} \quad \langle T^{coil}_{Pro} \rangle = \begin{bmatrix} 0.593 & 0.316 & -0.001 \\ 0.117 & -0.503 & -0.003 \\ 0.001 & 0.007 & -0.315 \end{bmatrix}$$

$$\langle T^{helix} \rangle = \begin{bmatrix} 0.020 & -0.424 & -0.905 \\ -0.425 & 0.816 & -0.391 \\ 0.905 & 0.393 & -0.164 \end{bmatrix} \quad \langle T^{coil}_{Ala-Pro} \rangle = \begin{bmatrix} 0.50 & 0.14 & 0.81 \\ -0.031 & -0.73 & 0.098 \\ 0.67 & -0.084 & -0.32 \end{bmatrix} \quad \langle T^{coil}_{Gly-Pro} \rangle = \begin{bmatrix} 0.47 & 0.20 & 0 \\ 0.019 & -0.00021 & 0 \\ 0 & 0 & 0.038 \end{bmatrix} \quad \langle T^{coil}_{Pro-Pro} \rangle = \begin{bmatrix} 0.423 & 0.474 & 0.548 \\ -0.589 & -0.221 & 0.582 \\ 0.640 & -0.667 & 0.186 \end{bmatrix}$$

N 060   "Why Cyclic Peptides? Complementary Approaches to Conformations"
Deber, C. M.; Madison, V.; Blout, E. R. *Acc. Chem. Res.* **1976**, *9*, 106.

Approaches are described in this paper to the deduction of peptide conformation in solution by the use of three techniques: ^{13}C nuclear magnetic resonance, conformational energy calculations, and circular dichroism. Sections of this paper illustrate the types of conformational information that can be derived from each of the individual methods for the synthetic and naturally occurring cyclic peptides. For complete conformational analysis, however, a combination of techniques is usually necessary.

N 061   "Helix-Coil Stability Constants for the Naturally Occurring Amino Acids in Water. X. Tyrosine Parameters from Random Poly(hydroxybutylglutamine-*co*-L-tyrosine)"
Scheule, R. K.; Cardinaux, F.; Taylor, G. T.; Scheraga, H. A. *Macromolecules* **1976**, *9*, 23.

The synthesis and characterization of water-soluble random copolymers containing L-tyrosine with N^5-(3-hydroxypropyl)-L-glutamine are described, and the thermally induced helix-coil transitions of these copolymers in water are studied. The incorporation of L-tyrosine is found to increase the helix content of the polymers at all temperatures. The *Zimm-Bragg* parameters σ and *s* for the helix-coil transition in poly(L-tyrosine) in water are deduced from an analysis of the melting curves of the copolymers. The large value of σ indicates that, in water, tyrosine has a tendency to promote helix-coil boundaries at all temperatures; the values of *s* indicate that this residue enhances helix growth at low temperature and reduces it at high temperatures. Values of the *Zimm-Bragg* parameters *s* for poly(L-tyrosine) in water are (computed with the *Allegra* theory with σ = 66 × 10^{-4}): *s* = 1.12 (273 K), 1.07 (283 K), 1.02 (293 K), 0.96 (303 K), 0.88 (313 K), 0.77 (323 K), and 0.65 (333 K).

N 062   "Statistical Mechanical Treatment of Protein Conformation. I. Conformational Properties of Amino Acids in Proteins"
Tanaka, S.; Scheraga, H. A. *Macromolecules* **1976**, *9*, 142.

A statistical mechanical (one-dimensional Ising model) treatment, based on the dominance of short range interactions, is developed in this series of papers; it is intended as an improvement over empirical prediction schemes for obtaining approximate initial conformations of proteins. In the first paper, the statistical weights for a two-state model (α-helical and other conformations) and for a three-state model (α-helical, extended, and other conformations) are evaluated from x-ray data on 16 native proteins. The method for evaluating the statistical weights is presented. Asymmetric α-helical nucleation parameters are also evaluated for the 20 naturally occurring amino acids. On the basis of these statistical weights, the conformational properties of the 20 naturally occurring amino acids are discussed. The statistical weights evaluated from x-ray data are also discussed in comparison with experimental results on the helix-coil transition in poly(amino acids) in solution. The predominant role of short-range interactions, and some possible long-range effects in determining the statistical weights, are discussed in conjunction with the mechanism of protein folding.

N 063   "Statistical Mechanical Treatment of Protein Conformation. II. A Three-State Model for Specific-Sequence Copolymers of Amino Acids"
Tanaka, S.; Scheraga, H. A. *Macromolecules* **1976**, *9*, 159.

A one-dimensional three-state Ising model [involving α-helical (α), extended (ε), and coil (or other) (c) states] for specific-sequence copolymers of amino acids is formulated on order to treat the conformational states of proteins. This model involves four parameters ($\omega_{h,i}$, $v_{h,i}$, $v_{\varepsilon,i}$, and $v_{c,i}$), and required a 4 × 4 matrix for generating statistical weights. Some problems in applying this model to a specific-sequence copolymer of amino acids are discussed. A nearest-neighbor approximation for treating this three-state model is also formulated; it requires a 3 × 3 matrix, in which the same four parameters appear, but only three parameters ($\omega_h{}^*$, $v_h{}^*$, and $v_\varepsilon{}^*$) are required if relative statistical weights are used. The relationship between the present three state model and models of the helix-coil transition is discussed. Then, the three-state model (3 × 3 matrix treatment) is incorporated into an earlier (*Tanaka-Scheraga*) model of helix-coil transition, in which asymmetric nucleation of helical sequences is taken into account. A method for calculating molecular averages and conformational-sequence probabilities is described. In this paper, these calculations are performed with the nearest-neighbor model, and without the feature of asymmetric nucleation. Finally, it is indicated how the three-state model and the methods for computing P(i/n/{ρ}) can be applied to predict protein conformation.

"Statistical Mechanical Treatment of Protein Conformation. III. Prediction of Protein Conformation Based on a Three-State Model"
Tanaka, S.; Scheraga, H. A. *Macromolecules* 1976, *9*, 168.

The method proposed for the evaluation of statistical weights in paper I (**N 062**), and the three-state model formulated in paper II (**N 063**), are used to develop a procedure to predict the backbone conformations of proteins, based on the concept of the predominant role played by the short-range interactions in determining protein conformation. Conformational probability profiles, in which the probabilities of formation of three consecutive α-helical conformations (triad) and of four consecutive extended conformations (tetrad) are defined relative to their average values over the whole molecule, are calculated for 19 proteins, of which 16 had been used in paper I to evaluate the set of statistical weights of the 20 naturally occurring amino acids. By comparing these conformational probability profiles to experimental x-ray observations, the following results are obtained: 80% of the α-helical regions and 72% of the extended conformational regions are predicted correctly for the 19 proteins. The percentage of residues predicted correctly is in the range of 53 to 90% for the α-helical conformation and in the range of 63 to 88% for the extended conformation for the 19 proteins in the two-state models [α-helical (α) and other (c) states, and extended (ε) and other (c) states]. In the three state model, the percentage of residues predicted correctly is in the range of 47 to 77% for 19 proteins.

In **N 063**, a three state model is formulated that is applicable to homopolymers and to specific-sequence copolymers of amino acids. In order to calculate the partition function and molecular averages, a 4×4 statistical weight matrix for the ith residue is formulated to correlate the states of residues $i-1$, i, and $i+1$ of the polymer chain: [rows and columns, corresponding to the states of bonds $i-1/i$ and $i/i+1$, respectively, are: $c / c \cup c \cup \alpha$, $\varepsilon / c \cup \varepsilon \cup \alpha$, $\alpha / c \cup \varepsilon$, α / α.]

$$W_i = \begin{bmatrix} v_c & v_\varepsilon & v_h & v_h \\ v_c & v_\varepsilon & v_h & v_h \\ v_c & v_\varepsilon & 0 & 0 \\ 0 & 0 & v_h & \omega_h \end{bmatrix}_j$$

An approximate nearest-neighbor interaction model, in which the states of residues $i-1$ and i are correlated, is also introduced; a 3×3 matrix is required in this model. Both the 4×4 and 3×3 matrix formulations involve four parameters ($v_{c,i}$, $\omega_{h,i}$, $v_{\varepsilon,i}$, and $v_{h,i}$). By choosing the c state as a standard state, the number of statistical weights is reduced to three (relative to that of the c state), viz. $\omega_{h,i}^*$, $v_{\varepsilon,i}^*$, and $v_{h,i}^*$. Hence, the 3×3 matrix for the nearest-neighbor interaction model can be written as: Here, the subscript i is placed on the matrix instead of on the individual statistical weights, and j designates the species of amino acid in the ith residue of the protein.

$$W_i = \begin{bmatrix} 1 & v_{\varepsilon,j}^* & (v_{h,j}^*)^2/\omega_{h,j}^* \\ 1 & v_{\varepsilon,j}^* & (v_{h,j}^*)^2/\omega_{h,j}^* \\ 1 & v_{\varepsilon,j}^* & \omega_{h,j}^* \end{bmatrix}_j$$

Definition of the eight helical states of the ith residue:

Conformational state (ϕ_i, ψ_i)	Hydrogen bond on C_iO_i	N_iH_i	State
c	No	No	1
c	No	Yes	2
c	Yes	No	3
c	Yes	Yes	4
h	No	No	5
h	No	Yes	6
h	Yes	No	7
h	Yes	Yes	8

Relative statistical weights of the ith amino acid used in the present scheme:

Amino acid j	Relative statistical weights			Amino acid j	Relative statistical weights		
	$\omega_{h,j}^*$	$v_{\varepsilon,j}^*$	$v_{h,j}^*$		$\omega_{h,j}^*$	$v_{\varepsilon,j}^*$	$v_{h,j}^*$
Ala	1.549	0.500	0.030	Leu	1.343	0.702	0.034
Arg	0.468	0.298	0.026	Lys	0.726	0.253	0.025
Asn	0.304	0.152	0.023	Met	1.000	0.667	0.033
Asp	0.481	0.130	0.023	Phe	0.727	0.318	0.026
Cys	0.444	0.556	0.031	Pro	0.315	0.333	0.027
Gln	0.795	0.432	0.029	Ser	0.336	0.234	0.025
Glu	1.188	0.250	0.025	Thr	0.488	0.464	0.029
Gly	0.226	0.195	0.024	Trp	1.105	0.421	0.028
His	0.535	0.233	0.025	Tyr	0.262	0.410	0.028
Ile	0.891	0.587	0.032	Val	1.028	0.789	0.036

N 065 "Statistical Mechanical Treatment of Protein Conformation. 4. A Four-State Model for Specific-Sequence Copolymers of Amino Acids"
Tanaka, S.; Scheraga, H. A. *Macromolecules* **1976**, *9*, 812.

One-dimensional short-range interaction models for specific-sequence copolymers of amino acids are developed in this series of papers. In this paper, the earlier three-state model [involving helical (h), extended (ε), and coil (or other) (c) states] is extended to a four state model by preserving the h and ε states, introducing the chain-reversal state (R and S), and redefining the c state. This model involves six parameters (ω_h, v_h, v_ε, v_R, v_S, and v_c) and requires 6 × 6 statistical weight matrix. A nearest-neighbor approximation of the four-state model is also formulated; it requires a 5 × 5 matrix, involving the same six parameters. By expressing the statistical weights relative to that of the ε state, only five parameters (ω_h^*, v_h^*, v_R^*, v_S^*, and v_c^*) are required in both the 6 × 6 and 5 × 5 matrices. These statistical weights, and the four-state model, are used to develop a procedure to predict the backbone conformations of proteins. Since the prediction of helical and extended conformations is carried out by the procedure described in **N 062** to **N 064**, particular attention is focussed on chain-reversal conformations.

Statistical weights for the four state model:

| Amino acid j | Relative statistical weights | | | | | Amino acid j | Relative statistical weights | | | | |
	$\omega_{h,j}^*$	$v_{h,j}^*$	$v_{R,j}^*$	$v_{S,j}^*$	$v_{c,j}^*$		$\omega_{h,j}^*$	$v_{h,j}^*$	$v_{R,j}^*$	$v_{S,j}^*$	$v_{c,j}^*$
Ala	1.683	0.077	0.505	0.293	0.356	Ser	0.608	0.031	0.442	0.419	0.331
Arg	0.889	0.067	0.322	0.322	0.244	Thr	0.528	0.104	0.358	0.292	0.264
Asn	0.800	0.033	0.317	0.900	0.667	Trp	0.815	0.111	0.167	0.426	0.259
Asp	1.705	0.159	0.784	1.148	1.045	Tyr	0.449	0.038	0.237	0.378	0.205
Cys	0.537	0.024	0.207	0.402	0.268	Val	0.663	0.018	0.156	0.181	0.147
Gln	0.860	0.100	0.420	0.300	0.300						
Glu	2.419	0.070	0.477	0.407	0.442						
Gly	0.638	0.032	0.479	0.915	1.202						
His	1.194	0.161	0.290	0.645	0.290						
Ile	0.747	0.061	0.131	0.121	0.212						
Leu	1.069	0.069	0.203	0.246	0.198						
Lys	1.553	0.105	0.526	0.539	0.487						
Met	1.083	0.000	0.271	0.187	0.292						
Phe	1.020	0.059	0.157	0.471	0.235						
Pro	0.400	0.108	0.738	0.092	0.138						

$$W_i = \begin{bmatrix} v_c^* & v_h^{*2}/\omega_h^* & 1 & v_R^* & 0 \\ v_c^* & \omega_h^* & 1 & v_R^* & 0 \\ v_c^* & v_h^{*2}/\omega_h^* & 1 & v_R^* & 0 \\ 0 & 0 & 0 & 0 & v_S^* \\ v_c^* & v_h^{*2}/\omega_h^* & 1 & v_R^* & 0 \end{bmatrix}_i$$

Rows and columns, corresponding to bonds i-1 and i, resp., are given as: c, h, ε, R, S.

N 066 "Helix-Coil Stability Constants for the Naturally Occurring Amino Acids in Water. 11. Lysine Parameters from Random Poly(hydroxybutylglutamine-*co*-L-lysine)"
Dygert, M. K.; Taylor, G. T.; Cardinaux, F.; Scheraga, H. A. *Macromolecules* **1976**, *9*, 794.

The synthesis and characterization of water-soluble random copolymers containing L-lysine with N^5-(4-hydroxybutyl)-L-glutamine, and the thermally induced helix-coil transitions of these copolymers in water, are described. The incorporation of L-lysine is found to decrease the helix content of the polymers at neutral pH. The *Zimm-Bragg* parameters σ and *s* for the helix-coil transition in poly(L-lysine) in water are deduced from an analysis of the melting curves. The computed values of *s* indicate that, in the temperature range of 273 to 333 K, lysine has a tendency to destabilize helical sequences. Values of the *Zimm-Bragg* parameters *s* for poly(L-lysine) in water or 0.1 N KCl at neutral pH are (computed according to *Lifson/Allegra*; σ = 1.0 × 10^{-4}): *s* = 0.832/0.857 (273 K), 0.899/0.909 (283 K), 0.934/0.939 (293 K), 0.944/0.947 (303 K), 0.937/0.939 (313 K), 0.924/0.926 (323 K), and 0.909/0.911 (333 K).

N 067   "Unperturbed Dimensions for Homopolypeptides and Sequential Copolypeptides Cross-Linked via a Disulfide Bond"
Mattice, W. L. *Macromolecules* **1977**, *10*, 511.

RIS theory, in the form appropriate for branched molecules, is used to calculate the mean-square unperturbed radius of gyration, $<s^2>_o$, for cross-linked polyglycine, poly(L-alanine), poly(L-proline), poly(L-alanyl-D-alanine), poly(L-prolyl-L-prolylglycine), poly(L-prolyl-L-alanylglycine), poly(glycyl-L-alanyl-L-proline), and poly(L-alanyl-L-alanylglycine). The central amino acid residue in each polypeptide chain is replaced by the L-cysteinyl residue involved in cross-link formation. Each cross-linked molecule is considered to contain two trifunctional branch points, the α-carbon atoms of the two L-cysteinyl residue. Random flight statistics provide a poor estimate for g, defined as the ratio of $<s^2>_o$ for branched and linear polypeptides containing the same flight statistics and RIS theory merge as the molecular weight becomes infinite. Deviations of g from its random flight value correlate with the size of the characteristic ratio, $<s^2>_o / n_p l_p^2$, for the linear polypeptides. The number of peptide bonds is n_p, and l_p denotes the distance between neighboring α-carbon atoms. Random flight statistics perform better in estimating the change in $<s^2>_o$ accompanying the cross-linking of the two polypeptide chains than it does in the estimation of g.

N 068   "Unperturbed Dimensions of Disordered Proteins Containing an Interchain Disulfide Cross-Link"
Mattice, W. L. *Macromolecules* **1977**, *10*, 516.

Mean-square radii of gyration, $<s^2>_o$, are calculated for several proteins crosslinked via an interchain disulfide bond. Thirty different polypeptide chains are used. Characteristic ratios tend to be smaller for cross-linked proteins than for the uncross-linked chains, although exceptions to this generalization do exist. Random flight statistics tend to overestimate the value of g, defined as the ratio of $<s^2>_o$ for the cross-linked protein to $<s^2>_o$ for an analogous linear polypeptide chain containing the same number of amino acid residues. The parameter f_i, defined as the ratio of the $<s^2>_o$ for the ith uncross-linked polypeptide chain and the cross-linked protein, is usually more accurately estimated by random flight statistics than g. When the cross-link connects two chains of identical amino acid sequence, the values of f_i obtained via random flight statistics are within 6% of those provided by RIS theory.

The following model requires two pages.
Therefore, this side has been left blank

"Rotational Isomeric State Treatment of the Cystine Residue. Configuration Partition Function and Its Relationship to the Optical Activity Exhibited by the Disulfide Bond"

Mattice, W. L. *J. Am. Chem. Soc.* **1977**, *99*, 2324.

The configuration partition function for two polypeptide chains cross-linked via a single cystinyl residue is formulated. This objective is achieved using RIS theory, in the form appropriate for branched molecules, to combine the configuration partition function for the cystinyl residue with appropriate representations for the configuration partition function of uncross-linked polypeptide chains. Rotational states, as well as preliminary estimates of the corresponding conformational energies, are obtained from consideration of semiempirical energy computations. These sequences -Cys-X- and -Cys-Pro- (X ≠ Pro) lead to different results. The *a priori* probabilities computed for the two rotational states of the disulfide bond are found to be in excellent agreement with the sign, magnitude, and temperature dependence of the optical activity exhibited by the lowest energy electronic transition in simple derivatives of L-cystine.

The C–S and S–S bond lengths are assigned the values 187 and 204 pm, respectively. The C–C–S bond angle is 113°, and the C–S–S bond angle is 107°. All other bond lengths and bond angles are identical with those used by *Brant et al.* (*J. Mol. Biol.* **1967**, *23*, 47). Torsional potentials for rotation about the N–C$^\alpha$ and C$^\alpha$–C bonds are also taken from *Brant et al.*, while the torsional potential for the C$^\alpha$–C$^\beta$ bond is that of *Abe et al.* (*J. Am. Chem. Soc.* **1966**, *88*, 631). The torsional potential for the C$^\beta$–S bond is threefold, with minima at 180° and ± 60°. For the S–S bond, the torsional potential is that used by *Allinger et al.* (*J. Am. Chem. Soc.* **1976**, *98*, 2741). Peptide bonds are maintained in the planar *trans* conformation.

$$U_i^1 = U_j^2 = U_j^4 = U_j^5 = [1], \; i < n_1, \; j > 1 \quad U_{n_1}^1 = [w_1 \; w_2 \; w_3 \; w_4] \quad U_1^2 = col\,(1,1,1,1)$$

$$U_1^3 = \begin{bmatrix} 1 & 1 & s_3 & 0 & 0 & 0 & 0 & 0 & 0 & 0 & 0 \\ 0 & 0 & 0 & 1 & 1 & s_3 & 0 & 0 & 0 & 0 & 0 \\ 0 & 0 & 0 & 0 & 0 & 0 & 1 & 1 & s_1 s_3 & 0 & 0 \\ 0 & 0 & 0 & 0 & 0 & 0 & 0 & 0 & 0 & 1 & 1 & s_2 s_3 \end{bmatrix} \quad U_3^3 = \begin{bmatrix} 1 & 1 \\ 1 & s_{16} \\ s_{16} & 1 \end{bmatrix} \quad U_4^3 = \begin{bmatrix} 1 & 1 & s_{16} \\ 1 & s_{16} & 1 \end{bmatrix}$$

$$U_2^3 = \begin{bmatrix} 1 & s_4 s_5 & s_7 \\ 1 & 1 & s_9 \\ 1 & s_{11} & s_{12} \\ 1 & s_4 & s_7 \\ 1 & s_8 & s_{10} \\ 1 & 1 & s_{13} \\ 1 & s_4 s_6 & s_7 \\ 1 & 1 & s_9 \\ 1 & s_{11} & s_{14} \\ 1 & s_4 & s_7 \\ 1 & 1 & s_9 \\ 1 & s_{11} & s_{15} \end{bmatrix}$$

$$U_5^3 = \begin{bmatrix} w_1 & w_1 & s_3 w_1 & w_2 & w_2 & s_3 w_2 & w_3 & w_3 & s_1 s_3 w_3 & w_4 & w_4 & s_2 s_3 w_4 \\ s_4 s_5 w_1 & w_1 & s_3 s_{11} w_1 & s_4 w_2 & s_8 w_2 & s_3 w_2 & s_4 s_6 w_3 & w_3 & s_1 s_3 s_{11} w_3 & s_4 w_4 & w_4 & s s_i b_2 s_3 s_{11} w_4 \\ s_7 w_1 & s_9 w_1 & s_3 s_{12} w_1 & s_7 w_2 & s_{10} w_2 & s_3 s_{13} w_2 & s_7 w_3 & s_9 w_3 & s_1 s_3 s_{14} w_3 & s_7 w_4 & s_9 w_4 & s_2 s_3 s_{15} w_4 \end{bmatrix}$$

$$U_1^4 = U_1^5 = U_1^4 \ominus U_1^5 = col\,(1,1,1,1,1,1,1,1,1,1,1)$$

For U^1_{n1} : columns : $\beta_1, \beta_2, \beta_3, \alpha_1$

For U^3_1 : rows : $\beta_1, \beta_2, \beta_3, \alpha_1$
columns : $\beta_1 t, \beta_1 g^+, \beta_1 g^-, \beta_2 t, \beta_2 g^+, \beta_2 g^-,$
$\beta_3 t, \beta_3 g^+, \beta_3 g^-, \alpha_1 t, \alpha_1 g^+, \alpha_1 g^-$

For U^3_2 : rows: ($\equiv$ columns of U^3_1)
columns: t, g^+, g^-

For U^3_5 : rows : t, g^+, g^-
columns: ($\equiv$ columns of U^3_1)

For U^3_3 : rows : t, g^+, g^-
columns: g^+, g^-

For U^3_4 : rows : g^+, g^-
columns: t, g^+, g^-

Minima in the L-alanyl residue conformational energy map:

energy [kJ mol^{-1}]	ϕ [°]	ψ [°]	$\langle\phi\rangle$ [°]	$\langle\psi\rangle$ [°]	statistical weight
0.00	100	330	94	328	0.415
0.88	100	280	92	275	0.173
1.13	30	330	26	312	0.363
3.81	110	130	107	128	0.033
5.57	10	120	15	119	0.010
9.80	230	240	236	248	0.004
9.96	250	350	247	348	0.002

States used for ϕ, ψ:

designation	$\langle\phi\rangle$ [°]	$\langle\psi\rangle$ [°]	statistical weight
β_1	94	328	w_1 (0.418)
β_2	26	312	w_2 (0.365)
β_3	92	275	w_3 (0.174)
α_1	86	126	w_4 (0.043)

Energies associated with the s_i:

statist. weights	estd. energy [kJ mol^{-1}]	statist. weights	estd. energy [kJ mol^{-1}]
s_1	2.5	s_9	0.8
s_2	8	s_{10}	4
s_3	0.8	s_{11}	8
s_4	0.8	s_{12}	2.1
s_5	2.9	s_{13}	6.3
s_6	4	s_{14}	20
s_7	−0.8	s_{15}	13
s_8	0.8	s_{16}	−0.4

States used for ϕ, ψ in Cys (Pro):

designation	energy [kJ mol^{-1}]	ϕ [°]	ψ [°]	$\langle\phi\rangle$ [°]	$\langle\psi\rangle$ [°]	statist. weight
β_1	0.00	100	320	97	311	0.49 (w_1)
β_2	0.59	20	310-320	30	310	0.51 (w_2)
	11.14	240	260	241	276	

Changes required for Cys (Pro):

$w_1 = 0.49$
$w_2 = 0.51$
$w_3 = w_4 = 0$
$E_{s5} = E_{s7} = E_{s8} = 0.0$
$E_{s12} = 6.3$ kJ mol^{-1}
$\langle\phi\rangle, \langle\psi\rangle = 97°, 311°$ for β_1
$\langle\phi\rangle, \langle\psi\rangle = 30°, 310°$ for β_2

N 070 "Helix-Coil Stability Constants for the Naturally Occurring Amino Acids in Water. XIV. Methionine Parameters from Random Poly(hydroxypropylglutamine-L-methionine)"
Hill, D. J. T.; Cardinaux, F.; Scheraga, H. A. *Biopolymers* **1977**, *16*, 2447.

Water-soluble, random copolymers containing L-methionine and N^5-(3-hydroxypropyl)-L-glutamine are prepared, fractionated, and characterized. The thermally induced helix-coil transitions of these copolymers in water are investigated, and it is found that incorporation of L-methionine increases the helix content of the polymers at all temperatures in the range of 273- 333 K. The *Zimm-Bragg* parameters σ and s for the helix-coil transition in poly(L-methionine) in water are deduced from an analysis of the melting curves. Values of the *Zimm-Bragg* parameters s for poly(L-methionine) in water are computed according to *Lifson* ($\sigma = 4.8 \times 10^{-3}$) / *Allegra* ($5.4 \times 10^{-3}$): s = 1.22/1.28 (273 K), 1.20/1.25 (283 K), 1.17/1.20 (293 K), 1.12/1.15 (303 K), 1.09/1.10 (313 K), 1.04/1.05 (323 K), and 1.01/1.00 (333 K).

N 071 "Helix-Coil Stability Constants for the Naturally Occurring Amino Acids in Water. 15. Arginine Parameters from Random Poly(hydroxybutylglutamine-*co*-L-arginine)"
Konishi, Y.; van Nispen, J. W.; Davenport, G.; Scheraga, H. A. *Macromolecules* **1977**, *10*, 1264.

Water-soluble, random copolymers containing L-arginine and N^5-(4-hydroxybutyl)-L-glutamine are prepared by copolymerization of the *N*-carboxy-α-amino acid anhydrides of N^δ-tert.-butoxycarbonyl-L-ornithine and γ-benzyl L-glutamate, followed by aminolysis with 4-amino-1-butanol, by removal of the *tert.*-butyloxycarbonyl protecting group, and by treatment with *O*-methylisourea. The copolymers are fractionated and characterized, and the thermally induced helix-coil transitions of these copolymers in water are studied at neutral pH in the presence and in the absence of KCl. The *Zimm-Bragg* parameters σ and s for the helix-coil transition in poly(L-arginine) in water are deduced from an analysis of the melting curves. The computed values of s indicate that L-arginine is a weak helix-making residue at low temperature and a weak helix-breaking residue at high temperature in aqueous solution. The results are found to be in good agreement with those obtained earlier in conformational analyses of arginyl residues in proteins. Values of the *Zimm-Bragg* parameters s for poly(L-arginine) in water or 0.2 N KCl at neutral pH are computed according to *Lifson* ($\sigma = 0.1 \times 10^{-5}$) / *Allegra* ($\sigma = 0.1 \times 10^{-5}$) / *Lifson* ($\sigma = 6.0 \times 10^{-5}$) / *Allegra* ($\sigma = 1.0 \times 10^{-5}$): s = 1.031/1.031/1.028/1.026 (273 K), 1.042/1.043/1.034/1.033 (293 K), 1.002/1.002/0.993/0.994 (313 K), and 0.929/0.930/0.938/0.940 (333 K).

N 072 "Helix-Coil Stability Constants for the Naturally Occurring Amino Acids in Water. 16. Aspartic Acid Parameters from Random Poly(hydroxybutylglutamine-*co*-L-aspartic acid)"
Kobayashi, Y.; Cardinaux, F.; Zweifel, B. O.; Scheraga, H. A. *Macromolecules* **1977**, *10*, 1271.

The synthesis and characterization of water-soluble, random copolymers containing L-aspartic acid with N^5-(4-hydroxybutyl)-L-glutamine, and the thermally induced helix-coil transitions of these copolymers in water and in 0.1 N KCl, are described. The incorporation of L-aspartic acid is found to decrease the helix content of the polymer at both high and low pH, in water and also in 0.1 N KCl. The *Zimm-Bragg* parameters σ and s for the helix-coil transition in poly(L-aspartic acid) in water and in 0.1 N KCl are deduced from an analysis of the melting curves. Corrections are made for the presence of a small amount of racemized aspartic acid, using data from random copolymers containing D-aspartic acid as the guest residue.

Values of the Zimm-Bragg parameter σ for poly(aspartic acid):

	Solvent conditions	Lifson	Allegra
L-Aspartic acid	pH 1.5, 0.1 N KCl	0.0166	0.0125
	pH 8.0, 0.1 N KCl	0.0105	0.0050
	pH 8.0, water	0.0070	0.0030
D-Aspartic acid	pH 1.5, 0.1 N KCl	0.0028	0.0010
	pH 8.0, 0.1 N KCl	0.0052	0.00105
	pH 8.0, water	0.0100	0.0040

Values of the *Zimm-Bragg* parameters s for pure poly(L-aspartic acid) at pH = 1.5 ($\sigma = 0.021$) / pH = 8 in 0.1 N KCl ($\sigma = 0.0070$) are: s = 0.83/0.74 (275 K), 0.78/0.68 (293 K), 0.70/0.59 (313 K), and 0.61/0.49 (333 K).

N 073   "Helix-Coil Stability Constants for the Naturally Occurring Amino Acids in Water. XII. Asparagine Parameters from Random Poly(hydroxybutylglutamine-co-L-asparagine)"
Matheson, R. R., Jr.; Nemenoff, R. A.; Cardinaux, F.; Scheraga, H. A. *Biopolymers* **1977**, *16*, 1567.

Water-soluble, random copolymers containing L-asparagine and N^5-(3-hydroxybutyl)-L-glutamine are prepared and characterized. The thermally induced helix-coil transitions of these copolymers in water are investigated. The incorporation of L-asparagine is found to decrease the helix content of the polymers in water at all temperatures. The *Zimm-Bragg* parameters σ and s for the helix-coil transition in poly(L-asparagine) in water are deduced from an analysis of the melting curves. Values of the *Zimm-Bragg* parameters s for poly(L-asparagine) in water are computed according to *Allegra* theory ($\sigma = 9.5 \times 10^{-6}$): s = 0.738 (273 K), 0.784 (293 K), 0.820 (313 K), and 0.847 (333 K).

N 074   "Helix-Coil Stability Constants for the Naturally Occurring Amino Acids in Water. XIII. The Presence of By-Products in Amino-Acid Analysis of Copolymers and Their Effect on the Guest Parameters; Recomputed Values of σ and s for L-Serine"
Van Nispen, J. W.; Hill, D. J.; Scheraga, H. A. *Biopolymers* **1977**, *16*, 1587.

Previously reported procedures for processing the amino acids from 6 N hydrochloric acid hydrolysis of poly[N^5-(4-hydroxybutyl)-L-glutamine], poly[N^5-(3-hydroxypropyl)-L-glutamine], and several random copolymers derived from these, lead to the formation of spurious products. These are isolated and characterized as the γ-ester of glutamic acid and the hydroxyalkyl amine, and chloro-alkyl amine hydrochloride. The former reduces the observed values for glutamic acid, but the latter has no effect on them. A method is used to avoid formation of these artifacts in the amino-acid analysis. Of all the copolymers studied previously in this series, the compositions of only those containing L-serine are in error as a result of the formation of the γ-ester. A redetermination of the amino-acid compositions of the copolymer fractions studied earlier leads to slightly revides values for the *Zimm-Bragg* parameters σ and s of serine. Values of the *Zimm-Bragg* parameters s for poly(L-serine) in water, computed according to *Allegra* theory ($\sigma = 1 \times 10^{-5}$) are: s = 0.70 ± 0.1 (273 K), 0.76 ± 0.1 (293 K), 0.79 ± 0.1 (313 K), and 0.73 ± 0.05 (333 K).

N 075   "Changes in Unperturbed Dimensions Accompanying Helix-Coil Transitions in Cross-Linked Homopolypeptides, with Special Reference to Poly(hydroxybutyl-L-glutamine)"
Mattice, W. L. *Macromolecules* **1978**, *11*, 15.

Mean-square unperturbed radii of gyration, $<s^2>_o$, are computed as functions of degree of polymerization and helix content for cross-linked polypeptides of the poly(L-alanine) type. *Zimm-Bragg* statistical weights are assigned values appropriate for poly(hydroxybutyl-L-alanine) in water. The g ($\equiv <s^2>_{o,\text{branched}} / <s^2>_{o,\text{linear}}$) for polypeptides with a finite degree of polymerization fall between the limits defined using random-flight statistics and rigid-rod behavior. For polypeptides of the molecular weight usually encountered, g varies strongly with helix content when helicity exceeds 20%. Partially helical polypeptides require substantially higher degrees of polymerization than do completely disordered polypeptides in order to attain the limiting g obtained from random-flight statistics. Calculation of $<s^2>_o$ for cross-linked polypeptides can be achieved using RIS theory in the form appropriate for branched molecules. Adopted procedures differ in only two important aspects from those described previously for disordered cross-linked polypeptides. First, the cross-link is represented by a single bond, 380 pm long, which is attached by a free joint to an appropriate α-carbon atom in each polypeptide chain. The second difference is that the statistical weight matrices for the amino acid residues must now allow for two states, helix and coil. The matrices are formulated using unity, σ, and s in the manner described previously (Mattice, W. L. *Macromolecules* **1975**, *8*, 644; *ibid.* **1976**, *9*, 48). Poly(hydroxybutyl-L-glutamine) in water has σ = 0.00068, while s decreased from 1.04 to 0.95 as the temperature increases from 273 to 373 K. The conformational energy map for disordered residues is that obtained by Brant et al. (*J. Mol. Biol.* **1967**, *23*, 47) for the L-alanyl residue, and the α-helix has ϕ,ψ = 133°, 122.8° and a value of 109.2° for the N–C$^\alpha$–C' angle.

N 076   "Helix-Coil Stability Constants for the Naturally Occurring Amino Acids in Water. 17. Threonine Parameters from Random Poly(hydroxybutlyglutamine-co-L-threonine)"
Hecht, M. H.; Zweifel, B. O.; Scheraga, H. A. *Macromolecules* **1978**, *11*, 545.

The synthesis and characterization of water-soluble, random copolymers containing L-threonine with N^5-(4-hydroxybutyl)-L-glutamine and the thermally induced helix-coil transitions of these copolymers in water are described. The incorporation of L-threonine is found to decrease the helix content of the copolymers in water at all temperatures considered. The *Zimm-Bragg* parameters σ and s for the helix-coil transition in poly(L-threonine) in water are deduced from an analysis of the melting curves. Values of the *Zimm-Bragg* parameters s for poly(L-threonine) in water are computed with the *Allegra* approximation, σ = 1.0 × 10^{-5}: s = 0.754 (313 K), 0.792 (323 K), 0.817 (333 K), and 0.836 (343 K).

N 077   "Unperturbed Dimensions of Crosslinked Histones Evaluated Using Random-Flight Statistics and Rotational Isomeric State Theory"
Mattice, W. L. *Biopolymers* **1979**, *18*, 225.

Unperturbed dimensions are computed via RIS theory for approximately 700 dimers obtained from histones 2A, 2B, and 4. Sets of statistical weights are chosen that yield either a low helicity or a helical content near 40%. These extremes correspond to helicities of histones in the commonly used acid/urea and sodium dodecyl sulfate systems, respectively. Mean features of the RIS results can be successfully reproduces by much simpler expressions based on random-flight statistics. For example, the two methods are about equally effective in predicting how the radius of gyration of a given type of crosslinked dimer should vary from that of the analogous linear polypeptide chain. This result is in marked contrast to that attained using crosslinked, partially helical homopolypeptides. The unexpected success of random-flight statistics with partially helical crosslinked histones is due to the suppression of long helical segments by helix-breaking amino acid residues.

Amino acid residue	σ	s
Pro	0	0
Gly	0.00001	0.60
Ser, Cys, Thr	0.0001	0.76
Lys, Arg, His	0.0001	0.94
Val, Ilu	0.0001	0.95
Glu, Asp, Gln, Asn	0.0006	0.96
Tyr	0.00066	0.99
Ala	0.0008	1.06

N 078   "Averaged Principal Moments of the Inertia Tensor for Unperturbed Poly(hydroxybutyl-L-glutamine) As It Passes through the Helix-Coil Transition"
Mattice, W. L. *Macromolecules* **1980**, *13*, 904.

Averaged principal moments, $<L_1^2> \geq <L_2^2> \geq <L_3^2>$, of the inertia tensor are obtained for polypeptides undergoing a helix-coil transition. Polypeptide chains containing 101, 201, 401, and 801 amino acid residues are studied. Conformational energy surfaces and *Zimm-Bragg* statistical weights used are those appropriate for aqueous poly(hydroxybutyl-L-glutamine) when it is unperturbed by long-range interactions. At all degrees of polymerization studied, $<L_1^2>$ and the mean-square radius of gyration pass through a minimum during the transition from random coil to α helix. Behaviour of the principal moments is related to the presence of short helical segments at low average helicity and long helical segments at high helicity. The conformational energy surface used for a nonhelical amino acid residue is that obtained by *Brant et al.* for unperturbed polypeptides bearing a $-CH_2-R$ side chain in the L configuration. Bond length and angles between bonds are those used by *Brant et al.* (*J. Mol. Biol.* 1967, *23*, 47). The α helix is constructed by using φ = 57.8° and ψ = 47.0°, dihedral angles being reported using the convention in which φ and ψ are each 180° in the completely extended chain.
The statistical weight matrix **U** used for an amino acid residue is given by:
$$U = \begin{bmatrix} 1 & \sigma S \\ 1 & S \end{bmatrix}$$
Nine different values of S are used for each n_p. Results shown in the original paper are calculated by using σ = 0.00068. The value for S for this polypeptide is about 1.044 at 273 K, and it falls to near 0.973 at 343 K.

"Helix End Effects in Block-Copolypeptides, Proteins and Protein-Detergent Complexes"
Mattice, W. L.; Srinivasan, G.; Santiago, G. *Macromolecules* **1980**, *13*, 1254.

Average configuration-dependent properties (mean-square radius of gyration, mean-square end-to-end distance, helical content, helix probability profile, and average number of amino acid residues in a helical segment) are evaluated for unperturbed partially helical polypeptides by using two weighting schemes. The weighting schemes differ with regard to whether end effects in a helical segment are assigned in like manner to amino acid residues at each end or are instead associated with only one amino acid residue. Molecules considered are block copolypeptides and specific sequence copolypeptides with amino acid sequences corresponding to those found in 44 proteins. Without exception, helix formation in the proteins is found to exhibit greater cooperativity if amino acid residues at each end of a helical segment contribute to the end effects. The helix probability profile for proteins is also found to be much smoother with this weighting scheme. Block copolypeptides show more dramatic changes in helicity, helix probability profile, and average number of amino acid residues in a helical segment. Unperturbed dimensions for both proteins and block copolypeptides are nearly independent of the weighting scheme adopted. Modification of the statistical weights used for arginyl, histidyl, and lysyl residues can produce helicities and unperturbed dimensions in reasonable agreement with helicities and dimensions deduced from CD spectra and viscosity measurements of complexes formed by reduced bovine serum albumin and sodium dodecyl sulfate.

Application of matrix methods to partially helical homopolypeptides is most easily achieved by using a 2 × 2 statistical weight matrix A. Columns index the state of amino acid residue i, rows index the state of amino acid residue $i-1$, and the order of indexing is coil (c), helix (h). The statistical weight for amino acid residue i is unity if it is not in a helical state. Its statistical weight is σs if it initiates a sequence of helical amino acid residues and s if it propagates an existing helix.

$$U_i = \begin{bmatrix} 1 & \sigma s \\ 1 & s \end{bmatrix}$$

Matrix schemes which assign end effects to each end of a helical segment are also proposed. A 3 × 3 statistical weight matrix B using this weighting scheme is used here. Row index states of amino acid residues $i-1$ and i, while columns index states of amino acid residues i and $i+1$. The order of indexing is c(cor h), hc, hh.

$$U_i = \begin{bmatrix} 1 & 0 & \sigma^{1/2}s \\ 1 & 0 & 0 \\ 0 & \sigma^{1/2}s & s \end{bmatrix}$$

σ and s values used in water at 303 K:

amino acid residue	$10^4 \sigma$	s
Pro	0	0
Gly	0.1	0.615
Asp	70	0.63
Ser(Cys)	0.1	0.78
Asn	0.1	0.806
Thr	0.1	0.836
Lys	1	0.946
Tyr	66	0.96

amino acid residue	$10^4 \sigma$	s
Val(Ile)	1	0.97
Glu(Gln)	6	0.97
Arg	0.01	1.028
Ala	8.4	1.056
Phe (His, Trp)	18	1.066
Leu	30	1.12
Met	51	1.14

N 080   "Conformational Properties of Central Nervous System Myelin Basic Protein, β-Endorphin, and β-Lipotropin in Water and in the Presence of Anionic Lipids"
Mattice, W. L.; Robinson, R. M. *Biopolymers* **1981**, *20*, 1421.

Conformational properties are examined for three proteins which are disordered when dissolved in water but become partially ordered in the presence of anionic lipids. When evaluated using matrix methods, the helical content of each protein is predicted to be vanishingly small in water, in agreement with experiment. Unperturbed root-mean-square radii of gyration are also evaluated for these proteins in water using generator matrices. Agreement between theory and experiment is found to be excellent. Attention is then directed to the changes induced upon interaction with anionic lipids or detergents. Computations predict an increase in helical content, with numerical results being in quite good agreement with experimental observations using several anionic lipids.

Helix-probability profiles are calculated using the method of *Zimm* and *Bragg*. A 3 × 3 statistical weight matrix is formulated:

$$U_i = \begin{bmatrix} 1 & 0 & \sigma^{1/2}s \\ 1 & 0 & 0 \\ 0 & \sigma^{1/2}s & s \end{bmatrix}$$

Row index states of amino acid residues $i-1$ and i, while columns index states of amino acid residues i and $i+1$. The order of indexing is c(c or h), hc, hh.

Statistical weights σ and s used for amino acid residues in water:

amino acid residue	$10^4 \sigma$	statistical weight, s, at			amino acid residue	$10^4 \sigma$	statistical weight, s, at		
		273 K	303 K	333 K			273 K	303 K	333 K
Ala	8	1.081	1.058	1.008	Asn	0.1	0.738	0.806	0.847
Asn	50	0.71	0.63	0.49	Pro	--	0	0	0
Glu, Gln	6	0.98	0.97	0.93	Arg	0.01	1.026	1.017	0.940
Phe, His, Trp	18	1.061	1.069	1.003	Ser	0.8	0.726	0.793	0.744
Gly	0.1	0.510	0.615	0.631	Thr	1	0.754	0.836	0.863
Lys	1	0.857	0.947	0.911	Val, Ile	66	0.85	0.97	1.06
Leu	33	1.10	1.14	1.09	Tyr	51	1.12	0.96	0.65
Met	54	1.28	1.15	1.00					

N 081   "Helix-Coil Stability Constants for the Naturally Occurring Amino Acids in Water. 19. Isoleucine Parameters from Random Poly(hydroxypropylglutamine-*co*-L-isoleucine)"
Fredrickson, R. A.; Chang, M. C.; Powers, S. P.; Scheraga, H. A. *Macromolecules* **1981**, *14*, 625.

The synthesis and characterization of water-soluble, random copolymers containing L-isoleucine with N^5-(3-hydroxypropyl)-L-glutamine, and the thermally induced helix-coil transitions of these copolymers in water, are described. The incorporation of L-isoleucine is found to increase the helix content of the copolymers in water at all temperatures considered. The *Zimm-Bragg* parameters σ and s for the helix-coil transition in poly(L-isoleucine) in water are deduced from an analysis of the melting curves. Values of the *Zimm-Bragg* parameters s for poly(L-isoleucine) in water are computed with the *Lifson* ($\sigma = 5.1 \times 10^{-3}$) / *Allegra* ($\sigma = 5.5 \times 10^{-3}$) approximation: $s = 1.20/1.25$ (273 K), 1.11/1.14 (293 K), 1.07/1.08 (313 K), and 1.01/1.01 (333 K).

N 082 "Helix-Coil Stability Constants for the Naturally Occurring Amino Acids in Water. 20. Reinvestigation of Valine Parameters from Random Poly(hydroxypropylglutamine-*co*-L-valine)"

Chang, M. C.; Fredrickson, R. A.; Powers, S. P.; Scheraga, H. A. *Macromolecules* **1981**, *14*, 633.

Even though L-valine and L-isoleucine are both β-branched amino acids, the effect of temperature on their helix-forming tendency differs markedly. Since this apparent anomaly might have arisen from significant nonrandom incorporation of L-valine into the copolymer used previously to determine its helix-coil stability constants, this copolymer is reinvestigated here, using a different method of synthesis to avoid formation of large, nonrandom blocks of L-valine. Within experimental error, the same temperature dependence of the *Zimm-Bragg* parameters s for L-valine is obtained as found with the polymer synthesized in earlier work. The difference in the behavior between L-valine and L-isoleucine is therefore a real one. Values of the *Zimm-Bragg* parameters s for poly(L-valine) in water are computed with the *Lifson* ($\sigma = 1 \times 10^{-5}$) / *Allegra* ($\sigma = 1 \times 10^{-4}$) / *Allegra* ($\sigma = 1 \times 10^{-5}$) approximation: s = 0.87/0.87/0.88 (273 K), 0.95/0.95/0.95 (293 K), 0.98/0.97/0.98 (313 K), and 1.04/1.02/1.04 (333 K).

N 083 "Local Configuration of Poly(L-proline) in Dilute Solution"

Darsey, J. A.; Mattice, W. L. *Macromolecules* **1982**, *15*, 1626.

Representative poly(L-proline) chains containing peptide bonds in the *trans* configuration are generated using a conformational energy surface that successfully reproduces the unperturbed dimensions in dilute solution. Representative chains are also generated using selected portions of that surface in order to assess the influence of various features on the local conformation, and approach of the X component of the end-to-end vector to its asymptotic limit is characterized for these same cases. In dilute solution, the overall configuration for chains containing as few as 70 prolyl residues is that of a random coil. Within the chain can be found a few short sequences that adopt a threefold left-handed helical structure reminiscent of that seen in the solid state. Threefold helices containing more than two turns are not seen in chains representative of the ensemble found in dilute solution.

"Stability of the Cross-Linked Tropomyosin Dimer: Cross-Link Effect on the Cooperativity of the Ordering Process and on the Maximum in the Helix Probability Profile"
Mattice, W. L.; Skolnick, J. *Macromolecules* **1982**, *15*, 1088.

A configuration partition function is formulated for a partially helical, cross-linked, in-register dimer composed of two polypeptide chains, A and B, of identical degree of polymerization. A cross-link is present between residues A_x and B_x. The three important features incorporated in the present model are as follows: *(1)* Conformational flexibility is allowed in the cross-linking unit, with certain of the cross-link configurations giving rise to nonalignment of helices propagating away from the cross-link site. *(2)* Cross-linking of the helices may be accompanied by deformation of the helices near the cross-link site. *(3)* There is assumed to be a vanishingly small probabiliity that a sequence of random coil residues connecting two helical segments will adopt a configuration that causes the helical segments to be collinear. Application is made to the case of cross-linked tropomyosin.

The states potentially available to each residue are denoted c, h, and h*. States denoted by h* differ from those denoted by h in that they merit an additional weighting factor that arises from helix-helix interaction in the cross-linked dimer. For a single chain, a 3 × 3 matrix U_i^{single} is used in which rows index the states of residue $i-1$ and i, while columns index the states of residues i and $i+1$. The order of indexing is c(c *or* h), hc, hh. The required statistical weight matrices for the A_iB_i pair are of dimensions 10 × 10 and are of different formulation depending upon whether $i<x$, $i=x$, and $i>x$ (see: below). For indexing, see: original paper.

Statistical weights σ and s, used in water at 303 K:

amino acid residue	$10^4 \sigma$	s
Gly	0.1	0.615
Asp	50	0.63
Ser(Cys)	0.8	0.793
Trp	70	1.06
Asn	0.1	0.806
Thr	0.1	0.836
Lys	1	0.947
Tyr	66	0.96
Val	1	0.97
Ile	55	1.11
Glu(Gln)	6	0.97
Arg	0.01	1.017
Ala	8.4	1.056
Phe (His)	18	1.069
Leu	33	1.14
Met	54	1.15

$$U_i^{single} = \begin{bmatrix} 1 & 0 & \sigma^{1/2}s \\ 1 & 0 & 0 \\ 0 & \sigma^{1/2}s & s \end{bmatrix}$$

$$U_i = \begin{bmatrix} U_B \otimes U_A & & & & & & & 0 \\ 0 & 0\ 0\ 0 & \sigma_A^{1/2}\sigma_B^{1/2}s_As_Bw & \sigma_B^{1/2}s_As_Bw & 0 & \sigma_A^{1/2}s_As_Bw & 0 & s_As_Bw \end{bmatrix}_{i, i > x}$$

$$U_x = \begin{bmatrix} U_B \otimes U_A & & & & \sigma_A^{1/2}\sigma_B^{1/2}s_As_{B^{(i)}} \\ & & & & 0 \\ & & & & \sigma_B^{1/2}\sigma_B^{1/2}s_As_{B^{(i)}} \\ & & & & 0 \\ & & & & 0 \\ & & & & \sigma_A^{1/2}s_As_{B^{(i)}} \\ & & & & 0 \\ & & & & 0 \\ 0 & 0\ 0\ 0 & \sigma_A^{1/2}\sigma_B^{1/2}s_As_{B^{(i)}} & \sigma_B^{1/2}s_As_{B^{(i)}}\ 0\ \sigma_A^{1/2}s_As_{B^{(i)}}\ 0 & s_As_{B^{(i)}} \end{bmatrix}_x$$

$$U_i = \begin{bmatrix} U_B \otimes U_A & \sigma_A^{1/2}\sigma_B^{1/2}s_As_Bw \\ & 0 \\ & \sigma_B^{1/2}s_As_Bw \\ & 0 \\ & 0 \\ & \sigma_A^{1/2}s_As_Bw \\ & 0 \\ & 0 \\ 0 & s_As_Bw \end{bmatrix}_{i, i < x}$$

N 085 "Representative Configurations of Unperturbed Poly(L-alanine) Chains"
Erie, D.; Darsey, J. A.; Mattice, W. L. *Macromolecules* **1983**, *16*, 910.

Representative unperturbed poly(L-alanine) chains are generated using Monte Carlo methods and the conformational energy surface obtained by *Brant, Miller,* and *Flory* (*J. Mol. Biol.* **1967**, *23*, 47). Representative poly(L-alanine) chains have the overall character of a random coil. They contain no evidence whatsoever of α-helical character at the local level. Instead, these are occasional stretches that resemble the conformation adopted in the pleated sheet. Propagation of the minimum-energy configuration would result in a left-handed helix with three residues per turn. In the representative chains, however, sequences that approximate a left-handed helix with three residues per turn occur much less frequently and are considerably shorter than sequences containing about two residues per turn. This situation arises because most unperturbed L-alanyl residues adopt conformations that are distributed over a rather broad region of low conformational energy. At one corner of this low-energy region lie conformations that can propagate a left-handed helix with three residues per turn. A longer traverse through the broad region of low conformational energy is found for a contour line on which lie conformations that would propagate a helix with two residues per turn.

N 086 "Helix Initiation and Propagation by (Hydroxyethyl)-L-glutaminyl Residues in Water"
Hawkins, E. R.; Robinson, R. M.; Mattice, W. L. *Macromolecules* **1983**, *16*, 158.

Random copolypeptides composed of (hydroxyethyl)-L-glutaminyl and (hydroxybutyl)-L-glutaminyl residues are synthesized by aminolysis of high molecular weight poly(γ-benzyl L-glutamate). Circular dichroism spectra measured in water have the shapes and intensities expected for partially helical polypeptides. *Zimm-Bragg* statistical weights for the (hydroxyethyl)-L-glutaminyl residue in water are deduced by matrix methods and the σ, *s* reported by *von Dreele et al.* (*Macromolecules* **1971**, *4*, 408) for the (hydroxybutyl)-L-glutaminyl residue. Ability of theory to reproduce experimental helicities is not sensitive to the value assigned to σ for the (hydroxyethyl)-L-glutaminyl residue as long as the value is small. With $\sigma = 1 \times 10^{-5}$, *s* for the (hydroxyethyl)-L-glutaminyl residue is found to decrease from 0.945 to 0.928 as the temperature rises from 277 to 337 K.

σ (L-glutaminyl) : 3.3×10^{-5}
σ ((hydroxyethyl)-L-glutaminyl) : 1×10^{-5}
σ ((hydroxypropyl)-L-glutaminyl) : 2.2×10^{-4}
σ ((hydroxybutyl)-L-glutaminyl) : 6.7×10^{-4}

$$U = \begin{bmatrix} 1 & \sigma S \\ 1 & S \end{bmatrix}$$

N 087 "Energy Parameters in Polypeptides. 9. Updating of Geometrical Parameters, Nonbonded Interactions, and Hydrogen Bond Interactions for the Naturally Occurring Amino Acids"
Némethy, G.; Pottle, M. S.; Scheraga, H. A. *J. Phys. Chem.* **1983**, *87*, 1883.

Some of the parameters that are used in the computer program ECEPP (empirical conformational energy program for peptides) of *Momany et al.* (*J. Phys. Chem.* **1975**, *79*, 2361) are updated. The changes are based on experimental information that has become available since 1975.

"Mean-Square Hydrophobic Moment for Partially Helical Polypeptides"
Hamed, M. M.; Mattice, W. L. *Biopolymers* **1984**, *23*, 201.

A mean-square helical hydrophobic moment, $<h^2>$, is defined for polypeptides in analogy to the mean-square dipole moment, $<\mu^2>$, for polymer chains. For a freely jointed polymer chain, $<\mu^2>$ is given by $\Sigma\, m_i^2$, where m_i denotes the dipole moment associated with bond i. In the absence of any correlations in the hydrophobic moments of individual amino acid residues in the helix, $<h^2>$ is specified by $\Sigma\, H_i^2$, where H_i denotes the hydrophobicity of residue i. Matrix-generation schemes are formulated that permit rapid evaluation of $<h^2>$ and $<H^2>$. The behaviour of $<h^2> / <H^2>$ is illustrated by calculations performed for model sequential copolypeptides.

The statistical weight matrix used for amino acid residue i, $\mathbf{U}_i$, is:

$$\mathbf{U}_i = \begin{bmatrix} 1 & 0 & \sigma^{1/2}s \\ 1 & 0 & 0 \\ 0 & \sigma^{1/2}s & s \end{bmatrix}_i$$

rows of $\mathbf{U}_i$ index states of amino acid residues i-1 and i, columns index states of amino acid residues i and $i+1$, and the order of indexing is $c(c \cup h)$, hc, hh.

The matrix required to transform from the coordinate system of virtual bond $i+1$ to the coordinate system of virtual bond i in the α-helix is denoted by $\mathbf{T}_h$. Virtual-bond vector i extends from the C^α atom of amino acid residue i–1 to the C^α of amino acid residue i. Using the dihedral angles ($\phi,\psi = -57.4°, -47.5°$) found for helical poly(L-alanine) in the solid state, $\mathbf{T}_h^1$ results; if the α-helix are constructed from standard peptide units, with all $\phi,\psi = -47°, -57.2°$, $\mathbf{T}_h^2$ results.

$$\mathbf{T}_h^1 = \begin{bmatrix} 0.011 & -0.549 & -0.836 \\ -0.291 & 0.798 & -0.529 \\ 0.957 & 0.249 & -0.152 \end{bmatrix} \qquad \mathbf{T}_h^2 = \begin{bmatrix} 0.020 & -0.424 & -0.905 \\ -0.425 & 0.816 & -0.391 \\ 0.905 & 0.393 & -0.164 \end{bmatrix}$$

N 089   "Mean-Square Helical Hydrophobic Moments in Partially Ordered Proteins"
Hamed, M. M.; Mattice, W. L. *Biopolymers* **1984**, *23*, 1057.

Helical hydrophobic moment ratios, $<h^2>/<H^2>$, are evaluated for 34 polypeptides under conditions where the helix content is dictated solely by the short-range interactions operative in aqueous media. The mean-square helical hydrophobic moment is denoted by $<h^2>$, and $<H^2>$ is the averaged of the squared hydrophobicities. This ratio would be one in absence of any correlation in the hydrophobicities of amino acid residues in helices.

Calculations are performed using the generator matrix formalism described in detail by *Hamed* and *Mattice* (**N 088**). The specific expressions used in the present work are:

$$Z = J^*U_1U_2...U_nJ_u \qquad\qquad <H^2> = Z^{-1} J^*U^*_1U^*_2...U^*_n \mathrm{col}(0,J_u)$$

$$U_i = \begin{bmatrix} 1 & 0 & \sigma^{1/2}s \\ 1 & 0 & 0 \\ 0 & \sigma^{1/2}s & s \end{bmatrix}_i \qquad U^*_i = \begin{bmatrix} U & H_i^2 U' \\ 0 & U \end{bmatrix}_i \qquad \text{where} \qquad U'_i = \begin{bmatrix} 0 & 0 & \sigma^{1/2}s \\ 0 & 0 & 0 \\ 0 & \sigma^{1/2}s & s \end{bmatrix}_i \qquad T_h^1 = \begin{bmatrix} 0.011 & -0.549 & -0.836 \\ -0.291 & 0.798 & -0.529 \\ 0.957 & 0.249 & -0.152 \end{bmatrix}$$

where Z denotes the configuration partition function, $\sigma^{1/2}s$ and s are statistical weights for amino acid residues at the termini or interior, respectively, of helical segments, J^* denotes [10...0], and J_u denotes a column whose transpose is [110]. The transformation matrix, T_h, assumes the α-helix has the conformation found for helical poly(L-alanine) in the solid state.

Hydrophobicities and statistical weights for amino acids in water:

Residue	Hydrophobicity	$\sigma \cdot 10^4$	s	Residue	Hydrophobicity	$\sigma \cdot 10^4$	s
Ala	0.3	8	1.858	Lys	−1.8	1	0.947
Arg	−1.4	0.1	1.017	Met	0.4	54	1.15
Asn	−0.5	0.1	0.806	Phe	0.5	18	1.069
Asp	−0.6	50	0.63	Pro	−0.3	0	0
Cys	0.9	0.1	0.78	Ser	−0.1	0.1	0.78
Gln	−0.7	32	0.94	Thr	−0.2	0.1	0.836
Glu	−0.7	6	0.97	Trp	0.3	78	1.06
Gly	0.3	0.1	0.615	Tyr	−0.4	66	0.96
His	−0.1	0.1	0.68	Val	0.6	1	0.97
Ile	0.7	55	1.11				
Leu	0.5	33	1.14				

N 090   "Matrix Formulation of the Transition from a Statistical Coil to an Intramolecular Antiparallel β Sheet"
Mattice, W. L.; Scheraga, H. A. *Biopolymers* **1984**, *23*, 1701.

A tractible matrix formulation is developed for the formation of intramolecular antiparallel β sheets in a homopolymer chain molecule. The formulation is applicable to chains with a finite degree of polymerization. It can readily be extended to treat specific-sequence heteropolymers. Individual sheets may contain any number of strands, the number of residues per strand can range upward from two. The weighting scheme utilizes two end-effect parameters, denoted by τ and δ. The first parameter is associated with each residue that does not have a partner in a preceding strand, and the latter is associated with each β bend. A third parameter, t, is associated with every residue in the sheet. Conditions are described which lead to the formation of different types of sheets: *(1)* "sheets" comprised of isolated extended strands; *(2)* cross-β fibers in which a sheet contains a large number of very short strands; *(3)* fibers in which a few very long strands run parallel to the fiber axis; *(4)* sheets comprised of several strands in which the average strand contains five residues. The fourth type of sheet resembles those found in globular proteins. It is formed when τ and δ are both small, with the ratio, τ/δ, being slightly less than one.

In formulating the terms of the partition function, it is convenient to group them according to the maximum number of residues permitted in a strand, which is denoted by I. The ultimate interest is in the limiting conformational behaviour as I approaches n.

Using **0** to denote a rectangular null submatrix, and leaving zero elements blank in **U**, it is obtained:

For $I = 2$: $Z = $ row $(1,0)$ **U**^{n} col $(1,0,1,0,1)$, where

$$U = \begin{bmatrix} 1 & \tau t & & & \\ & & \tau t & & \\ 1 & & & \delta t & \\ & & & & t \\ 1 & & & \delta t & \end{bmatrix}$$

For $I = 3$: $Z = $ row $(1,0)$ **U**^{n} col $(1,0,1,1,0,0,1,r,1)$, where

$$U = \begin{bmatrix} 1 & \tau t & & & & & & \\ & & \tau t & & & & & \\ 1 & & & \tau t & \delta t & & & \\ 1 & & & & \delta t & & & \\ & & & & & t & & \\ & & & & & & t & \\ 1 & & & r\tau t & \delta t & & & \\ r & & & & r\delta t & & & t \\ 1 & & & & \delta t & & & \end{bmatrix}$$

For $I = 4$: $Z = $ row $(1,0)$ **U**^{n} col $(1,0,1,1,1,0,0,0,1,r,r,1,r,1)$, where

$$U = \begin{bmatrix} 1 & \tau t & & & & & & & & & \\ & & \tau t & & & & & & & & \\ 1 & & & \tau t & & \delta t & & & & & \\ 1 & & & & \tau t & & \delta t & & & & \\ 1 & & & & & & \delta t & & & & \\ & & & & & & & t & & & \\ & & & & & & & & t & & \\ & & & & & & & & & t & \\ 1 & & & r\delta t & & \delta t & & & & & \\ r & & & & & r\delta t & & & t & & \\ r & & & & & & r\delta t & & & t & \\ 1 & & & r\tau t & & & \delta t & & & & \\ r & & & & & & t\delta t & & & & t \\ 1 & & & & & & \delta t & & & & \end{bmatrix}$$

A generalization of the matrix system to arbitrary I ($I > 2$) is also provided.

N 091   "Suppression of the Statistical Coil State during the α ↔ β Transition in Homopolypeptides"
Mattice, W. L.; Scheraga, H. A. *Biopolymers* **1984**, *23*, 2879.

A matrix formulation of the conformational partition function is used to examine helix ↔ sheet transitions in homopolyamino acids. α-Helices are weighted by *Zimm-Bragg* parameters σ and *s*. Antiparallel β-sheets with tight bends are weighted by the parameters t, δ, and τ, where t is the propagation parameter. In addition, each bend contributes a factor δ, and each residue in the sheet that does not have a partner in the preceding strand contributes a factor τ. The helix can be the dominant conformation in a long chain only if two conditions are satisfied simultaneously: *(i)* $s > 1$, and *(ii)* either $s > t$, or σ, δ, and τ are assigned values that inflict a greater penalty on antiparallel sheets than on helices.

N 092   "Helix-Coil Stability Constants for the Naturally Occurring Amino Acids in Water. 22. Histidine Parameters from Random Poly(hydroxybutylglutamine-*co*-L-histidine)"
Sueki, M.; Lee, S.; Powers, S. P.; Denton, J. B.; Konishi, Y.; Scheraga, H. A. *Macromolecules* **1984**, *17*, 148.

The synthesis and characterization of water-soluble random copolymers containing L-histidine with N^5-(4-hydroxybutyl)-L-glutamine, and the thermally induced helix-coil transitions of these copolymers in water and in aqueous 0.5 N KCl solution, are described. The incorporation of both charged and uncharged L-histidine is found to decrease the helix content of the polymer in water, even in the presence of 0.5 N KCl. The *Zimm-Bragg* parameters σ and *s* for the thermally induced helix-coil transition of charged and uncharged poly(L-histidine) in water and in 0.5 N KCl are deduced from an analysis of the melting curves of the copolymers. Values of the *Zimm-Bragg* parameters *s* for poly(L-histidine) in water are computed with the *Allegra* theory; pH 3.0 (σ = 1 × 10^{-5}) / pH 9.0 (σ = 2.1 × 10^{-2}): *s* = 0.69/0.98 (273 K), 0.69/0.85 (293 K), 0.66/0.68 (313 K), and 0.59/0.55 (333 K).

N 093   "Role of Interstrand Loops in the Formation of Intramolecular Cross-β-Sheets by Homopolyamino Acids"
Mattice, W. L.; Scheraga, H. A. *Biopolymers* **1985**, *24*, 565.

A matrix treatment of the formation of intramolecular anti-parallel β-sheets from a statistical coil is extended to incorporate interstrand loops of arbitrary size. The behavior of the model is compared with a simpler version in which all pairs of contiguous strands are connected by β-sheets. When large interstrand loops are allowed, there are many more types of sheets than is the case when all contiguous strands must be connected by tight or β-bends. For this reason, the larger interstrand loops make it easier to introduce the initial sheet into a statistical coil, and the sheet content is enhanced in the early stages of sheet formation (*i.e.*, at small values of the growth parameter *t*). As the transition continues (*i.e.*, as *t* increases), a stage will be reached where occupancy of the statistical coil state is negligible because nearly all residues are in sheets or interstrand loops.

N 094   "Stabilization of Short Helices by Intramolecular Cluster Formation"
Mattice, W. L. *Biopolymers* **1985**, *24*, 2231.

The intramolecular formation of multiple clusters of interacting helices is characterized in a homopolymer. The configuration partition function permits the formation of clusters in which the number of interacting helices may be as large as the greatest integer in $n/2$, where n denotes the number of amino acid residues in the chain. The theoretical formulation has its origin in a recent, tractable matrix expression (**N 090**) for the configuration partition function for intramolecular antiparallel β-sheet formation. Reassignment of the expression for one of the $n(n+3)/2$ elements in the sparse statistical weight matrix, along with a simple change in notation, converts that treatment into a matrix formulation of the configuration partition function for a chain containing multiple clusters of interacting antiparallel helices. The five statistical weights used are δ, f_l, ω, and the *Zimm-Bragg* σ and *s*. Each tight bend that connects two interacting helices contributes a factor of δ, f_l is used in the weight for larger loops between interacting helices, and ω arises from helix-helix interaction. The influence of the helix-helix interaction is illustrated by two helix-coil transitions in a chain with $n = 156$ and σ = 0.001.

N 095   "Dominance of Irregular Structures in the Formation of Intramolecular Antiparallel β Sheets by Homopolyamino Acids"
Mattice, W. L.; Lee, E.; Scheraga, H. A. *Can. J. Chem.* **1985**, *63*, 140.

A matrix formulation of the conformational partition function is used to assess the influence of irregular structures on the formation of intramolecular antiparallel β-sheets. An antiparallel sheet is considered to be irregular if any pair of contiguous strands has an unequal number of residues. The regular structures in the model consist of antiparallel sheets in which every strand contains the same number of residues. The regular structures in the model consist of antiparallel sheets in which every strand contains the same number of residues. Aside from a growth parameter t, the model contains two parameters, δ and τ, that account for the influence of edge effects.

N 096   "Helical Hydrophobic Moment Profiles in α and β Tropomyosin"
Hamed, M. M.; Mattice, W. L. *Int. J. Biol. Macromol.* **1985**, *7*, 15.

Averaged helical hydrophobic moment ratios are evaluated in order to assess the potential of amphiphilic regions contributing to the helix-helix interaction responsible for stabilization of tropomyosin dimers. These ratios yield profiles that are higher in the amino-terminal half than in the carboxyl-terminal half of α and β tropomyosin chains. The higher profiles found in the amino-terminal half of α tropomyosin may contribute to the greater stability of the dimer in this region.

N 097   "Similar matrix expressions describe configuration partition functions for intrachain formation of antiparallel β sheets and interacting α helices"
Mattice, W. L. *Macromolecules* **1985**, *18*, 1345.

With the changes described in the present reference, the methodology developed for the treatment of intramolecular antiparallel sheets can directly applied to the study of systems of interacting helices.

N 098   "Long-Range Aspects of the Formation of Intramolecular Antiparallel β Sheets"
Mattice, W. L.; Scheraga, H. A. *in: Math. and Comput. in Biomed. Appl.*, J. Eisenfeld, C. DeLisi (eds.), Elsevier, North Holland, **1985**, 13.

Conditions for the formation of intramolecular antiparallel β sheets with tight bends are evaluated for model peptides of a sequence $A_xBCA_{36-2x}BCA_x$. Residues A, B, and C are chosen so that bend formation can be achieved without penalty by the BC sequence, but is accompanied by a large penalty for all other sequences of residues. Nevertheless, there are values of the statistical weights and x for which the favored conformations contain three bends. The origin of the third bend becomes apparent upon examination of bend-formation profiles generated by matrix methods.

N 099   "Helix Breakers in Block Copolypeptides"
Mattice, W. L. *Biopolymers* **1986**, *25*, 1449.

The ability of residue of type Y to disrupt the helix formed by residues of type X is studied in X-Y-X-block copolypeptides. The degree of polymerization of each block is so large that it can be considered to be infinite. Matrix methods are used to obtain a general expression for the helix content of the central residue in the Y-block. The resulting expression is specialized to the case where statistical weight matrix is of dimensions 2×2, with elements, 1, 1, σs, and s. The behavior is evaluated for physically realistic values of σ and s. Two useful generalizations emerge: (1) The ability of the Y-block to disrub the helix is determined primarly by the value of s for residues of type Y. Helix disruption does not correlate well with the helix content of a Y-homopolymer of infinite degree of polymerization. (2) In contrast, the ability of the X-block to resist the influence of the Y-block is determined primarily by the helix content of the X-block. It does not correlate well with s for residues of type X. A random-coil region in a Y-block that continues into the following X-block is more stable if residues of type X have a large value of σ because there is helix initiation in the X-block. It is this influence of σ that prevents a correlation between s and the ability of the X-block to resist the disruptive influence of the Y-block.

"Triangular Matrix Representation of Dimensionless Helical Hydrophobic Moment Ratios"
Maroun, R. C.; McCord, R. W.; Mattice, W. L. *Int. J. Biol. Macromol.* **1986**, *8*, 73.

If the hydrophobicity of the *i*th residue in an α-helix is denoted by H_i, one can define a corresponding vector, $\mathbf{h}_i = a_i H_i$. Here, a_i is a unit vector directed radially from the helix axis through C_i^α. In the absence of any correlation in the $\mathbf{h}_i$, it is $h^2 = \Sigma H_i^2$, where the sum extends over all N residues in the helix. The dimensionless ratio $h^2/\Sigma H_i^2$ has the value 1 when the $\mathbf{h}_i$ are uncorrelated. However, if hydrophobic residues occur on the opposite surface of a helix from the hydrophilic residues, $h^2/\Sigma H_i^2 > 1$. The $h^2 / \Sigma H_i^2$ for the $n(n-1)/2$ helices that might occur in a chain of *n* residues are conveniently displayed as the elements in a symmetric $n \times n$ matrix.

With *Janin's* hydrophobicities (*Nature* **1979**, *277*, 491), ΣH_i^2 is simply calculated as $\Sigma H_i^2 = H_j^2 + H_{j+1}^2 + \ldots + H_k^2$. The geometry assumed for the α-helix is that reported for poly(L-alanine) in the solid state (Arnott, S.; Dover, S. D. *J. Mol. Biol.* **1967**, *30*, 209). The unit twist, *t*, is 99.57 degrees. The *a priori* probability for a helix that initiates at residue j and terminates at resiedue k is denoted by p_{jk}. It was computed via $p_{jk} = Z^{-1} [1\ 0]\ \mathbf{U}_1 \mathbf{U}_2 \ldots \mathbf{U}_{j-1} \mathbf{V}_{jk} \mathbf{U}_{k+1} \ldots \mathbf{U}_n$ col (1 0 1), where $Z = [1\ 0]\ \mathbf{U}_1 \mathbf{U}_2 \ldots \mathbf{U}_n$ col (1 0 1). Z denotes the configuration partition function, 0 is a null row or column of the dimension required so that matrices are conformable for multiplication, and the $\mathbf{U}_i$ are 8×8 statistical weight matrices in which the elements are 1, 0, $\sigma^{1/6}s$, *s*, or ω. Using x for $\sigma^{1/6}s$, the statistical weight matrix for all residues except Pro is $\mathbf{U}_a$. Matrices $\mathbf{U}_b$, $\mathbf{U}_c$ and $\mathbf{U}_d$ are used for Pro when the preceeding residue is Gly, Pro, or anything else, respectively.

$$\mathbf{U}_a = \begin{bmatrix} 1 & x & 0 & 0 & 0 & 0 & 0 \\ 0 & 0 & x & 0 & 0 & 0 & 0 \\ 0 & 0 & 0 & x & 0 & 0 & 0 \\ 0 & 0 & 0 & 0 & s & x & 0 & 0 \\ 0 & 0 & 0 & 0 & s & x & 0 & 0 \\ 0 & 0 & 0 & 0 & 0 & 0 & x & 0 \\ 0 & 0 & 0 & 0 & 0 & 0 & 0 & x \\ 1 & x & 0 & 0 & 0 & 0 & 0 \end{bmatrix} \quad \mathbf{U}_b = \begin{bmatrix} 1 & w & 0 & 0 & 0 & 0 & 0 \\ 0 & 0 & w & 0 & 0 & 0 & 0 \\ 0 & 0 & 0 & w & 0 & 0 & 0 \\ 0 & 0 & 0 & 0 & 0 & 0 & 0 \\ 0 & 0 & 0 & 0 & 0 & 0 & 0 \\ 0 & 0 & 0 & 0 & 0 & 0 & 0 \\ 0 & 0 & 0 & 0 & 0 & 0 & 0 \\ 1 & w & 0 & 0 & 0 & 0 & 0 \end{bmatrix} \quad \mathbf{U}_c = \begin{bmatrix} 1 & w & 0 & 0 & 0 & 0 & 0 \\ 0 & 0 & 0 & 0 & 0 & 0 & 0 \\ 0 & 0 & 0 & 0 & 0 & 0 & 0 \\ 0 & 0 & 0 & 0 & 0 & 0 & 0 \\ 0 & 0 & 0 & 0 & 0 & 0 & 0 \\ 0 & 0 & 0 & 0 & 0 & 0 & 0 \\ 0 & 0 & 0 & 0 & 0 & 0 & 0 \\ 0 & 0 & 0 & 0 & 0 & 0 & 0 \end{bmatrix} \quad \mathbf{U}_d = \begin{bmatrix} 1 & w & 0 & 0 & 0 & 0 & 0 \\ 0 & 0 & 0 & 0 & 0 & 0 & 0 \\ 0 & 0 & 0 & 0 & 0 & 0 & 0 \\ 0 & 0 & 0 & 0 & 0 & 0 & 0 \\ 0 & 0 & 0 & 0 & 0 & 0 & 0 \\ 0 & 0 & 0 & 0 & 0 & 0 & 0 \\ 0 & 0 & 0 & 0 & 0 & 0 & 0 \\ 1 & w & 0 & 0 & 0 & 0 & 0 \end{bmatrix}$$

The effect of these statistical weight matrices is to require that Pro occur in a non-helical region or at one of the first *m* positions in a helix, where *m* = 3 if the sequence is Gly-Pro, and *m* = 1 otherwise. The statistical weight for Pro is ω if it is one of the first *m* residues in a helix. The estimate for ω is $10^{-2/3}$. All other residues contribute a statistical weight of $\sigma^{1/6}s$ if they are one of the initial three or final three residues in a helix. Intervening non-prolyl helical residues contribute a statistical weight of *s*. The only non-zero element in $\mathbf{V}_{jk}$ is the statistical weight of the helix that initiates at residue j and terminates at residue k. This statistical weight is $(\sigma_j \sigma_{j+1} \sigma_j + 2\sigma_{k-2} \sigma_{k-1} \sigma_k)^{1/6} s_j s_{j+1} \ldots s_k$. It occurs in the last column of the first row of $\mathbf{V}_{jk}$. The σ, *s* for Cys are taken to be identical with those for Ser. Statistical weights for the remaining 18 residues are those determined for "host-guest" copolypeptides in water at 298 K (**N 092**).

N 101   "Theoretical Study of the Thermoelastic Properties of Elastin Model Chains"
DeBolt, L. C.; Mark, J. E. *Polymer* **1987**, *28*, 416.

The origin of the elastic behaviour of the bioelastomer elastin was a matter of some controversy for a long time, as was the degree of stable secondary structure present in the native state. Available data on the primary sequence of the protein chains constituting the elastin network indicate a significant incidence of three repeat peptides of length four, five and six, with all containing a Pro-Gly pair within them. The present investigation involves the calculation of the temperature coefficient of the mean-square end-to-end distance $<r^2>_0$ of random-coil model chains composed exclusively of one of the three repeat peptides. This configuration-dependent property is directly related to the experimentally obtained energetic component of the elastic force. The values of $d (\ln <r^2>_0) / dT$ for each model chain are evaluated through the use of semiempirical energy calculations on terminally-blocked residues present in the aforementioned repeat peptides, and subsequent use of established matrix techniques for calculating $<r^2>_0$ for polypeptides within a scheme involving independently rotating virtual bonds. Three levels of hydrogen bonding inclusion are investigated to help establish the importance of this contribution. The small positive value of the energetic component of the elastic force for elastin in the literature is found to be adequately reproduced in these calculations on the random-coil model chains.

N 102   "Basis for Large Differences in the Cooperativity of the Formation of Antiparallel β-Sheets and Clusters of Interacting α-Helices in Isolated Chains"
Mattice, W. L.; Tilstra, L. *Biopolymers* **1987**, *26*, 203.

Configuration partition functions that describe the intramolecular formation of antiparallel β-sheets and clusters of antiparallel interacting α-helices are very nearly of the same form. They can be interconverted by a simple change in notation and the addition of one weighting factor for each cluster of interacting α-helices. This extra weighting factor is the *Zimm-Bragg* σ which must be less than one. When it is assigned a reasonable numerical value, it plays an important role in the determination of the nature of the transition from the disordered chain to the ordered structure. It causes the formation of clusters of interacting α-helices to be more cooperative than the formation of antiparallel β-sheets in isolated chains.

N 103   "A Comparison of the CHARMM, AMBER, and ECEPP Potentials for Peptides. II ϕ-ψ Maps for *N*-Acetyl Alanine *N*'-Methyl Amide: Comparisons, Contrasts and Simple Experimental Tests"
Roterman, I. K.; Lambert, M. H.; Gibson, K. D.; Scheraga, H. A. *J. Biomol. Struc. Dyn.* **1989**, *7*, 421.

ϕ-ψ maps of *N*-acetyl alanine *N*'-methyl amide are computed using the CHARMM potential, the all-atom AMBER potential, and the ECEPP/2 potential, before and after adiabatic relaxation. The ϕ-ψ maps are subjected to three simple comparisons with experiment. *(1)* The maps are used to predict the characteristic ratio for poly(L-alanine), and the results are compared with experimental findings. The agreement with experiment is acceptable for ECEPP, and for CHARMM after adiabatic relaxation, marginal for AMBER after adiabatic relaxation, and unsatisfactory for CHARMM or AMBER without adiabatic relaxation. *(2)* Deviations of bond angles from their equilibrium values, in energy-minimized conformations, are compared with values deduced from crystals of terminally blocked amino acids. With both the CHARMM and AMBER potentials using flexible geometry, one or more excessive deviations is observed in the C_7^{ax} local minimum. *(3)* Distributions of ϕ and ψ among residues other than glycine and proline, taken from the coordinates of high-resolution crystals of 16 non-homologous proteins, are plotted for those residues that have zero, one, two, or three hydrogen bonds, respectively, involving a backbone atom. Comparison of the plots with the ϕ-ψ maps generated using CHARMM or AMBER with flexible geometry shows that, in the plots based on the X-ray data, far more residues have ϕ-ψ values in the $α_R$ region, and far fewer have ϕ-ψ values in the C_7^{ax} region, than would be expected from those calculated maps. Comparison of the plots with the maps generated using ECEPP or AMBER with fixed geometry shows much better agreement; however, some discrepancies remain. It is concluded that none of these potentials leads to predictions that are completely compatible with all the experimental results.

"Deuterium NMR Analysis of Poly(γ-benzyl L-glutamate) in the Lyotropic Liquid-Crystalline State: Orientational Order of the α-Helical Backbone and Conformation in the Pendant Side Chain"

Abe, A.; Yamazaki, T. *Macromolecules* **1989**, *22*, 2138.

Deuterium NMR studies are performed on variously labeled PBLG samples in the lyotropic liquid-crystalline state. The orientational order S of the α-helical backbone is estimated from the observed quadrupolar splittings $\Delta\nu$ of the N—D and C^α—D bonds. The relative orientation of the side chains with respect to the backbone is elucidated by the rotational isomeric state analysis of the $\Delta\nu/S$ data observed in the β-, γ-, and ζ-methylene and the p-benzyl deuterons. The most preferred conformation of the side chain is found to be $\chi_1\chi_2\chi_3\chi_4\chi_5$ = ttttt (31%) in 1,4-dioxane, ttttt (18%) in chloroform, and g⁻tttt (20%) in m-cresol at 303 K. The analysis also indicates that the steric overlaps between adjacent side chains play an important role in determining the preferrence of conformations in the PBLG system. The lateral dimensions of the α-helical PBLG rod, expresses by the radius of gyration, is estimated to be 6.5 ± 0.1 Å, being substantially larger than those reported using the SAXS method (3.5 - 4.8 Å).

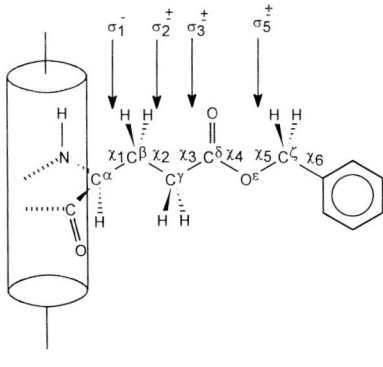

Bond Length [pm]	Valence Angles [°]
Main Chain:	*Main Chain:*
C^α-C* : 153.0	N-C^α-C* : 110.3
C*-N : 132.5	C*-N-C^α : 121.0
N-C^α : 145.3	C^α-C*-N : 115.0
C*=O : 123.0	C^αC*-O : 121.3 (120.5)
C^α-C^β : 153.0	N-C*-O : 123.7 (124.5)
N-D : 100.0	C*-N-D : 123.2 (124.0)
C^α-D^α : 110.0	C^α-N-D : 115.8 (115.0)
	C*-C^α-C^β : 114.0 (111.0)
	N-C^α-C^β : 115.0 (111.0)
	N-C^α-D^α : 108.9 (111.6)
Side Chain:	*Side Chain:*
C-C : 153.0	C^α-C^β-C^γ : 112.0
C-D : 110.0	C^β-C^γ-V^δ : 115.0
C^δ-O^ϵ : 137.0	C^γ-C^δ-O^ϵ : 111.4
O^ϵ-C^ζ : 144.0	C^δ-O^ϵ-C^ζ : 118.3
C^ζ-C^{ar} : 145.0	O^ϵ-C^ζ-C^{ar} : 108.0
	C^α-C^β-D^β : 109.2
	C^β-C^γ-D^γ : 108.4
	O^ϵ-C^ζ-D^ζ : 110.2

Values of the statistical weight parameters for the side chain flanking the α-helical backbone of PBLG:

Parameter	1,4-dioxane			chloroform			m-cresol		
	303K	327 K	353K	303K	327K	353K	303K	327K	353K
σ_1^- :	0.429	0.463	0.502	0.435	0.477	0.516	1.357	1.385	1.447
σ_2^+ :	0.000	0.049	0.093	0.000	0.062	0.135	0.746	0.761	0.664
σ_2^- :	0.000	0.009	0.075	0.000	0.000	0.003	0.180	0.254	0.191
σ_3^+ :	0.598	0.850	1.200	0.610	0.770	0.880	0.590	0.545	0.480
σ_3^- :	0.000	0.040	0.100	0.620	1.120	1.680	0.020	0.380	0.650
σ_5^+ :	0.210	0.260	0.300	0.370	0.420	0.460	0.245	0.320	0.660
σ_5^- :	0.210	0.260	0.300	0.370	0.420	0.460	0.070	0.360	0.660

Torsion angles [°]:

Side Chain:
NC^α-C^β-C^γ : 180.4 , −75.0
$C^\alpha C^\beta$-$C^\gamma C^\delta$: 180.0 , ±67.5
$C^\beta C^\gamma$-$C^\delta O^\epsilon$: 180.0 , ±57.3
$C^\delta O^\epsilon$-$C^\zeta C^{ar}$: 180.0 , ±76.0

Main Chain:
C*-N-C^αC* (φ) : −62.5 (−68.1)
NC^α-C*N (ψ) : −42.3 (−38.1)
C^αC*-NC^α (ω) : 180.0

Polynucleotide chains and related compounds:

N 105   "Theory of "Melting" of the Helical Form in Double Chains of the DNA Type"
Zimm, B. H. *J. Chem. Phys.* **1960**, *33*, 1349.

Previous theoretical treatments of the transition between the helical and random forms of the desoxyribose nucleic acid (DNA) molecule are extended to include formally the explicit consideration of the dissociation into two separate chains and the consideration of the effects of the ends of the chains. An approximate form for the fraction of the base pairs that are bonded is obtained in terms of two parameters, a stability constant for base pairing and a constant representing the interaction of adjacent base pairs. The matrix method of statistical mechanics proves to be adaptable to this problem. Some numerical examples are worked out for very long molecules, for which case it is found that the effect of concentration is small.

N 106   "Simplified Theory of the Helix-Coil Transition in DNA Based on a Grand Partition Function"
Lifson, S.; Zimm, B. H. *Biopolymers* **1963**, *1*, 15.

The statistical mechanics of the helix-coil transition in DNA is treated by means of a grand partition function. This method is simpler than methods used before, although it is suitable only in the limit of great chain length. Expressions for the number of base pairs bonded are obtained and shown to be identical to those obtained by other methods in the limit of great chain length. The physical meaning of the grand partition function is also discussed.

N 107   "Theory of the Helix-Coil Transition in DNA Considered as a Copolymer"
Lifson, S. *Biopolymers* **1963**, *1*, 25.

The grand-partition-function theory of the preceeding paper (**N 106**) is used to derive expressions for the number of base pairs bonded in DNA with explicit consideration of the copolymeric nature of the DNA. The following features of the DNA molecule are taken into account: *(1)* the DNA's from different sources have different ratios of a-t and g-c pairs; *(2)* the a-t pair is weaker than the g-c pair; *(3)* the nearest-neighbor frequencies of the two kinds of base pairs are nonrandom; *(4)* the stacking energies (nearest-neighbor energies) of the various combinations of pairs may be different. An ensemble is constructed in which the features *(1)* through *(4)* are introduced by means of statistical weights of the various pairs and combinations of pairs, and an expression for the corresponding partition function is written. Expressions are derived for the number of bonded base pairs and the number of helical sequences.

N 108   "On the Theory of Order-Disorder Transition and Copolymer Structure of DNA"
Lifson, S.; Allegra, G. *Biopolymers* **1964**, *2*, 65.

The statistical mechanical theory of the helix-coil transition of DNA is improved by introducing approximate normalization factors for the unnormalized statistical weights of finding a given molecule of the assembly in a given microscopic state.

"Spatial Configurations of Polynucleotide Chains. I. Steric Interactions in Polyribonucleotides: A Virtual Bond Model"

Olson, W. K.; Flory, P. J. *Biopolymers* **1972**, *11*, 1.

A simplified scheme for treating the spatial configurations of polynucleotide chains is developed using the RIS approximation and statistical mechanical methods applicable to linear systems of interacting subunits. As a consequence of geometric constraints imposed by the skeletal structure and of the severity of certain steric interactions, it is possible to represent the repeat unit comprising six skeletal bonds by two virtual bonds of fixed length. The configuration of the polynucleotide chain as a whole may be conveniently described by an alternating succession of these two virtual bonds. Moreover, analysis of steric interactions suggest that bond rotations governing the mutual orientation of a given pair of successive virtual bonds should be sensibly independent of the rotations affecting the mutual orientation of other pairs. The statistical mechanical treatment of configuration dependent properties is much simplified in consequence of this mutual independence.

Bond Length [pm]

General:
C-H : 100
O-H : 100
N-H : 100

Chain geometry:
$C_{4'}-C_{3'}$: 152
$C_{3'}-O_{3'}$: 147
$O_{3'}-P$: 156
$O_{5'}-C_{5'}$: 146
$C_{5'}-C_{4'}$: 154

Glycosidic geometry:
$O_{1'}-C_{1'}$: 145
$C_{2'}-C_{1'}$: 154
$C_{1'}-N_9$: 147
N_9-C_4 : 135
N_9-C_8 : 133

Virtual bond scheme:
l_a : 264
l_b : see table

Valence Angles [°]

109

$C_{4'}-C_{3'}-O_{3'}$: 110 ($\theta C_{3'}$)
$C_{3'}-O_{3'}-P$: 119 ($\theta O_{3'}$)
$O_{3'}-P-O_{5'}$: 104 (θP)
$P-O_{5'}-C_{5'}$: 121 ($\theta O_{5'}$)
$C_{5'}-C_{4'}-C_{3'}$: 112 ($\theta C_{4'}$)

$O_{1'}-C_{1'}-O_9$: 112
$C_{2'}-C_{1'}-N_9$: 106

$C_{1'}-N_9-C_4$: 127
$C_{1'}-N_9-C_8$: 127

Statistical Weights

$\sigma = 0.17$
$\rho = 0.80$
$\eta = 0.83$
$\alpha = 0.43$
$\beta = 0.62$
$\beta' = 0.43$
$\gamma = 0.87$
$\delta = 0.27$

$$U_a(\phi', \phi'') = \begin{bmatrix} 1 & 1 & 1 \\ \sigma & \sigma & \sigma\rho\beta' \\ \sigma & \sigma\beta & \sigma\rho\alpha \end{bmatrix}$$

$$U_b(\psi', \psi'') = \begin{bmatrix} 1 & \eta & \eta \\ 1 & \eta\gamma & \eta\delta \\ 1 & \eta\delta & \eta\gamma \end{bmatrix}$$

ω'	ω''	l_b	κ	Φ	ζ	Ψ	τ
263	0	461	54	26	19	74	15
263	35	425	64	25	34	99	16
263	55	397	69	27	43	114	14
263	85	351	74	34	62	137	3
333	35	492	38	29	48	158	13
333	55	477	39	37	81	160	42
333	85	451	38	51	128	153	82

see: Appendix B and Figure I of the original paper for definitions of angles ζ, τ, ξ, ν, an κ.

The angles ν and ξ are fixed at 31°, and 28°, respectively.

N 110 "Spatial Configurations of Polynucleotide Chains. II. Conformational Energies and the Average Dimensions of Polyribonucleotides"
Olson, W. K.; Flory, P. J. *Biopolymers* **1972**, *11*, 25.

Conformational energies are calculated for pairs of successive bond rotations within an internal residue of a polyribonucleotide chain. Contributions to these energies include bond torsional strain, van der Waals repulsions, London attractions, electrostatic interactions, and inductive interactions between nonbonded atoms in the nucleotide repeat unit. The average dimensions of unperturbed random-coil polyribonucleotide chains are then evaluated on the basis of energies thus estimated, using for this purpose the previously developed virtual bond treatment. The characteristic ratio C_∞ of the mean-square end-to-end distance calculated for polyribonucleotide chains in which all pentose rings are fixed in a $C_{3'}$-*endo* conformation is about 9; for chains consisting exclusively of $C_{2'}$-*endo* units it is about 25. These values are considerably greater than those obtained by giving equal weight to all conformations judged to be sterically allowed. Satisfactory agreement between the calculations here and experimental values from viscosity and light scattering studies is achieved by treating the chain as a random copolymer of $C_{3'}$-*endo* and $C_{2'}$-*endo* conformational isomers.

N 111 "Spatial Configurations of Polynucleotide Chains. III. Polydeoxyribonucleotides"
Olson, W. K.; Flory, P. J. *Biopolymers* **1972**, *11*, 57.

The virtual bond scheme set forth in preceding papers for treating the average properties of poly(riboadenylic acid) (poly rA) is here applied to the calculation of the unperturbed mean-square end-to-end distance of poly(deoxyriboadenylic acid) (poly dA). The modification in structure and in charge distribution resulting from the replacement of the hydroxyl group at $C_{2'}$ in the ribose residue by hydrogen in the desoxyribose produce only minor modifications in the conformational energies associated with the poly dA chain as compared to those found for poly rA. The main difference is manifested in the energy associated with rotations about the $C_{3'}-O_{3'}$ bond of the desoxyribose residue in the $C_{2'}$-*endo* conformation; accessible rotations are confined to the range between 0° and 30° relative to the *trans* conformation, whereas in the ribose unit the accessible regions comprise two ranges centered at approximately 35° and 85°. The characteristic ratio calculated on the basis of the conformational energy estimates is about 9 for the poly dA chain with all deoxyribose residues in the $C_{3'}$-*endo* conformation and about 21 with all residues in the $C_{2'}$-*endo* form. Satisfactory agreement is achieved between the theoretical values and experimental results on apurinic acid by treating the poly dA chain as a random copolymer of $C_{3'}$-*endo* and $C_{2'}$-*endo* conformational isomers present in a ratio of about 1 to 9.

N 112 "Backbone Conformations in Secondary and Tertiary Structural Units of Nucleic Acids. Constraint in the Phosphodiester Conformation"
Yathindra, N.; Sundaralingam, M. *Proc. Nat. Acad. Sci. USA* **1974**, *71*, 3325.

The possible backbone phosphodiester conformations in a dinucleotide monophosphate and a dinucleotide triphosphate are investigated by semiempirical energy calculations. Conformational energies are computed as a function of the rotations ω' and ω about the internucleotide P-O(3') and P-O(5') linkages, with the nucleotide residues themselves assumed to be in one of the preferred [C(3')-*endo*] conformations.

N 113   "Configurational Statistics of Polynucleotide Chains. A Single Virtual Bond Treatment"
Olson, W. K. *Macromolecules* **1975**, *8*, 272.

A simplified single virtual bond scheme is developed for the calculation of mean-square unperturbed dimensions in polynucleotide chains. As a consequence of the structural rigidity of the sugar residues in the chain, it is possible to represent the six chemical bonds comprising the chain backbone repeating unit by a single virtual bond (connecting successive phosphorus atoms). The mutual orientation of a pair of adjoining virtual bonds is determined by the angles of rotation about the phosphodiester bonds adjoining intervening phosphorus atoms and is independent of the orientation of all other virtual bonds in the chain. Computed values of chain dimensions based on the single virtual bond scheme are comparable to those calculated previously using a two virtual bond model which permits rotational flexibility in the sugar moieties of the chain.

Geometrical parameters of the virtual bond model:

Ring pucker	ψ'	l	κ	ϕ	ζ	τ	ψ
$C_{3'}$ endo	266	563	43	83	48	31	114
$C_{2'}$ exo	278	589	41	79	44	29	121
$C_{2'}$ endo	317	660	32	78	27	28	150
$C_{3'}$ exo	328	669	30	81	21	31	160

For definition of the angles, see original paper.

N 114   "Configuration-Dependent Properties of Randomly Coiling Polynucleotide Chains.
I. A Comparison of Theoretical Energy Estimates"
Olson, W. K. *Biopolymers* **1975**, *14*, 1775.

Various theoretical estimates of the conformational energy associated with polynucleotides in solution are compared with each other and also with the experimentally observed conformations found in X-ray crystallographic investigations of low-molecular weight nucleic acid analogs. In view of the disparities between these data, certain configuration-dependent properties (*i.e.*, the mean-square unperturbed end-to-end distance and the average vicinal NMR coupling constant) appropriate to randomly coiling polynucleotides described by either the energy estimates of by the crystallographically preferred conformations are also calculated and compared with the known solution behaviour of polynucleotide chains. Both the theoretical energy surfaces and the X-ray data show good agreement with the NMR coupling constant indications of the preferred rotations about the O–C and C–C bond of the chain backbone.

N 115 "Configuration-Dependent Properties of Randomly Coiling Polynucleotide Chains. II. The Role of the Phosphodiester Linkage"
Olson, W. K. *Biopolymers* **1975**, *14*, 1797.

The dependence of the unperturbed dimensions of randomly coiling polynucleotides on the rotations about the phosphodiester linkages of the chain is examined in order to understand the conformational discrepancies, set forth in **N 114**, regarding these angles (ω' and ω). Large values of the characteristic ratio, which agree with the experimental behavior of the chain, are obtained only if a sizeable proportion of the polymer residues have *trans* ω' values. The asymmetric torsional potential that is believed to arise from *gauche* effects associated with the P—O bonds is approximated using a hard core model.

N 116 "The Spatial Configuration of Ordered Polynucleotide Chains. I. Helix Formation and Base Stacking"
Olson, W. K. *Biopolymers* **1976**, *15*, 859.

A single virtual bond scheme set forth previously for the treatment of average properties of randomly coiled polynucleotides is applied to the calculation of helical parameters which characterize a regularly repeating polynucleotide molecule. Only a fraction of the enormous number of conformationally feasible helices fulfill the geometric criteria of vertical base stacking usually associated with ordered polynucleotide chains. Detailed examination of the nature and mode of base stacking feasible in a single helical backbone structure indicates that the handeness of a base stacking arrangement does not correlate either quantitatively or qualitatively with the handedness of the polymer backbone. A number of polynucleotide chains which exhibit lefthanded base stacking patterns in NMR and CD studies may, in fact, be righthanded helices.

N 117 "A Configurational Interpretation of the Axial Phosphate Spacing in Polynucleotide Helices and Random Coils"
Olson, W. K.; Manning, S. *Biopolymers* **1976**, *15*, 2391.

The structural implications arising from the observation that the charge density of a single-stranded randomly coiling polynucleotide chain is approximately equal to that of one strand of the familiar double helix are here examined. A computational scheme is described to obtain (using bond length, valence bond angles, and internal rotation angles) the mean phosphate-phosphate spacing parameter b along the chain axes of any single-stranded polynucleotide molecule. [The charge-spacing parameter b describing the polyelectrolyte chains treated by dilute solution theory is defined simply as the average distance between the projections of the charged groups of the polymer along the cylindrical axes of the rod-like structures comprising the chain backbone.] Attention is then focussed upon the computed interphosphate spacing associated with both the theoretical randomly coiling polynucleotide that reproduces the observed experimental unperturbed dimensions and the familiar single-stranded helix. The calculations clearly demonstrate that the parameter b only weakly reflects the spatial configuration of the chain. The approximate equivalence of the b values associated with the single-stranded helix and the unperturbed randomly coiling polynucleotide is not indicative of strong configurational similarities between the two forms. The torsion angles chosen here, $\phi' = 35°$, $\phi = 0°$, $\psi' = 266°$ (C(3')-*endo*) or $317°$ (C(2')-*endo*), $\psi = 0°$ (*trans*) or $240°$ (g^-), are consistent with the ranges of these angles observed in X-ray, NMR, and theoretical analyses.

Statistical weights $w_{\omega'\omega}$ of phosphodiester conformation in C(3')-endo and C(2')-endo polynucleotide chains:

$w_{\omega'\omega}$	tt	tg$^+$	tg$^-$	g$^+$t (Minimum)	g$^+$g$^+$	g$^+$g$^-$	g$^-$t	g$^-$g$^+$	g$^-$g$^-$
C(2')-*endo*	0.35	0.19	0.32	0.04	0.03	0.00	0.05	0.00	0.02
C(3')-*endo*	0.30	0.16	0.24	0.14	0.10	0.00	0.05	0.00	0.01

N 118   "Radius of Gyration, Asymmetry, and Head-Group Orientation in Unperturbed Lecithins"
Mattice, W. L. *J. Am. Chem. Soc.* **1979**, *101*, 7651.

Biologically interesting lipids derived from glycerol have available an enormous number ($>10^{20}$) of configurations if three states are assigned to each bond about which rotation occurs. Averaging over such a large number of configurations can be achieved using methods developed to treat branched polymers. Configuration-dependent properties for unperturbed lecitins are obtained via RIS theory from a representative sample generated by Monte Carlo methods. Necessary *a priory* and conditional probabilities are obtained from a treatment which incorporates first- and second-order interactions. This procedure is a logical extension of an earlier successful treatment of triglycerides.

nonhydrogen atoms in diacetylphosphorylcholine

Bond Length [pm]	Valence Angles [°]	Torsion Angles [°]	(for 298 K)
C-C : 153	C-C-C : 112	t : 180	
C-C' : 150	C-C-C' : 113.7		
C=C : 134	C-C=C : 125.0	g^+ : 60	
C-N : 151	C-C-N : 115		
C-O : 143	C-N-C : 109.5	g^- : -60	
C'-O : 135	C-C-O : 109		
C'-O' : 120	C-C'-O : 126.3		
O-P : 161	C-O-C' : 116.7		
O=P : 149	C-O-P : 118		
	O-P-O : 103		
	O-P=O : 108		
	O=P=O : 122		

$$U_2^3 = \begin{bmatrix} 1 & 0 & 1 \\ 1 & 1 & 0 \\ 1 & 0 & 0 \end{bmatrix} \quad U_3^3 = U_5^3 = \begin{bmatrix} 1 & 1 & 1 \\ 1 & 1 & 1 \\ 1 & 1 & 1 \end{bmatrix}$$

$$U_4^3 = \begin{bmatrix} 0 & 0.07 & 0.07 \\ 0.07 & 1 & 0.05 \\ 0.07 & 0.05 & 1 \end{bmatrix} \quad U_6^3 = \begin{bmatrix} 1 & 2 & 2 \\ 1 & 10 & 0 \\ 1 & 0 & 10 \end{bmatrix}$$

First-order a priori probability for population of rotational states by bonds 2 - 6 in branch 3:

bond	t	g^+	g^-
2	0.51	0.08	0.41
3	0.06	0.47	0.47
4	0.06	0.47	0.47
5	0.18	0.41	0.41
6	0.11	0.44	0.44

N 119   "The Flexible DNA Double Helix. I. Average Dimensions and Distribution Functions"
Olson, W. K. *Biopolymers* **1979**, *18*, 1213.

A scheme is developed to treat the spatial arrangements and properties of double-helical DNA in terms of the constituent atoms and bonds of the system. Heretofore the behavior of DNA in solution has been interpreted in terms of various artificial theoretical models. In this work the flexibility of the DNA double helix is taken to arise from minor perturbations of the rotation angles along the polynucleotide backbone. The rotational motions of the chain are limited to conformations within the B-DNA family of helices that permit base stacking. The disruptions of hydrogen bonding associated with these angular fluctuations are offered as a plausible description of the well-known breathing of double-stranded DNA. Radii of gyration, $<s^2>_o$, computed on the basis of this model are found to agree with experimental measurements.

N 120   "The Flexible DNA Double Helix. II. Superhelix Formation"
Olson, W. K. *Biopolymers* **1979**, *18*, 1235.

A simple super or s-virtual bond scheme is developed for the treatment of tertiary or superhelical structure in polynucleotide chains. The scheme is utilized to examine the enormous variety of tertiary structure that can be generated by regularly bending a B-DNA reference helix at the phosphodiester linkages.

N 121   "Configuration Statistics of Polynucleotide Chains. An Updated Virtual Bond Model to Treat Effects of Base Stacking"
Olson, W. K. *Macromolecules* **1980**, *13*, 721.

An updated virtual bond scheme is developed to include "long-range" effects of base stacking in the treatment of the spatial configurations of the polynucleotides. As a consequence of the relative rigidity of rotations about the C—O bonds (ϕ' and ϕ) of the sugar-phosphate backbone, it is possible to represent the six chemical bonds constituting each nucleotide repeating unit in terms of two virtual bonds of comparable magnitudes (spanning C—C—O—P bond fragments of the chain). The "long-range" (three-bond) interdependence of the alternate C—C and O—P bonds in the polynucleotide backbone somewhat complicates computations of chain averages compared to the previous treatments. Despite this complexity, the "long-range" interactions introduce a novel theoretical probe to deduce the solution conformation of the phosphodiester linkages ($\omega'\omega$) from the experimentally observed conformations of the C—C rotations ($\psi'\psi$). On the basis of hard-core conformational analysis and Karplus treatment of NMR coupling constants, the helix-coil transition was modeled of poly(rA) over the temperature range $-12°$ to $60°C$. For the purpose of evaluating the averaged transformation matrices three statistical weight matrices $U_2(\psi',\omega')$, $U_3(\omega',\omega)$, and $U_4(\omega,\psi)$ are introduced that relate the rotational interdependence of these angle pairs. It is also defined a premultiplication matrix $U_1(\psi')$ and a postmultiplication matrix $U_5(\psi)$ that reflect the rotational preferences of these two independent angles. With statistical weight matrices $U_1...U_5$ constructed in this manner, the partition function for each independent nucleotide unit is $Z = U_1U_2U_3U_4U_5$. The $U_1(\psi')$ matrix describing the sugar puckering through the ψ' rotation is given by the 1×2 row vector $U_1(\psi') = [F_{3E}\ F_{2E}]$, where F_{3E} represents the fractional population of ^3E sugar puckering and $F_{2E} = 1 - F_{3E}$ the amount of ^2E puckering. The F_{3E} parameter is varied in the calculations reported. Finally, the $U_5(\psi)$ matrix is given by the 3×1 column vector $U_5(\psi) = \text{col}(F_{gt}\ F_{tg}\ F_{gg})$. F_{gg} refers to the fractional population of $\psi = g^-$ conformers where $O_{5'}$ is in the so-called double *gauche* (gg) conformation with respect to atoms $O_{1'}$ and $O_{3'}$. F_{gt} is the statistical weight of the $\psi = t$ state where $O_{5'}$ is *gauche* to $O_{1'}$ and *trans* to $C_{3'}$ and F_{tg} the statistical weight for $\psi = g^+$ where $O_{5'}$ is *trans* to $O_{1'}$ and *gauche* to $C_{3'}$. F_{gg} is varied in the calculations with F_{gt} and F_{tg} determined by the relationships $F_{tg} + F_{gt} + F_{gg} = 1$ and $F_{tg} = F_{gt}/3$.

Bond Length [pm]	Valence Angles [°]
C-C : 152	O-P-O : 104.0 (θ^P)
C-O : 144	P-O-C : 121.5 ($\theta O_{5'} = \theta O_{3'}$)
O-P : 160	O-C-C : 110.0 ($\theta C_{5'} = \theta C_{3'}$)
	C-C-C : 116.0 ($\theta C_{4'}$)

$$U_2(\psi',\omega') = \begin{bmatrix} 0 & 1 & 1 & 1 & 1 & 0 \\ 1 & 1 & 1 & 0 & 0 & 1 \end{bmatrix}$$

$$U_3(\omega',\omega) = \begin{bmatrix} 1 & 1 & 1 & 1 & 1 & 1 & 1 & 1 \\ 1 & 1 & 1 & 1 & 0 & 0 & 1 & 1 \\ 1 & 1 & 1 & 0 & 0 & 0 & 0 & 1 \\ 1 & 1 & 0 & 0 & 0 & 0 & 0 & 1 \\ 1 & 0 & 0 & 0 & 0 & 0 & 0 & 1 \\ 1 & 0 & 0 & 0 & 0 & 0 & 1 & 1 \\ 1 & 0 & 0 & 0 & 0 & 1 & 1 & 1 \\ 1 & 1 & 0 & 0 & 0 & 1 & 1 & 1 \\ 1 & 1 & 1 & 1 & 1 & 1 & 1 & 1 \end{bmatrix}$$

$$U_4(\omega,\psi) = \begin{bmatrix} 1 & 1 & 0 \\ 1 & 1 & 1 \\ 1 & 1 & 0 \\ 1 & 1 & 0 \end{bmatrix}$$

Geometrical parameters of the virtual bond scheme:

Virtual Bond	ϕ or ϕ'	l	κ	Φ	ζ	τ	Ψ
a	0	395	13.5	0.0	90.0	90.0	0.0
b	30	388	28.6	21.5	16.0	27.5	42.0

(For definition of the angles, see original paper)

N 122   "The Poly(rU) Coil: A Minimum-Energy Model That Matches Experimental Observations"
Hingerty, B. E.; Olson, W. K. *Biopolymers* **1982**, *21*, 1167.

A model of the randomly coiling form of poly(rU) based on minimum-energy conformers of UpU is described. The blend of conformers is chosen to fit the C—C rotational populations derived in NMR studies of UpU and poly(rU) and to match the experimental unperturbed dimensions of the poly(rU) chain. In addition, estimates of loop closure based on the model are comparable to the sizes of loops most frequently seen in the model oligonucleotides. Approximately 60% of the conformers constituting the model are characterized by stacked, extended C2'-*endo* ω'ωψ = tg⁻g⁺ rotations.

N 123   "The Long-Range Stiffness and Local Mobility of Double Stranded DNA"
Olson, W. K. *in: Intramolecular Dynamics*, Jortner, J.; Pullmann, B (Eds.), D. Reidel Publishing Company, 1982, 525.

Theoretical conformational analysis provides a basis for understanding the unique features of double-stranded DNA in terms of its chemical architecture. The well-known stiffness of the chain as a whole derives from the sequence of heterocyclic bases, while the local mobility of the constituent nucleotides reflects the structural complexity of the sugar-phosphate backbone.

N 124   "Loop Formation in Polynucleotide Chains. I. Theory of Hairpin Loop Closure"
Marky, N. L.; Olson, W. K. *Biopolymers* **1982**, *21*, 2329.

A theoretical model to determine the probability of loop formation, based on an elaborated form of the *Jacobson-Stockmayer* theory of cyclization equilibria, is developed and used on RNA chains of homogeneous puckering and lengths up to 2^7 residues.

N 125   "Theory of Helix-Coil Transitions of α-Helical, Two-Chain, Coiled Coils"
Skolnick, J.; Holtzner, A. *Macromolecules* **1982**, *15*, 303.

A theory of the helix-coil transition for in-register, two-chain, α-helical, coiled coils such as tropomyosin and paramyosin is developed. The treatment differs from those fromulated previously for DNA- or collagen-like double helices; in the present treatment, isolated single chains and each of the two strands in the dimer may be partially helical. The fraction of helix is calculated in the two-chain, coiled coil as a function of *Zimm-Bragg* cooperativity (σ) and helix stability (s) parameters appropriate to single chains and of a new parameter ω that takes account of the enhanced helix stability in a paired chain via-à-vis an isolated chain. The importance of the quasi-repeating heptet in stabilizing the helix conformation in a two-chain, coiled coil is accounted for via a coarse-graining approximation. Singly cross-linked homopolypeptide chains and both singly cross-linked and non-cross-linked chains with the rabbit α-tropomyosin sequence are treated in detail and estimated of the stability per helical residue are given. Thus, the treatment differs from those developed previously for the helix-coil transition for simple polypeptides in that dimers are accounted for; and it differs from those developed previously for DNA- or collagen-like helices in that both isolated strands and double strands may be helical in the present work.

An approach is used which has the following features: *(1)* The theory is developed in terms of *Zimm-Bragg* parameters σ and s appropriate to each type of amino acid in the primary structure of an isolated, single chain. *(2)* It is assumed that cross-linking has no effect on σ and s of the cysteine residue. *(3)* The statistical weight of a randomly coiled residue is assigned a value of unity. *(4)* Since completely helical, two-chain, coiled coils are composed of parallel, in-register α-helices, the possibility of mismatched association between residues on different chains is excluded. Several models are formulated of the helix-coil transition in cross-linked dimers. The relevant statistical weight matrix for the ith pair of segments in terms of helix initiation (σ) and stability (s) parameters of chain j may be written:

$$U_i = \begin{bmatrix} 1 & \sigma_2 s_2 & \sigma_1 s_1 & \sigma_1 s_1 \sigma_2 s_2 \omega \\ 1 & s_2 & \sigma_1 s_1 & \sigma_1 s_1 s_2 \omega \\ 1 & \sigma_2 s_2 & s_1 & s_1 \sigma_2 s_2 \omega \\ 1 & s_2 & s_1 & s_1 s_2 \omega \end{bmatrix}$$

rows and columns: c(1)c(2), c(1)h(2), h(1)c(2), h(1)h(2)

Furthermore, defining the statistical matrix of the ith residue on isolated chain j by:

$$U_j = \begin{bmatrix} 1 & \sigma_j^i s_j^i \\ 1 & s_j^i \end{bmatrix}$$

and the three-dimensional identity matrix by $E = \begin{bmatrix} 1 & 0 & 0 \\ 0 & 1 & 0 \\ 0 & 0 & 1 \end{bmatrix}$ U_i now can be written in the compact form: $U_i = U_1^i \otimes U_2^i \begin{bmatrix} E & 0 \\ 0 & \omega^i \end{bmatrix}$

N 126   "Theoretical Probes of Nucleic Acid Conformation"
Olson, W. K. *in: Intramolecular Dynamics*, Jortner, J.; Pullmann, B (Eds.), D. Reidel Publishing Company, 1983, 217.

The computations described briefly in this paper illustrate the interrelationship between the local structure and macroscopic behavior of the DNA helix. Statistical mechanical studies help to identify the most likely morphological arrangements of the polynucleotide backbone and to understand the macroscopic flexibility of the DNA as a whole. Model building and potential energy calculations uncover the detailed local geometries of the chain and clarify the likely pathways between the multitude of allowed spatial forms.

N 127   "How Flexible Is the Furanose Ring? 1. A Comparison of Experimental and Theoretical Studies"
Olson, W. K.; Sussman, J. L. *J. Am. Chem. Soc.* **1982**, *104*, 270.

A series of statistical computations are carried out to test various theoretical potential energy estimates of furanose pseudorotation. Energy surfaces describing the flexibility of both ribose and deoxyribose are compared through a Boltzmann analysis with each other and also with published X-ray and NMR measurements of pseudorotation. No single theoretical approach is able to account simultaneously for the experimental properties of both ribose and deoxyribose. In the commonly occurring mononucleosides and -nucleotides the pseudorotational motions are decidedly stiff with the potential energy barrier somewhat higher for ribose than for deoxyribose. When incorporated into a polynucleotide backbone, this local stiffness is a major determinant of chain flexibility and a characteristic difference between RNA and DNA systems.

N 128   "How Flexible Is the Furanose Ring? 2. An Updated Potential Energy Estimate"
Olson, W. K. *J. Am. Chem. Soc.* **1982**, *104*, 278.

A potential energy function is developed to estimate the pseudorotational motions of ribose and 2'-deoxyribose sugars. In addition to standard nonbonded, torsional and valence angle strain contributions, an intrinsic *gauche* energy term is required to account for the puckering preferences and hindered ring flexibilities suggested by solid-state and solution studies. The *gauche* effect is also found to be an essential factor in reproducing the properties of 3'-deoxyribose and 2'-fluoro-2'-deoxyribose rings in solution. The extreme sensitivity of the potential energies to variations in ring geometry is helpful in understanding the disparities noted in the preceeding article (**N 127**) among earlier studies of furanose pseudorotation. Apparently minor deviations of valence bond angles from standard X-ray values are found to perturb the normal motions of the furanose drastically.

N 129   "Influence of L-Cystinyl Side-Chain Configurations on the Melting of Crosslinked α-Tropomyosin Dimers"
Maroun, R. C.; Mattice, W. L. *Biochim. Biophys. Acta* **1984**, *784*, 133.

The experimental melting profile for the rabbit α-tropomyosin dimer crosslinked at cysteine residue 190 is analyzed using matrix methods. The configuration partition function employed includes a term arising from interactions at the crosslink site. This term, denoted by ω, is found to be smaller than 1, implying that events at the crosslink site resist helix formation by dimer. A theoretical analysis of the conformational restrictions imposed on the crosslink provides a satisfactory estimate of ω at high temperatures. Agreement deteriorates at lower temperatures, perhaps as a consequence of difficulty in establishing a reliable value for ω from analysis of the low-temperature circular dichroism data.

N 130   "Loop Formation in Polynucleotide Chains. II. Flexibility of the Anticodon Loop of *t*RNA^{Phe}*"
Marky, N. L.; Olson, W. K. *Biopolymers* **1987**, *26*, 415.

The flexibility of hairpin loops containing *n* bases (residues) is examined using a theoretical model of oligonucleotide ring closure (**N 124**). The study is based on correlated probabilities of chain separation and terminal residue orientation. The probabilities are calculated standard statistical methods as functions of local conformational changes of the chain backbone. Model calculations suggest that the anticodon loop is a dynamic structure capable of assuming a variety of different spatial conformations.

Polysaccharide chains and related compounds:

N 131   "Mean Square Length and Mean Square Radius of Gyration of 1,4'-Polysaccharides and of Polybutadienes"
Eliezer, I.; Hayman, H. J. G. *J. Polym. Sci.* **1957**, *23*, 387.

The mean square length and mean square radius of gyration of 1,4'-polysaccharides are calculated by an extension of the method of *Eyring* and *Benoit*, the results being expressed in terms of the degree of polymerization and of certain parameters of physical significance connected with the rotations about the various single bonds. The calculations are based on the assumptions that no correlation exists between rotations about bonds separated by at least one pyranose ring and that the polysaccharide molecules are made up of pyranose rings in the chair form. The excluded volume effect is neglected throughout.

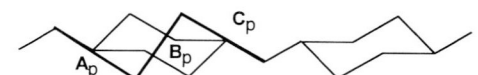

trans-type 1,4'-polysaccharide (e.g., cellulose)

The length of the virtual bonds:

Virtual Bond	Cellulose:	Pectic acid:	Amylose:
A	266 pm	168 pm	441 pm
B	266 pm	168 pm	293 pm
C	142 pm	282 pm	—

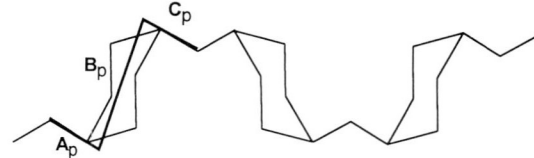

trans-type 1,4'-polysaccharide (e.g., pectic acid)

N 132   "The Radii of Gyration of Polysaccharides in Indifferent Solvents"
Burchard, W. *Makromol. Chem.* **1960**, *42*, 151.

The radii of gyration of amylose and cellulose are calculated by a method of *Lifson* for different glucose conformations. It is found that the radii of gyration differ widely and it can be determinated clearly by means of viscosity and light-scattering measurements which kind of glucose conformation is present in the polysaccharides. The effects of twisting upon the linear molecule is discussed.

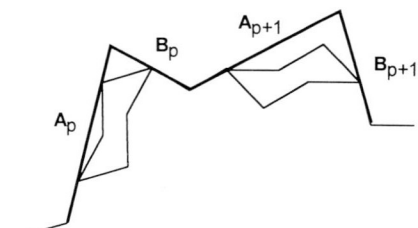

cis-type 1,4'-polysaccharide (e.g., amylose)

N 133   "Conformational Analysis of Cellobiose, Cellulose, and Xylan"
Rees, D. A.; Skerrett, R. J. *Carbohyd. Res.* **1968**, *7*, 334.

All of the likely conformations of cellobiose, cellulose, and xylan are explored systematically assuming the ring conformations and (C-1)-O-(C-4') angle for each pair of residues to be fixed and derivable from known crystal structures. The absolute van der Waals energies, but not the relative energies of different conformations, are sensitive to the choice of energy functions and atomic coordinates. The results lead to possible explanations of the known conformational stiffness of cellulose and its solubility properties in alkali. The characteristics of xylan conformations are compared with cellulose.

N 134   "Configurational Statistics of Polysaccharide Chains. Part I. Amylose"
Rao, V. S. R.; Yathindra, N.; Sundararajan, P. R. *Biopolymers* **1969**, *8*, 325.

The conformational energy of amylose molecule is computed as a function of rotations ϕ and ψ about interunit glycosidic bonds. The characteristic ratio $<r^2>_0 / nl_v^2$ of the mean-square end-to-end distance of the unperturbed chain and the square of the virtual bond length l_v = 420 pm calculated for amylose chain is about 6.9, which is in good agreement with the experimental value (6.6 ± 0.2). The value of this ratio calculated for free rotation about the inter unit glycosidic bond is 1.1. The conformational map reveals that the rotations of pyranose units about the interunit glycosidic bonds are highly restricted; this tends to lock these units in a preferred orientation. From the probability of occurrence of the glucose residue conformations is concluded that amylose exists as a random coil with short helical sequences in neutral aqueous solutions.

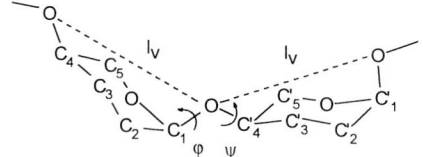

N 135   "Conformational Studies of β-D-1,4'-Xylan"
Sundararajan, P. R.; Rao, V. S. R. *Biopolymers* **1969**, *8*, 305.

The nonbonded interaction energy is computed for xylobiose and xylan as a function of the dihedral angles (ϕ, ψ). The energy maps indicate that interactions higher than the second neighbor are negligible. Of the four possibilities, the left-handed helical conformation with (ϕ,ψ) = (63°,25°) is of lowest energy. The allowed region map shows that the freedom of rotation of the monomer units in cellulose is more restricted than that of the monomer units in xylan, because of the presence of the -CH$_2$OH group in the former.

N 136   "Studies on the Helical Conformational of Amylose and Possible Interconversions"
Sundararajan, P. R.; Rao, V. S. R. *Biopolymers* **1969**, *8*, 313.

The nonbonded energy (van der Waals) is computed for isolated helical amylose chains as a function of the dihedral angles (ϕ,ψ), *i.e.*, the relative orientations of the glucose residues in the polysaccharide chain. In conformity with x-ray data, different helical conformations are proposed for different crystalline modifications of amylose.

N 137   "Conformational Analysis of Polysaccharides. Part II. Alternating Copolymers of the Agar-Carrageenan-Chondroitin type by Model Building in the Computer with Calculation of Helical Parameters"
Rees, D. A. *J. Chem. Soc. (B)* **1969**, 217.

Model building in the computer is used to analyze the conformational effects of steric interactions between atoms of the polymer skeleton for κ-carrageenan, ι-carrageenan, λ-carrageenan, agar, chondroitin, chondroitin sulfates, dermatan sulfate, keratan sulfate, hyaluronic acid, and related polysaccharides. Over 99% of the conformations are thus excluded and virtually all of the remainder for each polysaccharide lie close together. Predictions are also made from disaccharide crystal structures and checked against experimental results.

N 138   "The Conformational Energy of Maltose and Amylose"
Goebel, C. V.; Dimpfl, W. L.; Brant, D. A. *Macromolecules* **1970**, *3*, 644.

The conformational energies of helical amylose and dimeric skeletal unit maltose are calculated using structural models based upon the crystal structures of cyclohexaamylose and methyl β-maltoside. Potential functions which account for intramolecular van der Waals, Coulombic, and hydrogen-bonded interactions are employed. Intramolecular hydrogen bonding is found greatly to enhance the stability of certain helical conformations favoured also on the basis of van der Waals interactions. Coulombic interactions appear to make negligible contributions to the conformational energy within the domain of sterically allowed conformations.

N 139   "A Theoretical Interpretation of the Aqueous Solution Properties of Amylose and Its Derivatives"
Brant, D. A.; Dimpfl, W. L. *Macromolecules* **1970**, *3*, 655.

The observed dependence upon degree of polymerization and temperature of the unperturbed dimensions of two derivatives of amylose in aqueous solution are interpreted using the statistical mechanical theory of polymer configuration. The results are used to provide a description of the aqueous configuration of amylose. The model accounts for interdependence of bond rotations at each glycosidic bridge but assumes the rotations at each bridge to be independent of those at neighboring bridges due to suppression of intramolecular hydrogen bonding by solvation in aqueous medium. Structural models based on the crystal structures of cyclohexaamylose and methyl β-maltoside are employed. Configurational partition functions are established using approximate conformational energy calculations for an appropriately chosen chain segment. Excellent agrement is achieved between theory and experiment when the cyclohexaamylose structural geometry is used and for reasonable values of all strucutral and potential function parameters. It is concluded that amylose in aqueous solution is a statistical coil without identifiable helical character in the absence of complexing agents.

The numerical results for $<T>$ obtained with the "HRW" geometry set for $\theta = 114.00°$ and $114.75°$ ($\theta \equiv C_1-O-C_{4'}$ valence angle) at the temperatures 288, 298, 308, and 318 K using potential functions which include the dipolar terms and parameters a_{jk} which are evaluated by adding 10 pm to the conventional van der Waals radii of all atoms are given.

	288 K	298 K	308 K	318 K
$\theta = 144.00°$	$<T> = \begin{bmatrix} 0.619 & 0.628 & -0.413 \\ -0.767 & 0.516 & -0.255 \\ 0.073 & 0.487 & 0.801 \end{bmatrix}$	$<T> = \begin{bmatrix} 0.619 & 0.630 & -0.409 \\ -0.767 & 0.517 & -0.249 \\ 0.075 & 0.480 & 0.802 \end{bmatrix}$	$<T> = \begin{bmatrix} 0.618 & 0.631 & -0.405 \\ -0.767 & 0.517 & -0.243 \\ 0.077 & 0.473 & 0.803 \end{bmatrix}$	$<T> = \begin{bmatrix} 0.617 & 0.633 & -0.401 \\ -0.766 & 0.517 & -0.236 \\ 0.079 & 0.466 & 0.804 \end{bmatrix}$
$\theta = 144.75°$	$<T> = \begin{bmatrix} 0.607 & 0.645 & -0.396 \\ -0.775 & 0.509 & -0.229 \\ 0.075 & 0.459 & 0.808 \end{bmatrix}$	$<T> = \begin{bmatrix} 0.607 & 0.647 & -0.392 \\ -0.774 & 0.509 & -0.222 \\ 0.077 & 0.452 & 0.809 \end{bmatrix}$	$<T> = \begin{bmatrix} 0.606 & 0.649 & -0.388 \\ -0.774 & 0.509 & -0.216 \\ 0.079 & 0.445 & 0.810 \end{bmatrix}$	$<T> = \begin{bmatrix} 0.606 & 0.650 & -0.383 \\ -0.774 & 0.509 & -0.210 \\ 0.081 & 0.438 & 0.811 \end{bmatrix}$

N 140 "Configurational Statistics of Polysaccharide Chains. Part II. Cellulose"
Yathindra, N.; Rao, V. S. R. *Biopolymers* **1970**, *9*, 783.

The characteristic ratio of the unperturbed cellulose chain is computed as a function of the angle τ at the bridge oxygen atom and the degree of polymerization. Very high values of the order of 40 or more, depending on the angle at the bridge oxygen atom, are obtained for this ratio, indicating that cellulose chains are highly extended. The average dimensions of cellulose chains are found to be sensitive even for small changes in the angle at the bridge oxygen, and these chains attain the character of a random coil in very high molecular weight range. The large differences in the unperturbed dimensions of cellulosic chains observed in different solvents are attributed to the possible small changes in the angle τ caused by specific solvent interactions. The cellulose chain may thus be considered as a succession virtual bonds of length l_v = 545 pm and can be represented by (N-1) sets of rotational angles.

N 141 "Conformational Analysis of Polysaccharides. Part IV. Long-range Contacts in Some β-Glucans by Model Building in the Computer and the Influence of Oligosaccharide Conformation on Optical Rotation"
Rees, D. A.; Skerrett, R. J. *J. Chem. Soc. (B)* **1970**, 189.

Model building in the computer is used to compare the influence of remote contacts on the stereochemistry and properties of β-1,2-, β-1,3-, and β-1,4-glucans, including their optical rotation behavior, and to explore the possibility of chain folding in cellulose, chitin, and xylan derivatives.

N 142 "Conformational Analysis of Polysaccharides. Part V. The Characterization of Linkage Conformations (Chain Conformations) by Optical Rotation at a Single Wavelength. Evidence for Distortion of Cyclohexa-amylose in Aqueous Solution. Optical Rotation and Amylose Conformation"
Rees, D. A. *J. Chem. Soc. (B)* **1970**, 877.

A relationship is suggested between the optical rotation measured at a single wavelength for a di-, oligo-, or polysaccharide, and the conformation at the glycosidic linkage as expressed in torsion angles about the C—O and O—C bonds; it is based on *Kauzmann's* additivity prinziples for optical rotation, *Brewster's* empirical treatment in terms of screw patterns of polarizability, and *Whiffen's* earlier suggestions for carbohydrates. By use of conformation details from crystal structures with the optical rotations of component monosaccharide derivatives, it is then possible to predict the molecular rotations of β-cellobiose and α-lactose derivatives in water, and of cyclohexa-amylose in dimethyl sulfoxide, to within a few degrees.

N 143 "Conformational Energy Calculations on Alginic Acid I. Helix Parameters and Flexibility of the Homopolymers"
Whittington, S. G. *Biopolymers* **1971**, *10*, 1481.

Conformational energy maps are calculated for the 1-4-linked dimers of β-D-mannuronic acid and α-L-guluronic acid. Helix parameters are calculated for poly(mannuronic acid) and for poly(guluronic acid), which are in reasonable agreement with data from x-ray fiber diffraction studies of these polysaccharides. The flexibility of the homopolymers is investigated by calculating the characteristic ratios, *i.e.*, the ratio of the mean-square end-to-end length of the unperturbed chains to the product of the number of residues in the chain and the virtual bond length. The general conclusions are that both polymers are very stiff and extended, but that poly(mannuronic acid) is less extended than poly(guluronic acid).

N 144 "Conformational Energy Calculations on Alginic Acid II. Conformational Statistics of the Copolymers"
Whittington, S. G. *Biopolymers* **1971**, *10*, 1617.

Conformational energy maps are calculated for β-D-mannuronic acid (1-4) α-L-guluronic acid and for α-L-guluronic acid (1-4) β-D-mannuronic acid. These are used, together with maps previously calculated for the homomonomeric dimers, to estimate the characteristic ratios and Kuhn length of the alternating copolymer and of a stochastic copolymer similar in composition to that from *L. digitata*. The results show that the alternating copolymer is less extended than either homopolymer. Kuhn length calculated for the stochastic copolymer agree well with experimental results on high ionic strength solutions of alginate isolated from *L. digitata*.

N 145 "Polysaccharide Conformation. Part VI. Computer Model-Building for Linear and Branched Pyranoglycans. Correlations with Biological Function. Preliminary Assessment of Inter-Residue Forces in Aqueous Solution. Further Interpretation of Optical Rotation in Terms of Chain Conformation"
Rees, D. A.; Scott, W. E. *J. Chem. Soc. (B)* **1971**, 469.

Homopolymers of glucopyranose, galactopyranose, mannopyranose, xylopyranose, and arabinopyranose, with various positions and configurations of linkage, are compared by model-building in the computer in an attempt to formulate simple rules for conformational analysis. Regular conformations are very restricted by steric forces alone, and each polymer has one of four characteristic shapes: *Type A:* extended and ribbon-like; *Type B:* flexible and helical; *Type C:* rigid and crumpled; *Type D:* very flexible but, on the average, rather extended.

N 146 "Unperturbed Dimensions of Some 1-4-Linked Homopolysaccharides"
Whittington, S. G. *Macromolecules* **1971**, *4*, 569.

The characteristic ratios of a number of 1-4-linked homopolysaccharides are calculated using the methods develped by *Flory et al.*. It appears that the characteristic ratio is related to the freedom of rotation about the glycosidic bonds but also to the type of bonding (*e.g.*, axial-axial, equatorial-axial, etc.) between the monomers. The ring geometry is derived from the idealized xylose coordinates of *Settineri* and *Marchessault* (*J. Polym. Sci., Part C* **1965**, *11*, 253) and is assumed to be fixed in each calculation. The bridge parameters used are those obtained by *Chu* and *Jeffrey* (*Acta Crystallogr., Sect. B* **1968**, *24*, 830), *i.e.*, the C(1)—O bond length is 139.7 pm, the O—C(4') bond length is 144.6 pm, and the bridge angle is 116°. With a fixed ring geometry, so that the only variables are the dihedral angles about the glycosidic bonds, it is convenient to define a sequence of virtual bonds, l_i, between adjacent bridge oxygens.

N 147 "Configurational Statistics of Polysaccharide Chains. Part III. Linear β-(1→4') Xylan and Mannan"
Yathindra, N.; Rao, V. S. R. *Biopolymers* **1971**, *10*, 1891.

The characteristic ratio $<r^2>_0 / Nl_v^2$ of the β-D(1→4')-linked polysaccharides xylan and mannan are computed as a function of the angle τ at the bridge oxygen atom and the degree of polymerization N. The calculated values of the characteristic ratio are very high relative to their free rotational dimensions. The characteristic ratio of these polysaccharides converges to the asymptotic value at low degree of polymerization at higher τ values. The low values of the calculated characteristic ratio of xylan compared to cellulose and mannan for the same τ value indicate that the former is more flexible and assumes a compact configuration.

N 148 "Ionic Polysaccharides. V. Conformational Studies of Hyaluronic Acid, Cellulose, and Laminaran"
Cleland, R. L. *Biopolymers* **1971**, *10*, 1925.

Hindered rotation is studied for the disaccharides composed of basic β-glucopyranose units. The van der Waals interactions are calculated for the Lennard-Jones, Buckingham, and Kitaygorodsky interatomic potential functions. Values of the ratio of unperturbed to free-rotation root-mean-square end-to-end distance are calculated for chains composed of the unsolvated disaccharide repeating units.

N 149   "Conformational Statistics of 1,3- and 1,4-Linked Homopolysaccharides"
Whittington, S. G.; Glover, R. M. *Macromolecules* **1972**, *5*, 55.

Unperturbed dimensions are calculated for several 1,3- and 1,4-linked homopolysaccharides. The characteristic ratio depends strongly on the bonding geometry and on the degree of rotational freedom about the glycosidic bonds. It appears that unperturbed dimensions can be predicted qualitatively from a knowledge of the regular conformations of the polysaccharide.

N 150   "Conformational Statistics of Some Copolysaccharides"
Hallman, G. M.; Whittington, S. G. *Macromolecules* **1973**, *6*, 386.

Unperturbed dimensions are calculated for several polysaccharide copolymers. The effect of overall composition on the characteristic ratio is investigated for the three polymers: glucomannan (a 1,4-linked copolymer of β-D-glucose and β-D-mannose), alginic acid (a 1,4-linked copolymer of α-L-guluronic acid and β-D-mannuronic acid) and pectic acid (considered as a copolymer of α-D-galacturonic acid and α-L-rhamnose, in which the galacturonic acid units are linked 1,4 and the rhamnose units are linked 1,2). In each case the effect of sequence distribution is also considered. For Bernoullian sequence statistics the characteristic ratio is a monotonic function of the composition for glucomannan and for pectic acid but it passes through a minimum for alginic acid. For a fixed monomer composition the characteristic ratio is sensitive to the sequence statistics of the copolymer.

N 151   "A General Treatment of the Configurational Statistics of Polysaccharides"
Brant, D. A.; Goebel, K. D. *Macromolecules* **1975**, *8*, 522.

A treatment of the configurational statistics of polysaccharides is given in the isomeric state approximation. All classes of linear polysaccharides of specified chemical sequence are treated simultaneously. Chain tortuosity arising from torsional motions about the chemical bonds of the glycosidic linkages is recognized explicitly as is the possibility for conformational isomerism of the sugar residues. Valence angles and length are taken to be fixed at the equilibrium values, and pyranose residues in their chain conformation are treated as inflexible constituents of the skeletal structure. Pyranose and furanose forms capable of pseudorotation may be incorporated as rigid skeletal entities as well, provided suitable attention is given to the selection and interpretation of the conformational isomeric states included. Separation of the configuration energy into independent contributions is shown to be impossible in general. Methods are described for assessing the influence of neighbor interactions on the populations of the several conformers of the sugar residues. The relative conformational free energy of the flexible and chain form conformers of pyranose sugars is discussed, and appropriate measures of polysaccharide chain flexibility and stiffness are suggested.

N 152   "A Monte Carlo Study of the Amylosic Chain Conformation"
Jordan, R. C.; Brant, D. A.; Cesàro, A. *Biopolymers* **1978**, *17*, 2617.

Monte Carlo studies of the unperturbed amylosic chain conformation are carried out in the approximation of separable chain configuration energies. Sample chains of arbitrary chain length are generated so as to be distributed consistent with refined estimates of the configuration energy and thus suitable for evaluating of averages of the desired configuration dependent properties. The amylosic persistence vector and persistence length are calculated as a function of chain length for the chain model employed.

N 153 "Analysis of Cooperative Conformational Transitions in Cellulose and Amylose Tricarbanilates"
Hsu, B.; McWherter, C. A.; Brant, D. A.; Burchard, W. *Macromolecules* **1982**, *15*, 1350.

Dissolved cellulose tricarbanilate (CTC) and amylose tricarbanilate (ATC) both exhibit temperature and solvent-induced transitions between forms of differing stiffness and chain extension. The transition temperatures are strongly dependent on chain length, suggesting that the transitions are very cooperative. It is shown in this paper that the chain length dependences of the unperturbed dimensions of CTC and ATC also reflect the cooperative conformational transitions inferred from other kinds of measurements. The method of *Miller* and *Flory* (*J. Mol. Biol.* **1966**, *15*, 298) and realistic chain models for the two polymers are used to fit the observed chain length dependences of the unperturbed dimensions. *Zimm-Bragg* cooperativity parameters σ on the order of 10^{-5} and 10^{-6} for CTC and ATC, respectively, are required to match the slow convergence of the characteristic ratios to their asymptotic values at high degrees of polymerization. It was not possible to fit the data in question with any model that does not involve a cooperative transition from a more flexible form of the chain skeleton, which predominates at low degrees of polymerization, to a less flexible form, the stability of which is enhanced at high degrees of polymerization. The theoretical models providing the best fits to the unperturbed dimension data are used to calculate persistence length of 110 and 103 Å for CTC and ATC, respectively. The valence angle β at the oxygen of the glycosidic bridge between successive residues is taken to be 116.5°. The pyranoside ring of each substituted sugar residue is assumed to be rigid so as to render the "virtual bond" distance between successive glycosidic oxygens fixed at the values 547 and 440 pm, respectively, for the CTC and ATC chains. The value of σ is determined as $\sigma = 1.1 \times 10^{-5}$ for CTC and $\sigma = 3.0 \times 10^{-5}$ for ATC. The partition function is calculated by $Z_x = U_1 U_2 \ldots U_i \ldots U_x$, where

$$U_i = \begin{bmatrix} s & 1 \\ \sigma s & 1 \end{bmatrix} \qquad U_1 = \begin{bmatrix} \sigma s & 1 \end{bmatrix} \qquad U_x = \begin{bmatrix} 1 \\ 1 \end{bmatrix}$$

N 154 "Comparative Flexibility, Extension, and Conformation of Some Polysaccharide Chains"
Burton, B. A.; Brant, D. A. *Biopolymers* **1983**, *22*, 1769.

Realistic polymer chain models are developed for several polysaccharides to illustrate, using perspective drawings of representative chain conformations, the wide range of configuration, extension, and flexibility found in chains of the polysaccharide class. A method for incorporating the *gauche* or *exo*-anomeric effect into polysaccharide conformational energy functions is described, and a novel measure of directional correlation and pseudohelical persistence is utilized to help distinguish the differences in chain configuration observed among the polysaccharides compared.

N 155 "The Sequence Statistics and Solution Conformation of a Barley (1 → 3, 1 → 4)-β-D-Glucan"
Buliga, G. S.; Brant, D. A.; Fincher, G. B. *Carbohydr. Res.* **1986**, *157*, 139.

The sequence statistics and aqueous solution conformation of the 40° water-soluble (1 → 3, 1 → 4)-β-D-glucan is modeled realistically using the known sequence distribution of (1 → 3) and (1 → 4) linkages, theoretical conformational analysis, and the statistical mechanical theory of polymer-chain conformation. Chain flexibility in the 40° water-soluble β-glucan fraction is shown to arise principally from the isolated β-(1 → 3) linkages; blocks of two or more contiguous β-(1 → 3) linkages provide a source of additional flexibility which may influence the properties of barely β-glucan fractions containing a significant proportion of such sequences.

N 156 "The Influence of Side Chains on the Calculated Dimensions of Three Related Bacterial Polysaccharides"
Talashek, T. A.; Brant, D. A. *Carbohydr. Res.* **1987**, *160*, 303.

The effect of van der Waals interactions between side chains and backbone on the shape of three bacterial polysaccharides in solution is investigated. The three polymers, namely, gellan, welan, and rhamsan, share the same four-sugar backbone repeating unit. Gellan is unbranched, whereas welan and rhamsan display comblike branching. Consequently, the effect of chain branching on backbone conformation may be investigated. Van der Waals repulsive interactions of side chains and backbone serve to limit, somewhat, the range of conformational freedom of the welan backbone in comparison to that of gellan. Attractive side chain-backbone interactions, which may be as significant as 8 - 12 kJ · mol^{-1}, predominate over much of the accessible conformational space of the welan backbone. Despite the strength of these interactions, the unperturbed shape of welan in solution is calculated to be very similar to that of the unbranched gellan. Attractive side chain-backbone interactions in rhamsan have a modest influence on the conformational characteristics of the rhamsane backbone. The calculated, unperturbed conformation in solution is slightly more extended than that of gellan and welan, but the fundamental shape of the chain is changed only slightly. Significant differences in the physical properties of these polymers seem not to arise from differences in their random-coil conformations provoked by van der Waals interactions of side chain and backbone.

N 157 "Theoretical Interpretation of the Unperturbed Aqueous Solution Configuration of Pullulan"
Buliga, G. S.; Brant, D. A. *Int. J. Biol. Macromol.* **1987**, *9*, 77.

A structurally realistic theoretical model for the pullulan chain is developed and refined to be consistent with the experimentally measured unperturbed dimensions of aqueous pullulan and with the temperature dependence of the unperturbed dimensions. The model is based on the mean structural geometry for an α-D-glucose residue recommended by *Arnott* and *Scott* (*J. Chem. Soc., Perkin Trans. 2* **1972**, 324) and incorporates, in estimates of the conformational energy of the chain, a treatment of the *gauche*, or *exo*-anomeric, effect proposed by *Abe* and *Mark* (*J. Am. Chem. Soc.* **1976**, *98*, 6468, 6477). The valence angle β_4 [$C_1-O-C_{4'}$] at the glycosidic oxygen of the (1 → 4) linked dimers is varied within the range 115-119° actually observed in a variety of crystalline oligosaccharides. The glycosidic valence angle β_6 [$C_1-O-C_{6'}$] of the (1 → 6)-linked dimer is varied within the range 110-115°.

Received: November 1996

Author Index Volumes 101-131/132

Author Index Vols. 1-100 see Vol. 100

Adolf, D. B. see Ediger, M. D.: Vol. 116, pp. 73-110.
Aharoni, S. M. and *Edwards, S. F.*: Rigid Polymer Networks. Vol. 118, pp. 1-231.
Améduri, B., Boutevin, B. and Gramain, P.: Synthesis of Block Copolymers by Radical Polymerization and Telomerization. Vol. 127, pp. 87-142.
Améduri, B. and *Boutevin, B.*: Synthesis and Properties of Fluorinated Telechelic Monodispersed Compounds. Vol. 102, pp. 133-170.
Amselem, S. see Domb, A. J.: Vol. 107, pp. 93-142.
Andrady, A. L.: Wavelenght Sensitivity in Polymer Photodegradation. Vol. 128, pp. 47-94.
Andreis, M. and *Koenig, J. L.*: Application of Nitrogen-15 NMR to Polymers. Vol. 124, pp. 191-238.
Angiolini, L. see Carlini, C.: Vol. 123, pp. 127-214.
Anseth, K. S., Newman, S. M. and *Bowman, C. N.*: Polymeric Dental Composites: Properties and Reaction Behavior of Multimethacrylate Dental Restorations. Vol. 122, pp. 177-218.
Armitage, B. A. see O'Brien, D. F.: Vol. 126, pp. 53-58.
Arndt, M. see Kaminski, W.: Vol. 127, pp. 143-187.
Arnold Jr., F. E. and *Arnold, F. E.*: Rigid-Rod Polymers and Molecular Composites. Vol. 117, pp. 257-296.
Arshady, R.: Polymer Synthesis via Activated Esters: A New Dimension of Creativity in Macromolecular Chemistry. Vol. 111, pp. 1-42.

Bahar, I., Erman, B. and *Monnerie, L.*: Effect of Molecular Structure on Local Chain Dynamics: Analytical Approaches and Computational Methods. Vol. 116, pp. 145-206.
Baltá-Calleja, F. J., González Arche, A., Ezquerra, T. A., Santa Cruz, C., Batallón, F., Frick, B. and *López Cabarcos, E.*: Structure and Properties of Ferroelectric Copolymers of Poly(vinylidene) Fluoride. Vol. 108, pp. 1-48.
Barshtein, G. R. and *Sabsai, O. Y.*: Compositions with Mineralorganic Fillers. Vol. 101, pp.1-28.
Batallán, F. see Baltá-Calleja, F. J.: Vol. 108, pp. 1-48.
Barton, J. see Hunkeler, D.: Vol. 112, pp. 115-134.
Bell, C. L. and *Peppas, N. A.*: Biomedical Membranes from Hydrogels and Interpolymer Complexes. Vol. 122, pp. 125-176.
Bennett, D. E. see O'Brien, D. F.: Vol. 126, pp. 53-84.
Berry, G.C.: Static and Dynamic Light Scattering on Moderately Concentraded Solutions: Isotropic Solutions of Flexible and Rodlike Chains and Nematic Solutions of Rodlike Chains. Vol. 114, pp. 233-290.
Bershtein, V. A. and *Ryzhov, V. A.*: Far Infrared Spectroscopy of Polymers. Vol. 114, pp. 43-122.
Bigg, D. M.: Thermal Conductivity of Heterophase Polymer Compositions. Vol. 119, pp. 1-30.

Binder, K.: Phase Transitions in Polymer Blends and Block Copolymer Melts: Some Recent Developments. Vol. 112, pp. 115-134.
Bird, R. B. see Curtiss, C. F.: Vol. 125, pp. 1-102.
Biswas, M. and *Mukherjee, A.*: Synthesis and Evaluation of Metal-Containing Polymers. Vol. 115, pp. 89-124.
Boutevin, B. and *Robin, J. J.*: Synthesis and Properties of Fluorinated Diols. Vol. 102. pp. 105-132.
Boutevin, B. see Amédouri, B.: Vol. 102, pp. 133-170.
Boutevin, B. see Améduri, B.: Vol. 127, pp. 87-142.
Bowman, C. N. see Anseth, K. S.: Vol. 122, pp. 177-218.
Boyd, R. H.: Prediction of Polymer Crystal Structures and Properties. Vol. 116, pp. 1-26.
Bronnikov, S. V., *Vettegren, V. I.* and Frenkel, S. Y.: Kinetics of Deformation and Relaxation in Highly Oriented Polymers. Vol. 125, pp. 103-146.
Bruza, K. J. see Kirchhoff, R. A.: Vol. 117, pp. 1-66.
Burban, J. H. see Cussler, E. L.: Vol. 110, pp. 67-80.

Cameron, N. R. and *Sherrington, D. C.:* High Internal Phase Emulsions (HIPEs)-Structure, Properties and Use in Polymer Preparation. Vol. 126, pp. 163-214.
Candau, F. see Hunkeler, D.: Vol. 112, pp. 115-134.
Capek, I.: Kinetics of the Free-Radical Emulsion Polymerization of Vinyl Chloride. Vol. 120, pp. 135-206.
Carlini, C. and *Angiolini, L.*: Polymers as Free Radical Photoinitiators. Vol. 123, pp. 127-214.
Casas-Vazquez, J. see Jou, D.: Vol. 120, pp. 207-266.
Chen, P. see Jaffe, M.: Vol. 117, pp. 297-328.
Choe, E.-W. see Jaffe, M.: Vol. 117, pp. 297-328.
Chow, T. S.: Glassy State Relaxation and Deformation in Polymers. Vol. 103, pp. 149-190.
Chung, T.-S. see Jaffe, M.: Vol. 117, pp. 297-328.
Connell, J. W. see Hergenrother, P. M.: Vol. 117, pp. 67-110.
Criado-Sancho, M. see Jou, D.: Vol. 120, pp. 207-266.
Curro, J.G. see Schweizer, K.S.: Vol. 116, pp. 319-378.
Curtiss, C. F. and *Bird, R. B.*: Statistical Mechanics of Transport Phenomena: Polymeric Liquid Mixtures. Vol. 125, pp. 1-102.
Cussler, E. L., *Wang, K. L.* and *Burban, J. H.*: Hydrogels as Separation Agents. Vol. 110, pp. 67-80.

Dimonie, M. V. see Hunkeler, D.: Vol. 112, pp. 115-134.
Dodd, L. R. and *Theodorou, D. N.*: Atomistic Monte Carlo Simulation and Continuum Mean Field Theory of the Structure and Equation of State Properties of Alkane and Polymer Melts. Vol. 116, pp. 249-282.
Doelker, E.: Cellulose Derivatives. Vol. 107, pp. 199-266.
Domb, A. J., *Amselem, S.*, *Shah, J.* and *Maniar, M.*: Polyanhydrides: Synthesis and Characterization. Vol.107, pp. 93-142.
Dubrovskii, S. A. see Kazanskii, K. S.: Vol. 104, pp. 97-134.
Dunkin, I. R. see Steinke, J.: Vol. 123, pp. 81-126.

Economy, J. and *Goranov, K.*: Thermotropic Liquid Crystalline Polymers for High Performance Applications. Vol. 117, pp. 221-256.

Ediger M. D. and *Adolf, D. B.*: Brownian Dynamics Simulations of Local Polymer Dynamics. Vol. 116, pp. 73-110.
Edwards, S. F. see Aharoni, S. M.: Vol. 118, pp. 1-231.
Endo, T. see Yagci, Y.: Vol. 127, pp. 59-86.
Erman, B. see Bahar, I.: Vol. 116, pp. 145-206.
Ezquerra, T. A. see Baltá-Calleja, F. J.: Vol. 108, pp. 1-48.

Fendler, J.H.: Membrane-Mimetic Approach to Advanced Materials. Vol. 113, pp. 1-209.
Fetters, L. J. see Xu, Z.: Vol. 120, pp. 1-50.
Förster, S. and *Schmidt, M.*: Polyelectrolytes in Solution. Vol. 120, pp. 51-134.
Frenkel, S. Y. see Bronnikov, S. V.: Vol. 125, pp. 103-146.
Frick, B. see Baltá-Calleja, F. J.: Vol. 108, pp. 1-48.
Fridman, M. L.: see Terent'eva, J. P.: Vol. 101, pp. 29-64.
Ganesh, K. see Kishore, K.: Vol. 121, pp. 81-122.
Geckeler, K. E. see Rivas, B.: Vol. 102, pp. 171-188.
Geckeler, K. E.: Soluble Polymer Supports for Liquid-Phase Synthesis. Vol. 121, pp. 31-80.
Gehrke, S. H.: Synthesis, Equilibrium Swelling, Kinetics Permeability and Applications of Environmentally Responsive Gels. Vol. 110, pp. 81-144.
Godovsky, D. Y.: Electron Behavior and Magnetic Properties Polymer-Nanocomposites. Vol. 119, pp. 79-122.
González Arche, A. see Baltá-Calleja, F. J.: Vol. 108, pp. 1-48.
Goranov, K. see Economy, J.: Vol. 117, pp. 221-256.
Gramain, P. see Améduri, B.: Vol. 127, pp. 87-142.
Grosberg, A. and *Nechaev, S.*: Polymer Topology. Vol. 106, pp. 1-30.
Grubbs, R., Risse, W. and *Novac, B.*: The Development of Well-defined Catalysts for Ring-Opening Olefin Metathesis. Vol. 102, pp. 47-72.
van Gunsteren, W. F. see Gusev, A. A.: Vol. 116, pp. 207-248.
Gusev, A. A., Müller-Plathe, F., van Gunsteren, W. F. and *Suter, U. W.*: Dynamics of Small Molecules in Bulk Polymers. Vol. 116, pp. 207-248.
Guillot, J. see Hunkeler, D.: Vol. 112, pp. 115-134.
Guyot, A. and *Tauer, K.*: Reactive Surfactants in Emulsion Polymerization. Vol. 111, pp. 43-66.

Hadjichristidis, N. see Xu, Z.: Vol. 120, pp. 1-50.
Hall, H. K. see Penelle, J.: Vol. 102, pp. 73-104.
Hammouda, B.: SANS from Homogeneous Polymer Mixtures: A Unified Overview. Vol. 106, pp. 87-134.
Hedrick, J. L. see Hergenrother, P. M.: Vol. 117, pp. 67-110.
Heller, J.: Poly (Ortho Esters). Vol. 107, pp. 41-92.
Hemielec, A. A. see Hunkeler, D.: Vol. 112, pp. 115-134.
Hergenrother, P. M., Connell, J. W., Labadie, J. W. and *Hedrick, J. L.*: Poly(arylene ether)s Containing Heterocyclic Units. Vol. 117, pp. 67-110.
Hiramatsu, N. see Matsushige, M.: Vol. 125, pp. 147-186.
Hirasa, O. see Suzuki, M.: Vol. 110, pp. 241-262.
Hirotsu, S.: Coexistence of Phases and the Nature of First-Order Transition in Poly-N-isopropylacrylamide Gels. Vol. 110, pp. 1-26.
Hunkeler, D., Candau, F., Pichot, C., Hemielec, A. E., Xie, T. Y., Barton, J., Vaskova, V., Guillot, J., Dimonie, M. V., Reichert, K. H.: Heterophase Polymerization: A Physical and Kinetic Comparision and Categorization. Vol. 112, pp. 115-134.

Ichikawa, T. see Yoshida, H.: Vol. 105, pp. 3-36.
Ilavsky, M.: Effect on Phase Transition on Swelling and Mechanical Behavior of Synthetic Hydrogels. Vol. 109, pp. 173-206.
Inomata, H. see Saito, S.: Vol. 106, pp. 207-232.
Irie, M.: Stimuli-Responsive Poly(N-isopropylacrylamide), Photo- and Chemical-Induced Phase Transitions. Vol. 110, pp. 49-66.
Ise, N. see Matsuoka, H.: Vol. 114, pp. 187-232.
Ivanov, A. E. see Zubov, V. P.: Vol. 104, pp. 135-176.

Jaffe, M., Chen, P., Choe, E.-W., Chung, T.-S. and *Makhija, S.*: High Performance Polymer Blends. Vol. 117, pp. 297-328.
Jou, D., Casas-Vazquez, J. and *Criado-Sancho, M.*: Thermodynamics of Polymer Solutions under Flow: Phase Separation and Polymer Degradation. Vol. 120, pp. 207-266.

Kaetsu, I.: Radiation Synthesis of Polymeric Materials for Biomedical and Biochemical Applications. Vol. 105, pp. 81-98.
Kaminski, W. and *Arndt, M.*: Metallocenes for Polymer Catalysis. Vol. 127, pp. 143-187.
Kammer, H. W., Kressler, H. and *Kummerloewe, C.*: Phase Behavior of Polymer Blends - Effects of Thermodynamics and Rheology. Vol. 106, pp. 31-86.
Kandyrin, L. B. and *Kuleznev, V. N.*: The Dependence of Viscosity on the Composition of Concentrated Dispersions and the Free Volume Concept of Disperse Systems. Vol. 103, pp. 103-148.
Kaneko, M. see Ramaraj, R.: Vol. 123, pp. 215-242.
Kang, E. T., Neoh, K. G. and *Tan, K. L.*: X-Ray Photoelectron Spectroscopic Studies of Electroactive Polymers. Vol. 106, pp. 135-190.
Kazanskii, K. S. and *Dubrovskii, S. A.*: Chemistry and Physics of „Agricultural" Hydrogels. Vol. 104, pp. 97-134.
Kennedy, J. P. see Majoros, I.: Vol. 112, pp. 1-113.
Khokhlov, A., Starodybtzev, S. and *Vasilevskaya, V.*: Conformational Transitions of Polymer Gels: Theory and Experiment. Vol. 109, pp. 121-172.
Kilian, H. G. and *Pieper, T.*: Packing of Chain Segments. A Method for Describing X-Ray Patterns of Crystalline, Liquid Crystalline and Non-Crystalline Polymers. Vol. 108, pp. 49-90.
Kishore, K. and *Ganesh, K.*: Polymers Containing Disulfide, Tetrasulfide, Diselenide and Ditelluride Linkages in the Main Chain. Vol. 121, pp. 81-122.
Klier, J. see Scranton, A. B.: Vol. 122, pp. 1-54.
Kobayashi, S., Shoda, S. and *Uyama, H.*: Enzymatic Polymerization and Oligomerization. Vol. 121, pp. 1-30.
Koenig, J. L. see Andreis, M.: Vol. 124, pp. 191-238.
Kokufuta, E.: Novel Applications for Stimulus-Sensitive Polymer Gels in the Preparation of Functional Immobilized Biocatalysts. Vol. 110, pp. 157-178.
Konno, M. see Saito, S.: Vol. 109, pp. 207-232.
Kopecek, J. see Putnam, D.: Vol. 122, pp. 55-124.
Koßmehl, G. see Schopf, G.: Vol. 129, pp. 1-145.
Kressler, J. see Kammer, H. W.: Vol. 106, pp. 31-86.
Kirchhoff, R. A. and *Bruza, K. J.*: Polymers from Benzocyclobutenes. Vol. 117, pp. 1-66.
Kuchanov, S. I.: Modern Aspects of Quantitative Theory of Free-Radical Copolymerization. Vol. 103, pp. 1-102.
Kuleznev, V. N. see Kandyrin, L. B.: Vol. 103, pp. 103-148.

Kulichkhin, S. G. see Malkin, A. Y.: Vol. 101, pp. 217-258.
Kummerloewe, C. see Kammer, H. W.: Vol. 106, pp. 31-86.
Kuznetsova, N. P. see Samsonov, G. V.: Vol. 104, pp. 1-50.*Labadie, J. W.* see Hergenrother, P. M.: Vol. 117, pp. 67-110.

Lamparski, H. G. see O'Brien, D. F.: Vol. 126, pp. 53-84.
Laschewsky, A.: Molecular Concepts, Self-Organisation and Properties of Polysoaps. Vol. 124, pp. 1-86.
Laso, M. see Leontidis, E.: Vol. 116, pp. 283-318.
Lazár, M. and *Rychlý, R.*: Oxidation of Hydrocarbon Polymers. Vol. 102, pp. 189-222.
Lenz, R. W.: Biodegradable Polymers. Vol. 107, pp. 1-40.
Leontidis, E., de Pablo, J. J., Laso, M. and *Suter, U. W.*: A Critical Evaluation of Novel Algorithms for the Off-Lattice Monte Carlo Simulation of Condensed Polymer Phases. Vol. 116, pp. 283-318.
Lesec, J. see Viovy, J.-L.: Vol. 114, pp. 1-42.
Liang, G. L. see Sumpter, B. G.: Vol. 116, pp. 27-72.
Lin, J. and *Sherrington, D. C.*: Recent Developments in the Synthesis, Thermostability and Liquid Crystal Properties of Aromatic Polyamides. Vol. 111, pp. 177-220.
López Cabarcos, E. see Baltá-Calleja, F. J.: Vol. 108, pp. 1-48.

Majoros, I., Nagy, A. and *Kennedy, J. P.*: Conventional and Living Carbocationic Polymerizations United. I. A Comprehensive Model and New Diagnostic Method to Probe the Mechanism of Homopolymerizations. Vol. 112, pp. 1-113.
Makhija, S. see Jaffe, M.: Vol. 117, pp. 297-328.
Malkin, A. Y. and *Kulichkhin, S. G.*: Rheokinetics of Curing. Vol. 101, pp. 217-258.
Maniar, M. see Domb, A. J.: Vol. 107, pp. 93-142.
Matsumoto, A.: Free-Radical Crosslinking Polymerization and Copolymerization of Multivinyl Compounds. Vol. 123, pp. 41-80.
Matsuoka, H. and *Ise, N.*: Small-Angle and Ultra-Small Angle Scattering Study of the Ordered Structure in Polyelectrolyte Solutions and Colloidal Dispersions. Vol. 114, pp. 187-232.
Matsushige, K., Hiramatsu, N. and *Okabe, H.*: Ultrasonic Spectroscopy for Polymeric Materials. Vol. 125, pp. 147-186.
Mattice, W. L. see Rehahn, M.: Vol. 131/132, pp. 1-475
Mays, W. see Xu, Z.: Vol. 120, pp. 1-50.
Mikos, A. G. see Thomson, R. C.: Vol. 122, pp. 245-274.
Miyasaka, K.: PVA-Iodine Complexes: Formation, Structure and Properties. Vol. 108. pp. 91-130.
Monnerie, L. see Bahar, I.: Vol. 116, pp. 145-206.
Morishima, Y.: Photoinduced Electron Transfer in Amphiphilic Polyelectrolyte Systems. Vol. 104, pp. 51-96.
Müllen, K. see Scherf, U.: Vol. 123, pp. 1-40.
Müller-Plathe, F. see Gusev, A. A.: Vol. 116, pp. 207-248.
Mukerherjee, A. see Biswas, M.: Vol. 115, pp. 89-124.
Mylnikov, V.: Photoconducting Polymers. Vol. 115, pp. 1-88.

Nagy, A. see Majoros, I.: Vol. 112, pp. 1-11.
Narasinham, B., Peppas, N. A.: The Physics of Polymer Dissolution: Modeling Approaches and Experimental Behavior. Vol. 128, pp. 157-208.
Nechaev, S. see Grosberg, A.: Vol. 106, pp. 1-30.

Neoh, K. G. see Kang, E. T.: Vol. 106, pp. 135-190.
Newman, S. M. see Anseth, K. S.: Vol. 122, pp. 177-218.
Nijenhuis, K. te: Thermoreversible Networks. Vol. 130, pp. 1-252.
Noid, D. W. see Sumpter, B. G.: Vol. 116, pp. 27-72.
Novac, B. see Grubbs, R.: Vol. 102, pp. 47-72.
Novikov, V. V. see Privalko, V. P.: Vol. 119, pp. 31-78.

O'Brien, D. F., Armitage, B. A., Bennett, D. E. and *Lamparski, H. G.*: Polymerization and Domain Formation in Lipid Assemblies. Vol. 126, pp. 53-84
Ogasawara, M.: Application of Pulse Radiolysis to the Study of Polymers and Polymerizations. Vol.105, pp.37-80.
Okabe, H. see Matsushige, K.: Vol. 125, pp. 147-186.
Okada, M.: Ring-Opening Polymerization of Bicyclic and Spiro Compounds. Reactivities and Polymerization Mechanisms. Vol. 102, pp. 1-46.
Okano, T.: Molecular Design of Temperature-Responsive Polymers as Intelligent Materials. Vol. 110, pp. 179-198.
Onuki, A.: Theory of Phase Transition in Polymer Gels. Vol. 109, pp. 63-120.
Osad'ko, I.S.: Selective Spectroscopy of Chromophore Doped Polymers and Glasses. Vol. 114, pp. 123-186.

de Pablo, J. J. see Leontidis, E.: Vol. 116, pp. 283-318.
Padias, A. B. see Penelle, J.: Vol. 102, pp. 73-104.
Pascault, J.-P. see Williams, R. J. J.: Vol. 128, pp. 95-156.
Pasch, H.: Analysis of Complex Polymers by Interaction Chromatography. Vol. 128, pp. 1-46.
Penelle, J., Hall, H. K., Padias, A. B. and *Tanaka, H.*: Captodative Olefins in Polymer Chemistry. Vol. 102, pp. 73-104.
Peppas, N. A. see Bell, C. L.: Vol. 122, pp. 125-176.
Peppas, N. A. see Narasimhan, B.: Vol. 128, pp. 157-208.
Pichot, C. see Hunkeler, D.: Vol. 112, pp. 115-134.
Pieper, T. see Kilian, H. G.: Vol. 108, pp. 49-90.
Pospíšil, J.: Functionalized Oligomers and Polymers as Stabilizers for Conventional Polymers. Vol. 101, pp. 65-168.
Pospíšil, J.: Aromatic and Heterocyclic Amines in Polymer Stabilization. Vol. 124, pp. 87-190.
Priddy, D. B.: Recent Advances in Styrene Polymerization. Vol. 111, pp. 67-114.
Priddy, D. B.: Thermal Discoloration Chemistry of Styrene-co-Acrylonitrile. Vol. 121, pp. 123-154.
Privalko, V. P. and *Novikov, V. V.*: Model Treatments of the Heat Conductivity of Heterogeneous Polymers. Vol. 119, pp 31-78.
Putnam, D. and *Kopecek, J.*: Polymer Conjugates with Anticancer Acitivity. Vol. 122, pp. 55-124.

Ramaraj, R. and *Kaneko, M.*: Metal Complex in Polymer Membrane as a Model for Photosynthetic Oxygen Evolving Center. Vol. 123, pp. 215-242.
Rangarajan, B. see Scranton, A. B.: Vol. 122, pp. 1-54.
Reichert, K. H. see Hunkeler, D.: Vol. 112, pp. 115-134.
Rehahn, M., Mattice, W. L., Suter, U. W.: Rotational Isomeric State Models in Macromolecular Systems. Vol. 131/132, pp. 1-475

Risse, W. see Grubbs, R.: Vol. 102, pp. 47-72.
Rivas, B. L. and *Geckeler, K. E.*: Synthesis and Metal Complexation of Poly(ethyleneimine) and Derivatives. Vol. 102, pp. 171-188.
Robin, J. J. see Boutevin, B.: Vol. 102, pp. 105-132.
Roe, R.-J.: MD Simulation Study of Glass Transition and Short Time Dynamics in Polymer Liquids. Vol. 116, pp. 111-114.
Rozenberg, B. A. see Williams, R. J. J.: Vol. 128, pp. 95-156.
Ruckenstein, E.: Concentrated Emulsion Polymerization. Vol. 127, pp. 1-58.
Rusanov, A. L.: Novel Bis (Naphtalic Anhydrides) and Their Polyheteroarylenes with Improved Processability. Vol. 111, pp. 115-176.
Rychlý, J. see Lazár, M.: Vol. 102, pp. 189-222.
Ryzhov, V. A. see Bershtein, V. A.: Vol. 114, pp. 43-122.

Sabsai, O. Y. see Barshtein, G. R.: Vol. 101, pp. 1-28.
Saburov, V. V. see Zubov, V. P.: Vol. 104, pp. 135-176.
Saito, S., Konno, M. and *Inomata, H.*: Volume Phase Transition of N-Alkylacrylamide Gels. Vol. 109, pp. 207-232.
Samsonov, G. V. and *Kuznetsova, N. P.*: Crosslinked Polyelectrolytes in Biology. Vol. 104, pp. 1-50.
Santa Cruz, C. see Baltá-Calleja, F. J.: Vol. 108, pp. 1-48.
Sato, T. and *Teramoto, A.*: Concentrated Solutions of Liquid-Christalline Polymers. Vol. 126, pp. 85-162.
Scherf, U. and *Müllen, K.*: The Synthesis of Ladder Polymers. Vol. 123, pp. 1-40.
Schmidt, M. see Förster, S.: Vol. 120, pp. 51-134.
Schopf, G. and *Koßmehl, G.*: Polythiophenes - Electrically Conductive Polymers. Vol. 129, pp. 1-145.
Schweizer, K. S.: Prism Theory of the Structure, Thermodynamics, and Phase Transitions of Polymer Liquids and Alloys. Vol. 116, pp. 319-378.
Scranton, A. B., Rangarajan, B. and *Klier, J.*: Biomedical Applications of Polyelectrolytes. Vol. 122, pp. 1-54.
Sefton, M. V. and *Stevenson, W. T. K.*: Microencapsulation of Live Animal Cells Using Polycrylates. Vol.107, pp. 143-198.
Shamanin, V. V.: Bases of the Axiomatic Theory of Addition Polymerization. Vol. 112, pp. 135-180.
Sherrington, D. C. see Cameron, N. R. , Vol. 126, pp. 163-214.
Sherrington, D. C. see Lin, J.: Vol. 111, pp. 177-220.
Sherrington, D. C. see Steinke, J.: Vol. 123, pp. 81-126.
Shibayama, M. see Tanaka, T.: Vol. 109, pp. 1-62.
Shoda, S. see Kobayashi, S.: Vol. 121, pp. 1-30.
Siegel, R. A.: Hydrophobic Weak Polyelectrolyte Gels: Studies of Swelling Equilibria and Kinetics. Vol. 109, pp. 233-268.
Singh, R. P. see Sivaram, S.: Vol. 101, pp. 169-216.
Sivaram, S. and *Singh, R. P.*: Degradation and Stabilization of Ethylene-Propylene Copolymers and Their Blends: A Critical Review. Vol. 101, pp. 169-216.
Starodybtzev, S. see Khokhlov, A.: Vol. 109, pp. 121-172.
Steinke, J., Sherrington, D. C. and *Dunkin, I. R.*: Imprinting of Synthetic Polymers Using Molecular Templates. Vol. 123, pp. 81-126.
Stenzenberger, H. D.: Addition Polyimides. Vol. 117, pp. 165-220.
Stevenson, W. T. K. see Sefton, M. V.: Vol. 107, pp. 143-198.

Sumpter, B. G., Noid, D. W., Liang, G. L. and *Wunderlich, B.*: Atomistic Dynamics of Macromolecular Crystals. Vol. 116, pp. 27-72.
Suter, U. W. see Gusev, A. A.: Vol. 116, pp. 207-248.
Suter, U. W. see Leontidis, E.: Vol. 116, pp. 283-318.
Suter, U. W. see Rehahn, M.: Vol. 131/132, pp. 1-475
Suzuki, A.: Phase Transition in Gels of Sub-Millimeter Size Induced by Interaction with Stimuli. Vol. 110, pp. 199-240.
Suzuki, A. and *Hirasa, O.*: An Approach to Artifical Muscle by Polymer Gels due to Micro-Phase Separation. Vol. 110, pp. 241-262.

Tagawa, S.: Radiation Effects on Ion Beams on Polymers. Vol. 105, pp. 99-116.
Tan, K. L. see Kang, E. T.: Vol. 106, pp. 135-190.
Tanaka, T. see Penelle, J.: Vol. 102, pp. 73-104.
Tanaka, H. and *Shibayama, M.*: Phase Transition and Related Phenomena of Polymer Gels. Vol. 109, pp. 1-62.
Tauer, K. see Guyot, A.: Vol. 111, pp. 43-66.
Teramoto, A. see Sato, T.: Vol. 126, pp. 85-162.
Terent'eva, J. P. and *Fridman, M. L.*: Compositions Based on Aminoresins. Vol. 101, pp. 29-64.
Theodorou, D. N. see Dodd, L. R.: Vol. 116, pp. 249-282.
Thomson, R. C., Wake, M. C., Yaszemski, M. J. and *Mikos, A. G.*: Biodegradable Polymer Scaffolds to Regenerate Organs. Vol. 122, pp. 245-274.
Tokita, M.: Friction Between Polymer Networks of Gels and Solvent. Vol. 110, pp. 27-48.
Tsuruta, T.: Contemporary Topics in Polymeric Materials for Biomedical Applications. Vol. 126, pp. 1-52.

Uyama, H. see Kobayashi, S.: Vol. 121, pp. 1-30.

Vasilevskaya, V. see Khokhlov, A.: Vol. 109, pp. 121-172.
Vaskova, V. see Hunkeler, D.: Vol.:112, pp. 115-134.
Verdugo, P.: Polymer Gel Phase Transition in Condensation-Decondensation of Secretory Products. Vol. 110, pp. 145-156.
Vettegren, V. I.: see Bronnikov, S. V.: Vol. 125, pp. 103-146.
Viovy, J.-L. and *Lesec, J.*: Separation of Macromolecules in Gels: Permeation Chromatography and Electrophoresis. Vol. 114, pp. 1-42.
Volksen, W.: Condensation Polyimides: Synthesis, Solution Behavior, and Imidization Characteristics. Vol. 117, pp. 111-164.

Wake, M. C. see Thomson, R. C.: Vol. 122, pp. 245-274.
Wang, K. L. see Cussler, E. L.: Vol. 110, pp. 67-80.
Williams, R. J. J., Rozenberg, B. A., Pascault, J.-P.: Reaction Induced Phase Separation in Modified Thermosetting Polymers. Vol. 128, pp. 95-156.
Wunderlich, B. see Sumpter, B. G.: Vol. 116, pp. 27-72.

Xie, T. Y. see Hunkeler, D.: Vol. 112, pp. 115-134.
Xu, Z., Hadjichristidis, N., Fetters, L. J. and *Mays, J. W.*: Structure/Chain-Flexibility Relationships of Polymers. Vol. 120, pp. 1-50.

Yagci, Y. and *Endo, T.*: N-Benzyl and N-Alkoxy Pyridium Salts as Thermal and Photochemical Initiators for Cationic Polymerization. Vol. 127, pp. 59-86.

Yannas, I. V.: Tissue Regeneration Templates Based on Collagen-Glycosaminoglycan Copolymers. Vol. 122, pp. 219-244.

Yamaoka, H.: Polymer Materials for Fusion Reactors. Vol. 105, pp. 117-144.

Yaszemski, M. J. see Thomson, R. C.: Vol. 122, pp. 245-274.

Yoshida, H. and *Ichikawa, T.*: Electron Spin Studies of Free Radicals in Irradiated Polymers. Vol. 105, pp. 3-36.

Zubov, V. P., Ivanov, A. E. and *Saburov, V. V.*: Polymer-Coated Adsorbents for the Separation of Biopolymers and Particles. Vol. 104, pp. 135-176.

Printing: Saladruck, Berlin
Binding: Buchbinderei Lüderitz & Bauer, Berlin